VOLUME FOUR HUNDRED AND NINETY-NINE

Methods in
ENZYMOLOGY

Biology of Serpins

METHODS IN ENZYMOLOGY

Editors-in-Chief

JOHN N. ABELSON AND MELVIN I. SIMON

Division of Biology
California Institute of Technology
Pasadena, California

Founding Editors

SIDNEY P. COLOWICK AND NATHAN O. KAPLAN

VOLUME FOUR HUNDRED AND NINETY-NINE

METHODS IN ENZYMOLOGY

Biology of Serpins

EDITED BY

JAMES C. WHISSTOCK AND **PHILLIP I. BIRD**
Department of Biochemistry and Molecular Biology
Monash University Clayton
Victoria, Australia

AMSTERDAM • BOSTON • HEIDELBERG • LONDON
NEW YORK • OXFORD • PARIS • SAN DIEGO
SAN FRANCISCO • SINGAPORE • SYDNEY • TOKYO
Academic Press is an imprint of Elsevier

ELSEVIER

Academic Press is an imprint of Elsevier
525 B Street, Suite 1900, San Diego, CA 92101-4495, USA
225 Wyman Street, Waltham, MA 02451, USA
32 Jamestown Road, London NW1 7BY, UK

First edition 2011

Copyright © 2011, Elsevier Inc. All Rights Reserved.

No part of this publication may be reproduced, stored in a retrieval system or transmitted in any form or by any means electronic, mechanical, photocopying, recording or otherwise without the prior written permission of the publisher

Permissions may be sought directly from Elsevier's Science & Technology Rights Department in Oxford, UK: phone (+44) (0) 1865 843830; fax (+44) (0) 1865 853333; email: permissions@elsevier.com. Alternatively you can submit your request online by visiting the Elsevier web site at http://elsevier.com/locate/permissions, and selecting *Obtaining permission to use Elsevier material*

Notice

No responsibility is assumed by the publisher for any injury and/or damage to persons or property as a matter of products liability, negligence or otherwise, or from any use or operation of any methods, products, instructions or ideas contained in the material herein. Because of rapid advances in the medical sciences, in particular, independent verification of diagnoses and drug dosages should be made

For information on all Academic Press publications
visit our website at elsevierdirect.com

ISBN: 978-0-12-386471-0
ISSN: 0076-6879

Printed and bound in United States of America
11 12 13 14 10 9 8 7 6 5 4 3 2 1

Working together to grow
libraries in developing countries

www.elsevier.com | www.bookaid.org | www.sabre.org

ELSEVIER   BOOK AID International   Sabre Foundation

Contents

Contributors	xi
Preface	xix
Volumes in Series	xxi

1. Analysis of Serpin Secretion, Misfolding, and Surveillance in the Endoplasmic Reticulum — 1
Shujuan Pan, Michael J. Iannotti, and Richard N. Sifers

1. Introduction	2
2. Protocols	3
3. Concluding Remarks	14
Acknowledgments	14
References	14

2. Serpin–Enzyme Receptors: LDL Receptor-Related Protein 1 — 17
Dudley K. Strickland, Selen Catania Muratoglu, and Toni M. Antalis

1. Introduction	18
2. Purification of LRP1	18
3. Expression of Receptor Fragments	21
4. Ligand Binding to LRP1	23
5. Binding of Serpin–Enzyme Complexes to LRP1	26
6. Summary	28
Acknowledgments	28
References	28

3. The Role of Autophagy in Alpha-1-Antitrypsin Deficiency — 33
Tunda Hidvegi, Amitava Mukherjee, Michael Ewing, Carolyn Kemp, and David H. Perlmutter

1. Introduction	34
2. Methods for Detection of Autophagy in Cell Line Models	36
3. LC3 Conversion Assay	38
4. Assessing Autophagy by Analyzing Degradation of Long-Lived Proteins	41
5. Use of Autophagy-Deficient Cell Lines	42

6.	Pulse–Chase Labeling to Determine the Effect of Autophagy on Degradation of Individual Proteins	43
7.	Methods for Detection of Hepatic Autophagy *In Vivo*	45
8.	Methods for Detection of Hepatic Injury	48
	References	52

4. Serpins and the Complement System — 55

László Beinrohr, Thomas A. Murray-Rust, Leanne Dyksterhuis, Péter Závodszky, Péter Gál, Robert N. Pike, and Lakshmi C. Wijeyewickrema

1.	Introduction	56
2.	Human Plasma C1-Inhibitor	57
3.	Production of Recombinant C1-Inhibitor in Yeast	64
	Acknowledgments	73
	References	74

5. Use of Mouse Models to Study Plasminogen Activator Inhibitor-1 — 77

Paul J. Declerck, Ann Gils, and Bart De Taeye

1.	Introduction	78
2.	Role of PAI-1 in Different (Patho)Physiological Processes	80
3.	Mouse Models to Study the Role of PAI-1	82
4.	Mouse (Wild-Type) Models to Study the Role of PAI-1 in Cardiovascular Disease and Cancer	83
5.	Genetically Modified Mice to Study the Role of PAI-1 in Cardiovascular Disease and Cancer	87
6.	Conclusions	93
	References	94

6. Plasminogen Activator Inhibitor Type 2: Still an Enigmatic Serpin but a Model for Gene Regulation — 105

Robert L. Medcalf

1.	Introduction	106
2.	PAI-2 and the Plasminogen-Activating System	107
3.	General Features of PAI-2	108
4.	PAI-2 Gene Expression and Regulation	115
5.	Conclusions	122
6.	Methodology: Rapid Run-On Transcription Assay Protocol	123
	Acknowledgments	126
	References	126

7. **The SerpinB1 Knockout Mouse: A Model for Studying Neutrophil Protease Regulation in Homeostasis and Inflammation** — 135
   Charaf Benarafa
   1. Generation of *SerpinB1* Knockout Mice — 136
   2. Models of Lung Infection and Inflammation — 139
   3. Neutrophil Homeostasis — 142
   Acknowledgments — 146
   References — 146

8. **Investigating Maspin in Breast Cancer Progression Using Mouse Models** — 149
   Michael P. Endsley and Ming Zhang
   1. Introduction — 150
   2. Transgenic Mouse Models — 151
   3. Syngeneic Tumor Model — 157
   4. Conclusion — 161
   References — 162

9. **Hsp47 as a Collagen-Specific Molecular Chaperone** — 167
   Yoshihito Ishida and Kazuhiro Nagata
   1. Introduction — 168
   2. Hsp47 as a Collagen-Binding Protein in the ER — 169
   3. Interaction and Recognition of Collagen by Hsp47 — 170
   4. A Phenotype and Abnormal Collagen Maturation in Hsp47 Knockout Mice — 171
   5. Possible Roles of Hsp47 in Procollagen Maturation in the ER — 172
   6. Hsp47 Null Cells: A Tool for Studying the Fate of Misfolded Collagen — 174
   7. Regulation of Hsp47 Expression and Its Clinical Importance — 176
   References — 178

10. **Assays for the Antiangiogenic and Neurotrophic Serpin Pigment Epithelium-Derived Factor** — 183
    Preeti Subramanian, Susan E. Crawford, and S. Patricia Becerra
    1. Introduction — 184
    2. Purification of PEDF Protein — 185
    3. Techniques to Assay PEDF — 188
    4. Neurotrophic Assays — 194
    5. Antiangiogenic Assays — 197
    Acknowledgments — 202
    References — 202

11. The *Drosophila* Serpins: Multiple Functions in Immunity and Morphogenesis   205

Jean Marc Reichhart, David Gubb, and Vincent Leclerc

1.	Introduction	206
2.	The Range of *Drosophila* Serpin Functions	207
3.	Techniques for Analysis of Immune Response	214
	Acknowledgments	220
	References	220

12. Modeling Serpin Conformational Diseases in *Drosophila melanogaster*   227

Thomas R. Jahn, Elke Malzer, John Roote, Anastasia Vishnivetskaya, Sara Imarisio, Maria Giannakou, Karin Panser, Stefan Marciniak, and Damian C. Crowther

1.	Introduction	228
2.	Why *Drosophila* Models of Serpinopathies?	229
3.	Screening for Polymerogenic Mutations in Physiologically Important Fly Serpins	229
4.	First Steps in the Generation of a Human Serpinopathy Model in *Drosophila*	230
5.	Longevity Assays in Flies Expressing Human Serpins	233
6.	Behavioral Assays in Flies Expressing Human Serpins	234
7.	Microscopic Phenotyping of Flies Expressing Human Serpins	235
8.	Using Genetic Screens to Identify Genes That Have a Role in Generating, or Suppressing Serpin-Induced Phenotypes	236
9.	Conclusions	237
10.	Method 1: Generating Transgenic *Drosophila*	237
11.	Method 2: The *Drosophila* UAS/GAL4 Expression System	242
12.	Method 3: Chemical Mutagenesis and X-ray Mutagenesis	242
13.	Method 4: Screening for Mutations Caused by Chemical and X-ray Mutagenesis	245
14.	Method 6: Examination of the Eye Imaginal Disc	246
15.	Method 7: Pseudopupil Assay	248
16.	Method 8: Protein Extraction from Flies	250
17.	Method 9: Genetic Backcrossing	251
18.	Method 10: Longevity Assays	252
19.	Method 11: Locomotor Assays	253
20.	Method 12: Immunostaining of Fly Brains	253
21.	Method 13: P-element and RNAi Screening	254
22.	Method 15: Deletion Kit Screening	255
	References	255

13. Using *Caenorhabditis elegans* to Study Serpinopathies   259
Olivia S. Long, Sager J. Gosai, Joon Hyeok Kwak, Dale E. King, David H. Perlmutter, Gary A. Silverman, and Stephen C. Pak

1. Introduction   260
2. Considerations for Transgenesis   261
3. Microinjection   268
4. High-Content Drug Screening   273
Acknowledgments   279
References   279

14. Using *C. elegans* to Identify the Protease Targets of Serpins *In Vivo*   283
Sangeeta R. Bhatia, Mark T. Miedel, Cavita K. Chotoo, Nathan J. Graf, Brian L. Hood, Thomas P. Conrads, Gary A. Silverman, and Cliff J. Luke

1. Introduction   284
2. Methods for Identifying the Targets of Intracellular Serpins in *C. elegans*   286
Acknowledgments   297
References   297

15. Viral Serpin Therapeutics: From Concept to Clinic   301
Hao Chen, Donghang Zheng, Jennifer Davids, Mee Yong Bartee, Erbin Dai, Liying Liu, Lyubomir Petrov, Colin Macaulay, Robert Thoburn, Eric Sobel, Richard Moyer, Grant McFadden, and Alexandra Lucas

1. Introduction   302
2. From Viral Pathogenesis to Identification of Immunomodulatory Potential   305
3. Testing Biological Potential   307
4. Assessing Clinical Therapeutic Potential for a Viral Serpin in Clinical Trial: Trial of Serp-1 Treatment in Patients with Acute Coronary Syndrome and Stent Implant   319
References   326

16. Human SCCA Serpins Inhibit Staphylococcal Cysteine Proteases by Forming Classic "Serpin-Like" Covalent Complexes   331
Tomasz Kantyka and Jan Potempa

1. Introduction   332
2. Purification of Staphopains   333
3. Purification of GST–SCCA1 and GST–SCCA2 Fusion Proteins   333
4. Characterization of Inhibition   334

5.	Detection of Serpin–Enzyme Complex	338
6.	Determination of an Interaction Site	341
7.	Summary	343
	Acknowledgments	344
	References	344

17. Plants and the Study of Serpin Biology   347

Thomas H. Roberts, Joon-Woo Ahn, Nardy Lampl, and Robert Fluhr

1.	Introduction	348
2.	Detection of Serpins in Plant Extracts and Localization of Serpins in Plant Tissues and Cells	349
3.	Purification of Serpins from Plant Tissues	353
4.	Production of Recombinant Plant Serpins	360
5.	Analysis of Plant Serpin–Protease Interactions	360
	Acknowledgments	364
	References	364

Author Index   *369*
Subject Index   *397*

CONTRIBUTORS

Joon-Woo Ahn
Plant Systems Engineering Research Center, Korea Research Institute of Bioscience and Biotechnology (KRIBB), Yuseong-gu, Daejeon, South Korea

Toni M. Antalis
Center for Vascular and Inflammatory Diseases, and Department of Physiology, University of Maryland School of Medicine, Baltimore, Maryland, USA

Mee Yong Bartee
Department of Medicine, Divisions of Cardiovascular Medicine and Rheumatology, University of Florida, Gainesville, Florida, USA

S. Patricia Becerra
Section of Protein Structure and Function, National Eye Institute, NIH, Bethesda, Maryland, USA

László Beinrohr
Institute of Enzymology, Biological Research Center, Hungarian Academy of Sciences, Hungary

Charaf Benarafa
Theodor Kocher Institute, University of Bern, Bern, Switzerland

Sangeeta R. Bhatia
Department of Pediatrics, Children's Hospital of Pittsburgh, University of Pittsburgh School of Medicine, Pittsburgh, Pennsylvania, USA

Hao Chen
Department of Medicine, Divisions of Cardiovascular Medicine and Rheumatology, University of Florida, Gainesville, Florida, USA

Cavita K. Chotoo
Department of Cell Biology and Physiology, University of Pittsburgh School of Medicine, Pittsburgh, Pennsylvania, USA

Thomas P. Conrads
Department of Pharmacology and Chemical Biology, University of Pittsburgh Cancer Institute, University of Pittsburgh School of Medicine, Pittsburgh, Pennsylvania, USA

Susan E. Crawford
Department of Surgery and Pathology, NorthShore University Research Institute, Evanston, Illinois, USA

Damian C. Crowther
Department of Genetics, and Cambridge Institute for Medical Research, University of Cambridge, Cambridge, United Kingdom

Erbin Dai
Department of Medicine, Divisions of Cardiovascular Medicine and Rheumatology, University of Florida, Gainesville, Florida, USA

Jennifer Davids
Department of Medicine, Divisions of Cardiovascular Medicine and Rheumatology, University of Florida, Gainesville, Florida, USA

Bart De Taeye
Feinberg Cardiovascular Research Institute, Northwestern University, Chicago, Illinois, USA

Paul J. Declerck
Laboratory for Pharmaceutical Biology, Faculty of Pharmaceutical Sciences, Katholieke Universiteit Leuven, Campus Gasthuisberg, O&N2, Herestraat, Leuven, Belgium

Leanne Dyksterhuis
Department of Biochemistry and Molecular Biology, Monash University, Clayton, Victoria, Australia

Michael P. Endsley
Robert H. Lurie Comprehensive Cancer Center and Center for Genetic Medicine, Department of Molecular Pharmacology and Biological Chemistry, Northwestern University, Feinberg School of Medicine, Chicago, Illinois, USA

Michael Ewing
Department of Pediatrics, University of Pittsburgh School of Medicine, Children's Hospital of Pittsburgh of UPMC, Pittsburgh, Pennsylvania, USA

Robert Fluhr
Department of Plant Sciences, Weizmann Institute of Science, Rehovot, Israel

Péter Gál
Institute of Enzymology, Biological Research Center, Hungarian Academy of Sciences, Hungary

Maria Giannakou
Department of Genetics, University of Cambridge, Cambridge, United Kingdom

Ann Gils
Laboratory for Pharmaceutical Biology, Faculty of Pharmaceutical Sciences, Katholieke Universiteit Leuven, Campus Gasthuisberg, O&N2, Herestraat, Leuven, Belgium

Sager J. Gosai
Department of Pediatrics, Cell Biology and Physiology, University of Pittsburgh School of Medicine, Children's Hospital of Pittsburgh of UPMC, Pittsburg, Pennsylvania, USA

Nathan J. Graf
Department of Pediatrics, Children's Hospital of Pittsburgh, University of Pittsburgh School of Medicine, Pittsburgh, Pennsylvania, USA

David Gubb
Unidad de Genómica Funcional, CIC bioGUNE, Parque Tecnológico de Vizcaya, Vizcaya, España, Spain

Tunda Hidvegi
Department of Pediatrics, University of Pittsburgh School of Medicine, Children's Hospital of Pittsburgh of UPMC, Pittsburgh, Pennsylvania, USA

Brian L. Hood
Department of Pharmacology and Chemical Biology, University of Pittsburgh Cancer Institute, University of Pittsburgh School of Medicine, Pittsburgh, Pennsylvania, USA

Michael J. Iannotti
Cell and Molecular Biology Interdepartmental Graduate Program, Baylor College of Medicine, Houston, Texas, USA

Sara Imarisio
Department of Genetics, University of Cambridge, Cambridge, United Kingdom

Yoshihito Ishida
Laboratory of Molecular and Cellular Biology, Department of Molecular Biosciences, Faculty of Life Sciences, Kyoto Sangyo University, Kyoto, Japan

Thomas R. Jahn
Department of Genetics, and Department of Chemistry, University of Cambridge, Cambridge, United Kingdom

Tomasz Kantyka
Department of Microbiology, Faculty of Biochemistry, Biophysics, and Biotechnology, Jagiellonian University, Krakow, Poland

Carolyn Kemp
Department of Pediatrics, University of Pittsburgh School of Medicine, Children's Hospital of Pittsburgh of UPMC, Pittsburgh, Pennsylvania, USA

Dale E. King
Department of Pediatrics, Cell Biology and Physiology, University of Pittsburgh School of Medicine, Children's Hospital of Pittsburgh of UPMC, Pittsburg, Pennsylvania, USA

Joon Hyeok Kwak
Department of Pediatrics, Cell Biology and Physiology, University of Pittsburgh School of Medicine, Children's Hospital of Pittsburgh of UPMC, Pittsburg, Pennsylvania, USA

Nardy Lampl
Department of Plant Sciences, Weizmann Institute of Science, Rehovot, Israel

Vincent Leclerc
Université de Strasbourg, UPR 9022 CNRS, IBMC, 15 rue Descartes, Strasbourg, France

Liying Liu
Department of Medicine, Divisions of Cardiovascular Medicine and Rheumatology, University of Florida, Gainesville, Florida, USA

Olivia S. Long
Department of Pediatrics, Cell Biology and Physiology, University of Pittsburgh School of Medicine, Children's Hospital of Pittsburgh of UPMC, Pittsburg, Pennsylvania, USA

Alexandra Lucas
Department of Medicine, Divisions of Cardiovascular Medicine and Rheumatology, and Department of Molecular Genetics and Microbiology, University of Florida, Gainesville, Florida, USA

Cliff J. Luke
Department of Pediatrics, Children's Hospital of Pittsburgh, University of Pittsburgh School of Medicine, Pittsburgh, Pennsylvania, USA

Colin Macaulay
Viron Therapeutics, Inc., London, Ontario, Canada

Elke Malzer
Department of Genetics; Department of Medicine, and Cambridge Institute for Medical Research, University of Cambridge, Cambridge, United Kingdom

Stefan Marciniak
Department of Medicine, and Cambridge Institute for Medical Research, University of Cambridge, Cambridge, United Kingdom

Grant McFadden
Department of Molecular Genetics and Microbiology, University of Florida, Gainesville, Florida, USA

Robert L. Medcalf
Australian Centre for Blood Diseases, Monash University, Melbourne, Victoria, Australia

Mark T. Miedel
Department of Pediatrics, Children's Hospital of Pittsburgh, University of Pittsburgh School of Medicine, Pittsburgh, Pennsylvania, USA

Richard Moyer
Department of Molecular Genetics and Microbiology, University of Florida, Gainesville, Florida, USA

Amitava Mukherjee
Department of Pediatrics, University of Pittsburgh School of Medicine, Children's Hospital of Pittsburgh of UPMC, Pittsburgh, Pennsylvania, USA

Selen Catania Muratoglu
Center for Vascular and Inflammatory Diseases, and Department of Surgery, University of Maryland School of Medicine, Baltimore, Maryland, USA

Thomas A. Murray-Rust
Department of Biochemistry and Molecular Biology, Monash University, Clayton, Victoria, Australia

Kazuhiro Nagata
Laboratory of Molecular and Cellular Biology, Department of Molecular Biosciences, Faculty of Life Sciences, Kyoto Sangyo University, Kyoto, Japan

Stephen C. Pak
Department of Pediatrics, Cell Biology and Physiology, University of Pittsburgh School of Medicine, Children's Hospital of Pittsburgh of UPMC, Pittsburg, Pennsylvania, USA

Shujuan Pan
Department of Pathology and Immunology, Baylor College of Medicine, Houston, Texas, USA

Karin Panser
Department of Genetics, University of Cambridge, Cambridge, United Kingdom

David H. Perlmutter
Department of Pediatrics, and Department of Cell Biology and Physiology, University of Pittsburgh School of Medicine, Children's Hospital of Pittsburgh of UPMC, Pittsburgh, Pennsylvania, USA

Lyubomir Petrov
Robarts' Research Institute, University of Western Ontario, London, Ontario, Canada

Robert N. Pike
Department of Biochemistry and Molecular Biology, Monash University, Clayton, Victoria, Australia

Jan Potempa
Department of Microbiology, Faculty of Biochemistry, Biophysics, and Biotechnology, Jagiellonian University, Krakow, Poland, and Department of Oral Health and Rehabilitation, University of Louisville Dental School, Louisville, Kentucky, USA

Jean Marc Reichhart
Université de Strasbourg, UPR 9022 CNRS, IBMC, 15 rue Descartes, Strasbourg, France

Thomas H. Roberts
Department of Chemistry and Biomolecular Sciences, Macquarie University, North Ryde, Australia

John Roote
Department of Genetics, University of Cambridge, Cambridge, United Kingdom

Richard N. Sifers
Department of Pathology and Immunology; Cell and Molecular Biology Interdepartmental Graduate Program; Department of Molecular and Cellular Biology, and Department of Molecular Physiology and Biophysics, Baylor College of Medicine, Houston, Texas, USA

Gary A. Silverman
Department of Pediatrics, and Department of Cell Biology and Physiology, University of Pittsburgh School of Medicine, Children's Hospital of Pittsburgh of UPMC, Pittsburgh, Pennsylvania, USA

Eric Sobel
Department of Medicine, Divisions of Cardiovascular Medicine and Rheumatology, University of Florida, Gainesville, Florida, USA

Dudley K. Strickland
Center for Vascular and Inflammatory Diseases; Department of Physiology, and Department of Surgery, University of Maryland School of Medicine, Baltimore, Maryland, USA

Preeti Subramanian
Section of Protein Structure and Function, National Eye Institute, NIH, Bethesda, Maryland, USA

Robert Thoburn
Department of Medicine, Divisions of Cardiovascular Medicine and Rheumatology, University of Florida, Gainesville, Florida, USA

Anastasia Vishnivetskaya
Department of Genetics, University of Cambridge, Cambridge, United Kingdom

Lakshmi C. Wijeyewickrema
Department of Biochemistry and Molecular Biology, Monash University, Clayton, Victoria, Australia

Péter Závodszky
Institute of Enzymology, Biological Research Center, Hungarian Academy of Sciences, Hungary

Ming Zhang
Robert H. Lurie Comprehensive Cancer Center and Center for Genetic Medicine, Department of Molecular Pharmacology and Biological Chemistry, Northwestern University, Feinberg School of Medicine, Chicago, Illinois, USA

Donghang Zheng
Department of Medicine, Divisions of Cardiovascular Medicine and Rheumatology, University of Florida, Gainesville, Florida, USA

Preface

Serpins comprise a family of proteins that has fascinated investigators for over 60 years. Besides highlighting the importance of the homeostatic regulation of proteolysis, serpin research has yielded precious insights into protein–protein interactions and protein conformational change, and the relationship between protein misfolding and disease. Further, the family contains striking examples of the use of a common protein fold for different biological purposes: besides protease inhibition, serpins are used as hormone precursors, hormone carriers, and protein folding chaperones. Insights into the flexibility of the serpin fold are pointing the way toward development of therapeutics that will ameliorate some common human afflictions.

At the time of writing, there are more than 1000 serpins, mostly appearing as predicted proteins derived from genome sequencing projects. The majority of these are anticipated to be protease inhibitors, based upon the presence of conserved motifs in the crucial reactive center loop region. It is, however, evident that serpins occur in all kingdoms of life and perform a wide array of evolutionary and biological roles (including important functions outside the inhibition of proteases). This diversity is reflected in the broad scope and multiple systems covered by chapters in these volumes. The challenge for the field is to elucidate the role of each particular serpin in its own niche, and this will typically require examination of structure, tissue distribution and cellular localization, identification of partners or targets, and assessment of physiological significance via forward or reverse genetics. Examples of these approaches will be found within these pages.

What then are the "big questions" or achievements remaining in serpin research? We would nominate (in no particular order) the precise physiological mechanism of serpin polymerization and associated cytotoxicity, the roles of noninhibitory serpins and the contribution of conformational change to their function, and the development of additional targeted antiserpin therapeutics. Finally, as we move into the era of the 1000 genomes project and the routine sequencing of human genomes for medical purposes, we expect that a greater number of serpin mutations will be associated with a wide range of human diseases. The next generation of serpin researchers will make great inroads into all these problems.

We acknowledge and thank all the authors for contributing chapters that neatly illustrate the state of play in our understanding of serpins and the cutting-edge approaches from a variety of disciplines that are being used to expand the frontiers of our knowledge. We hope that these chapters in these two volumes will prove a very useful resource for both seasoned serpinologists and new hands.

In closing, we dedicate these volumes to our teachers, mentors, colleagues, and students, who continue to inspire us in our personal journeys of discovery.

March 2011
JAMES C. WHISSTOCK AND PHILLIP I. BIRD

METHODS IN ENZYMOLOGY

VOLUME I. Preparation and Assay of Enzymes
Edited by SIDNEY P. COLOWICK AND NATHAN O. KAPLAN

VOLUME II. Preparation and Assay of Enzymes
Edited by SIDNEY P. COLOWICK AND NATHAN O. KAPLAN

VOLUME III. Preparation and Assay of Substrates
Edited by SIDNEY P. COLOWICK AND NATHAN O. KAPLAN

VOLUME IV. Special Techniques for the Enzymologist
Edited by SIDNEY P. COLOWICK AND NATHAN O. KAPLAN

VOLUME V. Preparation and Assay of Enzymes
Edited by SIDNEY P. COLOWICK AND NATHAN O. KAPLAN

VOLUME VI. Preparation and Assay of Enzymes *(Continued)*
Preparation and Assay of Substrates
Special Techniques
Edited by SIDNEY P. COLOWICK AND NATHAN O. KAPLAN

VOLUME VII. Cumulative Subject Index
Edited by SIDNEY P. COLOWICK AND NATHAN O. KAPLAN

VOLUME VIII. Complex Carbohydrates
Edited by ELIZABETH F. NEUFELD AND VICTOR GINSBURG

VOLUME IX. Carbohydrate Metabolism
Edited by WILLIS A. WOOD

VOLUME X. Oxidation and Phosphorylation
Edited by RONALD W. ESTABROOK AND MAYNARD E. PULLMAN

VOLUME XI. Enzyme Structure
Edited by C. H. W. HIRS

VOLUME XII. Nucleic Acids (Parts A and B)
Edited by LAWRENCE GROSSMAN AND KIVIE MOLDAVE

VOLUME XIII. Citric Acid Cycle
Edited by J. M. LOWENSTEIN

VOLUME XIV. Lipids
Edited by J. M. LOWENSTEIN

VOLUME XV. Steroids and Terpenoids
Edited by RAYMOND B. CLAYTON

VOLUME XVI. Fast Reactions
Edited by KENNETH KUSTIN

VOLUME XVII. Metabolism of Amino Acids and Amines (Parts A and B)
Edited by HERBERT TABOR AND CELIA WHITE TABOR

VOLUME XVIII. Vitamins and Coenzymes (Parts A, B, and C)
Edited by DONALD B. MCCORMICK AND LEMUEL D. WRIGHT

VOLUME XIX. Proteolytic Enzymes
Edited by GERTRUDE E. PERLMANN AND LASZLO LORAND

VOLUME XX. Nucleic Acids and Protein Synthesis (Part C)
Edited by KIVIE MOLDAVE AND LAWRENCE GROSSMAN

VOLUME XXI. Nucleic Acids (Part D)
Edited by LAWRENCE GROSSMAN AND KIVIE MOLDAVE

VOLUME XXII. Enzyme Purification and Related Techniques
Edited by WILLIAM B. JAKOBY

VOLUME XXIII. Photosynthesis (Part A)
Edited by ANTHONY SAN PIETRO

VOLUME XXIV. Photosynthesis and Nitrogen Fixation (Part B)
Edited by ANTHONY SAN PIETRO

VOLUME XXV. Enzyme Structure (Part B)
Edited by C. H. W. HIRS AND SERGE N. TIMASHEFF

VOLUME XXVI. Enzyme Structure (Part C)
Edited by C. H. W. HIRS AND SERGE N. TIMASHEFF

VOLUME XXVII. Enzyme Structure (Part D)
Edited by C. H. W. HIRS AND SERGE N. TIMASHEFF

VOLUME XXVIII. Complex Carbohydrates (Part B)
Edited by VICTOR GINSBURG

VOLUME XXIX. Nucleic Acids and Protein Synthesis (Part E)
Edited by LAWRENCE GROSSMAN AND KIVIE MOLDAVE

VOLUME XXX. Nucleic Acids and Protein Synthesis (Part F)
Edited by KIVIE MOLDAVE AND LAWRENCE GROSSMAN

VOLUME XXXI. Biomembranes (Part A)
Edited by SIDNEY FLEISCHER AND LESTER PACKER

VOLUME XXXII. Biomembranes (Part B)
Edited by SIDNEY FLEISCHER AND LESTER PACKER

VOLUME XXXIII. Cumulative Subject Index Volumes I-XXX
Edited by MARTHA G. DENNIS AND EDWARD A. DENNIS

VOLUME XXXIV. Affinity Techniques (Enzyme Purification: Part B)
Edited by WILLIAM B. JAKOBY AND MEIR WILCHEK

VOLUME XXXV. Lipids (Part B)
Edited by JOHN M. LOWENSTEIN

VOLUME XXXVI. Hormone Action (Part A: Steroid Hormones)
Edited by BERT W. O'MALLEY AND JOEL G. HARDMAN

VOLUME XXXVII. Hormone Action (Part B: Peptide Hormones)
Edited by BERT W. O'MALLEY AND JOEL G. HARDMAN

VOLUME XXXVIII. Hormone Action (Part C: Cyclic Nucleotides)
Edited by JOEL G. HARDMAN AND BERT W. O'MALLEY

VOLUME XXXIX. Hormone Action (Part D: Isolated Cells, Tissues, and Organ Systems)
Edited by JOEL G. HARDMAN AND BERT W. O'MALLEY

VOLUME XL. Hormone Action (Part E: Nuclear Structure and Function)
Edited by BERT W. O'MALLEY AND JOEL G. HARDMAN

VOLUME XLI. Carbohydrate Metabolism (Part B)
Edited by W. A. WOOD

VOLUME XLII. Carbohydrate Metabolism (Part C)
Edited by W. A. WOOD

VOLUME XLIII. Antibiotics
Edited by JOHN H. HASH

VOLUME XLIV. Immobilized Enzymes
Edited by KLAUS MOSBACH

VOLUME XLV. Proteolytic Enzymes (Part B)
Edited by LASZLO LORAND

VOLUME XLVI. Affinity Labeling
Edited by WILLIAM B. JAKOBY AND MEIR WILCHEK

VOLUME XLVII. Enzyme Structure (Part E)
Edited by C. H. W. HIRS AND SERGE N. TIMASHEFF

VOLUME XLVIII. Enzyme Structure (Part F)
Edited by C. H. W. HIRS AND SERGE N. TIMASHEFF

VOLUME XLIX. Enzyme Structure (Part G)
Edited by C. H. W. HIRS AND SERGE N. TIMASHEFF

VOLUME L. Complex Carbohydrates (Part C)
Edited by VICTOR GINSBURG

VOLUME LI. Purine and Pyrimidine Nucleotide Metabolism
Edited by PATRICIA A. HOFFEE AND MARY ELLEN JONES

VOLUME LII. Biomembranes (Part C: Biological Oxidations)
Edited by SIDNEY FLEISCHER AND LESTER PACKER

VOLUME LIII. Biomembranes (Part D: Biological Oxidations)
Edited by SIDNEY FLEISCHER AND LESTER PACKER

VOLUME LIV. Biomembranes (Part E: Biological Oxidations)
Edited by SIDNEY FLEISCHER AND LESTER PACKER

VOLUME LV. Biomembranes (Part F: Bioenergetics)
Edited by SIDNEY FLEISCHER AND LESTER PACKER

VOLUME LVI. Biomembranes (Part G: Bioenergetics)
Edited by SIDNEY FLEISCHER AND LESTER PACKER

VOLUME LVII. Bioluminescence and Chemiluminescence
Edited by MARLENE A. DELUCA

VOLUME LVIII. Cell Culture
Edited by WILLIAM B. JAKOBY AND IRA PASTAN

VOLUME LIX. Nucleic Acids and Protein Synthesis (Part G)
Edited by KIVIE MOLDAVE AND LAWRENCE GROSSMAN

VOLUME LX. Nucleic Acids and Protein Synthesis (Part H)
Edited by KIVIE MOLDAVE AND LAWRENCE GROSSMAN

VOLUME 61. Enzyme Structure (Part H)
Edited by C. H. W. HIRS AND SERGE N. TIMASHEFF

VOLUME 62. Vitamins and Coenzymes (Part D)
Edited by DONALD B. MCCORMICK AND LEMUEL D. WRIGHT

VOLUME 63. Enzyme Kinetics and Mechanism (Part A: Initial Rate and Inhibitor Methods)
Edited by DANIEL L. PURICH

VOLUME 64. Enzyme Kinetics and Mechanism
(Part B: Isotopic Probes and Complex Enzyme Systems)
Edited by DANIEL L. PURICH

VOLUME 65. Nucleic Acids (Part I)
Edited by LAWRENCE GROSSMAN AND KIVIE MOLDAVE

VOLUME 66. Vitamins and Coenzymes (Part E)
Edited by DONALD B. MCCORMICK AND LEMUEL D. WRIGHT

VOLUME 67. Vitamins and Coenzymes (Part F)
Edited by DONALD B. MCCORMICK AND LEMUEL D. WRIGHT

VOLUME 68. Recombinant DNA
Edited by RAY WU

VOLUME 69. Photosynthesis and Nitrogen Fixation (Part C)
Edited by ANTHONY SAN PIETRO

VOLUME 70. Immunochemical Techniques (Part A)
Edited by HELEN VAN VUNAKIS AND JOHN J. LANGONE

VOLUME 71. Lipids (Part C)
Edited by JOHN M. LOWENSTEIN

VOLUME 72. Lipids (Part D)
Edited by JOHN M. LOWENSTEIN

VOLUME 73. Immunochemical Techniques (Part B)
Edited by JOHN J. LANGONE AND HELEN VAN VUNAKIS

VOLUME 74. Immunochemical Techniques (Part C)
Edited by JOHN J. LANGONE AND HELEN VAN VUNAKIS

VOLUME 75. Cumulative Subject Index Volumes XXXI, XXXII, XXXIV–LX
Edited by EDWARD A. DENNIS AND MARTHA G. DENNIS

VOLUME 76. Hemoglobins
Edited by ERALDO ANTONINI, LUIGI ROSSI-BERNARDI, AND EMILIA CHIANCONE

VOLUME 77. Detoxication and Drug Metabolism
Edited by WILLIAM B. JAKOBY

VOLUME 78. Interferons (Part A)
Edited by SIDNEY PESTKA

VOLUME 79. Interferons (Part B)
Edited by SIDNEY PESTKA

VOLUME 80. Proteolytic Enzymes (Part C)
Edited by LASZLO LORAND

VOLUME 81. Biomembranes (Part H: Visual Pigments and Purple Membranes, I)
Edited by LESTER PACKER

VOLUME 82. Structural and Contractile Proteins (Part A: Extracellular Matrix)
Edited by LEON W. CUNNINGHAM AND DIXIE W. FREDERIKSEN

VOLUME 83. Complex Carbohydrates (Part D)
Edited by VICTOR GINSBURG

VOLUME 84. Immunochemical Techniques (Part D: Selected Immunoassays)
Edited by JOHN J. LANGONE AND HELEN VAN VUNAKIS

VOLUME 85. Structural and Contractile Proteins (Part B: The Contractile Apparatus and the Cytoskeleton)
Edited by DIXIE W. FREDERIKSEN AND LEON W. CUNNINGHAM

VOLUME 86. Prostaglandins and Arachidonate Metabolites
Edited by WILLIAM E. M. LANDS AND WILLIAM L. SMITH

VOLUME 87. Enzyme Kinetics and Mechanism (Part C: Intermediates, Stereo-chemistry, and Rate Studies)
Edited by DANIEL L. PURICH

VOLUME 88. Biomembranes (Part I: Visual Pigments and Purple Membranes, II)
Edited by LESTER PACKER

VOLUME 89. Carbohydrate Metabolism (Part D)
Edited by WILLIS A. WOOD

VOLUME 90. Carbohydrate Metabolism (Part E)
Edited by WILLIS A. WOOD

VOLUME 91. Enzyme Structure (Part I)
Edited by C. H. W. HIRS AND SERGE N. TIMASHEFF

VOLUME 92. Immunochemical Techniques (Part E: Monoclonal Antibodies and General Immunoassay Methods)
Edited by JOHN J. LANGONE AND HELEN VAN VUNAKIS

VOLUME 93. Immunochemical Techniques (Part F: Conventional Antibodies, Fc Receptors, and Cytotoxicity)
Edited by JOHN J. LANGONE AND HELEN VAN VUNAKIS

VOLUME 94. Polyamines
Edited by HERBERT TABOR AND CELIA WHITE TABOR

VOLUME 95. Cumulative Subject Index Volumes 61–74, 76–80
Edited by EDWARD A. DENNIS AND MARTHA G. DENNIS

VOLUME 96. Biomembranes [Part J: Membrane Biogenesis: Assembly and Targeting (General Methods; Eukaryotes)]
Edited by SIDNEY FLEISCHER AND BECCA FLEISCHER

VOLUME 97. Biomembranes [Part K: Membrane Biogenesis: Assembly and Targeting (Prokaryotes, Mitochondria, and Chloroplasts)]
Edited by SIDNEY FLEISCHER AND BECCA FLEISCHER

VOLUME 98. Biomembranes (Part L: Membrane Biogenesis: Processing and Recycling)
Edited by SIDNEY FLEISCHER AND BECCA FLEISCHER

VOLUME 99. Hormone Action (Part F: Protein Kinases)
Edited by JACKIE D. CORBIN AND JOEL G. HARDMAN

VOLUME 100. Recombinant DNA (Part B)
Edited by RAY WU, LAWRENCE GROSSMAN, AND KIVIE MOLDAVE

VOLUME 101. Recombinant DNA (Part C)
Edited by RAY WU, LAWRENCE GROSSMAN, AND KIVIE MOLDAVE

VOLUME 102. Hormone Action (Part G: Calmodulin and Calcium-Binding Proteins)
Edited by ANTHONY R. MEANS AND BERT W. O'MALLEY

VOLUME 103. Hormone Action (Part H: Neuroendocrine Peptides)
Edited by P. MICHAEL CONN

VOLUME 104. Enzyme Purification and Related Techniques (Part C)
Edited by WILLIAM B. JAKOBY

VOLUME 105. Oxygen Radicals in Biological Systems
Edited by LESTER PACKER

VOLUME 106. Posttranslational Modifications (Part A)
Edited by FINN WOLD AND KIVIE MOLDAVE

VOLUME 107. Posttranslational Modifications (Part B)
Edited by FINN WOLD AND KIVIE MOLDAVE

VOLUME 108. Immunochemical Techniques (Part G: Separation and Characterization of Lymphoid Cells)
Edited by GIOVANNI DI SABATO, JOHN J. LANGONE, AND HELEN VAN VUNAKIS

VOLUME 109. Hormone Action (Part I: Peptide Hormones)
Edited by LUTZ BIRNBAUMER AND BERT W. O'MALLEY

VOLUME 110. Steroids and Isoprenoids (Part A)
Edited by JOHN H. LAW AND HANS C. RILLING

VOLUME 111. Steroids and Isoprenoids (Part B)
Edited by JOHN H. LAW AND HANS C. RILLING

VOLUME 112. Drug and Enzyme Targeting (Part A)
Edited by KENNETH J. WIDDER AND RALPH GREEN

VOLUME 113. Glutamate, Glutamine, Glutathione, and Related Compounds
Edited by ALTON MEISTER

VOLUME 114. Diffraction Methods for Biological Macromolecules (Part A)
Edited by HAROLD W. WYCKOFF, C. H. W. HIRS, AND SERGE N. TIMASHEFF

VOLUME 115. Diffraction Methods for Biological Macromolecules (Part B)
Edited by HAROLD W. WYCKOFF, C. H. W. HIRS, AND SERGE N. TIMASHEFF

VOLUME 116. Immunochemical Techniques
(Part H: Effectors and Mediators of Lymphoid Cell Functions)
Edited by GIOVANNI DI SABATO, JOHN J. LANGONE, AND HELEN VAN VUNAKIS

VOLUME 117. Enzyme Structure (Part J)
Edited by C. H. W. HIRS AND SERGE N. TIMASHEFF

VOLUME 118. Plant Molecular Biology
Edited by ARTHUR WEISSBACH AND HERBERT WEISSBACH

VOLUME 119. Interferons (Part C)
Edited by SIDNEY PESTKA

VOLUME 120. Cumulative Subject Index Volumes 81–94, 96–101

VOLUME 121. Immunochemical Techniques (Part I: Hybridoma Technology and Monoclonal Antibodies)
Edited by JOHN J. LANGONE AND HELEN VAN VUNAKIS

VOLUME 122. Vitamins and Coenzymes (Part G)
Edited by FRANK CHYTIL AND DONALD B. MCCORMICK

Volume 123. Vitamins and Coenzymes (Part H)
Edited by Frank Chytil and Donald B. McCormick

Volume 124. Hormone Action (Part J: Neuroendocrine Peptides)
Edited by P. Michael Conn

Volume 125. Biomembranes (Part M: Transport in Bacteria, Mitochondria, and Chloroplasts: General Approaches and Transport Systems)
Edited by Sidney Fleischer and Becca Fleischer

Volume 126. Biomembranes (Part N: Transport in Bacteria, Mitochondria, and Chloroplasts: Protonmotive Force)
Edited by Sidney Fleischer and Becca Fleischer

Volume 127. Biomembranes (Part O: Protons and Water: Structure and Translocation)
Edited by Lester Packer

Volume 128. Plasma Lipoproteins (Part A: Preparation, Structure, and Molecular Biology)
Edited by Jere P. Segrest and John J. Albers

Volume 129. Plasma Lipoproteins (Part B: Characterization, Cell Biology, and Metabolism)
Edited by John J. Albers and Jere P. Segrest

Volume 130. Enzyme Structure (Part K)
Edited by C. H. W. Hirs and Serge N. Timasheff

Volume 131. Enzyme Structure (Part L)
Edited by C. H. W. Hirs and Serge N. Timasheff

Volume 132. Immunochemical Techniques (Part J: Phagocytosis and Cell-Mediated Cytotoxicity)
Edited by Giovanni Di Sabato and Johannes Everse

Volume 133. Bioluminescence and Chemiluminescence (Part B)
Edited by Marlene DeLuca and William D. McElroy

Volume 134. Structural and Contractile Proteins (Part C: The Contractile Apparatus and the Cytoskeleton)
Edited by Richard B. Vallee

Volume 135. Immobilized Enzymes and Cells (Part B)
Edited by Klaus Mosbach

Volume 136. Immobilized Enzymes and Cells (Part C)
Edited by Klaus Mosbach

Volume 137. Immobilized Enzymes and Cells (Part D)
Edited by Klaus Mosbach

Volume 138. Complex Carbohydrates (Part E)
Edited by Victor Ginsburg

VOLUME 139. Cellular Regulators (Part A: Calcium- and
Calmodulin-Binding Proteins)
Edited by ANTHONY R. MEANS AND P. MICHAEL CONN

VOLUME 140. Cumulative Subject Index Volumes 102–119, 121–134

VOLUME 141. Cellular Regulators (Part B: Calcium and Lipids)
Edited by P. MICHAEL CONN AND ANTHONY R. MEANS

VOLUME 142. Metabolism of Aromatic Amino Acids and Amines
Edited by SEYMOUR KAUFMAN

VOLUME 143. Sulfur and Sulfur Amino Acids
Edited by WILLIAM B. JAKOBY AND OWEN GRIFFITH

VOLUME 144. Structural and Contractile Proteins (Part D: Extracellular Matrix)
Edited by LEON W. CUNNINGHAM

VOLUME 145. Structural and Contractile Proteins (Part E: Extracellular Matrix)
Edited by LEON W. CUNNINGHAM

VOLUME 146. Peptide Growth Factors (Part A)
Edited by DAVID BARNES AND DAVID A. SIRBASKU

VOLUME 147. Peptide Growth Factors (Part B)
Edited by DAVID BARNES AND DAVID A. SIRBASKU

VOLUME 148. Plant Cell Membranes
Edited by LESTER PACKER AND ROLAND DOUCE

VOLUME 149. Drug and Enzyme Targeting (Part B)
Edited by RALPH GREEN AND KENNETH J. WIDDER

VOLUME 150. Immunochemical Techniques (Part K: *In Vitro* Models of B and T Cell Functions and Lymphoid Cell Receptors)
Edited by GIOVANNI DI SABATO

VOLUME 151. Molecular Genetics of Mammalian Cells
Edited by MICHAEL M. GOTTESMAN

VOLUME 152. Guide to Molecular Cloning Techniques
Edited by SHELBY L. BERGER AND ALAN R. KIMMEL

VOLUME 153. Recombinant DNA (Part D)
Edited by RAY WU AND LAWRENCE GROSSMAN

VOLUME 154. Recombinant DNA (Part E)
Edited by RAY WU AND LAWRENCE GROSSMAN

VOLUME 155. Recombinant DNA (Part F)
Edited by RAY WU

VOLUME 156. Biomembranes (Part P: ATP-Driven Pumps and Related Transport: The Na, K-Pump)
Edited by SIDNEY FLEISCHER AND BECCA FLEISCHER

VOLUME 157. Biomembranes (Part Q: ATP-Driven Pumps and Related Transport: Calcium, Proton, and Potassium Pumps)
Edited by SIDNEY FLEISCHER AND BECCA FLEISCHER

VOLUME 158. Metalloproteins (Part A)
Edited by JAMES F. RIORDAN AND BERT L. VALLEE

VOLUME 159. Initiation and Termination of Cyclic Nucleotide Action
Edited by JACKIE D. CORBIN AND ROGER A. JOHNSON

VOLUME 160. Biomass (Part A: Cellulose and Hemicellulose)
Edited by WILLIS A. WOOD AND SCOTT T. KELLOGG

VOLUME 161. Biomass (Part B: Lignin, Pectin, and Chitin)
Edited by WILLIS A. WOOD AND SCOTT T. KELLOGG

VOLUME 162. Immunochemical Techniques (Part L: Chemotaxis and Inflammation)
Edited by GIOVANNI DI SABATO

VOLUME 163. Immunochemical Techniques (Part M: Chemotaxis and Inflammation)
Edited by GIOVANNI DI SABATO

VOLUME 164. Ribosomes
Edited by HARRY F. NOLLER, JR., AND KIVIE MOLDAVE

VOLUME 165. Microbial Toxins: Tools for Enzymology
Edited by SIDNEY HARSHMAN

VOLUME 166. Branched-Chain Amino Acids
Edited by ROBERT HARRIS AND JOHN R. SOKATCH

VOLUME 167. Cyanobacteria
Edited by LESTER PACKER AND ALEXANDER N. GLAZER

VOLUME 168. Hormone Action (Part K: Neuroendocrine Peptides)
Edited by P. MICHAEL CONN

VOLUME 169. Platelets: Receptors, Adhesion, Secretion (Part A)
Edited by JACEK HAWIGER

VOLUME 170. Nucleosomes
Edited by PAUL M. WASSARMAN AND ROGER D. KORNBERG

VOLUME 171. Biomembranes (Part R: Transport Theory: Cells and Model Membranes)
Edited by SIDNEY FLEISCHER AND BECCA FLEISCHER

VOLUME 172. Biomembranes (Part S: Transport: Membrane Isolation and Characterization)
Edited by SIDNEY FLEISCHER AND BECCA FLEISCHER

VOLUME 173. Biomembranes [Part T: Cellular and Subcellular Transport: Eukaryotic (Nonepithelial) Cells]
Edited by SIDNEY FLEISCHER AND BECCA FLEISCHER

VOLUME 174. Biomembranes [Part U: Cellular and Subcellular Transport: Eukaryotic (Nonepithelial) Cells]
Edited by SIDNEY FLEISCHER AND BECCA FLEISCHER

VOLUME 175. Cumulative Subject Index Volumes 135–139, 141–167

VOLUME 176. Nuclear Magnetic Resonance (Part A: Spectral Techniques and Dynamics)
Edited by NORMAN J. OPPENHEIMER AND THOMAS L. JAMES

VOLUME 177. Nuclear Magnetic Resonance (Part B: Structure and Mechanism)
Edited by NORMAN J. OPPENHEIMER AND THOMAS L. JAMES

VOLUME 178. Antibodies, Antigens, and Molecular Mimicry
Edited by JOHN J. LANGONE

VOLUME 179. Complex Carbohydrates (Part F)
Edited by VICTOR GINSBURG

VOLUME 180. RNA Processing (Part A: General Methods)
Edited by JAMES E. DAHLBERG AND JOHN N. ABELSON

VOLUME 181. RNA Processing (Part B: Specific Methods)
Edited by JAMES E. DAHLBERG AND JOHN N. ABELSON

VOLUME 182. Guide to Protein Purification
Edited by MURRAY P. DEUTSCHER

VOLUME 183. Molecular Evolution: Computer Analysis of Protein and Nucleic Acid Sequences
Edited by RUSSELL F. DOOLITTLE

VOLUME 184. Avidin-Biotin Technology
Edited by MEIR WILCHEK AND EDWARD A. BAYER

VOLUME 185. Gene Expression Technology
Edited by DAVID V. GOEDDEL

VOLUME 186. Oxygen Radicals in Biological Systems (Part B: Oxygen Radicals and Antioxidants)
Edited by LESTER PACKER AND ALEXANDER N. GLAZER

VOLUME 187. Arachidonate Related Lipid Mediators
Edited by ROBERT C. MURPHY AND FRANK A. FITZPATRICK

VOLUME 188. Hydrocarbons and Methylotrophy
Edited by MARY E. LIDSTROM

VOLUME 189. Retinoids (Part A: Molecular and Metabolic Aspects)
Edited by LESTER PACKER

VOLUME 190. Retinoids (Part B: Cell Differentiation and Clinical Applications)
Edited by LESTER PACKER

VOLUME 191. Biomembranes (Part V: Cellular and Subcellular Transport: Epithelial Cells)
Edited by SIDNEY FLEISCHER AND BECCA FLEISCHER

VOLUME 192. Biomembranes (Part W: Cellular and Subcellular Transport: Epithelial Cells)
Edited by SIDNEY FLEISCHER AND BECCA FLEISCHER

VOLUME 193. Mass Spectrometry
Edited by JAMES A. MCCLOSKEY

VOLUME 194. Guide to Yeast Genetics and Molecular Biology
Edited by CHRISTINE GUTHRIE AND GERALD R. FINK

VOLUME 195. Adenylyl Cyclase, G Proteins, and Guanylyl Cyclase
Edited by ROGER A. JOHNSON AND JACKIE D. CORBIN

VOLUME 196. Molecular Motors and the Cytoskeleton
Edited by RICHARD B. VALLEE

VOLUME 197. Phospholipases
Edited by EDWARD A. DENNIS

VOLUME 198. Peptide Growth Factors (Part C)
Edited by DAVID BARNES, J. P. MATHER, AND GORDON H. SATO

VOLUME 199. Cumulative Subject Index Volumes 168–174, 176–194

VOLUME 200. Protein Phosphorylation (Part A: Protein Kinases: Assays, Purification, Antibodies, Functional Analysis, Cloning, and Expression)
Edited by TONY HUNTER AND BARTHOLOMEW M. SEFTON

VOLUME 201. Protein Phosphorylation (Part B: Analysis of Protein Phosphorylation, Protein Kinase Inhibitors, and Protein Phosphatases)
Edited by TONY HUNTER AND BARTHOLOMEW M. SEFTON

VOLUME 202. Molecular Design and Modeling: Concepts and Applications (Part A: Proteins, Peptides, and Enzymes)
Edited by JOHN J. LANGONE

VOLUME 203. Molecular Design and Modeling: Concepts and Applications (Part B: Antibodies and Antigens, Nucleic Acids, Polysaccharides, and Drugs)
Edited by JOHN J. LANGONE

VOLUME 204. Bacterial Genetic Systems
Edited by JEFFREY H. MILLER

VOLUME 205. Metallobiochemistry (Part B: Metallothionein and Related Molecules)
Edited by JAMES F. RIORDAN AND BERT L. VALLEE

Volume 206. Cytochrome P450
Edited by Michael R. Waterman and Eric F. Johnson

Volume 207. Ion Channels
Edited by Bernardo Rudy and Linda E. Iverson

Volume 208. Protein–DNA Interactions
Edited by Robert T. Sauer

Volume 209. Phospholipid Biosynthesis
Edited by Edward A. Dennis and Dennis E. Vance

Volume 210. Numerical Computer Methods
Edited by Ludwig Brand and Michael L. Johnson

Volume 211. DNA Structures (Part A: Synthesis and Physical Analysis of DNA)
Edited by David M. J. Lilley and James E. Dahlberg

Volume 212. DNA Structures (Part B: Chemical and Electrophoretic Analysis of DNA)
Edited by David M. J. Lilley and James E. Dahlberg

Volume 213. Carotenoids (Part A: Chemistry, Separation, Quantitation, and Antioxidation)
Edited by Lester Packer

Volume 214. Carotenoids (Part B: Metabolism, Genetics, and Biosynthesis)
Edited by Lester Packer

Volume 215. Platelets: Receptors, Adhesion, Secretion (Part B)
Edited by Jacek J. Hawiger

Volume 216. Recombinant DNA (Part G)
Edited by Ray Wu

Volume 217. Recombinant DNA (Part H)
Edited by Ray Wu

Volume 218. Recombinant DNA (Part I)
Edited by Ray Wu

Volume 219. Reconstitution of Intracellular Transport
Edited by James E. Rothman

Volume 220. Membrane Fusion Techniques (Part A)
Edited by Nejat Düzgüneş

Volume 221. Membrane Fusion Techniques (Part B)
Edited by Nejat Düzgüneş

Volume 222. Proteolytic Enzymes in Coagulation, Fibrinolysis, and Complement Activation (Part A: Mammalian Blood Coagulation Factors and Inhibitors)
Edited by Laszlo Lorand and Kenneth G. Mann

VOLUME 223. Proteolytic Enzymes in Coagulation, Fibrinolysis, and Complement Activation (Part B: Complement Activation, Fibrinolysis, and Nonmammalian Blood Coagulation Factors)
Edited by LASZLO LORAND AND KENNETH G. MANN

VOLUME 224. Molecular Evolution: Producing the Biochemical Data
Edited by ELIZABETH ANNE ZIMMER, THOMAS J. WHITE, REBECCA L. CANN, AND ALLAN C. WILSON

VOLUME 225. Guide to Techniques in Mouse Development
Edited by PAUL M. WASSARMAN AND MELVIN L. DEPAMPHILIS

VOLUME 226. Metallobiochemistry (Part C: Spectroscopic and Physical Methods for Probing Metal Ion Environments in Metalloenzymes and Metalloproteins)
Edited by JAMES F. RIORDAN AND BERT L. VALLEE

VOLUME 227. Metallobiochemistry (Part D: Physical and Spectroscopic Methods for Probing Metal Ion Environments in Metalloproteins)
Edited by JAMES F. RIORDAN AND BERT L. VALLEE

VOLUME 228. Aqueous Two-Phase Systems
Edited by HARRY WALTER AND GÖTE JOHANSSON

VOLUME 229. Cumulative Subject Index Volumes 195–198, 200–227

VOLUME 230. Guide to Techniques in Glycobiology
Edited by WILLIAM J. LENNARZ AND GERALD W. HART

VOLUME 231. Hemoglobins (Part B: Biochemical and Analytical Methods)
Edited by JOHANNES EVERSE, KIM D. VANDEGRIFF, AND ROBERT M. WINSLOW

VOLUME 232. Hemoglobins (Part C: Biophysical Methods)
Edited by JOHANNES EVERSE, KIM D. VANDEGRIFF, AND ROBERT M. WINSLOW

VOLUME 233. Oxygen Radicals in Biological Systems (Part C)
Edited by LESTER PACKER

VOLUME 234. Oxygen Radicals in Biological Systems (Part D)
Edited by LESTER PACKER

VOLUME 235. Bacterial Pathogenesis (Part A: Identification and Regulation of Virulence Factors)
Edited by VIRGINIA L. CLARK AND PATRIK M. BAVOIL

VOLUME 236. Bacterial Pathogenesis (Part B: Integration of Pathogenic Bacteria with Host Cells)
Edited by VIRGINIA L. CLARK AND PATRIK M. BAVOIL

VOLUME 237. Heterotrimeric G Proteins
Edited by RAVI IYENGAR

VOLUME 238. Heterotrimeric G-Protein Effectors
Edited by RAVI IYENGAR

VOLUME 239. Nuclear Magnetic Resonance (Part C)
Edited by THOMAS L. JAMES AND NORMAN J. OPPENHEIMER

VOLUME 240. Numerical Computer Methods (Part B)
Edited by MICHAEL L. JOHNSON AND LUDWIG BRAND

VOLUME 241. Retroviral Proteases
Edited by LAWRENCE C. KUO AND JULES A. SHAFER

VOLUME 242. Neoglycoconjugates (Part A)
Edited by Y. C. LEE AND REIKO T. LEE

VOLUME 243. Inorganic Microbial Sulfur Metabolism
Edited by HARRY D. PECK, JR., AND JEAN LEGALL

VOLUME 244. Proteolytic Enzymes: Serine and Cysteine Peptidases
Edited by ALAN J. BARRETT

VOLUME 245. Extracellular Matrix Components
Edited by E. RUOSLAHTI AND E. ENGVALL

VOLUME 246. Biochemical Spectroscopy
Edited by KENNETH SAUER

VOLUME 247. Neoglycoconjugates (Part B: Biomedical Applications)
Edited by Y. C. LEE AND REIKO T. LEE

VOLUME 248. Proteolytic Enzymes: Aspartic and Metallo Peptidases
Edited by ALAN J. BARRETT

VOLUME 249. Enzyme Kinetics and Mechanism (Part D: Developments in Enzyme Dynamics)
Edited by DANIEL L. PURICH

VOLUME 250. Lipid Modifications of Proteins
Edited by PATRICK J. CASEY AND JANICE E. BUSS

VOLUME 251. Biothiols (Part A: Monothiols and Dithiols, Protein Thiols, and Thiyl Radicals)
Edited by LESTER PACKER

VOLUME 252. Biothiols (Part B: Glutathione and Thioredoxin; Thiols in Signal Transduction and Gene Regulation)
Edited by LESTER PACKER

VOLUME 253. Adhesion of Microbial Pathogens
Edited by RON J. DOYLE AND ITZHAK OFEK

VOLUME 254. Oncogene Techniques
Edited by PETER K. VOGT AND INDER M. VERMA

VOLUME 255. Small GTPases and Their Regulators (Part A: Ras Family)
Edited by W. E. BALCH, CHANNING J. DER, AND ALAN HALL

VOLUME 256. Small GTPases and Their Regulators (Part B: Rho Family)
Edited by W. E. BALCH, CHANNING J. DER, AND ALAN HALL

VOLUME 257. Small GTPases and Their Regulators (Part C: Proteins Involved in Transport)
Edited by W. E. BALCH, CHANNING J. DER, AND ALAN HALL

VOLUME 258. Redox-Active Amino Acids in Biology
Edited by JUDITH P. KLINMAN

VOLUME 259. Energetics of Biological Macromolecules
Edited by MICHAEL L. JOHNSON AND GARY K. ACKERS

VOLUME 260. Mitochondrial Biogenesis and Genetics (Part A)
Edited by GIUSEPPE M. ATTARDI AND ANNE CHOMYN

VOLUME 261. Nuclear Magnetic Resonance and Nucleic Acids
Edited by THOMAS L. JAMES

VOLUME 262. DNA Replication
Edited by JUDITH L. CAMPBELL

VOLUME 263. Plasma Lipoproteins (Part C: Quantitation)
Edited by WILLIAM A. BRADLEY, SANDRA H. GIANTURCO, AND JERE P. SEGREST

VOLUME 264. Mitochondrial Biogenesis and Genetics (Part B)
Edited by GIUSEPPE M. ATTARDI AND ANNE CHOMYN

VOLUME 265. Cumulative Subject Index Volumes 228, 230–262

VOLUME 266. Computer Methods for Macromolecular Sequence Analysis
Edited by RUSSELL F. DOOLITTLE

VOLUME 267. Combinatorial Chemistry
Edited by JOHN N. ABELSON

VOLUME 268. Nitric Oxide (Part A: Sources and Detection of NO; NO Synthase)
Edited by LESTER PACKER

VOLUME 269. Nitric Oxide (Part B: Physiological and Pathological Processes)
Edited by LESTER PACKER

VOLUME 270. High Resolution Separation and Analysis of Biological Macromolecules (Part A: Fundamentals)
Edited by BARRY L. KARGER AND WILLIAM S. HANCOCK

VOLUME 271. High Resolution Separation and Analysis of Biological Macromolecules (Part B: Applications)
Edited by BARRY L. KARGER AND WILLIAM S. HANCOCK

VOLUME 272. Cytochrome P450 (Part B)
Edited by ERIC F. JOHNSON AND MICHAEL R. WATERMAN

VOLUME 273. RNA Polymerase and Associated Factors (Part A)
Edited by SANKAR ADHYA

VOLUME 274. RNA Polymerase and Associated Factors (Part B)
Edited by SANKAR ADHYA

VOLUME 275. Viral Polymerases and Related Proteins
Edited by LAWRENCE C. KUO, DAVID B. OLSEN, AND STEVEN S. CARROLL

VOLUME 276. Macromolecular Crystallography (Part A)
Edited by CHARLES W. CARTER, JR., AND ROBERT M. SWEET

VOLUME 277. Macromolecular Crystallography (Part B)
Edited by CHARLES W. CARTER, JR., AND ROBERT M. SWEET

VOLUME 278. Fluorescence Spectroscopy
Edited by LUDWIG BRAND AND MICHAEL L. JOHNSON

VOLUME 279. Vitamins and Coenzymes (Part I)
Edited by DONALD B. MCCORMICK, JOHN W. SUTTIE, AND CONRAD WAGNER

VOLUME 280. Vitamins and Coenzymes (Part J)
Edited by DONALD B. MCCORMICK, JOHN W. SUTTIE, AND CONRAD WAGNER

VOLUME 281. Vitamins and Coenzymes (Part K)
Edited by DONALD B. MCCORMICK, JOHN W. SUTTIE, AND CONRAD WAGNER

VOLUME 282. Vitamins and Coenzymes (Part L)
Edited by DONALD B. MCCORMICK, JOHN W. SUTTIE, AND CONRAD WAGNER

VOLUME 283. Cell Cycle Control
Edited by WILLIAM G. DUNPHY

VOLUME 284. Lipases (Part A: Biotechnology)
Edited by BYRON RUBIN AND EDWARD A. DENNIS

VOLUME 285. Cumulative Subject Index Volumes 263, 264, 266–284, 286–289

VOLUME 286. Lipases (Part B: Enzyme Characterization and Utilization)
Edited by BYRON RUBIN AND EDWARD A. DENNIS

VOLUME 287. Chemokines
Edited by RICHARD HORUK

VOLUME 288. Chemokine Receptors
Edited by RICHARD HORUK

VOLUME 289. Solid Phase Peptide Synthesis
Edited by GREGG B. FIELDS

VOLUME 290. Molecular Chaperones
Edited by GEORGE H. LORIMER AND THOMAS BALDWIN

VOLUME 291. Caged Compounds
Edited by GERARD MARRIOTT

VOLUME 292. ABC Transporters: Biochemical, Cellular, and Molecular Aspects
Edited by SURESH V. AMBUDKAR AND MICHAEL M. GOTTESMAN

VOLUME 293. Ion Channels (Part B)
Edited by P. MICHAEL CONN

VOLUME 294. Ion Channels (Part C)
Edited by P. MICHAEL CONN

VOLUME 295. Energetics of Biological Macromolecules (Part B)
Edited by GARY K. ACKERS AND MICHAEL L. JOHNSON

VOLUME 296. Neurotransmitter Transporters
Edited by SUSAN G. AMARA

VOLUME 297. Photosynthesis: Molecular Biology of Energy Capture
Edited by LEE MCINTOSH

VOLUME 298. Molecular Motors and the Cytoskeleton (Part B)
Edited by RICHARD B. VALLEE

VOLUME 299. Oxidants and Antioxidants (Part A)
Edited by LESTER PACKER

VOLUME 300. Oxidants and Antioxidants (Part B)
Edited by LESTER PACKER

VOLUME 301. Nitric Oxide: Biological and Antioxidant Activities (Part C)
Edited by LESTER PACKER

VOLUME 302. Green Fluorescent Protein
Edited by P. MICHAEL CONN

VOLUME 303. cDNA Preparation and Display
Edited by SHERMAN M. WEISSMAN

VOLUME 304. Chromatin
Edited by PAUL M. WASSARMAN AND ALAN P. WOLFFE

VOLUME 305. Bioluminescence and Chemiluminescence (Part C)
Edited by THOMAS O. BALDWIN AND MIRIAM M. ZIEGLER

VOLUME 306. Expression of Recombinant Genes in Eukaryotic Systems
Edited by JOSEPH C. GLORIOSO AND MARTIN C. SCHMIDT

VOLUME 307. Confocal Microscopy
Edited by P. MICHAEL CONN

VOLUME 308. Enzyme Kinetics and Mechanism (Part E: Energetics of Enzyme Catalysis)
Edited by DANIEL L. PURICH AND VERN L. SCHRAMM

VOLUME 309. Amyloid, Prions, and Other Protein Aggregates
Edited by RONALD WETZEL

VOLUME 310. Biofilms
Edited by RON J. DOYLE

VOLUME 311. Sphingolipid Metabolism and Cell Signaling (Part A)
Edited by ALFRED H. MERRILL, JR., AND YUSUF A. HANNUN

VOLUME 312. Sphingolipid Metabolism and Cell Signaling (Part B)
Edited by ALFRED H. MERRILL, JR., AND YUSUF A. HANNUN

VOLUME 313. Antisense Technology
(Part A: General Methods, Methods of Delivery, and RNA Studies)
Edited by M. IAN PHILLIPS

VOLUME 314. Antisense Technology (Part B: Applications)
Edited by M. IAN PHILLIPS

VOLUME 315. Vertebrate Phototransduction and the Visual Cycle (Part A)
Edited by KRZYSZTOF PALCZEWSKI

VOLUME 316. Vertebrate Phototransduction and the Visual Cycle (Part B)
Edited by KRZYSZTOF PALCZEWSKI

VOLUME 317. RNA–Ligand Interactions (Part A: Structural Biology Methods)
Edited by DANIEL W. CELANDER AND JOHN N. ABELSON

VOLUME 318. RNA–Ligand Interactions (Part B: Molecular Biology Methods)
Edited by DANIEL W. CELANDER AND JOHN N. ABELSON

VOLUME 319. Singlet Oxygen, UV-A, and Ozone
Edited by LESTER PACKER AND HELMUT SIES

VOLUME 320. Cumulative Subject Index Volumes 290–319

VOLUME 321. Numerical Computer Methods (Part C)
Edited by MICHAEL L. JOHNSON AND LUDWIG BRAND

VOLUME 322. Apoptosis
Edited by JOHN C. REED

VOLUME 323. Energetics of Biological Macromolecules (Part C)
Edited by MICHAEL L. JOHNSON AND GARY K. ACKERS

VOLUME 324. Branched-Chain Amino Acids (Part B)
Edited by ROBERT A. HARRIS AND JOHN R. SOKATCH

VOLUME 325. Regulators and Effectors of Small GTPases
(Part D: Rho Family)
Edited by W. E. BALCH, CHANNING J. DER, AND ALAN HALL

VOLUME 326. Applications of Chimeric Genes and Hybrid Proteins
(Part A: Gene Expression and Protein Purification)
Edited by JEREMY THORNER, SCOTT D. EMR, AND JOHN N. ABELSON

VOLUME 327. Applications of Chimeric Genes and Hybrid Proteins
(Part B: Cell Biology and Physiology)
Edited by JEREMY THORNER, SCOTT D. EMR, AND JOHN N. ABELSON

VOLUME 328. Applications of Chimeric Genes and Hybrid Proteins (Part C: Protein–Protein Interactions and Genomics)
Edited by JEREMY THORNER, SCOTT D. EMR, AND JOHN N. ABELSON

VOLUME 329. Regulators and Effectors of Small GTPases (Part E: GTPases Involved in Vesicular Traffic)
Edited by W. E. BALCH, CHANNING J. DER, AND ALAN HALL

VOLUME 330. Hyperthermophilic Enzymes (Part A)
Edited by MICHAEL W. W. ADAMS AND ROBERT M. KELLY

VOLUME 331. Hyperthermophilic Enzymes (Part B)
Edited by MICHAEL W. W. ADAMS AND ROBERT M. KELLY

VOLUME 332. Regulators and Effectors of Small GTPases (Part F: Ras Family I)
Edited by W. E. BALCH, CHANNING J. DER, AND ALAN HALL

VOLUME 333. Regulators and Effectors of Small GTPases (Part G: Ras Family II)
Edited by W. E. BALCH, CHANNING J. DER, AND ALAN HALL

VOLUME 334. Hyperthermophilic Enzymes (Part C)
Edited by MICHAEL W. W. ADAMS AND ROBERT M. KELLY

VOLUME 335. Flavonoids and Other Polyphenols
Edited by LESTER PACKER

VOLUME 336. Microbial Growth in Biofilms (Part A: Developmental and Molecular Biological Aspects)
Edited by RON J. DOYLE

VOLUME 337. Microbial Growth in Biofilms (Part B: Special Environments and Physicochemical Aspects)
Edited by RON J. DOYLE

VOLUME 338. Nuclear Magnetic Resonance of Biological Macromolecules (Part A)
Edited by THOMAS L. JAMES, VOLKER DÖTSCH, AND ULI SCHMITZ

VOLUME 339. Nuclear Magnetic Resonance of Biological Macromolecules (Part B)
Edited by THOMAS L. JAMES, VOLKER DÖTSCH, AND ULI SCHMITZ

VOLUME 340. Drug–Nucleic Acid Interactions
Edited by JONATHAN B. CHAIRES AND MICHAEL J. WARING

VOLUME 341. Ribonucleases (Part A)
Edited by ALLEN W. NICHOLSON

VOLUME 342. Ribonucleases (Part B)
Edited by ALLEN W. NICHOLSON

VOLUME 343. G Protein Pathways (Part A: Receptors)
Edited by RAVI IYENGAR AND JOHN D. HILDEBRANDT

VOLUME 344. G Protein Pathways (Part B: G Proteins and Their Regulators)
Edited by RAVI IYENGAR AND JOHN D. HILDEBRANDT

VOLUME 345. G Protein Pathways (Part C: Effector Mechanisms)
Edited by RAVI IYENGAR AND JOHN D. HILDEBRANDT

VOLUME 346. Gene Therapy Methods
Edited by M. IAN PHILLIPS

VOLUME 347. Protein Sensors and Reactive Oxygen Species (Part A: Selenoproteins and Thioredoxin)
Edited by HELMUT SIES AND LESTER PACKER

VOLUME 348. Protein Sensors and Reactive Oxygen Species (Part B: Thiol Enzymes and Proteins)
Edited by HELMUT SIES AND LESTER PACKER

VOLUME 349. Superoxide Dismutase
Edited by LESTER PACKER

VOLUME 350. Guide to Yeast Genetics and Molecular and Cell Biology (Part B)
Edited by CHRISTINE GUTHRIE AND GERALD R. FINK

VOLUME 351. Guide to Yeast Genetics and Molecular and Cell Biology (Part C)
Edited by CHRISTINE GUTHRIE AND GERALD R. FINK

VOLUME 352. Redox Cell Biology and Genetics (Part A)
Edited by CHANDAN K. SEN AND LESTER PACKER

VOLUME 353. Redox Cell Biology and Genetics (Part B)
Edited by CHANDAN K. SEN AND LESTER PACKER

VOLUME 354. Enzyme Kinetics and Mechanisms (Part F: Detection and Characterization of Enzyme Reaction Intermediates)
Edited by DANIEL L. PURICH

VOLUME 355. Cumulative Subject Index Volumes 321–354

VOLUME 356. Laser Capture Microscopy and Microdissection
Edited by P. MICHAEL CONN

VOLUME 357. Cytochrome P450, Part C
Edited by ERIC F. JOHNSON AND MICHAEL R. WATERMAN

VOLUME 358. Bacterial Pathogenesis (Part C: Identification, Regulation, and Function of Virulence Factors)
Edited by VIRGINIA L. CLARK AND PATRIK M. BAVOIL

VOLUME 359. Nitric Oxide (Part D)
Edited by ENRIQUE CADENAS AND LESTER PACKER

VOLUME 360. Biophotonics (Part A)
Edited by GERARD MARRIOTT AND IAN PARKER

VOLUME 361. Biophotonics (Part B)
Edited by GERARD MARRIOTT AND IAN PARKER

VOLUME 362. Recognition of Carbohydrates in Biological Systems (Part A)
Edited by YUAN C. LEE AND REIKO T. LEE

VOLUME 363. Recognition of Carbohydrates in Biological Systems (Part B)
Edited by YUAN C. LEE AND REIKO T. LEE

VOLUME 364. Nuclear Receptors
Edited by DAVID W. RUSSELL AND DAVID J. MANGELSDORF

VOLUME 365. Differentiation of Embryonic Stem Cells
Edited by PAUL M. WASSAUMAN AND GORDON M. KELLER

VOLUME 366. Protein Phosphatases
Edited by SUSANNE KLUMPP AND JOSEF KRIEGLSTEIN

VOLUME 367. Liposomes (Part A)
Edited by NEJAT DÜZGÜNEŞ

VOLUME 368. Macromolecular Crystallography (Part C)
Edited by CHARLES W. CARTER, JR., AND ROBERT M. SWEET

VOLUME 369. Combinational Chemistry (Part B)
Edited by GUILLERMO A. MORALES AND BARRY A. BUNIN

VOLUME 370. RNA Polymerases and Associated Factors (Part C)
Edited by SANKAR L. ADHYA AND SUSAN GARGES

VOLUME 371. RNA Polymerases and Associated Factors (Part D)
Edited by SANKAR L. ADHYA AND SUSAN GARGES

VOLUME 372. Liposomes (Part B)
Edited by NEJAT DÜZGÜNEŞ

VOLUME 373. Liposomes (Part C)
Edited by NEJAT DÜZGÜNEŞ

VOLUME 374. Macromolecular Crystallography (Part D)
Edited by CHARLES W. CARTER, JR., AND ROBERT W. SWEET

VOLUME 375. Chromatin and Chromatin Remodeling Enzymes (Part A)
Edited by C. DAVID ALLIS AND CARL WU

VOLUME 376. Chromatin and Chromatin Remodeling Enzymes (Part B)
Edited by C. DAVID ALLIS AND CARL WU

VOLUME 377. Chromatin and Chromatin Remodeling Enzymes (Part C)
Edited by C. DAVID ALLIS AND CARL WU

VOLUME 378. Quinones and Quinone Enzymes (Part A)
Edited by HELMUT SIES AND LESTER PACKER

VOLUME 379. Energetics of Biological Macromolecules (Part D)
Edited by JO M. HOLT, MICHAEL L. JOHNSON, AND GARY K. ACKERS

VOLUME 380. Energetics of Biological Macromolecules (Part E)
Edited by JO M. HOLT, MICHAEL L. JOHNSON, AND GARY K. ACKERS

VOLUME 381. Oxygen Sensing
Edited by CHANDAN K. SEN AND GREGG L. SEMENZA

VOLUME 382. Quinones and Quinone Enzymes (Part B)
Edited by HELMUT SIES AND LESTER PACKER

VOLUME 383. Numerical Computer Methods (Part D)
Edited by LUDWIG BRAND AND MICHAEL L. JOHNSON

VOLUME 384. Numerical Computer Methods (Part E)
Edited by LUDWIG BRAND AND MICHAEL L. JOHNSON

VOLUME 385. Imaging in Biological Research (Part A)
Edited by P. MICHAEL CONN

VOLUME 386. Imaging in Biological Research (Part B)
Edited by P. MICHAEL CONN

VOLUME 387. Liposomes (Part D)
Edited by NEJAT DÜZGÜNEŞ

VOLUME 388. Protein Engineering
Edited by DAN E. ROBERTSON AND JOSEPH P. NOEL

VOLUME 389. Regulators of G-Protein Signaling (Part A)
Edited by DAVID P. SIDEROVSKI

VOLUME 390. Regulators of G-Protein Signaling (Part B)
Edited by DAVID P. SIDEROVSKI

VOLUME 391. Liposomes (Part E)
Edited by NEJAT DÜZGÜNEŞ

VOLUME 392. RNA Interference
Edited by ENGELKE ROSSI

VOLUME 393. Circadian Rhythms
Edited by MICHAEL W. YOUNG

VOLUME 394. Nuclear Magnetic Resonance of Biological Macromolecules (Part C)
Edited by THOMAS L. JAMES

VOLUME 395. Producing the Biochemical Data (Part B)
Edited by ELIZABETH A. ZIMMER AND ERIC H. ROALSON

VOLUME 396. Nitric Oxide (Part E)
Edited by LESTER PACKER AND ENRIQUE CADENAS

VOLUME 397. Environmental Microbiology
Edited by JARED R. LEADBETTER

VOLUME 398. Ubiquitin and Protein Degradation (Part A)
Edited by RAYMOND J. DESHAIES

VOLUME 399. Ubiquitin and Protein Degradation (Part B)
Edited by RAYMOND J. DESHAIES

VOLUME 400. Phase II Conjugation Enzymes and Transport Systems
Edited by HELMUT SIES AND LESTER PACKER

VOLUME 401. Glutathione Transferases and Gamma Glutamyl Transpeptidases
Edited by HELMUT SIES AND LESTER PACKER

VOLUME 402. Biological Mass Spectrometry
Edited by A. L. BURLINGAME

VOLUME 403. GTPases Regulating Membrane Targeting and Fusion
Edited by WILLIAM E. BALCH, CHANNING J. DER, AND ALAN HALL

VOLUME 404. GTPases Regulating Membrane Dynamics
Edited by WILLIAM E. BALCH, CHANNING J. DER, AND ALAN HALL

VOLUME 405. Mass Spectrometry: Modified Proteins and Glycoconjugates
Edited by A. L. BURLINGAME

VOLUME 406. Regulators and Effectors of Small GTPases: Rho Family
Edited by WILLIAM E. BALCH, CHANNING J. DER, AND ALAN HALL

VOLUME 407. Regulators and Effectors of Small GTPases: Ras Family
Edited by WILLIAM E. BALCH, CHANNING J. DER, AND ALAN HALL

VOLUME 408. DNA Repair (Part A)
Edited by JUDITH L. CAMPBELL AND PAUL MODRICH

VOLUME 409. DNA Repair (Part B)
Edited by JUDITH L. CAMPBELL AND PAUL MODRICH

VOLUME 410. DNA Microarrays (Part A: Array Platforms and Web-Bench Protocols)
Edited by ALAN KIMMEL AND BRIAN OLIVER

VOLUME 411. DNA Microarrays (Part B: Databases and Statistics)
Edited by ALAN KIMMEL AND BRIAN OLIVER

VOLUME 412. Amyloid, Prions, and Other Protein Aggregates (Part B)
Edited by INDU KHETERPAL AND RONALD WETZEL

VOLUME 413. Amyloid, Prions, and Other Protein Aggregates (Part C)
Edited by INDU KHETERPAL AND RONALD WETZEL

VOLUME 414. Measuring Biological Responses with Automated Microscopy
Edited by JAMES INGLESE

VOLUME 415. Glycobiology
Edited by MINORU FUKUDA

VOLUME 416. Glycomics
Edited by MINORU FUKUDA

VOLUME 417. Functional Glycomics
Edited by MINORU FUKUDA

VOLUME 418. Embryonic Stem Cells
Edited by IRINA KLIMANSKAYA AND ROBERT LANZA

VOLUME 419. Adult Stem Cells
Edited by IRINA KLIMANSKAYA AND ROBERT LANZA

VOLUME 420. Stem Cell Tools and Other Experimental Protocols
Edited by IRINA KLIMANSKAYA AND ROBERT LANZA

VOLUME 421. Advanced Bacterial Genetics: Use of Transposons and Phage for Genomic Engineering
Edited by KELLY T. HUGHES

VOLUME 422. Two-Component Signaling Systems, Part A
Edited by MELVIN I. SIMON, BRIAN R. CRANE, AND ALEXANDRINE CRANE

VOLUME 423. Two-Component Signaling Systems, Part B
Edited by MELVIN I. SIMON, BRIAN R. CRANE, AND ALEXANDRINE CRANE

VOLUME 424. RNA Editing
Edited by JONATHA M. GOTT

VOLUME 425. RNA Modification
Edited by JONATHA M. GOTT

VOLUME 426. Integrins
Edited by DAVID CHERESH

VOLUME 427. MicroRNA Methods
Edited by JOHN J. ROSSI

VOLUME 428. Osmosensing and Osmosignaling
Edited by HELMUT SIES AND DIETER HAUSSINGER

VOLUME 429. Translation Initiation: Extract Systems and Molecular Genetics
Edited by JON LORSCH

VOLUME 430. Translation Initiation: Reconstituted Systems and Biophysical Methods
Edited by JON LORSCH

VOLUME 431. Translation Initiation: Cell Biology, High-Throughput and Chemical-Based Approaches
Edited by JON LORSCH

VOLUME 432. Lipidomics and Bioactive Lipids: Mass-Spectrometry–Based Lipid Analysis
Edited by H. ALEX BROWN

VOLUME 433. Lipidomics and Bioactive Lipids: Specialized Analytical Methods and Lipids in Disease
Edited by H. ALEX BROWN

VOLUME 434. Lipidomics and Bioactive Lipids: Lipids and Cell Signaling
Edited by H. ALEX BROWN

VOLUME 435. Oxygen Biology and Hypoxia
Edited by HELMUT SIES AND BERNHARD BRÜNE

VOLUME 436. Globins and Other Nitric Oxide-Reactive Protiens (Part A)
Edited by ROBERT K. POOLE

VOLUME 437. Globins and Other Nitric Oxide-Reactive Protiens (Part B)
Edited by ROBERT K. POOLE

VOLUME 438. Small GTPases in Disease (Part A)
Edited by WILLIAM E. BALCH, CHANNING J. DER, AND ALAN HALL

VOLUME 439. Small GTPases in Disease (Part B)
Edited by WILLIAM E. BALCH, CHANNING J. DER, AND ALAN HALL

VOLUME 440. Nitric Oxide, Part F Oxidative and Nitrosative Stress in Redox Regulation of Cell Signaling
Edited by ENRIQUE CADENAS AND LESTER PACKER

VOLUME 441. Nitric Oxide, Part G Oxidative and Nitrosative Stress in Redox Regulation of Cell Signaling
Edited by ENRIQUE CADENAS AND LESTER PACKER

VOLUME 442. Programmed Cell Death, General Principles for Studying Cell Death (Part A)
Edited by ROYA KHOSRAVI-FAR, ZAHRA ZAKERI, RICHARD A. LOCKSHIN, AND MAURO PIACENTINI

VOLUME 443. Angiogenesis: *In Vitro* Systems
Edited by DAVID A. CHERESH

VOLUME 444. Angiogenesis: *In Vivo* Systems (Part A)
Edited by DAVID A. CHERESH

VOLUME 445. Angiogenesis: *In Vivo* Systems (Part B)
Edited by DAVID A. CHERESH

VOLUME 446. Programmed Cell Death, The Biology and Therapeutic Implications of Cell Death (Part B)
Edited by ROYA KHOSRAVI-FAR, ZAHRA ZAKERI, RICHARD A. LOCKSHIN, AND MAURO PIACENTINI

VOLUME 447. RNA Turnover in Bacteria, Archaea and Organelles
Edited by LYNNE E. MAQUAT AND CECILIA M. ARRAIANO

VOLUME 448. RNA Turnover in Eukaryotes: Nucleases, Pathways and Analysis of mRNA Decay
Edited by LYNNE E. MAQUAT AND MEGERDITCH KILEDJIAN

VOLUME 449. RNA Turnover in Eukaryotes: Analysis of Specialized and Quality Control RNA Decay Pathways
Edited by LYNNE E. MAQUAT AND MEGERDITCH KILEDJIAN

VOLUME 450. Fluorescence Spectroscopy
Edited by LUDWIG BRAND AND MICHAEL L. JOHNSON

VOLUME 451. Autophagy: Lower Eukaryotes and Non-Mammalian Systems (Part A)
Edited by DANIEL J. KLIONSKY

VOLUME 452. Autophagy in Mammalian Systems (Part B)
Edited by DANIEL J. KLIONSKY

VOLUME 453. Autophagy in Disease and Clinical Applications (Part C)
Edited by DANIEL J. KLIONSKY

VOLUME 454. Computer Methods (Part A)
Edited by MICHAEL L. JOHNSON AND LUDWIG BRAND

VOLUME 455. Biothermodynamics (Part A)
Edited by MICHAEL L. JOHNSON, JO M. HOLT, AND GARY K. ACKERS (RETIRED)

VOLUME 456. Mitochondrial Function, Part A: Mitochondrial Electron Transport Complexes and Reactive Oxygen Species
Edited by WILLIAM S. ALLISON AND IMMO E. SCHEFFLER

VOLUME 457. Mitochondrial Function, Part B: Mitochondrial Protein Kinases, Protein Phosphatases and Mitochondrial Diseases
Edited by WILLIAM S. ALLISON AND ANNE N. MURPHY

VOLUME 458. Complex Enzymes in Microbial Natural Product Biosynthesis, Part A: Overview Articles and Peptides
Edited by DAVID A. HOPWOOD

VOLUME 459. Complex Enzymes in Microbial Natural Product Biosynthesis, Part B: Polyketides, Aminocoumarins and Carbohydrates
Edited by DAVID A. HOPWOOD

VOLUME 460. Chemokines, Part A
Edited by TRACY M. HANDEL AND DAMON J. HAMEL

VOLUME 461. Chemokines, Part B
Edited by TRACY M. HANDEL AND DAMON J. HAMEL

VOLUME 462. Non-Natural Amino Acids
Edited by TOM W. MUIR AND JOHN N. ABELSON

VOLUME 463. Guide to Protein Purification, 2nd Edition
Edited by RICHARD R. BURGESS AND MURRAY P. DEUTSCHER

VOLUME 464. Liposomes, Part F
Edited by NEJAT DÜZGÜNEŞ

VOLUME 465. Liposomes, Part G
Edited by NEJAT DÜZGÜNEŞ

VOLUME 466. Biothermodynamics, Part B
Edited by MICHAEL L. JOHNSON, GARY K. ACKERS, AND JO M. HOLT

VOLUME 467. Computer Methods Part B
Edited by MICHAEL L. JOHNSON AND LUDWIG BRAND

VOLUME 468. Biophysical, Chemical, and Functional Probes of RNA Structure, Interactions and Folding: Part A
Edited by DANIEL HERSCHLAG

VOLUME 469. Biophysical, Chemical, and Functional Probes of RNA Structure, Interactions and Folding: Part B
Edited by DANIEL HERSCHLAG

VOLUME 470. Guide to Yeast Genetics: Functional Genomics, Proteomics, and Other Systems Analysis, 2nd Edition
Edited by GERALD FINK, JONATHAN WEISSMAN, AND CHRISTINE GUTHRIE

VOLUME 471. Two-Component Signaling Systems, Part C
Edited by MELVIN I. SIMON, BRIAN R. CRANE, AND ALEXANDRINE CRANE

VOLUME 472. Single Molecule Tools, Part A: Fluorescence Based Approaches
Edited by NILS G. WALTER

VOLUME 473. Thiol Redox Transitions in Cell Signaling, Part A Chemistry and Biochemistry of Low Molecular Weight and Protein Thiols
Edited by ENRIQUE CADENAS AND LESTER PACKER

VOLUME 474. Thiol Redox Transitions in Cell Signaling, Part B Cellular Localization and Signaling
Edited by ENRIQUE CADENAS AND LESTER PACKER

VOLUME 475. Single Molecule Tools, Part B: Super-Resolution, Particle Tracking, Multiparameter, and Force Based Methods
Edited by NILS G. WALTER

VOLUME 476. Guide to Techniques in Mouse Development, Part A Mice, Embryos, and Cells, 2nd Edition
Edited by PAUL M. WASSARMAN AND PHILIPPE M. SORIANO

VOLUME 477. Guide to Techniques in Mouse Development, Part B Mouse Molecular Genetics, 2nd Edition
Edited by PAUL M. WASSARMAN AND PHILIPPE M. SORIANO

VOLUME 478. Glycomics
Edited by MINORU FUKUDA

VOLUME 479. Functional Glycomics
Edited by MINORU FUKUDA

VOLUME 480. Glycobiology
Edited by MINORU FUKUDA

VOLUME 481. Cryo-EM, Part A: Sample Preparation and Data Collection
Edited by GRANT J. JENSEN

VOLUME 482. Cryo-EM, Part B: 3-D Reconstruction
Edited by GRANT J. JENSEN

VOLUME 483. Cryo-EM, Part C: Analyses, Interpretation, and Case Studies
Edited by GRANT J. JENSEN

VOLUME 484. Constitutive Activity in Receptors and Other Proteins, Part A
Edited by P. MICHAEL CONN

VOLUME 485. Constitutive Activity in Receptors and Other Proteins, Part B
Edited by P. MICHAEL CONN

VOLUME 486. Research on Nitrification and Related Processes, Part A
Edited by MARTIN G. KLOTZ

VOLUME 487. Computer Methods, Part C
Edited by MICHAEL L. JOHNSON AND LUDWIG BRAND

VOLUME 488. Biothermodynamics, Part C
Edited by MICHAEL L. JOHNSON, JO M. HOLT, AND GARY K. ACKERS

VOLUME 489. The Unfolded Protein Response and Cellular Stress, Part A
Edited by P. MICHAEL CONN

VOLUME 490. The Unfolded Protein Response and Cellular Stress, Part B
Edited by P. MICHAEL CONN

VOLUME 491. The Unfolded Protein Response and Cellular Stress, Part C
Edited by P. MICHAEL CONN

VOLUME 492. Biothermodynamics, Part D
Edited by MICHAEL L. JOHNSON, JO M. HOLT, AND GARY K. ACKERS

VOLUME 493. Fragment-Based Drug Design
Tools, Practical Approaches, and Examples
Edited by LAWRENCE C. KUO

VOLUME 494. Methods in Methane Metabolism, Part A
Methanogenesis
Edited by AMY C. ROSENZWEIG AND STEPHEN W. RAGSDALE

VOLUME 495. Methods in Methane Metabolism, Part B
Methanotrophy
Edited by AMY C. ROSENZWEIG AND STEPHEN W. RAGSDALE

Volume 496. Research on Nitrification and Related Processes, Part B
Edited by Martin G. Klotz and Lisa Y. Stein

Volume 497. Synthetic Biology, Part A
Methods for Part/Device Characterization and Chassis Engineering
Edited by Christopher Voigt

Volume 498. Synthetic Biology, Part B
Computer Aided Design and DNA Assembly
Edited by Christopher Voigt

Volume 499. Biology of Serpins
Edited by James C. Whisstock and Phillip I. Bird

CHAPTER ONE

ANALYSIS OF SERPIN SECRETION, MISFOLDING, AND SURVEILLANCE IN THE ENDOPLASMIC RETICULUM

Shujuan Pan,[*] Michael J. Iannotti,[†] and Richard N. Sifers[*,†,‡,§]

Contents

1. Introduction — 2
2. Protocols — 3
    2.1. Transfection of mammalian cell lines — 3
    2.2. Monitoring newly synthesized protein fate — 4
    2.3. Identification of physically interacting proteins — 6
    2.4. Time course of protein interactions and role verification — 8
    2.5. Contribution of N-linked oligosaccharide processing events — 10
3. Concluding Remarks — 14
Acknowledgments — 14
References — 14

Abstract

Biological checkpoints are known to function in the cellular nucleus to monitor the integrity of inherited genetic information. It is now understood that posttranslational checkpoint systems operate in numerous biosynthetic compartments where they orchestrate the surveillance of encoded protein structures. This is particularly true for the serpins where opposing, but complementary, systems operate in the early secretory pathway to initially facilitate protein folding and then selectively target the misfolded proteins for proteolytic elimination. A current challenge is to elucidate how this posttranslational checkpoint can modify the severity of numerous loss-of-function and gain-of-toxic-function diseases, some of which are caused by mutant serpins. This chapter provides a description of the experimental methodology by which the fate of a newly synthesized serpin is monitored, and how the

[*] Department of Pathology and Immunology, Baylor College of Medicine, Houston, Texas, USA
[†] Cell and Molecular Biology Interdepartmental Graduate Program, Baylor College of Medicine, Houston, Texas, USA
[‡] Department of Molecular and Cellular Biology, Baylor College of Medicine, Houston, Texas, USA
[§] Department of Molecular Physiology and Biophysics, Baylor College of Medicine, Houston, Texas, USA

processing of asparagine-linked oligosaccharides helps to facilitate both the protein folding and disposal events.

1. Introduction

The conformational maturation and secretion of an active serpin involves a multistep process. Initially, cellular machinery orchestrate a series of requisite events that eventually direct the cotranslational insertion of the nascent polypeptide across the membrane of the rough endoplasmic reticulum (ER) and into its lumen (Cabral *et al.*, 2001). Transient physical interactions with a group of molecular chaperones promote correct protein folding by diminishing the frequency of unproductive folding events that can lead to inappropriate aggregation (Choudhury *et al.*, 1977; Fewell *et al.*, 2001).

Efficient conformational maturation functions as a quality control standard, permitting the progression of active proteins beyond the ER (Wu *et al.*, 2003). Noncompliance results in the ER retention of misfolded polypeptides and unassembled protein subunits, both of which exhibit prolonged physical interaction with molecular chaperones and are eventually targeted for intracellular degradation (Liu *et al.*, 1997). In recent years, a picture has emerged in which the addition of asparagine-linked oligosaccharides, and their covalent modification, functions to assist the productive folding of newly synthesized serpins by facilitating their physical interactions with a small family of lectins that contribute to the glycoprotein folding machinery (Cabral *et al.*, 2001; Ellgaard *et al.*, 1999).

Importantly, additional carbohydrate modifications are responsible for generating a tag that selectively targets the nonnative proteins for intracellular degradation (Cabral *et al.*, 2001; Molinari, 2007). A model has been proposed in which the timing of the latter glycan modification, relative to the duration of nonnative glycoprotein structure, functions as an underlying stochastic mechanism by which serpins that exhibit permanent structural defects are selectively targeted for intracellular disposal (Wu *et al.*, 2003). Later stages of the disposal process include retrotranslocation across the ER membrane and into the cytoplasm, polyubiquitination, and elimination by 26S proteasomes (Bonifacino and Weissman, 1998).

Almost all naturally occurring secretion-incompetent serpin variants misfold following biosynthesis (Sifers *et al.*, 1992), and therefore fail to pass the quality control checkpoint. The serpins have provided a model system to elucidate how glycoprotein processing contributes to these surveillance mechanisms. The following information describes the experimental strategies and methodologies designed to elucidate the involvement of asparagine-linked oligosaccharide processing in regulating the fate of a newly synthesized serpin.

2. Protocols

2.1. Transfection of mammalian cell lines

Because the biosynthesis of serpins is naturally expressed in only a few cell types, it will be necessary to transfect the serpin cDNA into a selected cultured mammalian cell line to generate a system for studying the protein's fate following biosynthesis. The wild-type cDNA can be utilized in the expression studies, or one that has undergone site-directed mutagenesis either to mimic a naturally occurring variant or to test a specific structural hypothesis. The cDNA is subcloned into any number of commercially available mammalian expression vectors (Invitrogen).

DNA restriction enzyme digestions, in combination with gel electrophoresis, can be used to validate the insert's proper orientation, which can be further verified by DNA sequencing analyses. Transfections can be performed by a number of commercially available reagents, and some vectors include a gene for antibiotic-based resistance if one intends to eventually select stably transfected clonal cell lines for experimentation. The expansion of selected cell lines is based on the relative level of recombinant serpin synthesized, and this is routinely measured by ECL Western blotting (Amersham Pharmacia; Le *et al.*, 1992; Sifers *et al.*, 1989). Alternatively, transient expression studies can be initiated within 24–48 h posttransfection.

To quantify the relative level of serpin expression, the growth media are collected and set on ice for the eventual detection of the secreted protein. For detection of the intracellular pool, cell monolayers are washed with ice cold PBS and then lysed by gently scraping with a rubber spatula in the smallest possible volume of $0.1\ M$ Tris–HCl, pH 7.5, containing $0.15\ M$ NaCl, 0.1% SDS, and a protease inhibitor cocktail (Sigma). Insoluble debris are removed from the cell lysate by centrifugation at $10,000 \times g$ for 15 min at 4 °C. Any intact cells that might be present in the collected media are removed by low-speed centrifugation. Equivalent aliquots of soluble cell lysates or media are heated with gel loading buffer and resolved by SDS–PAGE (10% polyacrylamide). Total protein is electrophoretically transferred to nitrocellulose (Le *et al.*, 1992), and the inclusion of commercially available prestained protein molecular weight markers during SDS–PAGE provides a method to detect successful electrophoretic transfer. The serpin's expression is then determined by Western blotting. Preblocking and signal detection are performed according to the manufacturer's instructions, except that it is sometimes helpful to include the blocking reagent during incubation with both the primary and secondary antibodies. Subsequent washing of the nitrocellulose, according to the manufacturer's instructions, can be supplemented with a final gentle 1 min room temperature wash with 0.1% SDS in phosphate-buffered saline (PBS) to eliminate any undesired background bands. The amount of commercially available antiserum against the serpin and the appropriate amount of conjugated secondary antibody are dependent on their titer, and are therefore empirically determined. Usually, optimal detection requires that both antibodies are diluted to the lowest concentration capable of generating a specific positive signal. Untransfected cells serve as negative control, whereas a human hepatoma cell line will often provide a positive control. Generally, transfected cell lines with the greatest quantity of detected recombinant serpin are selected for subsequent pulse–chase radiolabeling studies, but only after considering several potential caveats described below (discussed in Section 2.2.4).

2.2. Monitoring newly synthesized protein fate

The fate of the newly synthesized recombinant serpin can be monitored by a combination of metabolic pulse–chase radiolabeling and immunoprecipitation techniques. A cohort of the newly synthesized protein will either be secreted or degraded following prolonged intracellular retention. Methods to evaluate these fates are described below.

2.2.1. Metabolic pulse–chase radiolabeling

Metabolic pulse–chase radiolabeling, in combination with immunoprecipitation, provides a means to both monitor and quantify changes in the fate of newly synthesized recombinant serpins (Le et al., 1992; Liu et al., 1997, 1999). Washed monolayers of cells (of identical confluency) are incubated with the cell's regular growth media, deficient in methionine, and supplemented with a commercial source of [^{35}S]methionine for a period of time (the pulse) that will allow for incorporation of the radiolabel into only those proteins synthesized during that time frame. The amount of fetal bovine serum (FBS) must be substantially diminished to eliminate unlabeled methionine or replaced with commercially available dialyzed FBS. The amount of radiolabel and duration of the "pulse" must be determined empirically in response to either the original transfection efficiency (for transient expression studies) or overall expression levels (for stably transfected cell lines) determined in methods described above. As a rule, the shortest pulse period works best as it will allow you to begin monitoring the protein at its earliest stages following biosynthesis. If necessary to enhance the incorporation of radiolabel, the pulse can be preceded by a 15–30 min "methionine starvation" period in which the cells are preincubated with growth medium deficient in methionine.

At the end of the pulse, unincorporated radiolabel is removed from the monolayers by washing with 37 °C PBS. The radiolabeled cells are then subjected to a "chase" period that involves 37 °C incubation in regular growth medium supplemented with a sixfold excess of unlabeled methionine. At this (the zero chase) time point, and chosen subsequent time points (to be empirically determined), the fate of the radiolabeled serpin can be monitored by collecting the media in tubes that are subsequently set on ice. The remaining cell monolayer is washed with ice cold PBS, and soluble cell lysates are prepared by lysis with 0.1% SDS and centrifugation, as described above. Alternatively, one can prevent the solubilization of a significant amount of unwanted cellular protein by replacing the SDS with 0.5% NP-40 or Triton X-100 which are nonionic detergents.

2.2.2. Serpin immunoprecipitation and quantification

The serpin can be isolated from the cell lysate and media by immunoprecipitation (specifically immunoabsorption). An excess of serpin-specific antiserum (empirically determined) is added to each lysate or media sample and incubated with constant gentle rotation overnight at 4 °C. The immunocomplexes are then collected by the addition of beaded Protein A-agarose, or Protein G-agarose, followed by a 1-h incubation with constant rotation at 4 °C. Alternatively, it is often possible to shorten the overall process by prebinding the antiserum to the agarose beads (for 1 h at 4 °C) and then incubating this mixture with either the lysate or media sample for 1–2 h with constant gentle rotation at 4 °C. The usefulness of this alternative procedure must be empirically determined, but can often diminish nonspecific protein binding and/or prevent the dissociation of protein complexes of which the serpin might serve as a transient component (discussed in Section 2.3.1).

The absorbed immunocomplexes are then subjected to several 4 °C washes, performed at constant agitation (Eppendorf agitation mixer 5432), in 1 ml of 0.05 M Tris–HCl, pH 7.5, supplemented with 0.5% NP-40 and 0.5 M NaCl. A final wash is performed with PBS to remove excess salt prior to the release of proteins from the beads upon incubation for 5 min at 95 °C in gel loading buffer. The supernatants are collected after centrifugation, and the proteins are resolved by electrophoresis in SDS–PAGE. The resulting gels are soaked with fluorographic enhancers and vacuum-dried at 50 °C (Bio-Rad). Radiolabeled bands are specifically detected and quantified by densitometric scanning, or via phosphorimager analysis using related software. Efficient secretion of the serpin corresponds with its quantitative appearance in the media with concomitant loss from cell lysates.

2.2.3. Identification of the intracellular degradation system

In the absence of loop-sheet polymerization (Lomas *et al.*, 1992; Sifers *et al.*, 1992; Yu *et al.*, 1995), two fates usually await a newly synthesized serpin; either the radiolabeled molecules correctly fold and are secreted into the media, as assessed by methodology described above, or they misfold and are retained in the cell prior to intracellular proteolysis. This disposal process, which is usually accomplished by cytoplasmic 26S proteasomes, is often preceded by a characteristic lag period (Le *et al.*, 1990; Wu *et al.*, 1994) in which the nonnative molecules are specifically tagged in the ER lumen (discussed in Section 2.5) for retrograde translocation into the cytoplasm.

However, because misfolded molecules can become insoluble within cells and apparently disappear, detectable intracellular loss of radiolabeled molecules without concomitant secretion demands an additional level of experimentation to establish the involvement of intracellular proteolysis. This is usually accomplished by the capacity of specific proteolytic inhibitors to prevent the loss of protein. The activity of proteasomes can be inhibited with several commercially available compounds such as lactacystin or MG132 (Liu *et al.*, 1997, 1999; Qu *et al.*, 1977). In contrast, degradation by lysosomes can be interrupted with compounds such as chloroquine or ammonium chloride which function as lysosomotrophic amines (Le *et al.*, 1990; Wu *et al.*, 2007). Finally, the drug wortmannin (Sigma) has been used as an inhibitor of autophagy, as have commercially available siRNAs targeted against specific essential autophagic components (Teckman and Perlmutter, 2000). When working with an inhibitor, the compound is added to cell monolayers and then incubated for 30 min at 37 °C in regular growth medium prior to the pulse. During both the pulse and chase periods, the compound is maintained at that same concentration to maintain its effect.

When the data are quantified, this aforementioned combination of methodologies allows one to determine whether the serpin is subject to intracellular retention and identifies the responsible machinery. In contrast, the failure of these inhibitors to prevent the apparent loss of the radiolabeled molecules strengthens the argument that insolubility is the true culprit. Often, this conclusion can be verified by examining the cell pellet for the radiolabeled immunoreactive serpin (Graham *et al.*, 1990; Lindblad *et al.*, 2007).

2.2.4. Experimental caveats

It should be noted that different cell types may exhibit distinct capacities to handle the folding and retention of serpins. This is especially true in those instances in which the level of recombinant serpin expression exceeds either the folding or quality control capacity of the host cell. For example, saturation of the retention system might lead to the secretion of a misfolded protein. Likewise, adaptation to long-term culture conditions in which selective pressures are missing could have the same outcome, irrespective of the expression level. One can test these possibilities by first monitoring the fate of a terminally folded variant to ensure that it is completely retained by the host cell line, and is efficiently degraded by an inhibitable system.

2.3. Identification of physically interacting proteins

Numerous intracellular proteins help to facilitate the folding and degradation of newly synthesized serpins (Cabral *et al.*, 2000, 2002; Liu *et al.*, 1997; Wu *et al.*, 2003). Several of these exhibit a transient physical association with the serpin and can therefore be detected by coimmunoprecipitation (coimmunoabsorption) studies.

2.3.1. Protein coimmunoprecipitation and identification

Detection of the bound proteins can be accomplished by metabolic labeling. However, because many exhibit relatively long intracellular half-lives, it is necessary to subject the host cells to long-term (>6 h) metabolic radiolabeling in regular growth medium supplemented with [^{35}S]methionine prior to coimmunoprecipitation (Choudhury et al., 1997; Le et al., 1994).

To maintain protein interactions, cell monolayers must be lysed under low-stringency conditions. This involves lysis at 4 °C with the smallest volume of 0.1 M Tris–HCl, pH 7.5, 0.5% NP-40, 0.15 M NaCl, plus centrifugation as described above. Also, ER functions as an intracellular storage site for calcium ions, so it is advisable to include 1 mM CaCl$_2$ in the lysis buffer. One might also want to include apyrase (a commercially available ATPase) during cell lysis as an added ingredient that will help to maintain interactions with molecular chaperones that would normally dissociate in the presence of ATP.

Preformation of the immobilized Protein A beads–immunoglobulin complex, as described above, is an important step as it diminishes the duration of the subsequent incubation with the cell lysate. To maintain the low-stringency conditions, very gentle washing of the immunoprecipitates at 4 °C without agitation or a high NaCl concentration aids in maintaining the stability of any complexes. Bound protein complexes are dissociated by heating with gel loading buffer containing 1% SDS, as described above. Proteins are resolved by SDS–PAGE, and radiolabeled bands are detected by autoradiography as described above.

Mock-transfected cells serve as an essential negative control, as this lane of the gel will be used for the detection of all nonspecifically bound radiolabeled intracellular proteins. Identification of the specifically bound proteins (i.e., not appearing in the negative control lane) is initiated by observing their apparent molecular mass which involves a comparison of their relative electrophoretic mobility against known protein molecular weight markers. Based on these criteria, subsequent immunological detection of each bound protein can often be made by Western blotting with commercially available antisera against candidate molecular chaperones or known N-linked glycoprotein processing enzymes (Affinity Bioreagents; StressGen) that are of a similar mass. The suspected protein, either obtained commercially or detected as an intracellular protein from a crude cell lysate, should be included as a positive control to validate the identical nature of the electrophoretic mobilities.

2.3.2. Alternative methodologies

Alternative methods of protein identification include comparative partial peptide mapping (Le et al., 1994) or direct immunoprecipitation of the suspected intracellular protein with concomitant coimmunoprecipitation of the serpin (Choudhury et al., 1997; Liu et al., 1997). If sufficient quantities of the protein can be obtained by a large-scale immunoaffinity purification technique, often requiring the acquisition of cell lysates from multiple confluent culture dishes, then mass spectrometry can be utilized to identify the amino acid sequence of the bound protein and therefore its identity when compared against several known protein databases. Finally, in the event that conditions cannot be identified to adequately maintain all, or some, protein interactions, one might want to utilize commercially available chemical crosslinkers (Pierce Biochemicals) to maintain their stability. Either homo-functional or hetero-bifunctional crosslinkers can be utilized, depending on which work optimally with the serpin in question. It is noteworthy that noncleavable crosslinkers will increase the apparent molecular mass of protein-bound serpins in SDS–PAGE (Pind *et al.*, 1994). Although many of the aforementioned methodologies can still be used for the purposes of identification, associated problems can sometimes be overcome with the use of a cleavable crosslinker which can effectively dissociate the intracellular protein from the serpin (Pierce Biochemicals). Altogether, this proposed combination of steady-state metabolic radiolabeling, coimmunoprecipitation, and detection techniques should eventually identify the entire spectrum of proteins that detectably bind the newly synthesized serpin at all stages of its folding, retention, and/or degradation.

2.3.3. Experimental caveats

Finally, it should be mentioned that the identity of interacting intracellular proteins can differ among host cell lines. This apparent cell-type specificity might reflect either subtle developmental differences or distinct changes that accompany adaptation to the conditions of long-term cell culture.

2.4. Time course of protein interactions and role verification

The methodology described below can be used to assess the actual order, and time course, in which intracellular proteins (identified in Section 2.3) interact with a newly synthesized serpin.

2.4.1. Assessing the time course of detectable protein interactions

Having optimized the conditions for identifying bound intracellular proteins in steady-state radiolabeled host cells (Section 2.3), commercially available antisera against the identified proteins can be used to coimmunoprecipitate the newly synthesized serpin in pulse–chase experiments. This will allow one to establish the order in which the various intracellular proteins bind to, and are released from, the newly synthesized serpin. In addition, one might want to attempt to test antisera against suspected proteins that were not necessarily detected in the prior experiments.

Identical dishes of host cell monolayers are subjected to short-term (5–20 min) metabolic radiolabeling with [^{35}S]methionine, and then "chased" for selected periods in which the excess radiolabel is removed and dishes receive regular growth medium supplemented with a sixfold excess of unlabeled methionine. At the selected periods, the media is removed, and cell monolayers are collected and stored on ice. Soluble lysates are generated under low-stingency conditions, as described above. Again, to maintain the protein interactions, the selected antisera must be prebound to immobilized Protein A beads. Incubation with the cell lysate and subsequent washings must be performed according to the aforementioned low-stringency conditions at 4 °C. Bound protein complexes are dissociated in response to heating the bead-bound complexes with gel loading buffer containing 1% SDS, and resolved by SDS–PAGE. Quantification of signal intensity is usually performed by fluorography. In this procedure, the goal is to identify interactions with intracellular proteins by using the corresponding antisera to coimmunoprecipitate the serpin molecules that had been radiolabeled during the short pulse. Mock-transfected cells serve as negative control to validate all results.

2.4.2. Use of sedimentation velocity centrifugation for complex identification and enhancement of serpin coimmunoprecipitation

In some instances, the titer of the added antiserum might fail to quantitatively immunoprecipitate the corresponding protein, thereby diminishing the likelihood of its detection. In this situation, one can sequentially separate and then detect the complexes by prefractionating the cell lysate using sedimentation velocity centrifugation in buffered sucrose gradients (Liu et al., 1997). In many cases, the altered sedimentation of the complex can actually help diminish the content of the unbound protein that would otherwise compete in the immunoprecipitation, allowing for the quantitative coimmunoprecipitation of the target protein. In this procedure, the immunoprecipitates are generated under low-stringency conditions from each fraction, or from even- or odd-numbered fractions. The coimmunoprecipitated radiolabeled serpin is then detected following SDS–PAGE and quantified by fluorography.

2.4.3. RNAi technology as a means to validate participation in serpin processing

To further validate the suspected role played by a specific intracellular protein (identified above) in promoting a processing event, one can knockdown the corresponding mRNA with the use of RNAi technology (Termine et al., 2009). These are now commercially available, and several can be chosen to test for their optimal use. The experiment usually involves cotransfection with the serpin expression construct and the specific siRNA to ensure cotransfection of the same cells. Of course, it is important to verify that cotransfection has occurred and to establish (by Northern blotting or RT-PCR) that the siRNA does, in fact, successfully diminish the selected mRNA species in each experiment, or lower the corresponding protein. The influence of this manipulation on the fate of a newly synthesized serpin (in terms of altered secretion, degradation, and/or insolubility) can be identified by metabolic pulse–chase radiolabeling quantification of the serpin in the cell lysate and media, as described in Section 2.2.

2.4.4. Experimental caveats

The generated data from the low-stringency coimmunoprecipitation methodology should support those generated in preceding experiments. However, not all protein interactions, especially the weak and/or transient ones, will likely be detected. Also, one should be aware that the use of RNAi technology could force the host cell to employ a series of compensatory mechanisms that might abolish, or merely slow the kinetics of specific processing events. Alternatively, it is possible that compensatory changes in the machinery may not significantly alter the processing kinetics, but might change the rate of secretion, percent of molecules actually secreted, or result in the use of an alternative protein disposal system.

2.5. Contribution of N-linked oligosaccharide processing events

Asparagine(N)-linked oligosaccharides (N-linked glycans) are added cotranslationally to consensus sequences in nascent polypeptides as they are threaded through the Sec61 channel for translocation across the ER membrane and into its lumen. These polar appendages provide solubility (Hammond and Helenius, 1995) and are modified in a manner that can help facilitate serpin folding by promoting physical interactions with the specific lectin-like molecular chaperones, calnexin and calreticulin (Le *et al.*, 1994). In addition, subsequent opportunistic removal of terminal α1,2-linked mannose units can tag misfolded molecules for intracellular degradation. The methodologies described below can be used to determine which N-glycan processing events, if any, participate in the folding or intracellular degradation of the recombinant serpin expressed in the host cell line.

2.5.1. Assessment of total N-glycan number and role

Serpins contain two to four consensus sequences in the polypeptide backbone that promote covalent attachment of branched oligosaccharides. *In silico* predictions of the glycan number are often accurate, but should be verified by experimentation. The most common method is to pulse-radiolabel the cells and then incubate, for short periods, the NP-40-soluble cell lysate with a limiting amount of endoglycosidase H which cleaves between the two N-acetylglucosamine residues (Fig. 1.1). The number and pattern of immunoprecipitated radiolabeled serpin bands in SDS–PAGE, following short-term partial (rather than complete) digestion, will

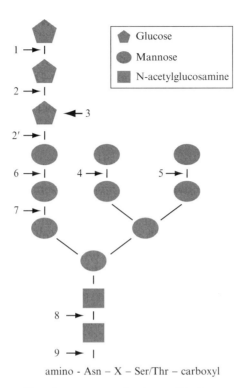

Figure 1.1 An ensemble of enzymes covalently modify the asparagine-linked oligosaccharides. The sites depicted are: (1) hydrolysis by glucosidase I, (2 and 2′) sequential hydrolytic steps by glucosidase II, (3) posttranslational addition of glucose by UDP-glucose:glycoprotein glycosyltransferase, (4) hydrolysis by ER mannosidase I, (5–7) hydrolysis by traditional Golgi-situated α1,2-mannosidases or following prolonged incubation with ER mannosidase I. The sites where endoglycosidase H cleaves the glycan within the chitobiose core (8) and the site where the addition of the N-linked glycan is inhibited by tunicamycin treatment (9) are depicted, also.

indicate the actual number of N-glycans attached during biosynthesis as the intermediate forms are generated. The fastest migrating band will represent the totally deglycosylated molecule. The identity of the fully deglycosylated band can be further verified by performing pulse-radiolabeling in the presence of tunicamycin, an inhibitor of N-linked glycosylation (Le et al., 1994; Liu et al., 1997), prior to immunoprecipitation and SDS–PAGE. For verification of site usage, site-directed mutagenesis of the serpin cDNA can be performed in which one, some, or all of the asparagine acceptors are mutated to encode glutamine (McCracken et al., 1989). This latter set of experiments can also be combined with metabolic pulse–chase radiolabeling and coimmunoprecipitation studies to determine whether any of the glycans are required for secretion, intracellular degradation, and/or physical interaction with specific intracellular proteins.

2.5.2. Biochemical assessment of serpin intracellular trafficking

Intracellular trafficking of the newly synthesized serpin can be assessed by indirect immunofluorescence microscopy in combination with a cycloheximide chase (Novoradovskaya *et al.*, 1998) or by direct immunofluorescence when the serpin cDNA has been fused in-frame with a fluorescent tag like green fluorescent protein (Novoradovskaya *et al.*, 1998). However, a biochemical method can be used for the same purpose. Initially following biosynthesis, all N-linked glycans are sensitive to digestion with endoglycosidase H (Calbiochem; Le *et al.*, 1990; Sifers *et al.*, 1989). However, resistance is eventually gained as the serpin traffics to the medial Golgi compartment where further modifications occur. This convenient gain of resistance is detected by the absence of altered band mobility in SDS–PAGE and can be used to determine the kinetics, or changes therein, at which the pulse-radiolabeled serpin has trafficked to the medial Golgi. Further, the methodology can be used to establish whether the serpin has already trafficked to the Golgi compartment prior to, or after, detected interactions with intracellular proteins (Section 2.4).

2.5.3. Establishment of N-glycan-mediated events

As stated above, the covalent modification of N-linked glycans can help to facilitate both arms of the posttranslational checkpoint as they pertain to serpins. These roles can be identified by the capacity of specific glycosidase inhibitors to block the corresponding biologic event in host cells. Briefly, as shown in Fig. 1.1, a branched 14-unit oligosaccharide (i.e., Glc3Man9GlcNAc2) is covalently linked to the Asn-X-Ser/Thr consensus sequences during translocation of nascent polypeptides into the ER lumen. The sequential cotranslational removal of the two outer glucose units by glucosidases I and II (Ellgaard *et al.*, 1999) generates a monoglycosylated asparagine-linked oligosaccharide (i.e., GlcMan9GlcNAc2). The appendage helps to facilitate proper protein folding upon recognition of the modified glycan by the lectin-like molecular chaperones, calnexin and calreticulin (Ellgaard and Helenius, 2001). Release from either lectin coincides with enzymatic removal of the remaining glucose unit by ER glucosidase II (Hebert *et al.*, 1995). UDP-glucose: glycoprotein glucosyltransferase (UGT) functions as a glycoprotein folding sensor that promotes posttranslational reassembly of nonnative glycoproteins with the glycoprotein folding machinery (Sousa *et al.*, 1992; Trombetta *et al.*, 1991). It accomplishes this task by catalyzing the transfer of a single glucose unit back to high mannose-type glycans asparagine-linked to nonnative proteins. Structurally mature glycoproteins are not recognized by UGT and are released from folding machinery to proceed down the secretory pathway. Modification by the α1,2-mannosidase ER mannosidase I (ERManI; Hosokawa *et al.*, 2003; Wu *et al.*, 2003), and perhaps the three traditional α1,2-mannosidases known to reside in the Golgi complex (Hosokawa et al., 2007), occurs in response to prolonged retention in the early secretory pathway rather than productive transport (Le et al., 1994; Liu et al., 1997) and generates the glycan-based component of a putative glycoprotein endoplasmic reticulum-associated degradation (GERAD) signal (Sifers, 2003; Wu et al., 2003).

The time of inhibitor application, in relation to the addition of [^{35}S] methionine, will determine whether cotranslational and/or posttranslational events are specifically interrupted for the radiolabeled serpin. Commercially available glycosidase inhibitors, and their sites of inhibition, are illustrated in Table 1.1. Changes in the fate of newly synthesized serpins and physical interactions with molecular chaperones, for example, are monitored by pulse–chase radiolabeling and coimmunoprecipitation, respectively. The cotranslational attachment of asparagine-linked oligosaccharides is inhibited by preincubating cells for 1 h at 37 °C in regular growth medium supplemented with tunicamycin (1–10 μg/ml; Sigma). Cotranslational inhibition of glucosidases I and II is accomplished during a similar preincubation period by adding castanospermine (0.2–0.5 mg/ml; Calbiochem) to cells.

Posttranslational removal of the single remaining glucose unit, by glucosidase II, requires the inclusion of either castanospermine (0.2–0.5 mg/ml) or 1-deoxynorjirimycin (Sigma) in only the chase media. A preincubation period of 1 h at 37 °C in regular growth medium, supplemented with either 1-deoxymannorjirimycin (0.5–1.0 mM; Calbiochem) or kifunensine (0.1–0.25 mM; Toronto Research Chemicals), and inclusion of the drug at the same concentration in all subsequent steps, is sufficient to inhibit the hydrolytic activities of ERManI and the three α1,2-mannosidases of the Golgi complex.

Table 1.1 Commonly used chemical inhibitors of asparagine-linked oligosaccharide modifications

Inhibitors	Target
Tunicamycin	N-linked glycosylation (9)
Castanospermine	ER glucosidase I (1)
1-Deoxynorjirimycin	ER glucosidases I and II (1, 2, 2′)
1-Deoxymannorjirimycin	α1,2-Mannosidases (4, 5, 6, 7)
Kifunensine	α1,2-Mannosidases (4, 5, 6, 7)
Enzyme	Target
Endoglycosidase H	Chitobiose core (8)

The inhibitors and enzyme are shown, as are the corresponding targets. The numbers in parentheses demonstrate the step inhibited in Fig. 1.1.

3. Concluding Remarks

Very little is known, or even appreciated, as to how biological systems that function exclusively at the level of encoded proteins can contribute to, or modify, gene expression. Conceivably, human biological variation at this level can influence the phenotypic expression of an individual's genotype (Cabral *et al.*, 2002). The devised methodologies outlined above describe various experimental strategies that have successfully elucidated the roles played by asparagine-linked oligosaccharide processing and recognition in serpin folding and quality control (Termine *et al.*, 2005). The results of each measured parameter will be dictated by both the serpin studied and the cell line into which it is expressed. The conceptual framework, and the associated techniques, should be amenable to the investigation of other asparagine-linked glycoproteins. It is suspected that this pursuit will eventually identify potential prognostic indicators for numerous diseases as well as sites for rationale therapeutic intervention.

ACKNOWLEDGMENTS

We acknowledge funding (to R. N. S.) by multiple grants from the National Institutes of Health (NIHLB and NIDDK), American Lung Association, American Heart Association, Alpha1-Foundation, and the Moran Foundation. S. P. was funded by a postdoctoral fellowship from the Alpha1-Foundation and a grant from the Baylor College of Medicine Digestive Disease Center (DK56338). M. J. I. is the recipient of a training grant from the Huffington Center on Aging, at Baylor College of Medicine.

REFERENCES

Bonifacino, J. S., and Weissman, A. M. (1998). Ubiquitin and the control of protein fate in the secretory and endocytic pathways. *Annu. Rev. Cell Dev. Biol.* **14,** 19–57.

Cabral, C. M., Choudhury, P., Liu, Y., and Sifers, R. N. (2000). Processing by endoplasmic reticulum mannosidases partitions a secretion-impaired glycoprotein into distinct disposal pathways. *J. Biol. Chem.* **275,** 25015–25022.

Cabral, C. M., Liu, Y., and Sifers, R. N. (2001). Dissecting glycoprotein quality control in the secretory pathway. *Trends Biochem. Sci.* **26**(10), 619–624.

Cabral, C. M., Liu, Y., Moremen, K. W., and Sifers, R. N. (2002). Organizational diversity among distinct glycoprotein ER-associated degradation programs. *Mol. Biol. Cell* **13,** 2639–2650.

Choudhury, P., Liu, Y., Bick, R. J., and Sifers, R. N. (1977). Intracellular association between udp-glucose:glycoprotein glucosyltransferase and an incompletely folded variant of α1-antitrypsin. *J. Biol. Chem.* **272,** 13446–13451.

Choudhury, P., Liu, Y., and Sifers, R. N. (1997). Quality control of protein folding: Participation in human disease. *News Physiol. Sci.* **12,** 162–165.

Ellgaard, L., and Helenius, A. (2001). ER quality control: Towards an understanding at the molecular level. *Curr. Opin. Cell Biol.* **13,** 431–437.
Ellgaard, L., Molinari, M., and Helenius, A. (1999). Setting the standards: Quality control in the secretory pathway. *Science* **286,** 1882–1888.
Fewell, S. W., Travers, K. J., Weissman, J. S., and Brodsky, J. L. (2001). The action of molecular chaperones in the early secretory pathway. *Annu. Rev. Genet.* **35,** 149–191.
Graham, K. S., Le, A., and Sifers, R. N. (1990). Accumulation of the insoluble PI Z variant of human α1-antitrypsin within the hepatic endoplasmic reticulum does not elevate the steady-state level of grp78/BiP. *J. Biol. Chem.* **265,** 20463–20468.
Hammond, E., and Helenius, A. (1995). Quality control in the secretory pathway. *Curr. Opin. Cell Biol.* **7,** 523–529.
Hebert, D. N., Foellmer, B., and Helenius, A. (1995). Glucose trimming and reglucosylation determine glycoprotein association with calnexin in the endoplasmic reticulum. *Cell* **81,** 425–433.
Hosokawa, N., Tremblay, L. O., You, Z., Herscovics, A., Wada, I., and Nagata, K. (2003). Enhancement of endoplasmic reticulum (ER) degradation of misfolded Null Hong Kong alpha1-antitrypsin by human ER mannosidase I. *J. Biol. Chem.* **278,** 26287–26294.
Hosokawa, N., You, Z., Tremblay, L. O., Nagata, K., and Herscovics, A. (2007). Stimulation of ERAD of misfolded null Hong Kong alpha1-antitrypsin by Golgi alpha1,2-mannosidases. *Biochem. Biophys. Res. Commun.* **362,** 626–632.
Le, A., Graham, K. S., and Sifers, R. N. (1990). Intracellular degradation of the transport-impaired human PI Z α1-antitrypsin variant. Biochemical mapping of the degradative event among compartments of the secretory pathway. *J. Biol. Chem.* **265,** 14001–14007.
Le, A., Ferrell, G. A., Dishon, D. S., Le, Q.-Q., and Sifers, R. N. (1992). Soluble aggregates of the human PI Z α1-antitrypsin variant are degraded within the endoplasmic reticulum by a mechanism sensitive to inhibitors of protein synthesis. *J. Biol. Chem.* **267,** 1072–1080.
Le, A., Steiner, J. L., Ferrell, G. A., Shaker, J. C., and Sifers, R. N. (1994). Association between calnexin and a secretion-incompetent variant of human α1-antitrypsin. *J. Biol. Chem.* **269,** 7514–7519.
Lindblad, D., Blomenkamp, K., and Teckman, J. (2007). Alpha-1-antitrypsin mutant Z protein content in individual hepatocytes correlates with cell death in a mouse model. *Hepatology* **46,** 1228–1235.
Liu, Y., Choudhury, P., Cabral, C. M., and Sifers, R. N. (1997). Intracellular disposal of incompletely folded human α1-antitrypsin involves release from calnexin and post-translational trimming of asparagine-linked oligosaccharides. *J. Biol. Chem.* **272,** 7946–7951.
Liu, Y., Choudhury, P., Cabral, C. M., and Sifers, R. N. (1999). Oligosaccharide modification in the early secretory pathway directs the selection of a misfolded glycoprotein for degradation by the proteasome. *J. Biol. Chem.* **274,** 5861–5867.
Lomas, D. A., Evans, D. L. I., Finch, J. T., and Carrell, R. W. (1992). The mechanism of Z α1-antitrypsin accumulation in the liver. *Nature* **357,** 605–607.
McCracken, A. A., Kruse, K. B., and Brown, J. L. (1989). Molecular basis for defective secretion of the Z variant of human alpha-1-protease inhibitor: Secretion variants having altered potential for salt bridge formation between amino acids 290 and 342. *Mol. Cell. Biol.* **9,** 1406–1414.
Molinari, M. (2007). N-glycan structure dictates extension of protein folding or onset of disposal. *Nat. Chem. Biol.* **3,** 313–320.
Novoradovskaya, N., Lee, J., Yu, Z. X., Ferrans, V. J., and Brantly, M. (1998). Inhibition of intracellular degradation increases secretion of a mutant form of α1-antitrypsin associated with a profound deficiency. *J. Clin. Invest.* **101,** 2693–2701.

Pind, S., Riordan, J. R., and Williams, D. B. (1994). Participation of the endoplasmic reticulum chaperone calnexin (p88, IP90) in the biogenesis of the cystic fibrosis transmembrane conductance regulator. *J. Biol. Chem.* **269,** 12784–12788.

Qu, D., Teckman, J. H., and Perlmutter, D. H. (1977). Review: Alpha 1-antitrypsin deficiency associated liver disease. *J. Gastroenterol. Hepatol.* **12**(5), 404–416.

Sifers, R. N. (2003). Cell biology: Protein degradation unlocked. *Science* **299,** 1330–1331.

Sifers, R. N., Brashears-Macatee, S., Kidd, V. J., Muensch, H., and Woo, S. L. C. (1989). A frameshift mutation results in a truncated α1-antitrypsin that is retained within the rough endoplasmic reticulum. *J. Biol. Chem.* **263,** 7330–7335.

Sifers, R. N., Finegold, M. J., and Woo, S. L. C. (1992). Molecular biology and genetics of α1-antirypsin deficiency. *Sem. Liv. Dis.* **12,** 301–310.

Sousa, M. C., Ferro-Garcia, M. A., and Parodi, A. J. (1992). Recognition of the oligosaccharide and protein moieties of glycoproteins by the UDP-glc:glycoprotein glucosyltransferase. *Biochemistry* **31,** 97–105.

Teckman, J. H., and Perlmutter, D. H. (2000). Retention of mutant alpha(1)-antitrypsin Z in endoplasmic reticulum is associated with an autophagic response. *Am. J. Physiol. Gastrointest. Liver Physiol.* **279,** G961–G974.

Termine, D., Wu, Y., Liu, Y., and Sifers, R. N. (2005). Alpha1-antitrypsin as a model for glycan function in the endoplasmic reticulum. *Methods* **35**(4), 348–353.

Termine, D. J., Moremen, K. W., and Sifers, R. N. (2009). The mammalian UPR boosts glycoprotein ERAD by suppressing the proteolytic down-regulation of ER mannosidase I. *J. Cell Sci.* **122,** 976–984.

Trombetta, S. E., Ganan, S., and Parodi, A. J. (1991). The UDP-Glc:glycoprotein glucosyltransferase is a soluble protein of the endoplasmic reticulum. *Glycobiology* **1,** 155–161.

Wu, Y., Whitman, I., Molmenti, E., Moore, K., Hippenmeyer, P., and Perlmutter, D. H. (1994). A lag in intracellular degradation of mutant 1-antitrypsin correlates with the liver disease phenotype in homozygous PiZZ α1-antitrypsin deficiency. *Proc. Natl. Acad. Sci. USA* **91,** 9014–9018.

Wu, Y., Swulius, M. T., Moremen, K. W., and Sifers, R. N. (2003). Elucidation of the molecular logic by which misfolded α1-antitrypsin is preferentially selected for degradation. *Proc. Natl. Acad. Sci. USA* **100**(14), 8229–8234.

Wu, Y., Termine, D. J., Swulius, M. T., Moremen, K. W., and Sifers, R. N. (2007). Human endoplasmic reticulum mannosidase I is subject to regulated proteolysis. *J. Biol. Chem.* **282,** 4841–4849.

Yu, M. H., Lee, K. N., and Kim, J. (1995). The Z type variation of human α1-antitrypsin causes a protein folding defect. *Nat. Struct. Biol.* **2,** 363–367.

CHAPTER TWO

Serpin–Enzyme Receptors: LDL Receptor-Related Protein 1

Dudley K. Strickland,[*,†,‡] Selen Catania Muratoglu,[*,‡] and Toni M. Antalis[*,†]

Contents

1. Introduction	18
2. Purification of LRP1	18
2.1. Isolation of full length LRP1 from tissue extracts	18
2.2. Isolation of soluble forms of LRP1 from plasma	21
3. Expression of Receptor Fragments	21
3.1. Expression of individual LDLa repeats in *Escherichia coli*	21
3.2. Expression of LRP1 fragments in cells	21
3.3. Expression of functional LRP1 "minireceptors"	22
4. Ligand Binding to LRP1	23
4.1. Determinants on LRP1 that bind to ligands	23
4.2. Assays to measure the binding of ligands to LRP1	24
5. Binding of Serpin–Enzyme Complexes to LRP1	26
5.1. Specificity of binding	26
5.2. Determinants on serpins that are responsible for binding to LRP1	26
5.3. Clearance of ^{125}I-labeled serpin–enzyme complexes in mice	27
6. Summary	28
Acknowledgments	28
References	28

Abstract

Early studies suggested the existence of an hepatic receptor that is involved in the clearance of serpin:enzyme complexes. Subsequent work has identified this receptor as the LDL receptor-related protein 1 (LRP1). LRP1 is a multifunctional receptor that serves to transport numerous molecules into the cell via endocytosis and also serves as a signaling receptor. LRP1 plays diverse roles in

[*] Center for Vascular and Inflammatory Diseases, University of Maryland School of Medicine, Baltimore, Maryland, USA
[†] Department of Physiology, University of Maryland School of Medicine, Baltimore, Maryland, USA
[‡] Department of Surgery, University of Maryland School of Medicine, Baltimore, Maryland, USA

biology, including roles in lipoprotein metabolism, regulation of protease activity, activation of lysosomal enzymes, and cellular entry of bacterial toxins and viruses. Deletion of the *Lrp1* gene leads to lethality in mice, revealing a critical, but as of yet undefined, role in development. Its identification as a receptor for serpin:enzyme complexes confirms a major role for LRP1 in regulating protease activity.

1. Introduction

The early work of Ohlsson *et al.* (1971), who investigated the clearance of trypsin–inhibitor complexes from the circulation, revealed the existence of a specific hepatic pathway responsible for removing complexes of proteases with their inhibitors. Subsequent work by Imber and Pizzo (1981) reinforced this concept by discovering that α_2-macroglobulin (α_2M) complexed with trypsin is rapidly removed by the liver, whereas the native form of the inhibitor is not removed from the circulation. In examining the specificity of these pathways, it was noted that various serpin:enzyme complexes could compete with one another for hepatic clearance (Fuchs *et al.*, 1984) but did not seem to alter the hepatic uptake of modified α_2M (Fuchs *et al.*, 1982), raising the possibility of different pathways for these molecules. However, upon isolation of the hepatic receptor responsible for clearing α_2M–protease complexes from the circulation (Ashcom *et al.*, 1990; Moestrup and Gliemann, 1989) and confirming its identity to the LDL receptor-related protein (LRP, now called LRP1; Kristensen *et al.*, 1990; Strickland *et al.*, 1990), we now know that this receptor is responsible for clearing both α_2M–protease complexes as well as serpin:enzyme complexes from the circulation (Kounnas *et al.*, 1996).

2. Purification of LRP1

2.1. Isolation of full length LRP1 from tissue extracts

LRP1 has been purified from placental tissue (Ashcom *et al.*, 1990) and from liver membrane extracts (Moestrup and Gliemann, 1989) by affinity chromatography over a Sepharose–α_2M–methylamine column.

2.1.1. Purification of α_2M and preparation of methylamine-reacted α_2M–Sepharose

1. α_2M is purified from human plasma employing the procedure detailed by Harpel (1976).

2. To prepare methylamine-activated α_2M, incubate native α_2M in 50 mM HEPES, 0.15 M NaCl, pH 8.0 with 100 mM methylamine at room temperature for 30 min.
3. Dialyze the α_2M:Me extensively against 0.1 M NaHCO$_3$, 0.5 M NaCl, pH 8.3.
4. After dialysis, couple to CNBr-activated Sepharose (Pharmacia Fine Chemical) as recommended by the manufacturer using 10 mg α_2M:Me/ml resin.
5. Allow protein to couple to resin for 2 h at room temperature using gentle mixing (end-over-end).
6. Remove solution, and replace with 0.1 M Tris, pH 8.0 for an additional 2 h at room temperature.
7. Wash resin with 0.1 M sodium acetate, 0.5 M NaCl, pH 5.0 and then with 0.1 M sodium bicarbonate, 0.5 M NaCl, pH 8.3.

2.1.2. Purification of LRP1 from human placenta

1. All procedures, unless otherwise is indicated, are carried out at 4 °C.
2. Wash fresh placenta with cold TBS (50 mM Tris, 150 mM NaCl, pH 7.4) and 200 mM sucrose. Remove the fetal membranes and umbilical cord, and grind the placenta in a meat grinder. Use the tissue immediately or store at -80 °C until needed.
3. Suspend the tissue in an equal volume of TBS containing 0.005% digitonin, 1 mM each of MgCl$_2$ and CaCl$_2$, along with proteinase inhibitors: 1 mM PMSF, 0.02 mg/ml leupeptin, and 0.02 mg/ml D-Phe-Pro-Arg–Ch$_2$Cl.
4. Stir for 15 min on ice, and then homogenize the mixture in a blender (three times for 30 s each). Following homogenization, centrifuge at $5000 \times g$ for 20 min.
5. Discard the supernatant, and suspend the pelleted tissue in an equal volume of extraction buffer (50 mM octyl-B-D-glucopyranoside in TBS containing 1 mM each of MgCl$_2$ and CaCl$_2$, 1 mM PMSF, 0.02 mg/ml leupeptin, 0.02 mg/ml D-Phe-Pro-Arg–Ch$_2$Cl). Stir in extraction buffer for 1 h at 4 °C.
6. Centrifuge the suspension at $5000 \times g$ for 20 min.
7. Remove the supernatant, and subject to additional centrifugation at $11,000 \times g$ for 20 min.
8. Apply the resultant supernatant to a 120 ml Sepharose CL-4B column, and collect the unabsorbed material.
9. Mix this material with 40–60 ml of α_2M:Me–Sepharose overnight at 4 °C using end-over-end mixing.
10. Wash the resin with eight column volumes of 25 mM octyl-β-D-glucopyranoside in TBS containing 1 mM CaCl$_2$, 1 mM MgCl$_2$, 1 mM PMSF.

11. Elute LRP1 from the column with TBS containing 20 mM EDTA, 25 mM octyl-β-D-glucopyranoside. Collect fractions and measure the absorbance at 280 nm to monitor protein. Analyze each fraction by SDS-PAGE, and pool fractions containing LRP1 (see Fig. 2.1).
12. Apply LRP1 containing fractions to a Mono Q anion exchange column (Pharmacia Fine Chemicals) at room temperature, previously equilibrated with 20 mM octyl-β-D-glucopyranoside in 50 mM Tris, pH 8.2, at a flow rate of 0.5 ml/min.
13. After washing the column with equilibration buffer, elute LRP1 with a linear gradient from 0 to 1 M NaCl over 60 min at a flow rate of 0.5 ml/min. Protein is monitored at 280 nm, and 0.5 ml fractions are collected. LRP1 elutes at approximately 0.55 M NaCl.
14. Fractions containing LRP1 are pooled, analyzed by SDS-PAGE, and dialyzed into PBS containing 20 mM octyl-β-D-glucopyranoside, and stored frozen at $-80\ °C$ until used. The concentration of LRP1 is determined by absorbance measurements at 280 nm using an $E_{1\%}^{280\ nm}$ of 13.5.

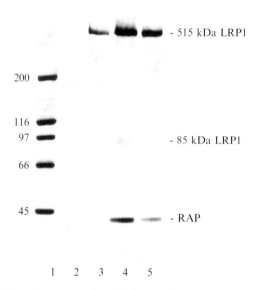

Figure 2.1 Affinity chromatography of placental extract over Sepharose–α_2M:Me. Fractions eluted from the α_2M:Me–Sepharose affinity column were assessed by SDS-PAGE on a 5–15% gradient gel with a 4% stacking gel using the Laemmli buffer system. Lane 1, standards; lanes 2–5, fractions eluted from the affinity column. From Ashcom et al. (1990).

2.2. Isolation of soluble forms of LRP1 from plasma

Soluble forms of LRP1 circulate in the plasma as a consequence of shedding (Quinn *et al.*, 1997), and Gaultier *et al.* (2008) have reported purification of soluble LRP1 using an affinity matrix in which GST-receptor-associated protein (RAP, a ligand of LRP1) was coupled to NHS-activated Sepharose 4 Fast Flow (GE Healthcare).

3. Expression of Receptor Fragments

3.1. Expression of individual LDLa repeats in *Escherichia coli*

LRP1 is composed of modules of β-propeller domains, EGF-repeats, and LDLa repeats (also called complement-type repeats or ligand-binding repeats). For structural studies, a number of LDLa repeats have been expressed in *E. coli* BL21 (Blacklow and Kim, 1996; Dolmer *et al.*, 1998) either as individual repeats or as fusion proteins with GST. In the case of the GST-fusion protein, GST is removed by thrombin digestion following purification. In all cases, the repeats need to be refolded. This can be accomplished as follows:

1. Dilute sample to approximately 0.2 mg/ml in 6 M guanidinium chloride, 50 mM Tris, 1 mM dithiothreitol, pH 8.5.
2. Dialyze against 50 mM Tris–HCl, pH 8.5, 10 mM CaCl$_2$, 1 mM GSH, and 0.5 mM GSSG for 24 h at room temperature under oxygen-free conditions as described (Blacklow and Kim, 1996).
3. Purify refolded LDLa repeats by reverse-phase HPLC as described (Blacklow and Kim, 1996).

3.2. Expression of LRP1 fragments in cells

Soluble minireceptors representing each of the four ligand-binding domains of LRP1 have been expressed in human glioblastoma U87 cells (Bu and Rennke, 1996) as well as in COS-1 cells (Lee *et al.*, 2006; Mikhailenko *et al.*, 2001). In all cases, coexpression of RAP greatly increased the yield of soluble domains found in the media.

1. Plate COS-1 cells in 100-mm dishes and grow them to approximately 50% confluence.
2. Transfect the cells in serum-containing medium with 30 μg of pSec-TagB carrying cDNA for various LRP1 fragments using the FuGENE 6

transfection reagent (Roche Molecular Biochemicals, Indianapolis) according to the manufacturer's protocol.
3. Twenty four hours after transfection, wash the cells, and change the medium to plain Dulbecco's modified Eagle's medium supplemented with 1% Nutridoma®-NS medium supplement (Roche Molecular Biochemicals).
4. Harvest this medium after 48 h of incubation, and detect soluble LRP1 fragments by immunoblot analysis using anti-*myc* antibody to detect recombinant proteins.

3.3. Expression of functional LRP1 "minireceptors"

Willnow *et al.* (1994) were the first to develop LRP1 "minireceptors" to identify regions on LRP1 that are involved in ligand binding. Since then, various "minireceptor" constructs covering all four clusters of ligand-binding repeats have been developed and used in functional studies to identify ligand-binding sites (Mikhailenko *et al.*, 2001; Obermoeller-McCormick *et al.*, 2001). These functional studies are best done in LRP1-deficient cells, such as the CHO 13-5-1 cell line (Fitzgerald *et al.*, 1995). To express LRP1 minireceptors and investigate ligand uptake, the following protocol is used:

1. Plate CHO 13-5-1 cells in 6-well plates (5×10^4 cells/well) 24 h prior to the transfection.
2. Transfect cells with 2 μg of DNA/well in 1.5 ml of serum-containing medium using FuGENE 6 transfection reagent (Roche Molecular Biochemicals).
3. Thirty-six to forty hours following transfection, wash cells with phosphate-buffered saline (PBS) which are ready to be used in the ligand internalization experiments.
4. To measure ligand uptake, incubate transfected CHO 13-5-1 cells transiently transfected with mini-LRP1 constructs for 3 h at 37 °C with ^{125}I-labeled ligands (5 nM).
5. After incubation, wash the cells with PBS and detach from plastic using 0.5 mg/ml trypsin, 0.5 mg/ml proteinase K, and 5 mM EDTA-containing buffer.
6. Internalized ^{125}I-labeled ligand is defined as radioactivity associated with the cell pellet.
7. Nonspecific uptake of ^{125}I-labeled ligand is determined by measuring ^{125}I-labeled ligand uptake in the presence of excess unlabeled ligand and is subtracted from the total internalization.
8. The cell numbers for each experimental condition are measured in parallel wells that do not contain radioactivity.

4. Ligand Binding to LRP1

4.1. Determinants on LRP1 that bind to ligands

LRP1 binds numerous ligands, including proteases, protease inhibitor complexes, apoE-enriched lipoproteins, matrix proteins, and certain growth factors. By far, the largest class of structurally related ligands includes protease-inhibitor complexes, and the list of all serpin–protease complexes that have been reported to bind to LRP1 is listed in Table 2.1. Most of the ligands that are recognized by LRP1 appear to bind to one of the clusters of LDLa repeats (or ligand-binding repeats) that are present in the extracellular domain of LRP1. LRP1 contains four such clusters, termed I–IV. Cluster I contains two LDLa, cluster II contains eight LDLa repeats, cluster III contains 10 LDLa repeats, while cluster IV contains 11 LDLa repeats.

Recent structural studies examining the interaction of RAP D3 domain with two LDLa repeats from the LDL receptor have generated a model for how ligands may interact with LRP1 (Fisher *et al.*, 2006). Prior work had established that the D3 domain of RAP contains a high-affinity LRP1 binding site, and random mutagenesis studies revealed a critical role for Lys270 and Lys256 in the binding interaction (Migliorini *et al.*, 2003). Fisher *et al.* (2006) obtained a crystal structure of the RAP D3 domain in complex with two repeats from the LDL receptor. The results reveal that Lys270 and Lys256 each interact with a single repeat from the LDL receptor. Within an individual LDLa repeats, three conserved, calcium-coordinating acidic residues encircle the lysine side chain. This electrostatic

Table 2.1 Serpin–enzyme complexes known to bind to LRP1

Complex	References
uPA:plasminogen activator inhibitor-I (PAI-1)	Nykjær *et al.* (1994)
tPA:PAI-1	Orth *et al.* (1994)
tPA:neuroserpin	Makarova *et al.* (2003)
Thrombin:antithrombin III	Kounnas *et al.* (1996)
Thrombin:heparin cofactor II	Kounnas *et al.* (1996)
Trypsin:α_1-antitrypsin	Kounnas *et al.* (1996)
Elastase: α_1-antitrypsin	Poller *et al.* (1995)
C1s:C1 inhibitor	Storm *et al.* (1997)
uPA:PAI-2	Croucher *et al.* (2006)
Thrombin:protease nexin-1	Knauer *et al.* (1997a,b)
Thrombin:protein C inhibitor	Kasza *et al.* (1997)
uPA:protein C inhibitor	Kasza *et al.* (1997)

interaction, combined with avidity effects resulting from the use of multiple sites, is thought to represent a model for ligand recognition by LRP1 and other LDL receptor family members.

4.2. Assays to measure the binding of ligands to LRP1

4.2.1. Quantitative measurements of ligand interaction using homologous ligand displacement experiments

Solid phase binding assays have been successful in measuring ligand association with LRP1 (Williams *et al.*, 1992). Quantitative measurements can be readily made employing homologous ligand displacement experiments. In this assay, trace levels of ^{125}I-labeled ligand is incubated with LRP1 immobilized in microtiter wells, and increasing concentrations of unlabeled ligand used to compete for binding. The data are analyzed by the program LIGAND (Munson and Rodbard, 1980), which has the advantage of fitting the nonspecific component as well.

1. Coat microtiter plates overnight at 4 °C with 100 μl of purified LRP1 (3–10 μg/ml) in 50 mM Tris, 150 mM NaCl, pH 7.4 (TBS) containing 5 mM Ca^{2+}.
2. Block the wells with 10 mg/ml BSA in TBS, 5 mM Ca^{2+} for 1 h at room temperature.
3. For high-affinity interactions such as the interaction of α_2M★ or RAP with LRP1, add 50–300 pM of ^{125}I-labeled α_2M★ (23 pCi/pg) or RAP in TBS, 5 mM Ca^{2+}, 30 mg/ml BSA to the wells in the presence of increasing concentrations (1–500 nM) of unlabeled ligands (α_2M★ or RAP).
4. Count aliquots of each stock solution to measure the total cpm added to each well.
5. Incubate overnight at 4 °C, and then wash the wells three times with TBS, 0.02% Tween 20.
6. Add 200 μl of 0.1 N NaOH to each well, and remove an aliquot (150 μl) for counting.
7. Analyze the data using the computer program LIGAND (Munson and Rodbard, 1980).

4.2.2. Using an ELISA to measure ligand binding to LRP1

Enzyme-linked immunosorbent assays have been extensively used to evaluate the binding of LRP1 to various proteins coated onto microtiter wells or to evaluate binding of proteins to LRP1 coated on microtiter wells.

1. For ELISAs, use 3 μg/ml of protein in 100 μl of coating buffer (50 mM Tris, pH 8.0, 150 mM NaCl (TBS), 5 mM CaCl$_2$) to coat microtiter plate wells (Linbro/Titertek, Flow Laboratories, Inc., McLean, VA) for 4 h at 37 °C or 18 h at 4 °C.

2. Incubate with 5% BSA in 250 μl of coating buffer for 1 h at 37 °C to block unbound sites. Then, add purified ligands or LRP1 to the wells at threefold dilutions in concentrations ranging from 150 to 0.06 nM in TBS containing 5% BSA, 0.05% Tween 20, 5 mM CaCl$_2$.
3. Following an 18-h incubation at 4 °C, wash the wells with TBS, 0.05% Tween, 5 mM CaC1$_2$ (TBS-Tween).
4. Detect bound ligand or receptor by incubating for 1 h at room temperature with antibodies (usually monoclonal) directed against the bound molecule. The antibodies are diluted in TBS-Tween.
5. After washing, add goat anti-mouse IgG conjugated to horseradish peroxidase and incubate for 1 h at room temperature.
6. After washing the wells, add the substrate 3,3′,5,5′-tetramethylbenzidine (Kirkegaard & Perry, Gaithersburg, MD), and measure the absorbance at 650 nm.

4.2.3. Surface Plasmon resonance measurements

Surface Plasmon resonance measurements have been useful for detecting binding of various ligands to LRP1 and other LDLR family members.

1. For these studies, activate the Biacore sensor chip (type CM5; Biacore AB) with a 1:1 mixture of 0.2 M N-ethyl-N_-(3-dimethylaminopropyl) carbodiimide and 0.05 M N-hydroxysuccinimide in water as described by the manufacturer.
2. Immobilize purified human LRP1 at the level of 3000 response units in a working solution of 10 μg/ml in 10 mM sodium acetate, pH 4.0. Flow over the chip at a rate of 5 μl/min.
3. Then block the remaining binding sites with 1 M ethanolamine, pH 8.5.
4. Wash out unbound protein with 0.5% SDS.
5. Use a second flow cell, similarly activated and blocked without immobilization of protein, as a negative control.
6. Use a flow cell with immobilized ovalbumin at the level of 500 response units as a control for nonspecific protein binding.
7. Perform all binding reactions in 10 mM HEPES, 0.15 M NaCl, pH 7.4 (HBS-P buffer; Biacore AB), containing 0.005% Tween 20.
8. Measure binding of ligands to LRP1 at 25 °C at a flow rate of 30 μl/min for 4 min, followed by 4 min of dissociation.
9. Subtract the bulk shift due to changes in refractive index measured on blank surfaces from the binding signal at each condition to correct for nonspecific signals.
10. Regenerate chip surfaces with subsequent 1-min pulses of 10 mM sodium acetate, pH 4.0, containing 1 M NaCl and 10 mM NaOH containing 1 M NaCl followed by 2 min of washing with running buffer to remove the high salt solution.

11. Binding of ligands is typically measured using twofold dilutions in HBS-P buffer over a range of concentrations (e.g., 0.6–50 nM).

5. Binding of Serpin–Enzyme Complexes to LRP1

5.1. Specificity of binding

Serpins can exist in a variety of conformational states, including the native serpin, the proteolytically modified form in which the inhibitory capacity is abolished, and finally as a stable proteinase-complexed form. For most serpins, including antithrombin III, heparin cofactor II, α_1-antitrypsin (Kounnas et al., 1996), and neuroserpin (Makarova et al., 2003), very little binding of the native or cleaved serpin to LRP1 occurs (Fig. 2.2). Thus, the only serpin form recognized by LRP1 is the stable proteinase-complexed form, consistent with the findings that only the proteinase-complexed forms of serpins are rapidly removed by the hepatic clearance pathway (Mast et al., 1991).

5.2. Determinants on serpins that are responsible for binding to LRP1

Proteinase cleavage of the exposed loop present in the serpin triggers a conformational change in the serpin and the formation of a covalent complex with the target proteinase. Studies investigating plasminogen activator

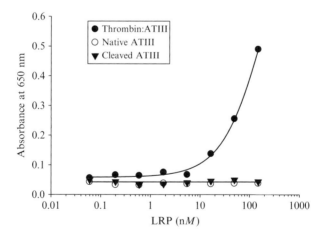

Figure 2.2 LRP1 appears specific for the enzyme:serpin complexes. Increasing concentrations of LRP1 were incubated with microtiter wells coated with thrombin:ATIII (closed circles), native ATIII (open circles), or cleaved ATIII (closed triangles). Following incubation, bound LRP was detected with monoclonal antibody 8G1 (Adapted from Kounnas et al., 1996).

inhibitor 1 (PAI-1) complexes with various proteinases have revealed that high-affinity LRP1 receptor binding is independent of the nature of the proteinase, since different PAI-1/proteinase complexes can cross-compete with one another, implying that the high-affinity receptor-binding epitope resides in the serpin alone (Stefansson *et al.*, 1998). However, the specific regions of serpins involved in receptor recognition at the molecular level are not well defined and appear to be cryptic in nature. While studies investigating the clearance of serpin–enzyme complexes originally implicated a pentapeptide sequence located at the COOH-terminal fragment of α_1-antitrypsin in receptor binding (Joslin *et al.*, 1991), mutation of this region in heparin cofactor II failed to diminish the binding, internalization, or degradation of thrombin:heparin cofactor II complexes (Maekawa and Tollefsen, 1996).

Mutagenesis studies have revealed some serpin epitopes that contribute to serpin interactions with LRP1. Basic residues clustered to one face of PAI-1, composed of parts of β-sheet-A and α-helix-D, appear to be involved in mediating PAI-1-proteinase binding to LRP1 (Rodenburg *et al.*, 1998; Skeldal *et al.*, 2006; Stefansson *et al.*, 1998). Alanine substitution of Lys-82 and Arg-120 reduced the ability of LRP1 to recognize PAI-1 complexed to urokinase plasminogen activator (uPA). Similarly, mutation of Arg-78 and Lys-124 to alanine also resulted in loss of binding of the complex to LRP1. Importantly, Stefansson *et al.* (1998) found that a PAI-1 molecule with Arg-76 mutated to glutamic acid within the heparin-binding domain abolished binding to LRP1.

For protease nexin 1 (PN-1), a region which separates β-sheet-6B and α-helix-B, corresponding to Pro-47 through Ile-58, appears responsible for interacting with LRP1 (Knauer *et al.*, 1997a,b). Thus a synthetic peptide representing this region (PHDNIVISPHGI) was shown to competitively inhibit the LRP1-dependent internalization of thrombin:PN1 complexes. An antibody prepared against this synthetic peptide inhibited PN1:thrombin complex degradation by 70%, but it had no effect on binding of the complex to cell surface heparins (Knauer *et al.*, 1999). In addition, mutagenesis within the corresponding region of PN-1 (His-48A and Asp-49A) reduced the catabolism rate of mutated PN-1 to 15% of wild type (Knauer *et al.*, 1999).

5.3. Clearance of ^{125}I-labeled serpin–enzyme complexes in mice

To measure the clearance of serpin–enzyme complexes from the circulation, the following protocol can be used:

1. Inject anesthetized mice with a bolus of 200 μl of serpin:enzyme complex (e.g., ATIII-^{125}I-thrombin, 100 n*M*) in the presence or absence of competitor (e.g., RAP 110 μ*M*) into the tail vein over a period of ~15 s.

2. Collect blood (40 µl) at selected time intervals following injection (1, 5, 10, and 20 min), by retro-orbital bleeding into 10 µl of 0.5 M EDTA.
3. Weigh the sample, and count for its ^{125}Iodine content.
4. The initial time point, taken 1 min after injection, is considered to represent 100% radioactivity in the circulation.
5. Examine the clearance of each preparation in two mice and average the results.

6. Summary

Substantial evidence exists confirming the role of LRP1 as the hepatic receptor is responsible for clearing serpin–enzyme complexes from the circulation. It should be pointed out that other members of the LDL receptor family are also able to bind serpin–enzyme complexes. These include the VLDL receptor (Argraves et al., 1995; Kasza et al., 1997) and LRP2/gp330 (Stefansson et al., 1995). Since these receptors are not expressed in the liver, LRP1 is mainly responsible for hepatic clearance of serpin–enzyme complexes from the circulation. While structural studies are beginning to reveal the molecular mechanisms by which ligands interact with this receptor family, the exact mechanisms by which serpin–enzyme complexes are recognized by LRP1 still need to be delineated. It is clear that LRP1 only binds serpin–enzyme complexes and does not bind to the native or cleaved serpin. Current evidence suggests that determinants present on the serpin contribute to binding, but it is likely that determinants on the protease also contribute to LRP1 binding as well.

ACKNOWLEDGMENTS

This work was supported by grants HL054710, HL050784, HL072929 (DKS), and CA098369, HL084387, DK081376 (TMA). SCM is supported by T32HL007698.

REFERENCES

Argraves, K. M., Battey, F. D., MacCalman, C. D., McCrae, K. R., Gåfvels, M., Kozarsky, K. F., Chappell, D. A., Strauss, J. F., and Strickland, D. K. (1995). The very low density lipoprotein receptor mediates the cellular catabolism of lipoprotein lipase and urokinase-plasminogen activator inhibitor type I complexes. *J. Biol. Chem.* **270,** 26550–26557.

Ashcom, J. D., Tiller, S. E., Dickerson, K., Cravens, J. L., Argraves, W. S., and Strickland, D. K. (1990). The human α2-macroglobulin receptor: Identification of a 420-kD cell surface glycoprotein specific for the activated conformation of α2-macroglobulin. *J. Cell Biol.* **110,** 1041–1048.

Blacklow, S. C., and Kim, P. S. (1996). Protein folding and calcium binding defects arising from familial hypercholesterolemia mutations of the LDL receptor. *Nat. Struct. Biol.* **3,** 758–762.
Bu, G., and Rennke, S. (1996). Receptor-associated protein is a folding chaperone for low density lipoprotein receptor-related protein. *J. Biol. Chem.* **271,** 22218–22224.
Croucher, D., Saunders, D. N., and Ranson, M. (2006). The urokinase/PAI-2 complex: A new high affinity ligand for the endocytosis receptor low density lipoprotein receptor-related protein. *J. Biol. Chem.* **281,** 10206–10213.
Dolmer, K., Huang, W., and Gettins, P. G. W. (1998). Characterization of the calcium site in two complement-like domains from the low-density lipoprotein receptor-related protein (LRP) and comparison with a repeat from the low-density lipoprotein receptor. *Biochemistry* **37,** 17016–17023.
Fisher, C., Beglova, N., and Blacklow, S. C. (2006). Structure of an LDLR-RAP complex reveals a general mode for ligand recognition by lipoprotein receptors. *Mol. Cell* **22,** 277–283.
Fitzgerald, D. J., Fryling, C. M., Zdanovsky, A., Saelinger, C. B., Kounnas, M., Winkles, J. A., Strickland, D., and Leppla, S. (1995). *Pseudomonas* exotoxin-mediated selection yields cells with altered expression of low-density lipoprotein receptor-related protein. *J. Cell Biol.* **129,** 1533–1541.
Fuchs, H. E., Shifman, M. A., and Pizzo, S. V. (1982). In vivo catabolism of α_1-proteinase inhibitor-trypsin, antithrombin III-thrombin, and α_2-macroglobulin-methylamine. *Biochem. Biophys. Acta* **716,** 151–157.
Fuchs, H. E., Michalopoulos, G. K., and Pizzo, S. V. (1984). Hepatocyte uptake of alpha 1-proteinase inhibitor-trypsin complexes in vitro: Evidence for a shared uptake mechanism for proteinase complexes of alpha 1-proteinase inhibitor and antithrombin III. *J. Cell. Biochem.* **25,** 231–243.
Gaultier, A., Arandjelovic, S., Li, X., Janes, J., Dragojlovic, N., Zhou, G. P., Dolkas, J., Myers, R. R., Gonias, S. L., and Campana, W. M. (2008). A shed form of LDL receptor-related protein-1 regulates peripheral nerve injury and neuropathic pain in rodents. *J. Clin. Invest.* **118,** 161–172.
Harpel, P. C. (1976). Human alpha2-macroglobulin. *Methods Enzymol.* **45,** 639–652.
Imber, M. J., and Pizzo, S. V. (1981). Clearance and binding of two electrophoretic "fast" forms of human alpha 2-macroglobulin. *J. Biol. Chem.* **256,** 8134–8139.
Joslin, G., Fallon, R. J., Bullock, J., Adams, S. P., and Perlmutter, D. H. (1991). The SEC receptor recognizes a pentapeptide neodomain of alpha 1- antitrypsin-protease complexes. *J. Biol. Chem.* **266,** 11282–11288.
Kasza, A., Petersen, H. H., Heegaard, C. W., Oka, K., Christensen, A., Dubin, A., Chan, L., and Andreasen, P. A. (1997). Specificity of serine proteinase/serpin complex binding to very- low-density lipoprotein receptor and alpha2-macroglobulin receptor/low-density-lipoprotein-receptor-related protein. *Eur. J. Biochem.* **248,** 270–281.
Knauer, M. F., Hawley, S. B., and Knauer, D. J. (1997a). Identification of a binding site in protease nexin I (PN1) required for the receptor mediated internalization of PN1-thrombin complexes. *J. Biol. Chem.* **272,** 12261–12264.
Knauer, M. F., Kridel, S. J., Hawley, S. B., and Knauer, D. J. (1997b). The efficient catabolism of thrombin-protease nexin 1 complexes is a synergistic mechanism that requires both the LDL receptor- related protein and cell surface heparins. *J. Biol. Chem.* **272,** 29039–29045.
Knauer, M. F., Crisp, R. J., Kridel, S. J., and Knauer, D. J. (1999). Analysis of a structural determinant in thrombin-protease nexin 1 complexes that mediates clearance by the low density lipoprotein receptor-related protein. *J. Biol. Chem.* **274,** 275–281.
Kounnas, M. Z., Church, F. C., Argraves, W. S., and Strickland, D. K. (1996). Cellular internalization and degradation of antithrombin III-thrombin, heparin cofactor

II-thrombin, and α_1-antitrypsin- trypsin complexes is mediated by the low density lipoprotein receptor-related protein. *J. Biol. Chem.* **271,** 6523–6529.

Kristensen, T., Moestrup, S. K., Gliemann, J., Bendtsen, L., Sand, O., and Sottrup-Jensen, L. (1990). Evidence that the newly cloned LRP is the α_2M receptor. *FEBS Lett.* **276,** 151–155.

Lee, D., Walsh, J. D., Mikhailenko, I., Yu, P., Migliorini, M., Wu, Y., Krueger, S., Curtis, J. E., Harris, B., Lockett, S., Blacklow, S. C., Strickland, D. K., *et al.* (2006). RAP uses a histidine switch to regulate its interaction with LRP in the ER and golgi. *Mol. Cell* **22,** 423–430.

Maekawa, H., and Tollefsen, D. M. (1996). Role of the proposed serpin-enzyme complex receptor recognition site in binding and internalization of thrombin-heparin cofactor II complexes by hepatocytes. *J. Biol. Chem.* **271,** 18604–18610.

Makarova, A., Mikhailenko, I., Bugge, T. H., List, K., Lawrence, D. A., and Strickland, D. K. (2003). The LDL receptor-related protein modulates protease activity in the brain by mediating the cellular internalization of both neuroserpin and neuroserpin: tPA complexes. *J. Biol. Chem.* **278,** 50250–50258.

Mast, A. E., Enghild, J. J., Pizzo, S. V., and Salvesen, G. (1991). Analysis of the plasma elimination kinetics and conformational stabilities of native, proteinase-complexed, and reactive site cleaved serpins: Comparison of α_1-proteinase inhibitor, α_1-antichymotrypsin, antithrombin III, α_2-antiplasmin, angiotensinogen, and ovalbumin. *Biochemistry* **30,** 1723–1730.

Migliorini, M. M., Behre, E. H., Brew, S., Ingham, K. C., and Strickland, D. K. (2003). Allosteric modulation of ligand binding to low density lipoprotein receptor-related protein by the receptor-associated protein requires critical lysine residues within its carboxyl-terminal domain. *J. Biol. Chem.* **278,** 17986.

Mikhailenko, I., Battey, F. D., Migliorini, M., Ruiz, J. F., Argraves, K., Moayeri, M., and Strickland, D. K. (2001). Recognition of alpha 2-macroglobulin by the low density lipoprotein receptor-related protein requires the cooperation of two ligand binding cluster regions. *J. Biol. Chem.* **276,** 39484–39491.

Moestrup, S. K., and Gliemann, J. (1989). Purification of the rat hepatic α2-macroglobulin receptor as an approximately 440 kDa single chain polypeptide. *J. Biol. Chem.* **264,** 15574–15577.

Munson, P. J., and Rodbard, D. (1980). LIGAND: A versatile computerized approach for characterization of ligand-binding systems. *Anal. Biochem.* **107,** 220–239.

Nykjær, A., Kjoller, L., Cohen, R. L., Lawrence, D. A., Gliemann, J., and Andreasen, P. A. (1994). Both pro-uPA and uPA:PAI-1 complex bind to the α_2-macroglobulin receptor/LDL receptor-related protein: Evidence for multiple independent contacts between the ligands and receptor. *Ann. NY Acad. Sci.* **737,** 483–485.

Obermoeller-McCormick, L. M., Li, Y., Osaka, H., Fitzgerald, D. J., Schwartz, A. L., and Bu, G. (2001). Dissection of receptor folding and ligand-binding property with functional minireceptors of LDL receptor-related protein. *J. Cell Sci.* **114,** 899–908.

Ohlsson, K., Ganrot, P. O., and Laurell, C. B. (1971). In vivo interaction beween trypsin and some plasma proteins in relation to tolerance to intravenous infusion of trypsin in dog. *Acta Chir. Scand.* **137,** 113–121.

Orth, K., Willnow, T., Herz, J., Gething, M. J., and Sambrook, J. (1994). Low density lipoprotein receptor-related protein is necessary for the internalization of both tissue-type plasminogen activator-inhibitor complexes and free tissue-type plasminogen activator. *J. Biol. Chem.* **269,** 21117–21122.

Poller, W., Willnow, T. E., Hilpert, J., and Herz, J. (1995). Differential recognition of α_1-antitrypsin-elastase and α_1-antichymotrypsin-cathepsin G complexes by the low density lipoprotein receptor-related protein. *J. Biol. Chem.* **270,** 2841–2845.

Quinn, K. A., Grimsley, P. G., Dai, Y. P., Tapner, M., Chesterman, C. N., and Owensby, D. A. (1997). Soluble low density lipoprotein receptor-related protein (LRP) circulates in human plasma. *J. Biol. Chem.* **272,** 23946–23951.

Rodenburg, K. W., Kjoller, L., Petersen, H. H., and Andreasen, P. A. (1998). Binding of urokinase-type plasminogen activator-plasminogen activator inhibitor-1 complex to the endocytosis receptors alpha2-macroglobulin receptor/low-density lipoprotein receptor-related protein and very-low-density lipoprotein receptor involves basic residues in the inhibitor. *Biochem. J.* **329,** 55–63.

Skeldal, S., Larsen, J. V., Pedersen, K. E., Petersen, H. H., Egelund, R., Christensen, A., Jensen, J. K., Gliemann, J., and Andreasen, P. A. (2006). Binding areas of urokinase-type plasminogen activator-plasminogen activator inhibitor-1 complex for endocytosis receptors of the low-density lipoprotein receptor family, determined by site-directed mutagenesis. *FEBS J.* **273,** 5143–5159.

Stefansson, S., Kounnas, M. Z., Henkin, J., Mallampalli, R. K., Chappell, D. A., Strickland, D. K., and Argraves, W. S. (1995). gp330 on type II pneumocytes mediates endocytosis leading to degradation of pro-urokinase, plasminogen activator inhibitor- 1 and urokinase-plasminogen activator inhibitor-1 complex. *J. Cell Sci.* **108,** 2361–2368.

Stefansson, S., Muhammad, S., Cheng, X. F., Battey, F. D., Strickland, D. K., and Lawrence, D. A. (1998). Plasminogen activator inhibitor-1 contains a cryptic high affinity binding site for the low density lipoprotein receptor-related protein. *J. Biol. Chem.* **273,** 6358–6366.

Storm, D., Herz, J., Trinder, P., and Loos, M. (1997). C1 inhibitor-C1s complexes are internalized and degraded by the low density lipoprotein receptor-related protein. *J. Biol. Chem.* **272,** 31043–31050.

Strickland, D. K., Ashcom, J. D., Williams, S., Burgess, W. H., Migliorini, M., and Argraves, W. S. (1990). Sequence identity between the α2-macroglobulin receptor and low density lipoprotein receptor-related protein suggests that this molecule is a multi-functional receptor. *J. Biol. Chem.* **265,** 17401–17404.

Williams, S. E., Ashcom, J. D., Argraves, W. S., and Strickland, D. K. (1992). A novel mechanism for controlling the activity of α_2-macroglobulin receptor/low density lipoprotein receptor-related protein. Multiple regulatory sites for 39-kDa receptor-associated protein. *J. Biol. Chem.* **267,** 9035–9040.

Willnow, T. E., Orth, K., and Herz, J. (1994). Molecular dissection of ligand binding sites on the low density lipoprotein receptor-related protein. *J. Biol. Chem.* **269,** 15827–15832.

CHAPTER THREE

THE ROLE OF AUTOPHAGY IN ALPHA-1-ANTITRYPSIN DEFICIENCY

Tunda Hidvegi,* Amitava Mukherjee,* Michael Ewing,* Carolyn Kemp,* *and* David H. Perlmutter*,[†]

Contents

1. Introduction	34
1.1. Autophagy in AT deficiency	35
2. Methods for Detection of Autophagy in Cell Line Models	36
2.1. Analysis of autophagy in cell lines by transmission electron microscopy	36
3. LC3 Conversion Assay	38
3.1. Sample preparation	39
3.2. Protein estimation by BCA assay	39
3.3. Gel electrophoresis	39
3.4. Immunoblotting	40
3.5. Detection of LC3 signal on membrane	40
3.6. Interpretation of LC3 blots	40
3.7. Use of lysosomal inhibitors to enhance the information provided by LC3 immunoblots	40
4. Assessing Autophagy by Analyzing Degradation of Long-Lived Proteins	41
4.1. Pulse–chase metabolic labeling for long-lived proteins	41
4.2. Putative inhibitors of autophagy	42
5. Use of Autophagy-Deficient Cell Lines	42
5.1. Cellular model	42
5.2. $Atg5^{-/-}$	42
5.3. $Atg5^{-/-}$ MEFs with regulated (Tet-Off) expression of Atg5	43
6. Pulse–Chase Labeling to Determine the Effect of Autophagy on Degradation of Individual Proteins	43
6.1. Pulse–chase protocol	43
6.2. Determination of total protein content by TCA precipitation	44
6.3. Immunoprecipitation to detect specific proteins	44

* Department of Pediatrics, University of Pittsburgh School of Medicine, Children's Hospital of Pittsburgh of UPMC, Pittsburgh, Pennsylvania, USA
[†] Department of Cell Biology and Physiology, University of Pittsburgh School of Medicine, Children's Hospital of Pittsburgh of UPMC, Pittsburgh, Pennsylvania, USA

7. Methods for Detection of Hepatic Autophagy *In Vivo*   45
   7.1. Analysis of hepatic autophagy *in vivo* by EM   45
   7.2. Analysis of autophagy by fluorescence microscopy using the GFP-LC3 mouse   46
   7.3. Use of autophagy-deficient mouse models   47
8. Methods for Detection of Hepatic Injury   48
   8.1. Mouse models   48
   8.2. Assessment of hepatic fibrosis by determination of hydroxyproline (OHP) content   50
   8.3. Assessment of injury/regenerative activity by BrdU labeling   51
References   52

Abstract

In the classical form of alpha-1-antitrypsin (AT) deficiency, a mutant protein accumulates in the endoplasmic reticulum of liver cells, causing hepatic fibrosis and hepatocellular carcinoma by a gain-of-toxic function mechanism. Autophagy is specifically activated by the accumulation of mutant AT, and the autophagy plays a key role in intracellular degradation of this mutant protein. Our recent study indicates that an autophagy enhancer drug can decrease the hepatic load of mutant AT and reduce hepatic fibrosis in a mouse model of AT deficiency. In this chapter, we discuss what is known about autophagy in AT deficiency and methods for characterizing autophagy in cell lines and animal models.

1. INTRODUCTION

With an incidence of 1 in 2000–3000 live births, the classical form of alpha-1-antitrypsin (AT) deficiency is the most common serpin deficiency. A point mutation that results in a substitution of lysine for glutamate at residue 342 renders the secretory glycoprotein prone to altered folding, polymerization, and aggregation early in the secretory pathway. The mutant ATZ protein accumulates in the endoplasmic reticulum of cells in which it is synthesized (Perlmutter, 2002). This is particularly problematic in the liver because AT is one of the most abundant of the secretory proteins of hepatocytes, and its expression is upregulated during the acute phase response. This is apparently part of the explanation for predisposition to hepatic cirrhosis and hepatocellular carcinoma in homozygotes for the ATZ allele by a gain-of-toxic function mechanism. Indeed, AT deficiency is the most common genetic cause of liver disease in infants and children. It is also a more common cause of severe liver disease in adults than previously recognized.

Because serum levels of AT are reduced by 85% in the deficiency and because the physiological function of AT is inhibition of neutrophil elastase and perhaps other neutrophil proteinases that can degrade the extracellular matrix of the lung, deficient individuals are predisposed to chronic obstructive pulmonary disease (COPD). This loss-of-function mechanism is exacerbated by cigarette smoking, especially because active oxygen intermediates that are released by smokers' mononuclear phagocytes can functionally inactivate AT by oxidation of its reactive site methionine. Nevertheless, several lines of evidence suggest that gain-of-toxic function mechanisms may also contribute to the pathogenesis of COPD (Perlmutter, 2011).

Cohort studies of a population identified by newborn screening in Sweden 40 years ago show that only 8% develop clinically significant liver disease (Bernspang et al., 2009). Perhaps even more important, this series of studies has provided the basis for the concept that genetic and/or environmental modifiers determine whether a deficient individual is susceptible to or protected from tissue damage due to the gain-of-toxic function mechanism (Perlmutter, 2002). We have theorized that these modifiers subtly affect cellular function in one of two ways: (1) an effect on intracellular degradation pathways could worsen the cytotoxicity from accumulation of mutant ATZ; (2) an effect on putative protective signaling pathways could increase the susceptibility or resistance of cells to the cytotoxicity of accumulated mutant ATZ (Perlmutter, 2002). To address this theory, we have first characterized the pathways for intracellular degradation of ATZ and found that the proteasomal and autophagic pathways play critical roles in disposal of ATZ (Perlmutter, 2011). Second, using systems that are ideally suited to characterizing signaling pathways, cell line and animal models which are engineered for inducible expression of ATZ, we have found that accumulation of ATZ in the ER activates the NFκB signaling pathway and the autophagic response but does not activate the unfolded protein response (Hidvegi et al., 2005, 2007; Kamimoto et al., 2006; Kruse et al., 2006a,b).

1.1. Autophagy in AT deficiency

Autophagy is an evolutionarily conserved degradative process in which intracellular constituents are eliminated in the lysosome. The process is mediated by vesicles that engulf cytosol, organelles, or parts of organelles as they form in the cytoplasm and the contents are degraded when these vesicles fuse with the lysosome. It has been classically characterized as a mechanism that permits cell survival during starvation or other stress states but recent studies have shown that it plays a critical role in disposal of damaged proteins, oxidized proteins, innate, and acquired immunity as well as tumor suppression. Moreover, it has become clear that constitutive autophagic activity plays a role in basal cellular proteostasis. By expressing mutant ATZ in an autophagy-deficient cell line, we have shown that autophagy plays a key role in disposal of ATZ. Autophagy is particularly important because it can dispose of the polymerized and aggregated forms of ATZ that cannot be degraded by the proteasomal pathway (Kamimoto et al., 2006). By breeding a mouse model with hepatocyte-specific inducible expression of ATZ to the GFP-LC3 mouse that produces green fluorescent autophagosomes, we found that accumulation of ATZ in the ER specifically activates the autophagic response (Kamimoto *et al.*, 2006).

Because autophagy appears to be specialized for disposal of insoluble polymers and aggregates of ATZ and is specifically activated when ATZ accumulates in the ER, we recently investigated the possibility that drugs which enhance autophagy could ameliorate the liver disease associated with AT deficiency (Hidvegi *et al.*, 2010). From a list of drugs which have been recently reputed to enhance autophagy, we selected carbamazepine (CBZ) because it has been used safely as an anticonvulsant and mood stabilizer for many years. The study showed that CBZ mediates an increase in intracellular degradation of ATZ in cell line models and decreases the hepatic ATZ load and hepatic fibrosis in a mouse model of AT deficiency. These results have led us to begin clinical trials of CBZ for severe liver disease due to AT deficiency and provide a proof in principle for the use of autophagy enhancer drugs in disorders caused by aggregation-prone proteins. This last notion has been further substantiated by recent results from high-content screening of drug libraries using a novel *Caenorhabditis elegans* model of AT deficiency. In a pilot screen, four of six hit compounds were known to be autophagy enhancers (Gosai *et al.*, 2010). Together, these observations have provided powerful evidence for the importance of autophagy in the pathobiology of AT deficiency.

2. Methods for Detection of Autophagy in Cell Line Models

2.1. Analysis of autophagy in cell lines by transmission electron microscopy

Transmission electron microscopy (EM) is the most-reliable and time-tested method for detection of autophagosomes. Autophagosomes are distinguished from other cellular vesicles by several ultrastructural criteria. These vesicles are encompassed by a ribosome-free double membrane and contain within their internal contents a variety of structures including debris, fragments of membranes, and membranous whirls. The lamellar accumulations of membranes and membrane fragments become more electron-dense as the so-called early autophagic vesicles mature to late/degradative autophagic vacuoles that are in various phases of fusing with lysosomes.

2.1.1. Plastic-embedded EM

i. Cells are grown to 85% confluence in 10-cm dishes and then trypsinized.
ii. By centrifugation cell pellets are obtained and then washed with cold PBS.

iii. 1% glutaraldehyde–0.1 M sodium cacodylate is used to fix the cells and then the cell block is embedded in polybed for ultrathin sectioning (Raposo et al., 1995).
iv. Cell blocks are mounted on specimen holders and frozen in liquid nitrogen before sectioning.
v. Ultrathin sections (60–80 nm) are cut at $-120\,°C$, using an Ultra-CutS cryomicrotome (Leica, Biel, Switzerland) and a diamond knife (Diatome, Biel, Switzerland). Ultrathin sections are collected with a mixture of methyl cellulose and 2.3 M sucrose (Liou and Slot, 1994).

2.1.2. Immuno EM

Postembedding immunogold labeling with thin cryosections is one of the most sensitive methods to immunolabel antigens for EM (Griffiths, 1993). Autophagic vacuoles can be delineated by immunogold labeling for LC3 which is generally present on autophagic membranes as well as inside the vacuoles (Jäger et al., 2004). After fusing with lysosomes, the autophagic vacuoles are generally defined by the presence of molecular markers of lysosomes such as lysosomal membrane protein (LAMP)-2 or cathepsin D as well as LC3 (Houwerzijl et al., 2009). An alternate approach is transfection of chimeric GFP Atg-5 in cells with localization of GFP-Atg5 immunogold EM using an anti-GFP antibody (Fujita et al., 2008) (Fig. 3.1).

i. Cells are grown to 85% confluence in 10-cm dishes and then trypsinized, pelleted, and washed with PBS.

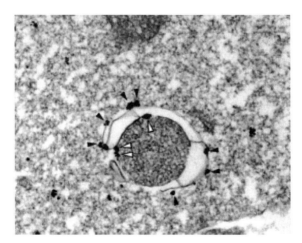

Figure 3.1 Ultrastructural localization of LC3 in HeLa cells by silver-enhanced immunogold labeling with anti-LC3 antibody. The open and closed arrowheads indicate immunogold labeling of LC3 associated with inner and outer membranes, respectively, of an autophagosome (reprinted with permission from Kabeya et al., 2000).

ii. Cells are then fixed with 2% paraformaldehyde (PFA)/0.2% glutaraldehyde in PBS, pH 7.2, at room temperature for 2 h.
iii. Cells are embedded in 10% gelatin (Slot *et al.*, 1989) but without fixation of the gelatin. After pelleting the cells, the gelatin is solidified on ice.
iv. Blocks for ultracryotomy are prepared and infused with 2.3 M sucrose for 2 days.
v. Labeling with the primary antibody is carried out for 2 h, and labeling with the appropriate species-specific secondary anti-IgG/gold conjugate (Jackson ImmunoResearch, West Grove, PA) is carried out for 1 h.
vi. Sections are stained with uranyl acetate and embedded in methyl cellulose. Specimens are viewed and photographed using a Zeiss 902 electron microscope.

2.1.3. Analysis of autophagy by DAMP EM

During the fusion of autophagosome and lysosome, the membrane of lysosome fuses with the outer membrane of the autophagosome. The resulting structure is called an autolysosome (Suzuki *et al.*, 2001). Enzymes present in lysosomes are generally activated under acidic pH conditions. The acidification status within lysosomal compartments can be analyzed by using DAMP (N-{3-[(2,4-dinitrophenylamino)propyl]}-N-(3-aminopropyl)-methylamine dihydrochloride). In acidic pH, the primary amino group of DAMP gets protonated and become positively charged. This positively charged DAMP then covalently link to proteins which are previously fixed with aldehydes.

i. Cells are incubated with 50 μM DAMP at 37 °C for 30 min in growth media to label intracellular acidic compartments before fixation for EM (Anderson *et al.*, 1984; Dunn, 1990).
ii. For quantitation of autophagic vacuoles, EM photomicrographs of 25 whole, intact cells containing a nucleus are examined. Quantitative grids of the appropriate size are superimposed on the photomicrographs, and the grid area occupied by nascent (AVi) and degradative (AVd) autophagic vacuoles is compared with the grid area occupied by the cytoplasm (Teckman and Perlmutter, 2000).

3. LC3 Conversion Assay

LC3 is an autophagosomal membrane-specific protein. LC3 is synthesized as a proform that is proteolytically cleaved to LC3-I, an isoform which is localized to the cytoplasm. When an autophagosome is forming, cytosolic LC3-I is converted to membrane-bound LC3-II, and in this process, its carboxyl terminus is modified by conjugation of phosphatidylethanolamine.

This modification leads to a change in electrophoretic mobility such that the LC3-II isoform migrates faster than LC3-I in SDS–PAGE. The ratio of LC3-II to LC3-I is now routinely used as a measure of autophagosome formation.

3.1. Sample preparation

i. Cells are plated and grown to confluence in 100 mm dishes.
ii. Media is removed and monolayers rinsed with PBS. Cells are scraped into 1 ml PBS and centrifuged to pellet at 4 °C.
iii. After removing supernatant, pellet is resuspended in 300 µl NP40 buffer in the presence of protease inhibitor cocktail.
iv. Samples are then incubated for 20–30 min on ice and centrifuged at 15,000 rpm for 15 min at 4 °C.
v. Supernatant is stored at -80 °C.

3.2. Protein estimation by BCA assay

i. Albumin standards (2000, 1000, 500, 250, 125, 62.5, and 31.25 µg/ml) are made from stock of (2 mg/ml) and transferred to microcentrifuge tubes. Serial dilutions were done using appropriate buffer.
ii. Working reagent is prepared by mixing 1 part reagent B to 50 parts reagent A (Pierce BCA protein Assay kit).
iii. Standards and blank are aliquoted (25 µl) into wells of a 96-well plate. Samples are diluted 1:10 in appropriate buffer and 25 µl is added to desired well. One working reagent (200 µl) is added to each well.
iv. The plate is covered and incubated at 56 °C for 20 min.
v. The plate is cooled down on bench top.

3.3. Gel electrophoresis

i. Lonza 15% precast acrylamide Tris–glycine gels (Fisher cat. # BMA-58110) are prepared using 1× Tris–glycine–SDS running buffer (10× TG–SDS from ISC BioExpress).
ii. Samples are prepared in 5× SDS sample buffer (+βME), and 50 µg of total protein is loaded in each lane. The gel is run for 110 min at 110 V for best separation.
iii. To transfer the proteins from gel to the membrane, PVDF membrane pieces are briefly soaked in methanol and rehydrated using 1× TG transfer buffer for 10 min. Transfer uses cold transfer buffer for 1 h at 400 mA.
iv. After transfer, the PVDF membrane is briefly rinsed in TBST (Tris buffer saline with 0.05% Tween-20) and blocked overnight in fresh TBST with

5% milk at 4 °C with gentle shaking. (*Note*: Prepare the TBST with fresh milk on the day of transfer to reduce background noise.)

3.4. Immunoblotting

i. Membrane is incubated in primary antibody (anti-LC3; NOVUS, cat. # NB100-2220) diluted 1:250 in 5% milk for 1 h at room temperature. Then the membrane is rinsed briefly in TBST. (*Note*: Fresh aliquot of antibody is thawed and diluted in 10 ml milk each time. Storing this antibody in milk for reuse does not yield consistent results.)
ii. Membrane is washed four times for 15 min each at room temperature in TBST with gentle shaking.
iii. Membrane is incubated in secondary anti-rabbit-HRP antibody diluted 1:50,000 in 5% milk for 30 min at room temperature with gentle shaking and washed four times for 15 min at room temperature in TBST.

3.5. Detection of LC3 signal on membrane

i. Membrane is incubated for 5 min in Pierce Supersignal West Dura substrate and exposed immediately to film.
ii. The best images usually require only 1–5 min exposure to film. If the background is too high, one to two additional washes will improve detection. If the signal is not detected within 30 min, Pierce Supersignal West Femto can be tried (Hidvegi *et al.*, 2010).

3.6. Interpretation of LC3 blots

LC3 immunoblots should be submitted to densitometry to determine the ratio of LC3-II to LC3-I. An increase in this ratio indicates increased autophagosomes. Use of lysosomal inhibitors in the design of the experiments may allow the investigator to determine whether the increase in autophagosomes is due to increased formation or decreased clearance (Fig. 3.2).

3.7. Use of lysosomal inhibitors to enhance the information provided by LC3 immunoblots

Immunoblotting for LC3 can provide even more information if separate cell monolayers are incubated in the absence or presence of lysosomal inhibitors, E64D and pepstatin A (20 µg/ml) for the past 4 h of an experiment. LC3-II itself is degraded by the lysosome, and, therefore, this additional experimental control allows one to distinguish between increased autophagosome formation versus decreased autophagosome clearance (Mizushima and

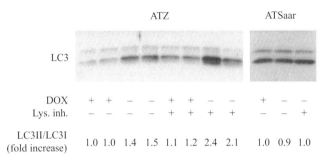

Figure 3.2 LC3 immunoblot of HeLa cells expressing human ATZ and ATSaar. Each cell line was incubated in the absence or presence of DOX for 4 weeks. Separate monolayers were incubated with lysosomal protease inhibitors (Lys inh), including E64d and pepstatin A (20 μg/ml) for the past 4 h prior to harvesting and homogenization. The homogenates were then subjected to immunoblot analysis for LC3. Densitometric values for the LC3-II to LC3-I ratio are shown at the bottom with the relative densitometric value in the presence of DOX but not Lys inh arbitrarily set as 1.0. The results show that there is an increase in the LC3-II to LC3-I ratio when DOX is removed in the case of ATZ and this is further increased in the presence of lysosomal inhibitors, indicating that here is an increase in autophagic flux. There was no increase in the LC3-II to LC3-I ratio when the ATSaar variant was induced (reprinted with permission from Hidvegi et al., 2010).

Yoshimuri, 2007). If the LC3-II to LC3-I ratio increases further when cells are treated with lysosomal inhibitors, one can conclude that there is increased autophagosome formation and increased autophagic flux. If the LC3-II to LC3-I ratio does not increase when cells are treated with lysosomal inhibitors when compared to the ratio in cells that have not been treated with lysosomal inhibitors, this scenario is most consistent with decreased autophagic clearance (Boland et al., 2008).

4. Assessing Autophagy by Analyzing Degradation of Long-Lived Proteins

Many long-lived proteins are degraded by autophagy, and so the function of autophagy can be assessed by measurement of the turnover of long-lived proteins in cell lines. This usually involves metabolic labeling and a pulse–chase.

4.1. Pulse–chase metabolic labeling for long-lived proteins

i. Confluent monolayers are labeled with (3H) leucine (2 μCi/ml) for 48 h at 37 °C to preferentially label long-lived proteins. This is called the pulse period.

ii. At the end of the pulse period, monolayers are rinsed thoroughly and then incubated with complete medium that contains a 200-fold molar excess of unlabeled leucine. This is called the chase period.
iii. At predetermined time points, individual monolayers are harvested and homogenized.
iv. Cell homogenates from different time points are subjected to precipitation with 10% TCA, and degradation is measured by the decay of TCA-precipitable protein over time.

4.2. Putative inhibitors of autophagy

The turnover of long-lived proteins is predominantly mediated by autophagy. However, the role of autophagy can be further evaluated by adding putative inhibitors of autophagy during the chase period.

i. Ammonium chloride (20 mM) inhibits autophagic degradation. However, it also inhibits other forms of lysosomal degradation and therefore is not specific for autophagy.
ii. 3-Methyladenine (10 mM) blocks formation of autophagosomes and fusion with lysosomes, and so it is often used (Seglen and Gordon, 1984). Nevertheless, because it works by inhibiting type I and type III PI3 kinase, it also cannot be considered specific for autophagy. This is also true for wortmannin and LY204002 (Teckman and Perlmutter, 2000).

5. USE OF AUTOPHAGY-DEFICIENT CELL LINES

5.1. Cellular model

Mouse embryonic fibroblasts (MEFs) deleted for Atg5 are deficient in the conventional autophagy pathway, and so these cells can be used to evaluate the specific role of autophagy in a cellular process.

5.2. Atg5$^{-/-}$

A 5.6-kb fragment containing the second and third exons of the Atg5 gene was replaced by the Neo-cassette creating a targeting vector to use for transfecting R1 ES cells by electroporation (Mizushima, 2001). The resulting Atg5$^{-/-}$ deficient ES cells were later used by the aggregation method (Wood, 1993) to create chimeric mice (Kuma, 2004). Then embryonic fibroblasts (13.5d) were separated, and immortalized cell lines were developed (Kuma, 2004). For the investigation of the role of autophagy in a serpinopathy, these Atg5$^{-/-}$ MEFs are compared to wild-type MEFs.

5.3. Atg5$^{-/-}$ MEFs with regulated (Tet-Off) expression of Atg5

The Atg5$^{-/-}$ MEF were used to generate a Tet-inducible Atg5 MEF cell line introducing a CAG-tTA regulator gene cassette and the TRE-mAtg5 (Hosokawa, 2006). Removal of doxycycline (DOX) results in Atg5 expression, while addition of DOX suppresses the expression of Atg5 and results in the absence of conventional autophagic activity. The advantage of this cell line is that the same cell lineage can be used to examine the impact of autophagy. There is no need for the wild-type MEFs. Further, the timing and amount of autophagy can be regulated by timing and dosing of DOX.

Each of these cell lines can be used to determine the effect of autophagy on a particular process. For instance, we were able to determine whether autophagy participates in the disposal of mutant ATZ by transfecting the human ATZ gene into the Atg5$^{-/-}$ MEFs (Kamimoto et al., 2006). Atg5$^{-/-}$ MEFs and wild-type MEFs with stable expression of mutant ATZ were compared to determine whether CBZ enhanced degradation of mutant ATZ via autophagy (Hidvegi et al., 2010).

6. Pulse–Chase Labeling to Determine the Effect of Autophagy on Degradation of Individual Proteins

Metabolic labeling using a pulse–chase design is the best method to determine the kinetics of degradation of individual proteins.

6.1. Pulse–chase protocol

i. Cell line is subcultured into separate wells of 24-well tissue culture plates until wells were fully covered with a monolayer of cells. Each sample is composed of three wells.

ii. After rinsing the wells twice with 0.5 ml/well HBSS, monolayers are incubated in pulse medium methionine 0.333 ml/well. The pulse media is DMEM without methionine supplemented with ^{35}S methionine 250 µCi/ml. The plates are then returned to the tissue culture incubator at 37 °C. The pulse period should be as short as possible but not so short that it becomes difficult to detect label incorporated into specific proteins.

iii. After the pulse period, the monolayers are rinsed twice with HBSS 0.5 ml/well as quickly as possible. One set of wells is given lysis buffer 0.333 ml/well and immediately put at −20 °C to provide the initial

time point (T0). The other wells are given complete DMEM 0.333 ml/ well and returned to the tissue culture incubator for the chase period.
iv. At predetermined time intervals, media is harvested and cells lysed to provide the extra- and intracellular samples for the time point.
v. Cell lysates are subjected to two freeze–thaw cycles and then harvested into centrifuge tubes. Cell lysates and extracellular fluid samples in microfuge tubes are spun in the microfuge for 10 min and then removed from any pelleted debris.

6.2. Determination of total protein content by TCA precipitation

i. From each sample, 5 µl is applied to a small piece of filter paper (Whatman Disc) previously labeled with the sample ID by pencil. Then 5 µl pulse medium is added to separate discs, and all of the discs are dried under a heat lamp for 15 min.
ii. The dried discs are collected in a large beaker, and 10% trichloroacetic acid (TCA) is added (∼10 ml/disc) and incubated at room temperature for 10 min with occasional stirring.
iii. The liquid is discarded into a "radioactive approved" sink and 5% TCA is added to the discs, the beaker covered with foil and the liquid heated in boiling water for 15 min. The liquid is discarded again into a "radioactive approved" sink, and the discs rinsed with 5% TCA and then with 100% ethanol, adding only enough liquid to cover them, then gently swirling for a few seconds, and discarding the liquid.
iv. The discs are dried under a heat lamp for 15 min, and then placed into scintillation vials one by one using forceps. Then a mixture of 130 µl distilled water and 700 µl solvable (tissue solubilizer) is added to each vial; the caps were secured and the vials shaken for 1 h at room temperature. Then 10 ml scintillation fluid is added to each sample, capped tightly again, and radioactivity was counted in a scintillation counter using the program for ^{35}S.

6.3. Immunoprecipitation to detect specific proteins

i. Aliquots of the cell lysates and extracellular fluid are subjected to immunoprecipitation for the specific protein being investigated. To determine the effect of a perturbation on the kinetics of degradation and/or secretion, one set of intracellular lysates and extracellular fluid samples from the different time points of a pulse–chase protocol will be compared to another set. The volume of the samples will be identical within each set of samples and will be matched with each other for total protein synthesis using the TCA counts for the time 0 intracellular sample.

ii. 30 μl Pansorbin/sample is centrifuged in microfuge tubes for 1 min to pellet Staphylococcus protein A, then the supernatant is aspirated, and the pellet loosened by vortexing.
iii. The samples are then added to the protein A pellet, vortexed well, and then incubated at 4 °C for 30 min to preclear. At the end of this time period, the samples are spun for 5 min at 4 °C and then the supernatant (precleared sample) is transferred to a new tube.
iv. An appropriate volume of antibody to the specific protein is added to each precleared sample, vortexed briefly and incubated overnight by rotating the tubes at 4 °C to form immune complexes.
v. The immune complexes are then captured by adding 30 μl washed Pansorbin to each sample and incubated at 4 °C for 1 h. Then the supernatant and pellet are separated by repeated centrifugation and consequent washing of the pellet first with lysis buffer containing BSA (once) and then with lysis buffer alone (four times). After the last wash, the pellet is spun again, so the last amount of supernatant residue can be carefully removed by aspiration with a pipette tip.
vi. The captured immune precipitate (pellet) is then reconstituted in 25 μl 2.5× loading buffer, boiled for 10 min, spun for 10 min, and loaded on 9% SDS–PAGE. The specific protein is then detected by fluorography on X-ray film by migration relative to known molecular mass markers.

7. Methods for Detection of Hepatic Autophagy *In Vivo*

EM is the gold standard for detecting autophagy in the liver *in vivo*.

7.1. Analysis of hepatic autophagy *in vivo* by EM

i. Liver tissue is fixed in 1% glutaraldehyde–0.1 M sodium cacodylate and embedded in polybed for ultrathin section transmission EM.
ii. Processed sections are viewed and photographed using a Zeiss 902 transmission electron microscope (Fig. 3.3).
iii. Quantification of autophagy is performed with grids superimposed on 10 photomicrographs of each specimen at 3000× and showing a complete hepatocyte with a nucleus. The area of cytoplasm occupied by autophagic vacuoles is then determined as a percentage of the total area of cytoplasm. The area occupied by fat droplets is excluded during the measurement (Teckman *et al.*, 2002).
iv. For immune EM, sections are embedded in 10% gelatin for ultrathin sectioning. Sections are incubated with primary antibody for 2 h and then with secondary antibody for 1 h.

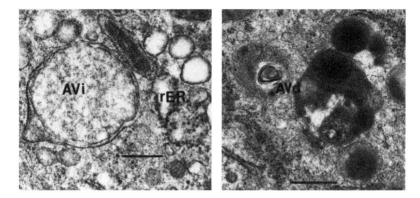

Figure 3.3 EM of human liver tissue of a patient with AT deficiency. *Left panel*: an early autophagic vesicle (AVi) adjacent to dilated rough ER; *right panel*: a more mature, degradative autophagic vesicle (AVd) as characterized by increased electron-dense debris within the internal contents. Scale bars 100 nm (reprinted with permission from Teckman and Perlmutter, 2000).

7.2. Analysis of autophagy by fluorescence microscopy using the GFP-LC3 mouse

Autophagosomes can be detected by fluorescence using the GFP-LC3 mouse because LC3 is specific for the membrane of autophagosomes. Previous studies have shown that vesicles with the appropriate ring-shaped structure can be detected in the cytoplasm of liver cells of the GFP-LC3 mouse only after starvation (Mizushima, 2004). Using this mouse model bred to a mouse with hepatocyte-specific inducible expression of mutant ATZ, we were able to show that accumulation of ATZ in the ER was sufficient to induce autophagy, that is, green fluorescent autophagosomes were detected in the cytoplasm of liver cells when ATZ expression was induced even though the mice were not starved (Kamimoto *et al.*, 2006). Staining with antibody to GFP enhances the detection of autophagosomes in liver from mice on the GFP-LC3 background (Fig. 3.4).

i. Mouse liver tissue is excised after humane euthanasia, cut into 6–8 × 6–8 mm pieces and placed into 2% buffered PFA, pH 7.2 for 1–2 h, then transferred into 30% sucrose for 24 h, changing the sucrose every 6–8 h. The samples are then wiped dry on filter paper and immersed into liquid nitrogen cooled 2-methylbutane for 30 s and liquid nitrogen for 10 s, and then placed into chilled containers and kept on dry ice (if used immediately), or at $-80\ °C$ until use.

ii. Cryostat sections (6 μm) of mouse liver tissue are obtained, mounted on glass slides, and kept at $-20\ °C$ until they were ready for use (for up to 2 weeks). The tissue is then rehydrated by washing the slides 2× in PBS.

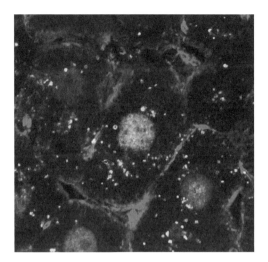

Figure 3.4 Activation of the autophagic response in the liver of the PiZ × GFP-LC3 mouse. The PiZ mouse has constitutive expression of human ATZ, and the GFP-LC3 mouse makes green fluorescent autophagosomes when subjected to starvation. Here, staining of a liver section from the PiZ × GFP-LC3 mouse with anti-GFP shows the green fluorescent autophagosomes in the absence of starvation. The section is also stained with dsRed anti-actin to show the plasma membrane of hepatocytes.

Blocking is performed in 2% BSA for 45 min, followed by five washing steps (PBS with 0.5% BSA).

iii. Rabbit anti-GFP is used as the first antibody. Antibody (100 μl 0.5 μg/ml) is applied on each section for 1 h at room temperature and then washed five times with PBS, 0.5% BSA.

iv. The secondary antibody is donkey anti-rabbit IgG-Alexa488 diluted in PBS, 0.5% BSA, and the incubation time is 1 h at room temperature. After washing the sections five times in PBS, 0.5% BSA and five times in PBS to remove traces of BSA, Hoechst stain (10 μg/ml) is added to stain the nuclei for 30 s at room temperature.

v. Finally, sections are washed another three times with PBS, and then a glass cover was adhered over the sections with gelvatol and incubated at 4 °C overnight in the dark.

7.3. Use of autophagy-deficient mouse models

The role of autophagy in physiologic and pathologic processes *in vivo* can be evaluated by using mouse models that are deleted for specific autophagy genes and therein rendered deficient in autophagy. Although homozygous deletion of beclin-1 (Atg6) in mice is lethal, the beclin 1-deficient mouse can be used in the heterozygote state. The tumor suppressor activity of autophagy has been demonstrated in this mouse even though its autophagy deficiency is partial (Qu *et al.*, 2003; Yue *et al.*, 2003). It has the advantage of partial deficiency of autophagy in all tissues. A mouse with conditional deletion of Atg7 has been generated and can be used for complete deficiency of autophagy mostly in the liver (Komatsu *et al.*, 2005). This mouse model was originally made with the promoter which permits deletion of the target gene in cells other than hepatocytes. A mouse with hepatocyte-specific conditional deletion of Atg7 has been made by using the albumin promoter (Matsumoto *et al.*, 2008). Both of these models can be used for investigating the effect of complete deficiency of autophagy in the liver although the latter model has the advantage of being exclusively liver specific.

8. Methods for Detection of Hepatic Injury

Several mouse models and methods for detection of hepatic injury have been used to investigate the role of autophagy in AT deficiency and recently to demonstrate the effect of autophagy enhancer drugs as potential therapeutic agents (Hidvegi *et al.*, 2010).

8.1. Mouse models

8.1.1. PiZ mouse

Germline transgenic mouse strains were generated using a genomic fragment of the entire human *alpha-1-antitrypsin Z* gene (~19 KB), including about 2 kb downstream and upstream flanking regions on C57Bl/6 background (Carlson *et al.*, 1989). This mouse is ideal for recapitulation of the hepatic pathology of AT deficiency. It is known to have abundant expression of ATZ in hepatocytes and other cell types that express ATZ (Fig. 3.4) in humans (Carlson *et al.*, 1989). In the liver, there are abundant ATZ-containing intrahepatocytic globules and inflammation that is characteristic of what is seen in the human liver (Carlson *et al.*, 1989; Rudnick *et al.*, 2004). We have found that the liver of the PiZ mouse (Fig. 3.5) resembles that of in humans with the classical form of AT deficiency in terms of regenerative activity, steatosis, dysplasia, mitochondrial injury, activation of autophagy, NFκB, and genes associated with fibrosis (Hidvegi *et al.*, 2005, 2007; Rudnick *et al.*, 2004; Teckman *et al.*, 2004; Teckman and Perlmutter, 2000). In our most recent study, we used Sirius Red staining and quantification of hydroxyproline (OHP) in the liver of these mice for the first time and found that there is also significant hepatic fibrosis, the most important hepatic histological marker of hepatic injury that occurs in the human disease. The hepatic OHP content is more than twofold higher than that in the background strain (Hidvegi *et al.*, 2010). Taken together, these

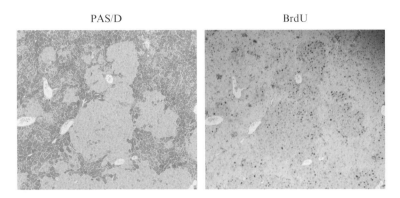

Figure 3.5 Hepatocellular proliferation in the liver of the PiZ mouse. *Left panel*: liver section stained with PAS/diastase (globules: magenta stain); *right panel*: liver section stained with BrdU, BrdU-labeled nuclei are almost exclusively found in cells that are devoid of globules indicating that globule-devoid hepatocytes are proliferating (reprinted with permission from Perlmutter, 2011). (For interpretation of the references to color in this figure legend, the reader is referred to the Web version of this chapter.)

observations indicate that the PiZ mouse is an appropriate model for the gain-of-toxic function mechanism that is responsible for liver damage in the classical form of AT deficiency. It is important to point out that the endogenous murine ortholog of AT is not knocked out in this mouse, so it does not have deficient serum levels of AT. In this perspective, it is not an exact phenocopy of the classical form of AT deficiency. In particular, it cannot be a model for the loss-of-function mechanisms associated with the classical form of AT deficiency.

8.1.2. Z mouse

As a first step to create a transgenic mouse strain that inducibly expresses the Z mutant of AT exclusively in the liver cells, we generated a "target" strain containing the coding region for ATZ downstream the Tet-responsive element of the pTRE vector (Clontech). These mice were created on the FVB/N background and then mated to a liver-specific promoter mouse (LAP, Jackson's Lab) to target ATZ expression to the liver. Germline littermates were then bred until ATZ expression could be detected in their liver. Line M5 proved to have a high on/off expression ratio and was bred further through 10 generations (Hidvegi *et al.*, 2005). Expression of ATZ is solely in hepatocytes and solely when DOX is removed from the drinking water. The advantage of this model is that expression of ATZ can be tightly controlled for experimental studies. However, it has relatively low levels of expression of ATZ and so there is minimal formation of intrahepatocytic globules and negligible fibrosis. Thus, it is not a good model of hepatic injury.

Each of these mice has been bred to the GFP-LC3 mouse to permit assessment of autophagy by detection of green fluorescent autophagosomes (Hidvegi *et al.*, 2010; Kamimoto *et al.*, 2006).

8.2. Assessment of hepatic fibrosis by determination of hydroxyproline (OHP) content

Determination of the total OHP content per microgram of dry liver tissue in the presence of large amounts of other amino acids is a quantitative measure of collagen deposition and fibrosis (Woessner, 1961).

8.2.1. Acid hydrolysis

i. Approximately 50–100 mg (wet weight) snap frozen liver tissue samples is incubated in open 1.7 ml Eppendorf tubes at 110 °C for 48 h, and then the dry weight is measured on an analytical scale. The tissue samples are then transferred into 2 ml glass vacuoles, and 1 ml 6 N HCl is added. Five alternating vacuum and nitrogen gas flush steps are used to evacuate oxygen, and then the vacuoles are closed by sealing the neck.
ii. The samples are then incubated at 110°°C for 24 h. After cooling off, the neck of vacuole is broken off and the samples air dried under a fume hood for at least 24 h to evaporate the acid.
iii. The dry samples are reconstituted by adding 2 ml PBS and placing them in hot water bath (60 °C) for 1 h. The top of the vacuoles is then sealed by Parafilm. The hydrolyzed sample solutions are then transferred into Eppendorf tubes and centrifuged at 10,000 rpm for 6 min to remove debris. The supernatant is then transferred into 2 ml screw cap tubes (the hydrolyzed samples can be stored at room temperature indefinitely).

8.2.2. OHP measurement

Detection of OHP is based on its oxidation to pyrrole by chloramine-T and conversion of pyrrole to a chromophore with p-dimethyl-aminobenzaldehyde (p-DMABA).

i. Stock solution is prepared freshly every time by dissolving 3 mg dried OHP in 300 μl PBS (10 μg/ml). Then 0, 100, 200, 300, 400, and 500 μl OHP were added to 1000, 900, 800, 700, 600, and 500 μl PBS, creating 0, 1, 2, 3, 4, and 5 μg/ml OHP solutions.
ii. Hydrolyzed liver tissue is diluted 1:10 with PBS (we found that about 40 samples can be measured safely at one time).
iii. 50 mM chloramine-T solution is prepared by adding 0.228 g chloramine-T to 6 ml distilled water and then 9 ml methyl cellusolve (ethylene-glycol monomethyl ether, 2-methoxyethanol) and 15 ml OHP-buffer (30 ml final volume) are added.

iv. 500 µl chloramine-T solution is added to each sample and standard, and the solution is then incubated at room temperature for 20 min.
v. During this incubation, p-DMABA solution can be made by adding 6 g p-DMABA to 30 ml methyl cellusolve (30 ml final volume). The solution is then vortexed and placed into a 60 °C water bath for 20 min to dissolve.
vi. 500 µl 3.15 M perchloric acid ($HCLO_4$) is added to each sample and standard, to destroy chloramine-T. Then all samples and standards are vortexed and incubated at room temperature for 5 min (extending the incubation time to 5–7 min does not affect the reaction).
vii. 500 µl 20% p-DMABA solution is added to each sample and standard, vortexed, and incubated at 60 °C (water bath) for 20 min. If it is necessary, the incubation can wait for maximum 30 min. The color of the reaction is stable for about 1 h.
viii. 200 ml of each sample and standard are pipetted into a flat bottom 96-well microplate (ELISA-plate) in triplicate, and absorbance is measured at 557 nm wavelength. The absorbances of the standards are then plotted against the OHP concentration for a linear curve. The OHP content of the samples is determined with an accounting for the dilution factor. Since the original volume of the sample lysates is 2 ml, the result is multiplied by 2 and divided by the dry weight of the tissue sample, resulting in a measure of microgram OHP/milligram dry tissue.

8.3. Assessment of injury/regenerative activity by BrdU labeling

Incorporation of BrdU in liver cells is a measure reflecting regenerative activity and therefore can be used as a marker of injury.

i. Mice undergo subcutaneous implantation of BrdU-filled osmotic pumps designed for sustained release (Alzet 2001; flow rate 20 µg/h).
ii. After implantation, the animals are allowed *ad libitum* food and water for 4 days.
iii. At the end of the 4 days, the mice are sacrificed and liver is harvested.
iv. Liver is fixed in buffered 10% PFA and then embedded in paraffin.
v. Liver sections are then stained for BrdU using the BrdU immunohistochemistry kit (Oncogene, Boston) (Fig. 3.5).
vi. The frequency of nuclear BrdU labeling is determined by examination of at least three random fields, magnification × 400 and at least 300 cells and nuclei in each tissue section.

REFERENCES

Anderson, R. G., Falck, J. R., Goldstein, J. L., and Brown, M. S. (1984). Visualization of acidic organelles in intact cells by electron microscopy. *Proc. Natl. Acad. Sci. USA* **81,** 4838–4842.

Bernspang, E., Carlson, J., and Pitulainen, E. (2009). The liver in 30-year-old individuals with alpha1-antitrypsin deficiency. *Scand. J. Gastroenterol.* **44,** 1349–1355.

Boland, B., Kumar, A., Lee, S., Platt, F. M., Wegiel, J., Yu, W. H., and Nixon, R. A. (2008). Autophagy induction and autophagosome clearance in neurons: Relationship to autophagic pathology in Alzheimer's disease. *J. Neurosci.* **28,** 6926–6937.

Carlson, J. A., Rogers, B. B., Sifers, R. N., Finegold, M. J., Clift, S. M., DeMayo, F. J., Bullock, D. W., and Woo, S. L. (1989). Accumulation of PIZ alpha1-antitrypsin causes liver damage in transgenic mice. *J. Clin. Invest.* **83,** 1183–1190.

Dunn, W. A. (1990). Studies on the mechanisms of autophagy: Maturation of the autophagic vacuole. *J. Cell Biol.* **110,** 1935–1945.

Fujita, N., Hayashi-Nishino, M., Fukumoto, H., Omor, H., Yamamoto, A., Noda, T., and Yoshimori, Y. (2008). An Atg4B mutant hampers the lipidation of LC3 paralogues and causes defects in autophagosome closure. *Mol. Biol. Cell* **19,** 4651–4659.

Gosai, S. J., Kwak, J. H., Luke, C. J., Long, O. S., King, D. E., Kovatch, K. J., Johnston, P. A., Shun, T. Y., Lazo, J. S., Perlmutter, D. H., Silverman, G. A., and Pak, S. C. (2010). Automated high-content live animal drug screening using *C. elegans* expressing the aggregation-prone serpin α1-antitrypsin Z. *PLOS One* **5**(11), 15460.

Griffiths, G. (1993). Fine Structure Immunocytochemistry. Springer-Verlag, Berlin, Heidelberg.

Hidvegi, T., Schmidt, B. Z., Hale, P., and Perlmutter, D. H. (2005). Accumulation of mutant α1-antitrypsin Z in the ER activates caspases-4 and -12, NFκB and BAP31 but not the unfolded protein response. *J. Biol. Chem.* **280,** 39002–39015.

Hidvegi, T., Mirnics, K., Hale, P., Ewing, M., Beckett, C., and Perlmutter, D. H. (2007). Regulator of G signaling 16 is a marker for the distinct endoplasmic reticulum stress state associated with aggregated mutant α1-antitrypsin Z in the classical form of α1-antitrypsin deficiency. *J. Biol. Chem.* **282,** 27769–27780.

Hidvegi, T., Ewing, M., Hale, P., Dippold, C., Beckett, C., Kemp, C., Maurice, N., Mukherjee, A., Goldbach, C., Watkins, S., Michalopoulos, G., and Perlmutter, D. H. (2010). An autophagy-enhancing drug promotes degradation of mutant α1-antitrypsin Z and reduces hepatic fibrosis. *Science* **329,** 229–232.

Hosokawa, N. H. (2006). Generation of cell lines with tetracyline-regulated autophagy and a role for autophagy in controlling cell size. *FEBS Lett.* **580,** 2623–2629.

Houwerzijl, E. J., Pol, H- Wd, Blom, N. R., van der Want, J. J. L., Wolf, J. D., and Vellenga, E. (2009). Erythroid precursors from patients with low-risk myelodysplasia demonstrate ultrastructural features of enhanced autophagy of mitochondria. *Leukemia* **23,** 886–891.

Jäger, S., Bucci, C., Tanida, I., Ueno, T., Kominami, E., Saftig, P., and Eskelinen, E.-L. (2004). Role for Rab7 in maturation of late autophagic vacuoles. *J. Cell Sci.* **117,** 4837–4848.

Kabeya, Y., Mizushima, N., Ueno, T., Yamamoto, A., Kirisako, T., Noda, T., Kominami, E., Ohsumi, Y., and Yoshimori, T. (2000). LC3, a mammalian homologue of yeast Apg8p, is localized in autophagosome membranes after processing. *EMBO J.* **19,** 5720–5728.

Kamimoto, T., Shoji, S., Mizushima, N., Umebayashi, K., Perlmutter, D. H., and Yoshimori, T. (2006). Intracellular inclusions containing mutant α1-antitrypsin Z are propagated in the absence of autophagic activity. *J. Biol. Chem.* **281,** 4467–4476.

Komatsu, M., Waguri, S., Ueno, T., Iwata, J., Murata, S., Tanida, I., Ezaki, J., Mizushima, N., Ohsumi, Y., Uchiyama, Y., Kominami, E., Tanaka, K., et al. (2005). Impairment of starvation-induced and constitutive autophagy in Atg7-deficient mice. *J. Cell Biol.* **169,** 425–434.

Kruse, K. B., Brodsky, J. L., and McCracken, A. A. (2006a). Characterization of an ERAD gene as VPS30/ATG6 reveals two alternative and functionally distinct protein quality control pathways: One for soluble A1PiZ and another for aggregates of A1PiZ. *Mol. Biol. Cell* **17,** 203–212.

Kruse, K. B., Dear, A., Kaltenbrun, E. R., Crum, B. E., George, P. M., Brennan, S. O., and McCracken, A. A. (2006b). Mutant fibrinogen cleared from the endoplasmic reticulum via endoplasmic reticulum-associated degradation and autophagy: An explanation for liver disease. *Am. J. Pathol.* **168,** 1300–1308.

Kuma, A. H. (2004). The role of autophagy during the early neonatal starvation period. *Nature* **432,** 1032–1036.

Liou, W., and Slot, J. W. (1994). Improved fine structure in immunolabeled cryosections after modifying the sectioning and pick-up conditions. *Proc. Int. Conf. Electr. Microsc.* **13,** 253–254.

Matsumoto, N., Ezaki, J., Komatsu, M., Takahashi, K., Mineki, R., Taka, H., Kikkawa, M., Fujimura, T., Takeda-Ezaki, M., Ueno, T., Tanaka, K., and Kominami, E. (2008). Comprehensive proteomics analysis of autophagy-deficient mouse liver. *Biochem. Biophys. Res. Commun.* **368,** 643–649.

Mizushima, N. Y. (2001). Dissection of autophagosome formation using APG5-deficient mouse embryonic stem cells. *J. Cell Biol.* **152,** 657–667.

Mizushima, N. Y. (2004). In vitro analysis of autophagy in response to nutrient starvation using transgenic mice expressing a fluorescent autophagosome marker. *Mol. Biol. Cell* **15,** 1101–1111.

Mizushima, N., and Yoshimuri, T. (2007). How to interpret LC3 immunoblotting. *Autophagy* **3,** 542–545.

Perlmutter, D. H. (2002). Liver injury in α1-antitrypsin deficiency: An aggregated protein induces mitochondrial injury. *J. Clin. Invest.* **110,** 233–238.

Perlmutter, D. H. (2011). Alpha-1-antitrypsin deficiency: Importance of proteasomal and autophagic degradative pathways in disposal of liver disease-associated protein aggregates. *Annu. Rev. Med.* **18**(62), 333–345.

Qu, X., Yu, J., Bhagat, G., Furuya, N., Hibshoosh, H., Troxel, A., Rosen, J., Eskelinen, E. L., Mizushima, N., Ohsumi, Y., Cattoretti, G., and Levine, B. (2003). Promotion of tumorigenesis by heterozygous disruption of the beclin 1 autophagy gene. *J. Clin. Invest.* **112,** 1809–1820.

Raposo, G., van Santen, H. M., Leijendekker, R., Geuze, J. H., and Ploegh, H. L. (1995). Misfolded major histocompatibility complex class I molecules accumulate in an expanded ER-Golgi intermediate compartment. *J. Cell Biol.* **131,** 1403–1419.

Rudnick, D. A., Liao, Y., An, J. K., Muglia, L. J., Perlmutter, D. H., and Teckman, J. H. (2004). Analyses of hepatocyte proliferation in a mouse model of alpha-1-antitrypsin deficiency. *Hepatology* **39,** 1048–1055.

Seglen, P. O., and Gordon, P. B. (1984). Amino acid control of autophagic sequestration and protein degradation in isolated rat hepatocytes. *J. Cell Biol.* **99,** 435–444.

Slot, J. W., Posthuma, G., Chang, L. Y., Crapo, J. D., and Geuze, H. J. (1989). Quantitative aspects of immunogold labeling in embedded and in nonembedded sections. *Am. J. Anat.* **185,** 271–281.

Suzuki, K., Kirisako, T., Kamada, Y., Mizushima, N., Noda, T., and Ohsumi, Y. (2001). The preautophagosomal structure organized by concerted functions of APG genes is essential for autophagosome formation. *EMBO J.* **20,** 5971–5981.

Teckman, J. H., and Perlmutter, D. H. (2000). Retention of mutant a1-antitrypsin Z in endoplasmic reticulum is associated with an autophagic response. *Am. J. Physiol. Gastrointest. Liver Physiol.* **279,** G961–G974.

Teckman, J. H., An, J. K., Loethen, S., and Perlmutter, D. H. (2002). Fasting in α-1 antitrypsin deficient liver: Constitutive activation of autophagy. *Am. J. Physiol. Gastrointest. Liver Physiol.* **283,** G1156–G1165.

Teckman, J. H., An, J. K., Blomenkamp, K., Schmidt, B., and Perlmutter, D. H. (2004). Mitochondrial autophagy and injury in the liver in alpha-1-antitrypsin deficiency. *Am. J. Physiol.* **286,** G851–G862.

Woessner, J. (1961). The determination of hydroxyproline in tissue and protein samples containing small proportions of this imino acid. *Arch. Biochem. Biophys.* **93,** 440–447.

Wood, S. A. (1993). Non-injection methods for the product ion of embryonic stem cell-embryo chimeras. *Nature* **365,** 87–89.

Yue, Z., Jin, S., Yang, C., Levine, A. J., and Heintz, N. (2003). Beclin 1, an autophagy gene essential for early embryonic development, is a haploinsufficient tumor suppressor. *Proc. Natl. Acad. Sci. USA* **100,** 15077–15082.

CHAPTER FOUR

SERPINS AND THE COMPLEMENT SYSTEM

László Beinrohr,* Thomas A. Murray-Rust,[†] Leanne Dyksterhuis,[†] Péter Závodszky,* Péter Gál,* Robert N. Pike,[†] and Lakshmi C. Wijeyewickrema[†]

Contents

1. Introduction	56
2. Human Plasma C1-Inhibitor	57
2.1. Purification of C1-inhibitor from human plasma	58
2.2. Functional activity of purified C1-inhibitor from human plasma	62
2.3. Method	64
2.4. Interpretation of data	64
3. Production of Recombinant C1-Inhibitor in Yeast	64
3.1. Generation of *P. pastoris* clones expressing C1-inhibitor	66
3.2. Expression of C1-inhibitor in *P. pastoris*	68
3.3. Purification of recombinant C1-inhibitor	71
Acknowledgements	73
References	74

Abstract

C1-inhibitor (serpin G1) is a 105 kDa inhibitor which functions as a major antiinflammatory protein in the body. It has its effects via inhibition of the proteases of the complement system and contact system of coagulation, as well as several direct effects mediated by its unique highly glycosylated N-terminal domain. The serpin controls a number of different proteases very efficiently and for some of these the function is augmented by the cofactor, heparin. Here, we describe the preparation of human plasma and recombinant C1-inhibitor and the basic methods required for their characterization, using the complement enzyme C1s as an example of a target enzyme.

---

* Institute of Enzymology, Biological Research Center, Hungarian Academy of Sciences, Hungary
[†] Department of Biochemistry and Molecular Biology, Monash University, Clayton, Victoria, Australia

1. Introduction

The complement system is a major component of the immune system, playing vital roles in the recognition of pathogens and altered human cells and thus protecting the body from numerous diseases. As part of its mechanisms of action, complement strongly upregulates inflammatory processes in the body. While this is an important part of its mechanism of action, it can also play roles in so-called inflammatory diseases, which are often characterized by a chronic inflammation component. The regulation of the complement system is therefore vital in order to maintain healthy body function.

The system can be activated by three separate, but highly connected pathways: classical, lectin, and alternative. Serpin G1 or C1-inhibitor is a major regulator of complement activation via the classical and lectin pathways of complement. The serpin has its effects by efficiently inhibiting the initiating proteases of the pathways. Other serpins, such as antithrombin, have also been shown to play a role in the regulation of some complement proteases.

The classical pathway of complement is initiated by the action of the C1 component, made up of the C1q recognition molecule and a tetramer composed of two copies of each of two proteases: C1r and C1s. Recognition of ligands, usually antibody–antigen complexes, by C1q results in autoactivation of C1r, which in turn activates C1s. The action of C1s on its cognate substrates, C4 and C2, results in the formation of the C4bC2a complex which functions as a C3 convertase to activate the rest of the pathway and hence the system as a whole. C1-inhibitor has been shown to inactivate both C1r and C1s using a typical serpin mechanism. The serpin inhibits C1r with an association rate (k_{ass}) of $\sim 3 \times 10^3 \, M^{-1} s^{-1}$ (Nilsson and Wiman, 1983) and C1s with a k_{ass} of $\sim 4 \times 10^4 \, M^{-1} s^{-1}$ (Wuillemin et al., 1997). Inhibition of the proteases dissociates them from the C1 complex (Sim et al., 1979) and thus terminates their action. Complexes of C1s with C1-inhibitor have been shown to be cleared from the circulation by binding to the low-density lipoprotein receptor-related protein (Storm et al., 1997). Heparin has been shown to accelerate the rate of association between both proteases and C1-inhibitor by up to 58-fold (Wuillemin et al., 1997). The mechanism involved has been postulated to involve a "charge sandwich," in which the protease and serpin both have positively charged interacting faces that are neutralized by the heparin-binding between them (Beinrohr et al., 2007). This is most likely a relevant event in inflammation because it is likely that most cells would be present at the site of such events and these cells are known to store heparin in their granules, the contents of which are released during inflammation. The rate of association between C1r/C1s and C1-inhibitor is maximally stimulated by high molecular weight forms of the nonphysiological molecule, dextran sulfate (rate enhanced by 130-fold for C1s by high molecular weight dextran sulfate; Wuillemin et al., 1997). Thus, it is clear that the two initiating proteases of the classical pathway are efficiently regulated by C1-inhibitor under conditions that are likely to occur during inflammatory processes.

Much less work has been undertaken on the regulation of the activity of the proteases of the lectin pathway by C1-inhibitor, but this pathway does appear to be efficiently controlled by the serpin. Activation of the lectin pathway involves recognition of carbohydrate patterns on the surface of microbes or altered cells by molecules such as mannose-binding lectin (MBL) or the ficolins. The carbohydrates recognized are often high in mannose sugars, unlike the sialic acid residues coating normal mammalian cells. The lectin recognition molecules are usually arranged into higher order, disulfide linked oligomers and are usually complexed with dimers of the MBL-associated serine proteases (MASPs; Thiel, 2007). There are three such MASPs: MASP-1, MASP-2, and MASP-3. Splice variants of these proteins also exist and, in fact, MASP-1 and MASP-3 are splice variants of the same gene, with only their serine protease domains differing from each other. MASP-2 has been shown to be responsible for the direct activation of complement via the lectin pathway as this enzyme is capable of autoactivating upon engagement of the recognition molecule with ligands, and is additionally capable of activation of the C4 and C2 substrates required to form the same C3 convertase complex as is formed in classical pathway activation. It is therefore of note in this context that C1-inhibitor has a very high association rate constant for MASP-2 even in the absence of heparin (Kerr et al., 2007). The roles of MASP-1 and MASP-3 in the complement system are less clear at present. C1-inhibitor has an association rate constant of $6.2 \times 10^3\ M^{-1} s^{-1}$ for MASP-1, which is not enhanced in the presence of heparin (Dobó et al., 2009). MASP-1 is more efficiently inhibited by antithrombin in the presence of heparin ($k_{ass} = 4 \times 10^4\ M^{-1} s^{-1}$), which appears to be due to effects of heparin on antithrombin and not any effects due to binding to MASP-1. MASP-3 is not inhibited by C1-inhibitor (Zundel et al., 2004) and thus far no serpin has yet been found which inhibits this enzyme.

Here, we describe the methods for the preparation and characterization of human plasma and recombinant forms of C1-inhibitor for use in further studying this serpin. The prepared protein can be used for a variety of tasks including characterization of kinetic mechanisms of interaction with target proteases and for structural studies using the serpin.

2. Human Plasma C1-Inhibitor

C1-inhibitor is an acute-phase protein that circulates in blood at levels of around $0.25\ g\ l^{-1}$, which can increase up to 2.5-fold during inflammation (Liu et al., 2007). There are a few different methods of purification from human plasma published, the latest in 1991 by Pilatte *et al.*, which is a three-step procedure that includes PEG fractionation, jacalin–agarose affinity chromatography, and hydrophobic interaction chromatography on phenyl-sepharose. The PEG precipitation is used to remove IgA. Jacalin is a D-galactose-specific lectin from the jackfruit, *Artocarous integrifolia* that interacts with only a few serum proteins and has been used to isolate functionally active C1-inhibitor without any degradation (Hiemstra *et al.*, 1987); the presence of a relatively high concentration of salt (0.5 M NaCl) in all the buffers also minimizes nonspecific binding. C1-inhibitor is markedly hydrophilic protein (Pilatte *et al.*, 1989), thus under the conditions employed, C1-inhibitor is the only protein not retained by the phenyl-sepharose column. This procedure has major advantages over those used previously, first, because it includes only two chromatographic steps. Secondly, since the C1-inhibitor pool is cleanly and predictably separated from the unwanted proteins by differential elution conditions in both chromatographic steps, no antigenic or functional assays are required to define the desired peaks. Thirdly, only the final product is dialyzed while all previously practiced methods of purification require several buffer changes. In this chapter, we therefore use this method and add a further anion-exchange step, to ensure purity.

2.1. Purification of C1-inhibitor from human plasma

1. 240 ml of human, citrated, fresh frozen plasma (kindly provided by Australian Red Cross Blood Service) was thawed at 37 °C, and then protease inhibitors: p-nitrophenyl-p'-guanido benzoate (NPGB), ethylenediaminetetraacetic acid (EDTA), and soybean trypsin inhibitor (SBTI) were added to achieve concentrations of 25, 10, and 50 mM, respectively (Sigma Chemical Co., St Louis, MO, USA). After this step, all procedures were performed at 4 °C and polypropylene or polycarbonate containers were used to minimize activation of the contact system.
2. Inhibitor-treated plasma was brought to a final concentration of 21.4% (w/v) polyethylene glycol 3350 (Sigma Chemical Co.). The solid powder was added to the plasma with constant stirring and allowed to equilibrate for 1 h. The precipitate was removed by centrifugation in an Avanti J26CPI centrifuge (Beckman Coulter, Brea, CA) at 10,000×g for 30 min.
3. The PEG concentration in the supernatant was adjusted to a final concentration of 45% (w/v) by addition of solid PEG 3350, equilibrated for l h with stirring and centrifuged at 10,000×g for 30 min to recover the precipitated proteins.
4. The precipitate recovered after 45% (w/v) PEG treatment was solubilized in 240 ml phosphate-buffered isotonic saline (PBS) containing

10 mM EDTA and 25 mM NPGB and applied to a 1.5 cm diameter column containing 10 ml of jacalin-agarose (Thermo Scientific, Rockford, IL, USA) equilibrated in the same buffer.

5. Following application of the sample, the column was washed with PBS containing 10 mM EDTA, 25 mM NPGB, and 0.5 M NaCl until the absorbance at 280 nm of the effluent dropped to, and was constant at 0.3. The bound proteins were eluted using 200 ml of 0.125 M melibiose (Sigma Chemical Co.) in the same buffer. All fractions were collected and the absorbance at 280 nm recorded for each fraction was used to plot a chromatographic profile (Fig. 4.1A). The fractions obtained after the addition of melibiose were analyzed using 10% SDS-PAGE (Laemmli, 1970). Briefly, 30 ml of each fraction was solubilized by addition of 10 ml volume of reducing sample buffer, and boiling for 5 min. Reducing sample buffer comprised 0.1 ml distilled water; 2 ml 50% (v/v) glycerol; 2.5 ml 0.5 M Tris–HCl, pH 6.8; 4 ml 10% (w/v) SDS; 0.4 ml 0.1% (w/v) bromophenol blue; and 1 ml 2-mercaptoethanol (Sigma Chemical Co.). The samples were run with Fermentas PageRuler Plus prestained molecular weight markers (Thermo Fisher Scientific, Waltham, MA) in a Mini Protean III cell (Bio-Rad). The gels were stained for protein using Coomassie Blue R-250 (Sigma Chemical Co.).

6. C1-inhibitor was identified in all fractions by Western blot analysis (Fig. 4.1B). For immunoblotting, proteins in the SDS-PAGE gels (as above) were rinsed in transfer buffer [25 mM Tris, 192 mM glycine, pH 8.3, containing 20% (v/v) methanol], overlayed with nitrocellulose (Schleicher & Schuell Bioscience, Keene, NH, USA), and sandwiched between two sheets of Whatman #1 filter paper in the Bio-Rad transfer unit. A constant voltage (100 V) was applied for 1 h with the transfer apparatus on ice. The nitrocellulose was then blocked in 3% (w/v) skim milk in TS buffer (0.01 M Tris–HCl, 0.15 M sodium chloride, pH 7.4) for 1 h on an orbital shaker, and then goat anti-C1-inhibitor (Sigma Chemical Co.) was applied, diluted in the same solution, for 16 h. The membrane was washed extensively with TS buffer containing 1% (v/v) Tween-20, and incubated with horseradish peroxidase (HRP)-conjugated rabbit anti-goat antibody (Chemicon International, Temecula, CA, USA) in the blocking solution. After further washing in the TS buffer/Tween-20, blots were developed using the ECL chemiluminescence reagent and exposed to X-ray film (Amersham-Pharmacia, Little Chalfont, Buckinghamshire, UK).

7. Fractions containing C1-inhibitor were pooled and adjusted to 0.4 M $(NH_4)_2SO_4$ by addition of a 4 M solution, and loaded onto a 5 ml HiTrap phenyl column (GE Healthcare Life Sciences, Piscataway, NJ), which was equilibrated in PBS containing 0.4 M $(NH_4)_2SO_4$. The column was washed with 10 column volumes of the same buffer.

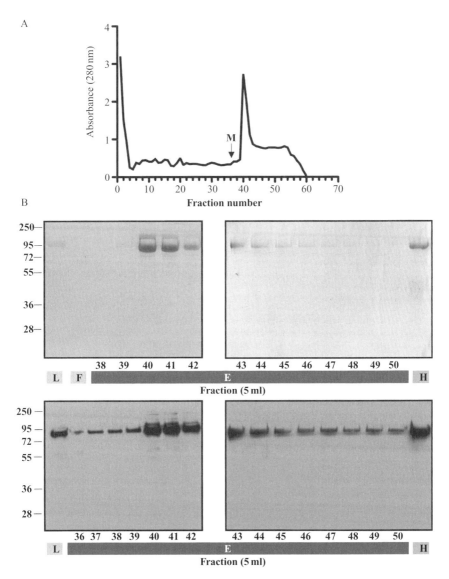

Figure 4.1 *Jacalin–agarose affinity chromatography of the PEG precipitate from human plasma.* Bound proteins were eluted with 0.125 M melibiose (beginning with M). (A) Fractions (5 ml) were collected and the values for absorbance at 280 nm were plotted. (B) Fractions were analyzed by 10% SDS-PAGE, under reducing conditions. The gels were stained with Coomassie Blue R-250 (upper panels) or transferred to nitrocellulose and Western blotted with goat anti-C1-inhibitor antibody (bottom panels). Fractions representing the load (L), the flow-through (F), or proteins eluting after applying the melibiose (E) are labeled accordingly. Purified human C1-inhibitor (H) was used as a positive control for both.

8. The material not retained by the column (~300 ml) was pooled, concentrated (~10 ml), and buffer exchanged into 20 mM Tris–HCl, pH 7.0 using a VivaFlow 20 (Sartorius Stedim Biotech S.A., Aubagne Cedex, France). This was loaded on to a Mono Q 5/50 GL column (GE Healthcare Life Sciences), equilibrated in the same buffer using a flow rate of 1 ml min^{-1}. The column was washed with 20 mM Tris–HCl, pH 7.0 and bound proteins eluted using a 20 ml linear gradient to 500 mM sodium chloride, collecting 0.5 ml fractions. Fractions were analyzed by SDS-PAGE and Western blotting (Fig. 4.2).

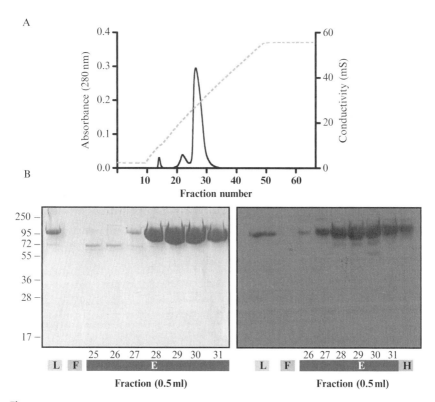

Figure 4.2 *Purification of C1-inhibitor by Mono Q FPLC anion-exchange chromatography.* (A) Elution profile of C1-inhibitor from Mono Q FPLC. Purified C1-inhibitor was loaded onto a Mono Q GL 5/50 column equilibrated in 20 mM Tris–HCl, pH 7.0. An increasing linear gradient of NaCl from 0 to 1 M was applied in the same buffer. (B) SDS-PAGE analysis of the purified C1-inhibitor, under reducing conditions. The gels were stained with Coomassie blue (left panel) or transferred to nitrocellulose and Western blotted with goat anti-C1-inhibitor antibody (right panel), and visualized using peroxidase-coupled rabbit anti-goat IgG and ECL reagent. Fractions representing the load (L), the flow-through (F), or proteins eluting after applying the salt gradient are labeled accordingly. Commercially available human C1-inhibitor (H) was used as a positive control for the Western blot.

2.2. Functional activity of purified C1-inhibitor from human plasma

2.2.1. Stoichiometry of inhibition as analyzed by SDS-PAGE

C1-inhibitor and C1s were incubated at concentrations of 0–5 and 2 µM, respectively, for 1 h at 37 °C, in a total volume of 30 ml in buffer containing 50 mM Tris, 150 mM NaCl, pH 7.3. The reactions were stopped by addition of 10 ml of reducing SDS-PAGE sample buffer. All samples were boiled for 5 min and analyzed by SDS-PAGE (Fig. 4.3). The gels were stained with Coomassie Brilliant Blue R-250.

2.2.2. Stoichiometry of inhibition analyzed using enzyme assay

The stoichiometry of protease inhibition by C1-inhibitor was measured by incubating 75 nM C1s with plasma C1-inhibitor at concentrations ranging from 0 to 67.5 nM. All reactions were carried out in 90 µl (final volume). After a 2 h 45 min incubation at 37 °C in reaction buffer [50 mM Tris, pH 7.4, 150 mM NaCl, 0.02% (w/v) PEG 8000, 0.02% (w/v) sodium azide], the residual C1s activity was detected by the addition of the substrate, Boc-Leu-Gly-Arg-AMC (Bachem, Germany). Fluorescence was measured with excitation at 360 nm (±10) and emission at 460 nm (±10) in white 96-well plates using a Labtech Fluostar Optima microplate reader (BMG Labtech, Australia). The SI value corresponds to the abscissa intercept of the linear regression analysis of fractional velocity (Fig. 4.4).

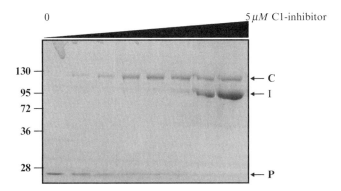

Figure 4.3 *SDS-PAGE analysis of the C1-inhibitor and C1s interaction.* C1s (2 µM) and increasing concentrations of C1-inhibitor (0–5 µM) were incubated at 37 °C for 1 h and the reaction products analyzed by 10% SDS-PAGE, under reducing conditions. The presence of C1s (P), native C1-inhibitor (I), and complexed C1s and C1-inhibitor (C) are indicated.

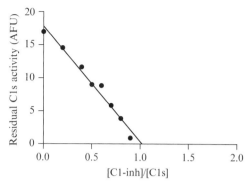

Figure 4.4 *Stoichiometry of C1-inhibitor and C1s inhibition determined from residual protease activity with increasing C1-inhibitor concentrations.* The stoichiometry of inhibition of C1s is taken as the *x*-intercept of the linear regression when residual protease activity is plotted against the ratio of C1-inhibitor to protease. A value of ∼1 is normal for most inhibitory serpins with a cognate target protease.

2.2.3. Measurement of association rate constant for C1-inhibitor with C1s

The association rate constant (k_{ass}) is a measure of the rate at which free protease interacts with free inhibitor and, as serpins are irreversible inhibitors of proteases, this is a key measure of the efficiency with which a serpin inhibits a target protease. The k_{ass} value is usually determined indirectly, by measuring residual protease activity, which is monitored using a fluorescent peptide substrate using discontinuous or continuous methods; we describe the latter here. For the discontinuous method, protease and inhibitor are incubated together in the absence of a fluorescent reporter target for several hours and aliquots are removed at various time points to be measured in the presence of the reporter substrate. This method is carried out under first order conditions with respect to the serpin and thus requires the use of excess inhibitor (5- to 10-fold greater than the C1s concentration). The continuous method is usually also carried out under first order conditions, but the method described here was carried out under essentially second order conditions (equimolar concentrations of enzyme and inhibitor). This was due to the balance between the high amounts of enzyme needed to rapidly and reproducibly report residual activity and the association rate with the serpin. Use of higher serpin–enzyme ratios would have resulted in reactions too rapid to be measured using the substrate and methods employed. This procedure has been found to be valid in our laboratories provided the SI value is close to one, which in this case it is. This method involves continuously monitoring residual activity in real time starting with delivery of the protease to wells containing reporter substrate and inhibitor.

2.3. Method

All experiments were performed in duplicate in a 96-well plate. Wells were blocked for 10 min by addition of 100 µl 0.5% (w/v) BSA in fluorescence activity buffer [FAB; 50 mM Tris, 150 mM NaCl, 0.2% (w/v) PEG 8000, 0.02% (w/v) sodium azide] and rinsed once with FAB alone. Care was taken to remove all residual liquid to avoid dilution of subsequent reactions. Fifty microliters of 1 mM Boc-Leu-Gly-Arg-AMC and 25 µl of 40 nM C1-inhibitor in FAB were coincubated in each well for 10 min at 37 °C. At the same time, 40 nM C1s in FAB was incubated for 10 min at 37 °C. Twenty-five microliters of C1s was added to each well immediately prior to measurement. The fluorescence liberated by the activity of C1s was measured using the microplate reader set to a temperature of 37 °C. Filters on the microplate reader were set to allow transmission of excitation at 360 nm and emission at 460 nm and excitation used 10 flashes. Mixing was performed for 1–2 s prior to the start of measurement only. Each cycle was between 30 and 40 s in duration and 250 cycles were used.

2.4. Interpretation of data

Measurements for all cycles in all wells were copied into GraphPad Prism. Nonlinear regression [see Eq. (4.1)] was used to fit the curves and to estimate k_{obs}. The k_{ass} value was calculated using Eq. (4.2) to take into account the presence of the peptide substrate in the reaction. The K_m value for C1s cleavage of Boc-Leu-Gly-Arg-AMC is 500 µM.

$$Y = Y_{init} + Y_{max}(1 - \exp[-E \times k \times X]) \quad (4.1)$$

- Y = fluorescence at given time
- Y_{init} = initial fluorescence
- Y_{max} = fluorescence at completion of reaction
- E = [enzyme] (M)
- k = estimate of k_{obs}
- X = time (s)

$$k_{ass} = k_{obs} \times (1 + [S]/K_m) \quad (4.2)$$

3. PRODUCTION OF RECOMBINANT C1-INHIBITOR IN YEAST

Mutations in the coding region of the *C1-inh* gene can compromise the structural and functional integrity of the protein. It is important therefore to study the structural and functional consequences of a given mutation (Cugno *et al.*, 2009). Since the number of different naturally occurring C1-inh mutants is quite high, it is not really feasible to isolate and purify them from the patient's sera. Therefore, recombinant expression of C1-inh and its potential variants is of great practical importance (Beinrohr *et al.*, 2008). While recombinant C1-inh produced in mammalian or insect cells is the most authentic protein, the low yields of protein provided by the expression systems are not sufficient for crystallization or detailed functional studies (Wolff *et al.*, 2001). Another obvious choice would be expression in *Escherichia coli*, which is the most widely used expression system, but C1-inhibitor cannot be expressed as an active protein in *E. coli* because the disulfide bonds that are crucial to its structure are inefficiently formed in the reducing environment of the bacterium's cytosol. Only trace amounts of active C1-inhibitor were detected when expression was conducted in a special *E. coli* strain capable of disulfide bond formation (Lamark *et al.*, 2001). Previous studies in our laboratory showed that *in vitro* refolding of sophisticated proteins produced in an insoluble form in *E. coli* is a feasible approach in difficult cases (Ambrus *et al.*, 2003). Indeed, we did manage to refold and purify a sufficient amount of C1-inhibitor to be visible on SDS-PAGE (Beinrohr, doctoral thesis, 2009). The protein was visualized as a single band, which indicated that it was not modified by glycosylation, and formation of covalent complexes was observed with target proteases. However, the method fell short in terms of the yield of protein obtained.

A compromise between the simplicity of *E. coli* and the complexity of mammalian cells are the eukaryotic yeasts. *Pichia pastoris*, for example, is an industrial methylotrophic yeast that is now established for the production of recombinant proteins (for a review of *P. pastoris* expression systems, see Cereghino and Cregg, 2000). *Pichia* has several advantages over other protein expression systems, including the high growth rates and high expression rates found in bacterial systems, the posttranslational modifications needed for proper eukaryotic protein production and the added benefit of being highly cost effective. The principal advantage of *P. pastoris* over other yeast strains, such as *Saccharomyces cerevisiae*, can be attributed to its preference for a respiratory rather than a fermentative mode of growth, resulting in high cell densities, reaching an astonishing abundance of 20–50% (v/v), without accumulating the toxic level of ethanol and acetic acid that restrain cultures of *S. cerevisiae*. In addition, for the purposes of recombinant protein expression, a tight and strong inducible promoter is advisable, which is available in *P. pastoris*. *P. pastoris* can grow solely on methanol as carbon source. Due to the toxicity of methanol, its utilization occurs only when all other carbon sources are missing. The enzyme responsible for oxidation of methanol to formaldehyde, the first step in methanol consumption, is alcohol oxidase 1 (AOX1). It has low affinity for methanol and oxygen, therefore large quantities are expressed, and AOX1 levels can reach ~20% of total cell protein. Therefore, the AOX1 promoter is frequently used for the controlled expression of foreign genes.

A pioneering study described expression of active C1-inhibitor at yields exceeding milligrams in *P. pastoris* (Bos *et al.*, 2003). Recently, we have modified and optimized the expression system and used the material to crystallize C1-inhibitor for which large amounts of homogeneous material were required (Beinrohr *et al.*, 2007).

3.1. Generation of *P. pastoris* clones expressing C1-inhibitor

3.1.1. Cloning

Many variations of the *P. pastoris* expression system have been developed. We used the classic system, which is based on the histidine auxotroph *P. pastoris* strain GS115 (Invitrogen) with pPic9 vectors (Invitrogen) that contain the *HIS4* gene as a metabolic marker (Invitrogen, K1710-01). The cloning was performed according to standard procedures (Sambrook *et al.*, 1989). Since natural C1-inhibitor is an extracellular protein, it must also be secreted by *P. pastoris*, otherwise no posttranslational modifications occur. This can be achieved by fusing C1-inhibitor to the α-factor signal sequence terminating in a KEX2 protease cleavage site, which cleaves the signal peptide during secretion. Here, we describe the cloning of the hexahistidine tagged serpin domain of C1-inhibitor, which can readily be adapted to other constructs as well.

1. A cDNA clone of C1-inhibitor was obtained as described (Bock *et al.*, 1986). The allele Val458Met may be used, which seems functionally identical to the more abundant Val458 variant (Cumming *et al.*, 2003). Numbering is as described previously (Bock *et al.*, 1986), beginning with the first amino acid of the mature protein. Optional: you may find it useful to eliminate the single *Eco*RI restriction site from the native C1-inhibitor gene.
2. The gene was amplified using PCR between codons 97 and 478, including a hexahistidine tag at the N-terminus of the protein coded in the oligonucleotide used for PCR. A *Sna*BI restriction site was placed at the N-terminus and an *Eco*RI site at the C-terminus. Other restriction sites may be used, but C1-inhibitor seems sensitive to the placing of the hexahistidine-tag: in our hands, a C-terminal tag did not work well.
3. The PCR product was inserted into the pPic9 or pPic9K vector (Invitrogen) between the *Sna*BI and *Eco*RI sites. As a result, this construct was fused to the α-mating factor signal peptide and followed by the sequence ...KR|EAEA<u>YV</u>HHHHHHTGSF... (the translated *Sna*BI site is underlined and the KEX2 signal peptidase cleavage site marked with "|").

3.1.2. Isolation of recombinant *P. pastoris* clones

To introduce the gene into *P. pastoris*, we used the following protocol adapted from Wu and Letchworth (2004). The gene is inserted via homologous recombination. Cleaving the expression plasmid at specific sites can target and stimulate the recombination. The digested DNA is then electroporated into *P. pastoris* strain GS115 similarly to the protocol in Wu and Letchworth (2004). It was found that recombination at the 5′ part of the *AOX1* gene or in the *HIS4* gene often produces nonexpressing clones. Instead, a double recombination event ("omega insertion") at both the 5′ and 3′ *AOX1* sites produces expressing clones in every case, but requires a time-consuming two-step selection procedure. This recombination event excises the native *AOX1* gene. The *P. pastoris* can still grow on methanol, because there is a second alcohol oxidase (AOX2), which has a low activity promoter compared to AOX1. Therefore, these clones with phenotype Methanol Utilization Slow (Mut^S) require minimal amounts of methanol during induction, lessening the toxicity for the cells. Strains with functional *AOX1* (e.g., wild types) are designated Mut$^+$.

1. The expression vector was linearized with *Dra*I to facilitate double homologous recombination at the *AOX1* gene. The DNA was prepared using standard miniprep kits. It was not necessary to desalt the digestion reaction and the maximal amount of DNA attainable in the restriction digestion was used and complete digestion was required.
2. A starter culture of GS115 was grown overnight at 30 °C with shaking at 240 rpm in 20 ml Yeast Peptone Dextrose (YPD) medium: 1% (w/v) yeast extract, 2% (w/v) peptone, and 2% (w/v) glucose. 1–10 ml of the overnight culture was then used to seed 100 ml cultures in YPD, which were grown at 30 °C with shaking at 240 rpm, till the absorbance at 600 nm reached 0.6–2.0. This usually took a few hours.
3. The cells were centrifuged at 2000 rpm for 5 min and resuspended in 25–50 ml of 0.6 M sorbitol, 0.1 M LiSO$_4$, 10 mM HEPES, pH 7.5. Freshly dissolved dithiotreitol (DTT) was added to yield a concentration of 10 mM, following which the cells were incubated for 30 min at room temperature.
4. The cells were centrifuged again (2000 rpm, 5 min), resuspended in ice-cold 1 M sorbitol and washed two times using 1 M sorbitol. Finally, the cells were resuspended to yield a density of ∼50% (v/v; usually in a 1–2 ml volume) and aliquoted into 100 μl batches.
5. A maximum of 2–3 μl digested DNA was added to the cells, which were chilled on ice, following which the mixture was electroporated in a 2-mm gapped electroporation cell (Bio-Rad) at 1500 V for 5 ms (ECM399, BTX). Immediately after the pulse, 900 μl of 1 M sorbitol was added.
6. All cells were streaked on Minimal Dextrose (MD) plates: 10% (v/v) 10× Yeast Nitrogen Base (YNB) solution, 2% (w/v) glucose, 400 μg l^{-1}

biotin, 0.1% (v/v) 1000× trace metals solution, and 1.5% (w/v) agarose for plates. 10× YNB solution: 100 g l^{-1} (NH$_4$)$_2$SO$_4$, 20.5 g l^{-1} MgSO$_4$·7 H$_2$O, 20 g l^{-1} KH$_2$PO$_4$, 2 g l^{-1} NaCl, 2.7 g l^{-1} CaCl$_2$·2 H$_2$O. 1000× trace metals solution: 500 mg l^{-1} H$_3$BO$_3$, 63 mg l^{-1} CuSO$_4$·5 H$_2$O, 100 mg l^{-1} KI, 340 mg l^{-1} FeCl$_3$·6 H$_2$O, 630 mg l^{-1} MnCl$_2$·4 H$_2$O, 230 mg l^{-1} NaMoO$_4$·2 H$_2$O, 720 mg l^{-1} ZnSO$_4$·7 H$_2$O. The plates were allowed to air-dry and incubated for several days at 30 °C.

7. Approximately 40 of the clones that appeared were spotted onto MD and minimal methanol (MM) plates simultaneously. MM media: 10% (v/v) 10× YNB, 0.5% (v/v) methanol, 400 µg l^{-1} biotin, 0.1% (v/v) 1000× trace metals solution, and 1.5% (w/v) agarose for plates. Mut$^+$ and Mut^S recombinants were discriminated by observing the clones for at least a week (200 µl methanol was added into the plate lids every 2 days). On MM plates, the Mut^S recombinants were visibly smaller than Mut$^+$ ones, while the clones were identical on MD plates. About 5% of the total clones were Mut^S.

8. The isolated clones were suspended in 1 M sorbitol, frozen in liquid nitrogen, and then kept at −70 °C for long-term storage.

3.2. Expression of C1-inhibitor in *P. pastoris*

P. pastoris can grow solely on methanol as a carbon source. Methanol is toxic, so its utilization occurs only when all other carbon sources are depleted. Since growth of *P. pastoris* on methanol is slower than on other carbon sources, two-phase expression protocols are used (Invitrogen, K1710-01). In the first phase, cell mass is generated on glycerol as a carbon source, a "normal" substrate. Glucose is not recommended, because it represses the AOX promoters. When the sufficient cell mass is achieved, and the glycerol carbon source depleted, induction is initiated by adding methanol to the cell culture.

3.2.1. Expression of C1-inhibitor in flasks

1. A 20–40 ml starter culture in YPD medium was grown overnight at 30 °C with shaking at 240 rpm. The culture was expanded in 0.5–4 l of Buffered Medium with Glycerol (BMGY): 1% (w/v) yeast extract, 2% (w/v) peptone, 10% (v/v) 10× YNB, 0.1 M KH$_2$PO$_4$, 2% glycerol. This culture was grown at 29 °C with shaking at 240 rpm over 2 days. Typically, the Erlenmeyer flasks were loaded to 12.5% and 25% of their nominal volume during induction and cell mass generation, respectively.

2. The cells were centrifuged (2000 rpm, 5 min) and resuspended in half of the volume of the original culture using fresh BMMY medium without methanol: 1% (w/v) yeast extract, 2% (w/v) peptone, 10% (v/v) 10× YNB, 0.1 M KH$_2$PO$_4$. The culture was further incubated with shaking

for at least 0.5 h at 25 °C to deplete residual carbon sources from the yeast extract and peptone. Then, 0.5% (v/v) methanol was added to induce expression and culturing was continued for 3–4 days at 25 °C with shaking at 240 rpm, with 0.5% (v/v) methanol added twice daily. Optionally, phenylmethylsulfonyl fluoride (PMSF) may be added to the methanol solution, to achieve a final concentration of 0.1 mM PMSF in the medium. This may help reduce unwanted proteolysis.

3. Ultimately, $\sim$5 mg l^{-1} protein should be obtained, as estimated following SDS-PAGE analysis with staining using Coomassie Blue R-250 (Fig. 4.5A). The sample should additionally be deglycosylated with Endoglycosidase H due to the heterogeneity of glycosylation in the material obtained.

3.2.2. Fermentor-based C1-inhibitor expression

We optimized the fermentation protocols to increase the amount of C1-inhibitor produced. We found that in chemically defined media, C1-inhibitor is extremely sensitive to proteolysis at the reactive center loop (Beinrohr, doctoral thesis, 2009). This is in sharp contrast with the C1-inhibitor from shake-flask cultures, where no proteolysis was detectable (Fig. 4.5A). The major source of the digesting proteases is the yeast itself, with a small percentage of dead cells most likely releasing proteases that are able to digest the recombinant protein, especially in high-density cultures (Sinha et al., 2005). In addition, serpins are inherently prone to be attacked by proteases at their RCL. To prevent unwanted proteolysis, inclusion of peptone, a peptidic medium component was essential to provide alternative substrates for these proteases (Wu et al., 2008). Moreover, a broad spectrum serine protease inhibitor (PMSF) was added as it is mainly serine proteases that are active at the pH of the expression medium. The third optimization step called for addition of a carbon source (sorbitol, in addition to methanol) to enhance cell viability by providing a more convenient fuel than the toxic methanol (Ramon et al., 2007). The optimized procedure, namely fermentation in rich media in the presence of protease inhibitor and a mild carbon fueling strategy resulted in the production of completely intact C1-inhibitor (Fig. 4.5B). The yields rose to $\sim$20–50 mg l^{-1} culture supernatant, which was considered enough for most purposes (e.g., crystallization attempts). Please note that individual fermentation parameters must be verified experimentally for a given setup. Therefore, the recipe below should serve as a guideline only.

1. A large scale expression in a fermentor with a 2-l vessel (Biostat B, B. Braun) was conducted, using a fed-batch technique in complete medium. A starter culture was grown for 2 days in 40-ml YPD media.

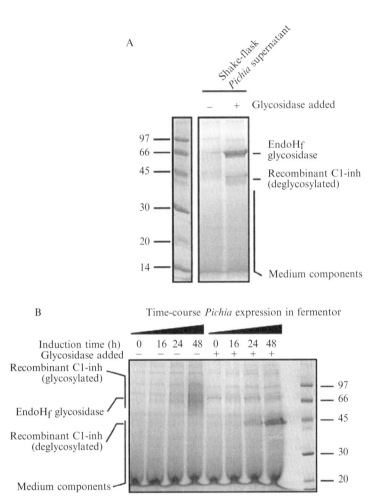

Figure 4.5 *Expression of C1-inhibitor.* (A) The *Pichia pastoris* supernatant contains intact protein after expression by the yeast. Samples were obtained from shake-flask cultures; 20–50 μl was loaded onto SDS-PAGE. The material displays high homogeneity, when glycosylation is removed. (B) Time-course of expression of C1-inhibitor using the optimized procedure for the fermentor. The material remains intact for extended periods of time and the yields are considerably higher than those obtained from shake-flasks.

The cells were centrifuged (2000 rpm, 5 min) and the spent medium was discarded. The fermentor was loaded with 1.5 l of the batch medium [1% (w/v) yeast extract, 2% (w/v) peptone, 10% (v/v) 10× YNB solution, 0.1 M KH_2PO_4, 4% (v/v) glycerol], which was inoculated with the starter culture.

2. The parameters were carefully adjusted to avoid dissolved oxygen (DO) limitations (maintaining levels at 10–50%) and antifoam (Struktol J633, Schill & Seilacher) was added only when necessary. Initial parameters were stirring: 750–900 rpm, temperature 28 °C, pH 5.5 with 25% (w/v) ammonia, and aeration 1.5 l min^{-1}. After depletion of batch glycerol (after 16–22 h), the wet cell weight (wcw) typically reaches 120 g l^{-1}.
3. The culture was continuously fed with 50% (v/v) glycerol using a pump (101 U/I, Watson-Marlow) and tubing (MasterFlex 913.A008.016). The feed was adjusted so that the DO remained in the 10–20% range. The stirring speed was raised to 900–990 rpm, the temperature was lowered to 25 °C and the pH was adjusted to 6.0 during this period. Additionally, 0.25% (w/v) yeast extract and 0.5% (w/v) peptone were added to replenish the spent medium.
4. Induction was commenced with a 25% (v/v) methanol, 40% (w/v) sorbitol mixture when the wcw reached 200 g l^{-1} (usually in 4–8 h). The DO is relatively unresponsive to this mixed carbon source, since the *Pichia* strain is Mut^S. A useful feed rate that maximizes expression without toxicity must be verified experimentally. In our setup, we used an arbitrary feeding rate of 1 overnight, which was raised to 2–3 the next day. The feed was shut off twice daily and PMSF (1.5 ml of a 0.2 M stock in methanol) was added. The DO dropped due to the consumption of methanol and therefore the automatic feed was only continued once the DO returned to previous levels. This process was continued for 2–3 days.

3.3. Purification of recombinant C1-inhibitor

The chromatography procedures were essentially performed according to the manufacturer's instructions (GE Healthcare). It should be noted that the expressed protein from *P. pastoris* appears as a broad smear between ∼50 and ∼100 kDa (Fig. 4.6A). The heterogeneity is due to "hyperglycosylation," the attachment of highly variable and long carbohydrates to the protein. Endoglycosidase H digestion (New England Biolabs) can be used to cleave the glycans off, but certain constructs (e.g., full-length ones) may not be fully deglycosylated.

1. A supernatant from the culture was centrifuged at 10,000 rpm for 10 min, then filtered through a glass fiber filter (GF 3362, Schleicher & Schuell). PMSF (0.2 M stock in methanol) and Na$_2$EDTA (0.5 M stock, pH 8.0) were added to achieve a final concentration of 0.5 and 25 mM, respectively. The pH of the clear supernatant obtained was adjusted to 6.0 using NaOH or HCl, after which it was stored at -70 °C.
2. The buffer of the supernatant was exchanged using a Sephadex G-25 Coarse 5 × 63 cm column (GE Healthcare) into 25 mM NaH$_2$PO$_4$,

pH 6.0. NaCl (4 M stock) and imidazole (1 M stock) were added to final concentrations of 0.4 M and 5 mM, respectively and the pH was adjusted to 7.5. If dialysis is preferred, extra care should be taken to ensure that no trace of EDTA remains in the supernatant.

3. The C1-inhibitor was then further purified using affinity purification on a Ni-NTA Superflow 2.5 × 5.5 cm column (Qiagen) equilibrated in 5 mM imidazole, 0.5 M NaCl, 25 mM NaH$_2$PO$_4$, pH 7.5. The column was washed using 20 mM imidazole, 0.5 M NaCl, 25 mM NaH$_2$PO$_4$, pH 7.5, following which the protein was eluted using 0.2 M imidazole, 0.5 M NaCl, 25 mM NaH$_2$PO$_4$, pH 7.5. Please note that all forms of C1-inhibitor copurified at this step.

4. The fractions containing C1-inhibitor were collected, pooled, and the buffer was exchanged using a fine Sephadex G-25 Fine column (2.6 × 26 cm, GE Healthcare) into 20 mM MOPS, 0.1 mM Na$_2$EDTA, pH 7.0. Dialysis may also be used.

5. The sample was loaded onto a Source 15S cation exchange column (1 × 8 cm, GE Healthcare) equilibrated in 20 mM MOPS, 0.1 mM Na$_2$EDTA, pH 7.0. Bound protein was eluted using a 10 column volume gradient from 0 to 0.5 M NaCl in the same buffer. The active and inactive C1-inhibitor forms were separated at this step (Fig. 4.6B). Please note that this purification step may only work for native and truncated "wild-type" C1-inhibitor forms, but not necessarily on mutants of the molecule.

6. *Optional step.* If higher homogeneity is required, quantitative and preparative deglycosylation is feasible using Endoglycosidase H. Pool the protein fractions obtained from the previous cation exchange step, and carry out the deglycosylation at 4 °C using 1000–4000 units of enzyme Endo H$_f$ (New England Biolabs). The reaction may be monitored using SDS-PAGE. The reaction is usually completed within a week, with over 90% efficiency. If the reaction is performed at higher temperatures, active C1-inhibitor may precipitate. If required, the deglycosylated protein can be purified away from the minute amounts of glycosylated material. The protein was diluted ∼2- to 10-fold using 20 mM MOPS, 0.1 mM Na$_2$EDTA, pH 7.0 and purified again using cation exchange chromatography as in step 5. The protein peaks eluted should be separated into several fractions and the individual fractions checked using SDS-PAGE to ascertain that they are sufficiently pure. Impure fractions should be discarded. For crystallization of C1-inhibitor, an additional gel filtration chromatography step on a Superdex 75 HiLoad 16/60, prep grade column (GE Healthcare) in 0.1 M NaCl, 20 mM MOPS, 1 mM Na$_2$EDTA, pH 7.2, is recommended. The eluted fractions should be measured for their absorbance at 280 nm and then concentrated to ∼10–14 mg ml^{-1} (estimated using a calculated extinction coefficient of 0.64 ml mg^{-1} cm^{-1}). Different C1-inhibitor forms and constructs may

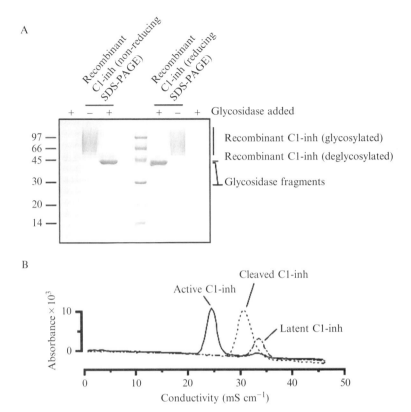

Figure 4.6 *Purification of C1-inhibitor.* (A) Purity of C1-inhibitor, after affinity purification and cation exchange. The extensive smear of C1-inhibitor is due to hyperglycosylation, evidenced by the effect of glycosidase on the molecular weight. (B) Elution profile of deglycosylated C1-inhibitor forms in cation exchange chromatography under the described conditions. Absorbance was measured at 280 nm. The active and inactive forms separate well, but the inactive forms do not. The profile has similar characteristics for different constructs, for example, glycosylated or full-length forms.

precipitate at such high concentrations. For long-term storage, samples were frozen in liquid nitrogen and stored at $-70\,°C$.

ACKNOWLEDGEMENTS

Pilot fermentations were kindly performed by Bálint Kupcsulik, Dávid Domonkos, and Kálmán Könczöl. This work was supported by the Ányos Jedlik Grant NKFP_07_1-MASPOK07 of the Hungarian National Office for Research and Technology, the Hungarian Scientific Research Fund Grant OTKA NK77978, a Baross grant and a Program Grant from the Australian National Health and Medical Research Council.

REFERENCES

Ambrus, G., Gál, P., Kojima, M., Szilágyi, K., Balczer, J., Antal, J., Gráf, L., Laich, A., Moffatt, B. E., Schwaeble, W., Sim, R. B., and Závodszky, P. (2003). Natural substrates and inhibitors of mannan-binding lectin-associated serine protease-1 and -2: A study on recombinant catalytic fragments. *J. Immunol.* **170,** 1374–1382.

Beinrohr, L., Harmat, V., Dobó, J., Lörincz, Z., Gál, P., and Závodszky, P. (2007). C1 inhibitor serpin domain structure reveals the likely mechanism of heparin potentiation and conformational disease. *J. Biol. Chem.* **282,** 21100–21109.

Beinrohr, L., Dobó, J., Závodszky, P., and Gál, P. (2008). C1, MBL-MASPs and C1-inhibitor: Novel approaches for targeting complement-mediated inflammation. *Trends Mol. Med.* **14,** 511–521.

Beinrohr, L. Doctoral thesis. (2009). Structural basis of C1-inhibitor specificity and deficiency. Eötvös Loránd University. http://www.doktori.hu/index.php?menuid=193&vid=4196&lang=EN.

Bock, S. C., Skriver, K., Nielsen, E., Thogersen, H. C., Wiman, B., Donaldson, V. H., Eddy, R. L., Marrinan, J., Radziejewska, E., Huber, R., Huber, R., Shows, T. B., Magnusson, S., *et al.* (1986). Human C1 inhibitor: Primary structure, cDNA cloning, and chromosomal localization. *Biochemistry* **25,** 4292–4301.

Bos, I. G., de Bruin, E. C., Karuntu, Y. A., Modderman, P. W., Eldering, E., and Hack, C. E. (2003). Recombinant human C1-inhibitor produced in Pichia pastoris has the same inhibitory capacity as plasma C1-inhibitor. *Biochim. Biophys. Acta* **1648,** 75–83.

Cereghino, J. L., and Cregg, J. M. (2000). Heterologous protein expression in the methylotrophic yeast Pichia pastoris. *FEMS Microbiol. Rev.* **24,** 45–66.

Cugno, M., Zanichelli, A., Foieni, F., Caccia, S., and Cicardi, M. (2009). C1-inhibitor deficiency and angioedema: Molecular mechanisms and clinical progress. *Trends Mol. Med.* **15,** 69–78.

Cumming, S. A., Halsall, D. J., Ewan, P. W., and Lomas, D. A. (2003). The effect of sequence variations within the coding region of the C1 inhibitor gene on disease expression and protein function in families with hereditary angio-oedema. *J. Med. Genet.* **40,** e114.

Dóbo, J., Harmat, V., Beinrohr, L., Sebestyén, E., Závodszky, P., and Gál, P. (2009). MASP-1, a promiscuous complement protease: Structure of its catalytic region reveals the basis of its broad specificity. *J. Immunol.* **183,** 1207–1214.

Hiemstra, P. S., Gorter, A., Stuurman, M. E., Van Es, L. A., and Daha, M. R. (1987). The IgA-binding lectin jacalin induces complement activation by inhibition of C-1-inactivator function. *Scand. J. Immunol.* **26,** 111–117.

Kerr, F. K., Thomas, A. R., Wijeyewickrema, L. C., Whisstock, J. C., Boyd, S. E., Kaiserman, D., Matthews, A. Y., Bird, P. I., Thielens, N. M., Rossi, V., and Pike, R. N. (2007). Elucidation of the substrate specificity of the MASP-2 protease of the lectin complement pathway and identification of the enzyme as a major physiological target of the serpin, C1-inhibitor. *Mol. Immunol.* **45,** 670–677.

Laemmli, U. K. (1970). Cleavage of structural proteins during the assembly of the head of bacteriophage T4. *Nature* **227,** 680–685.

Lamark, T., Ingebrigtsen, M., Bjornstad, C., Melkko, T., Mollnes, T. E., and Nielsen, E. W. (2001). Expression of active human C1 inhibitor serpin domain in Escherichia coli. *Protein Expr. Purif.* **22,** 349–358.

Liu, D., Lu, F., Qin, G., Fernandes, S. M., Li, J., and Davis, A. E., 3rd. (2007). C1 inhibitor-mediated protection from sepsis. *J. Immunol.* **179,** 3966–3972.

Nilsson, T., and Wiman, B. (1983). Kinetics of the reaction between human C1-esterase inhibitor and C1r or C1s. *Eur. J. Biochem.* **129,** 663–667.

Pilatte, Y., Hammer, C. H., Frank, M. M., and Fries, L. F. (1989). A new simplified procedure for C1 inhibitor purification. A novel use for jacalin-agarose. *J Immunol. Methods* **120,** 37–43.

Ramon, R., Ferrer, P., and Valero, F. (2007). Sorbitol co-feeding reduces metabolic burden caused by the overexpression of a Rhizopus oryzae lipase in Pichia pastoris. *J. Biotechnol.* **130,** 39–46.

Sambrook, J., Fritsch, E. F., and Maniatis, T. (1989). *Molecular Cloning: A Laboratory Manual* Cold Spring Harbor, New York, USA.

Sim, R. B., Arlaud, G. J., and Colomb, M. G. (1979). C1 inhibitor-dependent dissociation of human complement component C1 bound to immune complexes. *Biochem. J.* **179,** 449–457.

Sinha, J., Plantz, B. A., Inan, M., and Meagher, M. M. (2005). Causes of proteolytic degradation of secreted recombinant proteins produced in methylotrophic yeast Pichia pastoris: Case study with recombinant ovine interferon-tau. *Biotechnol. Bioeng.* **89,** 102–112.

Storm, D., Herz, J., Trinder, P., and Loos, M. (1997). C1 inhibitor-C1s complexes are internalized and degraded by the low density lipoprotein receptor-related protein. *J. Biol. Chem.* **272,** 31043–31050.

Thiel, S. (2007). Complement activating soluble pattern recognition molecules with collagen-like regions, mannan-binding lectin, ficolins and associated proteins. *Mol. Immunol.* **44,** 3875–3888.

Wolff, M. W., Zhang, F., Roberg, J. J., Caldwell, E. E., Kaul, P. R., Serrahn, J. N., Murhammer, D. W., Linhardt, R. J., and Weiler, J. M. (2001). Expression of C1 esterase inhibitor by the baculovirus expression vector system: Preparation, purification, and characterization. *Protein Expr. Purif.* **22,** 414–421.

Wu, S., and Letchworth, G. J. (2004). High efficiency transformation by electroporation of Pichia pastoris pretreated with lithium acetate and dithiothreitol. *Biotechniques* **36,** 152–154.

Wu, D., Hao, Y. Y., Chu, J., Zhuang, Y. P., and Zhang, S. L. (2008). Inhibition of degradation and aggregation of recombinant human consensus interferon-alpha mutant expressed in Pichia pastoris with complex medium in bioreactor. *Appl. Microbiol. Biotechnol.* **80,** 1063–1071.

Wuillemin, W. A., te Velthuis, H., Lubbers, Y. T., de Ruig, C. P., Eldering, E., and Hack, C. E. (1997). Potentiation of C1 inhibitor by glycosaminoglycans: Dextran sulfate species are effective inhibitors of in vitro complement activation in plasma. *J. Immunol.* **159,** 1953–1960.

Zundel, S., Cseh, S., Lacroix, M., Dahl, M. R., Matsushita, M., Andrieu, J. P., Schwaeble, W. J., Jensenius, J. C., Fujita, T., Arlaud, G. J., and Thielens, N. M. (2004). Characterization of recombinant mannan-binding lectin-associated serine protease (MASP)-3 suggests an activation mechanism different from that of MASP-1 and MASP-2. *J. Immunol.* **172,** 4342–4350.

CHAPTER FIVE

Use of Mouse Models to Study Plasminogen Activator Inhibitor-1

Paul J. Declerck,* Ann Gils,* *and* Bart De Taeye[†]

Contents

1. Introduction	78
1.1. PAI-1, a unique serpin	80
2. Role of PAI-1 in Different (Patho)Physiological Processes	80
2.1. PAI-1 in cardiovascular disease	80
2.2. PAI-1 and cancer	81
3. Mouse Models to Study the Role of PAI-1	82
4. Mouse (Wild-Type) Models to Study the Role of PAI-1 in Cardiovascular Disease and Cancer	83
4.1. Cardiovascular disease	83
4.2. Cancer	87
5. Genetically Modified Mice to Study the Role of PAI-1 in Cardiovascular Disease and Cancer	87
5.1. Cardiovascular disease	88
5.2. Cancer	92
6. Conclusions	93
References	94

Abstract

Plasminogen activator inhibitor-1 (PAI-1) is the main inhibitor of tissue-type plasminogen activator (t-PA) and urokinase-type plasminogen activator (u-PA) and therefore plays an important role in the plasminogen/plasmin system. PAI-1 is involved in a variety of cardiovascular diseases (mainly through inhibition of t-PA) as well as in cell migration and tumor development (mainly through inhibition of u-PA and interaction with vitronectin).

PAI-1 is a unique member of the serpin superfamily, exhibiting particular unique conformational and functional properties. Since its involvement in various biological and pathophysiological processes PAI-1 has been the subject of many *in vivo* studies in mouse models. We briefly discuss structural and

---

* Laboratory for Pharmaceutical Biology, Faculty of Pharmaceutical Sciences, Katholieke Universiteit Leuven, Campus Gasthuisberg, O&N2, Herestraat, Leuven, Belgium
[†] Feinberg Cardiovascular Research Institute, Northwestern University, Chicago, Illinois, USA

physiological differences between human and mouse PAI-1 that should be taken into account prior to extrapolation of data obtained in mouse models to the human situation. The current review provides an overview of the various models, with a focus on cardiovascular disease and cancer, using wild-type mice or genetically modified mice, either deficient in PAI-1 or overexpressing different variants of PAI-1.

1. INTRODUCTION

The serpin plasminogen activator inhibitor-1 (PAI-1) is an important component of the plasminogen/plasmin system as it is the main inhibitor of tissue-type plasminogen activator (t-PA) and urokinase-type plasminogen activator (u-PA). Consequently, PAI-1 plays an important role in cardiovascular diseases (mainly through inhibition of t-PA) and in cell migration and tumor development (mainly through inhibition of u-PA and interaction with vitronectin; Fig. 5.1). In blood, the balance between blood clot formation (the coagulation system) and their dissolution/degradation (the fibrinolytic process) is tightly regulated. In the fibrinolytic process, the t-PA-mediated pathway of plasmin generation plays a crucial role. Upon activation, the active enzyme plasmin serves to degrade the insoluble fibrin meshwork into soluble degradation products, resulting in dissolution and resorption of the blood clot (Rijken and Lijnen, 2009). The u-PA-mediated pathway of the plasminogen system is primarily involved in phenomena such as cell migration and tissue remodeling, occurring in many biological processes. Indeed, activation of the proMMPs by plasmin leads to degradation of extracellular matrix (ECM) components, a prerequisite for endothelial, smooth muscle, and inflammatory cells to migrate to distant sites. Therefore, the plasminogen system has been suggested to play a role in atherosclerosis, aneurysm formation, tumor growth, metastasis, and infection (reviewed in Duffy et al., 2008; Medcalf, 2007).

PAI-1 was first characterized, almost three decades ago, as a PAI produced by cultured bovine endothelial cells (Loskutoff et al., 1983) and was later found to be expressed by many different cell types in various tissues (Simpson et al., 1991). The expression of PAI-1 is regulated by a variety of cytokines, growth factors, hormones, and endotoxins (reviewed in Andreasen et al., 1990; Irigoyen et al., 1999; Vaughan, 2005). In 1986, four groups independently described the isolation of the full-length cDNA encoding human PAI-1 (Andreasen et al., 1986; Ginsburg et al., 1986; Ny et al., 1986; Pannekoek et al., 1986). PAI-1 is secreted as a 379 or 381 (depending on N-terminal heterogeneity) amino acid single-chain glycoprotein lacking cysteines, with a molecular weight of approximately 50 kDa. Three potential N-glycosylation sites have been identified in

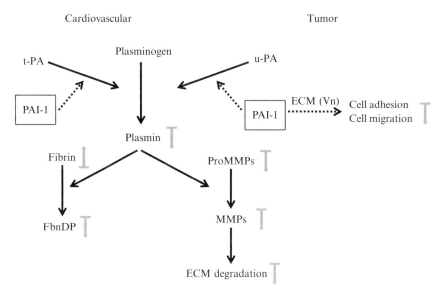

Figure 5.1 Role of PAI-1 in the plasminogen/plasmin system and the extracellular matrix. Solid arrows indicate activation, dotted lines indicate inhibition, gray arrows indicate the ultimate effect subsequent to an increase in PAI-1. PAI-1, plasminogen activator inhibitor-1; t-PA, tissue-type plasminogen activator; u-PA, urokinase-type plasminogen activator; FbnDP, fibrin degradation products; MMP, matrix metalloproteases; ECM, extracellular matrix; Vn, vitronectin.

human PAI-1 (Asn^{209}, Asn^{265}, and Asn^{329}) and two of these are glycosylated in vivo (Asn^{209} and Asn^{265}; Gils et al., 2003; Xue et al., 1998). However, the glycosylation of PAI-1 has been shown not to be a prerequisite for its activity (Lawrence et al., 1989). Also the cDNAs encoding PAI-1 from vervet monkey (Meissenheimer et al., 2006), bovine (Mimuro et al., 1989), porcine (Bijnens et al., 1997), mouse (Prendergast et al., 1990), rat (Zeheb and Gelehrter, 1988), and rabbit (Hofmann et al., 1992) have been cloned and reveal a nucleic acid identity of 81–97% and an amino acid identity of 78–97% with human PAI-1. Mouse and human PAI-1 share 78% amino acid identity and 82% amino acid homology (Prendergast et al., 1990). When considering mouse models to study PAI-1 and its pharmacological modulation, it is important to realize that, despite the high degree of homology, PAI-1 of human and mouse origin exhibits subtle but important structural differences that may result in a different susceptibility toward PAI-1 modulating agents (Dewilde et al., 2010; Gils et al., 2009).

In blood, PAI-1 occurs in two different pools: that is, plasma and platelets (Chmielewska et al., 1983; Erickson et al., 1984). The amount of circulating PAI-1 in human plasma is low (0–60 ng/ml). This PAI-1 is mainly in the active form and is probably secreted from the endothelial cells of vessel walls (Loskutoff et al., 1989). The main blood pool of PAI-1 exists in platelets (200–300 ng/ml), but only about 10% of this is in the active form (Booth et al., 1988; Declerck et al., 1988a; Kruithof et al., 1987). In the ECM, PAI-1 mainly occurs as an active form bound to vitronectin (Declerck et al., 1988b; Owensby et al., 1991; Seiffert et al., 1990). PAI-1 levels in mouse plasma are around 2 ng/ml (Declerck et al., 1995; Nagai et al., 2007), thus 5- to 10-fold lower compared to those in human plasma. However, PAI-1 levels in mouse platelets are 500-fold lower compared to those in human platelets (Kawasaki et al., 2000). It is obvious that also these differences need to be taken into account when evaluating mouse models to study PAI-1.

1.1. PAI-1, a unique serpin

Like most *serpins*, PAI-1 has a highly ordered structure consisting of three β-sheets (A, B, and C), nine α-helices (A through I), and a reactive center loop (RCL) containing 26 residues that are designated as P_{16}–$P_{10'}$ providing a bait peptide bond (P_1–$P_{1'}$) that mimics the normal substrate of the target protease.

PAI-1 is unique amongst *serpins* because of its functional flexibility. Although PAI-1 is synthesized in an active inhibitory form, under normal physiological conditions, it converts spontaneously into a nonreactive stable form (Hekman and Loskutoff, 1985) with an apparent half-life of ~2 h at 37 °C. This nonreactive *latent* form does not interact with the target proteases. The latent form can be partially reactivated by denaturing agents and subsequent refolding (Hekman and Loskutoff, 1985) and also *in vivo* reactivation of latent PAI-1 has been observed (Vaughan et al., 1990). Whereas in plasma and in the ECM, PAI-1 occurs mainly in the active conformation, in platelets, PAI-1 occurs mainly in the latent conformation.

2. ROLE OF PAI-1 IN DIFFERENT (PATHO) PHYSIOLOGICAL PROCESSES

Numerous *in vitro* experimental as well as epidemiological and clinical studies in humans have demonstrated that PAI-1 plays a role in many (patho)physiological processes such as fibrinolysis, thrombosis, restenosis, atherosclerosis, obesity, apoptosis, cell adhesion, cell migration, wound healing, angiogenesis, inflammation, chemotaxis, and fibrosis (reviewed in Gramling and Church, 2010; Lijnen, 2005).

2.1. PAI-1 in cardiovascular disease

A deficient fibrinolytic response may be caused by impaired release of t-PA by the vessel wall, by abnormalities in the function or synthesis of plasminogen, or by increased PAI-1 levels (reviewed in Declerck et al.,

1994; Juhan-Vague et al., 1995). The most commonly encountered abnormality of fibrinolysis is the result of excessive PAI-1 levels.

The link between cardiovascular disease and impaired fibrinolytic systems has been thoroughly investigated in humans. Evidence is provided of a link between increased PAI-1 levels and myocardial infarction (MI; Hamsten et al., 1985; Meade et al., 1993), coronary artery disease (Meade et al., 1993), deep vein thrombosis (DVT; Nilsson et al., 1985; Schulman and Wiman, 1996; Siemens et al., 1999; Wiman et al., 1985), restenosis (Binder et al., 2007), aneurysm formation (Sandford et al., 2007), and angiogenesis (Binder et al., 2007). Elevated PAI-1 levels have also been associated with an unfavorable outcome in numerous cases of sepsis (Kornelisse et al., 1996; Mesters et al., 1996). Further, PAI-1 expression was found to be increased in atherosclerotic vessel walls and atherosclerotic lesions (Arnman et al., 1994; Fay et al., 2007; Lupu et al., 1993; Narayanaswamy et al., 2000; Robbie et al., 1996; Schneiderman et al., 1992). Despite the link in these abovementioned studies, some are in contradiction with other studies. For example, some studies could not confirm the correlation between PAI-1 and DVT (reviewed in Prins and Hirsh, 1991). In addition, Juhan-Vague et al. (1996) showed that associations between PAI-1 levels and risk of cardiovascular events disappeared after adjustment for factors characterizing insulin resistance (i.e., BMI, triglycerides and HDL cholesterol, hyperglycemia, and hyperinsulinemia; Juhan-Vague et al., 1996). The latter was also observed by Thogersen et al. (1998) and Folsom et al. (2001). Eventually, these findings formed the basis of the hypothesis that elevated levels of PAI-1 form the link between obesity, insulin resistance, and the risk of cardiovascular events (including atherosclerosis; Alessi et al., 1997; Nordt et al., 1998; Schafer et al., 2003).

Beside patients with increased PAI-1 levels, also persons afflicted with PAI-1 deficiency have been identified (Dieval et al., 1991; Fay et al., 1992, 1997; Lee et al., 1993). PAI-1 deficiency is manifested by a hyperfibrinolytic state. Interindividual variations in plasma PAI-1 concentrations are also correlated to a polymorphism in the *PAI-1* gene, however, conflicting evidence exists regarding the relation between PAI-1 polymorphism and risk of arterial and venous thrombosis (reviewed by Francis, 2002).

2.2. PAI-1 and cancer

The effect of PAI-1 on cancer is related to both its inhibitory properties (i.e., reducing the extent of plasmin generation and subsequent events) and its vitronectin-binding properties. Clinical data on many different human cancers, including breast (Janicke et al., 1993; Knoop et al., 1998), gastric (Nekarda et al., 1994), ovarian (Konecny et al., 2001; Kuhn et al., 1999), bladder (Becker et al., 2010), and oral cancers (Gao et al., 2010), have shown that high levels of PAI-1 are a reliable marker of poor prognostic outcomes in patients (reviewed in Binder and Mihaly, 2008; Durand *et al.*, 2004; McMahon and Kwaan, 2008). This is in line with studies reporting that PAI-1 promotes, rather than inhibits invasion and metastasis. PAI-1 activity has also been shown to inhibit apoptosis (thereby enhancing tumor growth) in both cancer and noncancer cell lines (Kwaan *et al.*, 2000). In addition, the absence of host PAI-1 retards tumor development (Gutierrez *et al.*, 2000), prevents cancer invasion, and was proven to be a modest inhibitor of endothelial sprouting in an *ex vivo* study (Brodsky *et al.*, 2001).

In these processes, the effect of PAI-1 is not only mediated by its capacity to inhibit protease activity but also by virtue of its interaction with vitronectin (Stefansson *et al.*, 2001). Vitronectin plays an important role in the attachment of cells to their surrounding matrix and may participate in the regulation of cell differentiation, proliferation, and morphogenesis. PAI-1 interacts with vitronectin with high affinity, competing for binding with integrins and the u-PA receptor (u-PAR). Therefore, PAI-1 bound to vitronectin prevents the association of integrins to vitronectin and downregulates cell adhesion and migration. Similarly, by competing with u-PAR for binding to vitronectin, it inhibits u-PA-dependent cell adhesion and migration (reviewed in Czekay and Loskutoff, 2004).

3. Mouse Models to Study the Role of PAI-1

Numerous mouse models have been developed to study the role of PAI-1 *in vivo* (see below). These models are based either on (a) the identification of a relationship between disease and PAI-1 levels, (b) localization of PAI-1 at the site of a diseased tissue area, (c) transgenic mice either overexpressing wild-type (WT) mouse or human PAI-1, (d) transgenic mice expressing a stable variant of mouse (threefold stabilization) or human (80-fold stabilization) PAI-1, or (e) mice in which the *PAI-1* gene has been inactivated. However, some of these models have also been used to evaluate PAI-1 as a putative therapeutic target (Li and Lawrence, submitted for publication).

In all these studies, it should be realized that

1. the structure of mouse PAI-1 differs slightly from human PAI-1 and subsequently not all inhibitors identified for inhibition of human PAI-1 act on mouse PAI-1 (Dewilde *et al.*, 2010; Gils *et al.*, 2003, 2009);
2. even though quite similar, there are differences between the mouse and human plasminogen/plasmin system (Lijnen *et al.*, 1994; Matsuo *et al.*, 2007);
3. the levels of PAI-1 in plasma are fivefold lower, and the levels of PAI-1 in platelets are 500-fold lower in mice.

Therefore, extrapolation of data obtained in mouse models to the human situation may not always be fully justified.

Even though data obtained with the various models described below should be interpreted with some caution as to the applicability in humans, they have provided a wealth of information on the general as well as specific aspects of the role of PAI-1 *in vivo* in mammals.

4. MOUSE (WILD-TYPE) MODELS TO STUDY THE ROLE OF PAI-1 IN CARDIOVASCULAR DISEASE AND CANCER

The differences between the human and the mouse plasminogen activation systems are minor but must be kept in mind when designing and interpreting specific experiments, especially *in vivo* experiments. In addition, the various techniques used to trigger vascular thrombosis *in vivo* differ from each other and do not adequately model the more chronic and subtle forms of injury that result in vascular disease in humans. Moreover, vessel size and blood flow, which are important determinants of vascular function and remodeling, differ significantly between mice and humans. Lijnen *et al.* (1994) also showed that the fibrinolytic system in mouse is more resistant to activation than the human system first due to the relative resistance of mouse plasminogen by mouse t-PA and second due to the shorter half-life of mouse t-PA due to an apparent inhibition of mouse t-PA by inhibitors other than PAI-1. In addition, in most of the mouse models, thrombus formation is triggered within a normal blood vessel whereas pathologic thrombosis in humans usually occurs within a diseased vascular segment.

Larger and smaller vessels of both the arterial and venous systems as well as different vascular beds have been used to study thrombosis in mice. The carotid, the testicular, or the femoral artery are mostly used to study arterial thrombosis whereas the jugular vein, the inferior vena cava (IVC), and the femoral vein are mostly used to study venous thrombosis. In addition, thrombosis has been studied in the microvascular beds of the mesentery, the cremaster muscle, and the ear (Whinna, 2008).

4.1. Cardiovascular disease

No mouse strain develops thrombosis spontaneously. However, the mouse has some physiological and genetic characteristics that make it an extremely useful tool to evaluate venous thrombosis. Essential in developing a thrombosis mouse model is the reproducibility of the thrombus size. A thrombosis model in which blood is able to flow past the developing thrombus would make it possible to study the effect of a therapeutic agent on thrombosis and fibrinolysis. To analyze either a genetic or pharmacological intervention, consistent times to occlusion are essential in thrombosis models. Many variables (i.e., depth of anesthesia, maintenance of physiological body temperature under anesthesia, strain background, age and gender of experimental mice, pregnancy) can influence the outcome of thrombosis and should therefore be considered when using thrombosis models (Westrick et al., 2007).

Several mouse models have been used to study *venous thrombosis*. The *photochemical injury model* involves an intravenous administration of a photoreactive dye. Administration of Rose Bengal is followed by illumination of either the jugular vein or the IVC with a green light laser (540 nm) from a xenon lamp equipped with a heat-absorbing filter. The photochemical reaction triggers the production of singlet oxygen and other oxygen species that damage the vascular endothelium and initiate thrombus formation (Eitzman et al., 2000a; Kikuchi et al., 1998; Matsuno et al., 1991). In the *IVC stasis or ligation model*, a nonreactive suture ligature is placed around the IVC just below the renal veins to obtain complete blood stasis. This model is useful to study interactions between the vein wall and the occlusive thrombus and for assessing the progression from acute to chronic inflammation (Myers et al., 2002). The *electrolyte stasis model* induces thrombosis at the site of electrolysis (Cooley et al., 2005). Thrombus formation can be the result of either direct electrical injury (heating) or a free radical induced injury to the vein wall. Heat applied to the vein wall may lead to protein denaturation, which is not desirable. In the *IVC stenosis model*, a silk ligature is placed around the IVC. The ligature, combined with temporary endothelial compression with hemoclips, produces a laminar thrombus (Singh et al., 2002). The *mechanical injury model* uses an external mechanical force to damage the endothelium. Pierangeli et al. described a model in which forceps were used to deliver a standardized pinch. The mechanical injury model is usually combined with fiber optic technology to transilluminate the vein to visualize and record thrombosis (Pierangeli and Harris, 1994; Pierangeli et al., 1995). Diaz et al. (2010) recently developed a deep venous thrombosis model that generates consistent venous thrombi in the presence of continuous blood flow by combining the use of an electric current with endothelial cell activation. In their *electrolytic inferior vena cava model (EIM model)*, a 25G stainless-steel needle attached to a 30G silver coated copper wire is inserted into the exposed caudal IVC and positioned against the anterior wall (anode). Another wire is implanted subcutaneously completing the circuit (cathode), and for 15 min, a 250-µA current is applied to establish consistent thrombus formation. The direct current results in the formation of toxic products of electrolysis that erode the endothelial surface of the *vena cava* thereby promoting a thrombogenic environment and subsequent thrombus formation.

A number of these venous thrombosis models in mice have been used to study the role of PAI-1. McDonald *et al.* (2010) applied the mouse IVC stasis model on young (2 months old, weighing 20–25 g) and old (11 months old, weighing 29–40 g) male C57BL/6J mice to evaluate the effect of aging on venous thrombosis. They concluded that aging was associated with increased thrombus size in mice. Old thrombosed animals had a significant increase in circulating PAI-1 activity levels compared to young thrombosed mice. Their donor/recipient experiments suggest that the host age and the physiological environment are important factor in stasis thrombus resolution. Cleuren *et al.* (2009) evaluated the effect of nutritionally induced obesity and short-term oral estrogen administration, either alone or in combination, on venous thrombosis using a mouse IVC thrombosis model in 5-week-old female C57BL/6J mice. Mice were fed either a standard fat diet (SFD) or a high-fat diet (HFD). After 14 weeks, either 1 µg ethinylestradiol per day or vehicle was given orally for 8 days. Subsequently, the IVC was ligated upon deposition of a Prolene thread. By removing the Prolene thread, slow flow was restituted triggering platelet activation, stasis, and thrombus formation. PAI-1 plasma levels increased significantly in diet-induced obese mice compared to the SFD group whereas oral estrogens did not alter PAI-1 levels, neither in mice on an SFD nor in mice on an HFD. They concluded that nutritionally induced obesity causes a trend toward an increased thrombus weight in the IVC model. They also found that under obese conditions, short-term oral estrogen administration results in a significant lower thrombogenicity. This illustrates that the finding that oral estrogens in woman are prothrombotic cannot simply be translated into the mouse context.

In all these models, thrombosis is induced by techniques that induce vascular injury and subsequent tissue factor exposure. Alternatively, exogenous tissue factor can be injected intravenously (i.v.) to induce thrombi. Van de Craen *et al.* evaluated five monoclonal antibodies that are able to inhibit both glycosylated and nonglycosylated vitronectin-bound mouse PAI-1 (Van de Craen *et al.*, 2011). In this model, female Swiss mice were used, and monoclonal antibodies were administered via the tail vein. Five minutes after the antibody injection, a suboptimal concentration of t-PA was injected i.v., immediately followed by i.v. injection of tissue factor to evoke thromboembolism. Mice were evaluated after 15 min and gained a negative score in case of paralysis of one or more limbs or in case of death. For all five antibodies evaluated, a statistically significant decrease in the proportion of paralyzed or dead mice (i.e., 41–50% paralyzed or dead) could be detected compared to control mice to which a control antibody was administered (i.e., 71% paralyzed or dead). MA-33H1F7 (Debrock and Declerck, 1997) was also evaluated in the same thromboembolism model because of the differential reactivity that MA-33H1F7 has to vitronectin-bound PAI-1 compared to the newly evaluated antibodies. Administration of MA-33H1F7 resulted also in a decrease in the proportion of paralyzed or dead mice (i.e., 53% vs. 71%, for MA-33H1F7 and control antibody, respectively). However, this decrease did not reach statistical significance. It was concluded that reactivity toward vitronectin-bound PAI-1 is an important aspect to achieve stronger PAI-1 inhibition *in vivo*.

The role of PAI-1 in *arterial thrombosis* has been studied in mice using the *photochemical injury model* (as described for the venous thrombosis model), the *ferric chloride (FeCl$_3$) injury model*, or the *carotid artery ligation and cuff placement* model.

Kurz *et al.* (1990) first demonstrated that topical application of a solution of ferric chloride to the adventitial surface of an artery rapidly induces formation of a thrombus that typically progresses to complete vascular occlusion. The rate of thrombus formation depends on the concentration of ferric chloride and on the size of the filter paper. Usually, a 3-mm disk of filter paper soaked in (35–50%) ferric chloride is used. Kurz *et al.* demonstrated that the produced thrombus is composed of platelets and red blood cells enmeshed in a fibrin network. Rhodamine 6G can be injected i.v. to monitor arterial occlusion using a fluorescence microscope. Izuhara *et al.* (2008) described the formation of thrombi in the testicular artery upon ferric chloride exposure in male C57BL/6J mice. The time from endothelial damage by ferric chloride to occlusion of the testicular arteries by large thrombi prolonged from 12.7 ± 2.7 to 56.3 ± 8.9 min upon administration of an oral PAI-1 inhibitor (TM5007).

Numaguchi *et al.* (2009) investigated the *in vivo* interaction between the angiotensin IV receptor (AT4R) and PAI-1 by assessing both the acute and chronic occlusive thrombus formation and fibrinolysis in the artery of insulin-regulated aminopeptidase (IRAP) knockout mice and WT male C57Bl/6J mice treated with the PAI-1 inhibitor T-686. To assess the effects of PAI-1 inhibition on acute thrombosis, T-686 was given orally for 7 days to WT mice. Subsequently, mice were subjected to carotid artery injury using ferric chloride. Complete thrombotic occlusion was observed in 100% of (nontreated) WT mice but in only 72% and 87% of IRAP$^{-/-}$ mice and T-686-treated WT mice, respectively. Moreover, in occluded arteries, the mean time for occlusion was significantly longer in T-686-treated mice (11.7 ± 2.1 min) and IRAP$^{-/-}$ mice (14.8 ± 2.8 min) than in WT mice (8.6 ± 1.7 min). For the assessment of chronic occlusive thrombus formation and fibrinolysis in the artery, a new model combining carotid artery ligation with perivascular cuff placement was developed. This model is characterized by chronic blood flow cessation and endothelial cell injury limited within an intracuff lesion. This allows to quantitate thrombus formation in the arteries. T-686 was administered for 7 days before and 28 days after surgery. The authors observed that although the administration of T-686 had a significant effect on thrombus formation in the acute phase, T-686 failed to suppress nuclear factor kappa B (NFκB) activation and inflammatory responses whereas both NFκB activation and inflammation were decreased in the arteries of IRAP$^{-/-}$ mice. They hypothesized that the angiotensin IV-AT4R pathway could regulate both thrombus formation through PAI-1 induction and inflammation through NFκB activation and leukocyte infiltration.

The role of PAI-1 in chronic *lung fibrosis* can be tested using bleomycin-induced pulmonary fibrosis through intratracheal instillation of bleomycin; Izuhara *et al.* (2008) used this model to demonstrate in male C57BL/6J mice that bleomycin significantly increases the lung hydroxyproline content and the PAI-1 activity levels. Subsequently, the oral PAI-1 inhibitor (TM5007) was found to reduce bleomycine-induced PAI-1 increase and fibrosis. In a recent study, Senoo *et al.* (2010) instilled intranasally either PAI-1 small interfering RNA (siRNA) or nonspecific siRNA into C57Bl6/J mice 1, 4, 8, 11, 17, and 20 days after the intratracheal administration of bleomycin. From this study, it could be concluded that lung specific suppression of PAI-1 results in the prevention of pulmonary fibrosis and the improvement of survival.

4.2. Cancer

PAI-1 has an impact on both proteolytic activity and cell migration during angiogenesis. The role of PAI-1 in tumorigenesis is highly controversial. Using an *in vivo* matrigel tumor angiogenesis model, Leik *et al.* (2006) showed that treatment with the oral PAI-1 inhibitor PAI-039 (tiplaxtinin) reduced angiogenesis although PAI-039 was not able to inhibit vitronectin-bound PAI-1. However, other studies have indicated that the PAI-1 source (tumor vs. host), the spatial localization of PAI-1, and the different tumor models define the outcome of the experiment. In addition, mainly transgenic mice have been used to study the role of PAI-1 in cancer (see below).

5. Genetically Modified Mice to Study the Role of PAI-1 in Cardiovascular Disease and Cancer

The ability to generate genetically modified mice has led to the generation of PAI-1 deficient animals (Carmeliet *et al.*, 1993a). In addition, several models were generated overexpressing PAI-1 under the control of a variety of promoters, leading to quantitative, spatial, and temporal differences in PAI-1 expression. Overall, the various models significantly contributed to the elucidation of the *in vivo* role for PAI-1 in cardiovascular disease and cancer.

5.1. Cardiovascular disease

5.1.1. Thrombosis

As expected, genetic disruption of the *PAI-1* gene in mice resulted in a mild hyperfibrinolytic state and a greater resistance to venous thrombosis, however, without leading to impaired hemostasis (Carmeliet *et al.*, 1993b). Using a model of L-NAME induced thrombosis, it was shown that PAI-1 deficient mice were largely protected from the development of hepatic vein thrombosis (Smith *et al.*, 2006). To further elucidate the role of PAI-1 in vascular thrombosis, different injury models have been used in combination with the genetic deficiency. Using ferric chloride ($FeCl_3$)-induced carotid artery injury, significantly smaller residual thrombus was observed 24 h after injury induction in PAI-1$^{-/-}$ versus WT mice, even though time to occlusion was similar (Farrehi *et al.*, 1998; Fay *et al.*, 1999). In addition, reperfusion during thrombolytic therapy with t-PA occurred more efficiently in PAI-1$^{-/-}$ animals (Zhu *et al.*, 1999). Of interest, while a role for PAI-1 in preventing premature thrombus dissolution was confirmed and ascribed to the local synthesis of PAI-1 in the thrombus/vessel wall after injury, Konstantinides *et al.* (2001) showed that time to occlusion was significantly increased in PAI-1$^{-/-}$ mice. Further, while it was confirmed that PAI-1$^{-/-}$ mice have a longer time to occlusion and a shorter total duration of vascular occlusion than their WT counterparts, they have more unstable thrombi, associated with an increased number of emboli (Koschnick *et al.*, 2005). Interestingly, combined PAI-1 and vitronectin-deficient animals had the same thrombotic phenotype as mice deficient in one of the genes suggesting that both proteins influence thrombus stability by regulating a common pathway (Koschnick *et al.*, 2005).

In more acute studies of photochemically induced arterial and venous thrombosis, using rose bengal activated by a green laser light, it was shown that vascular occlusion occurred faster and lasted longer in WT than in PAI-1$^{-/-}$ animals (Eitzman *et al.*, 2000a; Matsuno *et al.*, 1999) and thrombus size was significantly larger (Kawasaki *et al.*, 2000). Administering recombinant PAI-1 to PAI-1$^{-/-}$ animals restored thrombus development (Kawasaki *et al.*, 2000). Depletion and reconstitution of PAI-1$^{-/-}$ mice with PAI-1$^{+/+}$ platelets did not restore a normal thrombotic response whereas this was the case for PAI-1$^{+/+}$ mice receiving PAI-1$^{-/-}$ platelets, suggesting that thrombosis is controlled by vascular rather than by platelet PAI-1 (Kawasaki *et al.*, 2000). Considering the 500-fold higher levels of PAI-1 in human versus mouse platelets, this does not necessarily indicate that platelet PAI-1 is unimportant for human biology.

Since cardiovascular diseases are often present on a background of hypercholesterolemia, the apoE$^{-/-}$ mouse has become a useful tool to study the role of PAI-1 for thrombotic vascular occlusion during the development and progression of atherosclerotic lesions (Nakashima *et al.*,

1994). Using FeCl$_3$-induced carotid artery injury in atherosclerosis-prone apoE$^{-/-}$ mice fed a so-called western diet, it was shown that vascular PAI-1 expression significantly increases with hypercholesterolemia resulting in a prothrombotic phenotype. Deletion of the *PAI-1* gene reversed this prothrombotic tendency and reduced neointimal growth after injury (Schafer *et al.*, 2003).

Besides the PAI-1 deficient mouse, transgenic mice overexpressing PAI-1 have been generated. Several models support a contribution of excess PAI-1 to thrombotic disorders. Transgenic mice expressing a stable variant of active human PAI-1 under the control of the preproendothelin-1 promoter developed spontaneous coronary arterial thrombosis, often associated with subendocardial infarction (Eren *et al.*, 2002). Whereas a related mouse strain expressing a human PAI-1 variant with impaired vitronectin-binding displayed a similar degree of coronary arterial thrombosis, this was not the case in a third strain lacking the antiproteolytic activity, underscoring the importance of the RCL domain in the regulation of vascular fibrinolysis (Eren *et al.*, 2007). Transgenic animals expressing the native form of human PAI-1 under the control of the metallothionein I promoter developed venous but no arterial occlusions (Erickson *et al.*, 1990). The former occurred in the tail and hind feet, consisted of cellular thrombi and resolved spontaneously at young age. In contrast, for transgenic animals overexpressing mouse PAI-1 under the control of the CMV or adipocyte-specific aP2 promoter, no reports were made regarding thrombotic disorders (Eitzman *et al.*, 1996b; Lijnen *et al.*, 2003). Finally, a recent study using mice expressing stable mouse PAI-1 under control of the hybrid CMV/chicken β-actin promoter reported no apparent thrombosis (Fahim *et al.*, 2009). Whereas obviously the choice of promoter can influence the differences in observed venous and arterial thrombosis, a cross-species difference in PAI-1 function might also contribute since thrombosis is observed only in models overexpressing human PAI-1. Alternatively, it is also important to note that the "stable" mouse PAI-1 variant used by Fahim *et al.* (2009) exhibited only a threefold higher stability compared to its WT counterpart whereas studies using the stable human variant used a human variant with a 80-fold higher stability compared to its WT counterpart (Eren *et al.*, 2002).

Cerebral hemorrhage and stroke are two common complications associated with disturbed fibrinolysis. Using ligation of the left middle cerebral artery (MCA) to induce focal cerebral ischemic infarction (FCI), it was shown that PAI-1 deficiency increases infarct size (Nagai *et al.*, 1999). Using transgenic mice overexpressing mouse PAI-1 under control of the aP2 promoter in two different stroke models, a permanent MCA ligation model and a transient photochemically induced thrombotic MCA model, it was established that high levels of PAI-1 reduce infarct volume in the permanent model but enhance it in the MCA thrombosis model (Nagai *et al.*, 2005). The enhancement in the thrombosis model most likely is related to the inhibition of t-PA-mediated thrombolysis of microthrombi. The protective effect can potentially be related to the inhibition of the neurotoxic effects of t-PA (Nagai *et al.*, 2005). Using the MCA occlusion model, it was shown that whereas obesity led to a more deleterious outcome of thrombotic ischemic stroke in WT mice, this was not the case in PAI-1$^{-/-}$ animals suggesting that the increased PAI-1 plasma levels typical for obesity play a functional role in the development of thrombotic ischemic stroke (Nagai *et al.*, 2007).

5.1.2. Fibrosis

Cardiac fibrosis is the abnormal accumulation of ECM in the heart. Generally, mechanisms causing fibrosis are related to increased ECM synthesis or decreased ECM degradation. As such, it has been well established that increased PAI-1 expression predisposes to fibrosis in several organs and tissues. However, its role in cardiac fibrosis remains controversial. PAI-1 deficiency has been shown to protect against cardiac fibrosis after MI by coronary ligation (Takeshita *et al.*, 2004) and against hypertension and coronary perivascular fibrosis induced by NOS inhibition (Kaikita *et al.*, 2001, 2002). Mice overexpressing PAI-1 under the control of the aP2 promoter displayed increased cardiac fibrosis associated with an acute MI induced by coronary occlusion (Zaman *et al.*, 2009). In contrast, some studies suggest that decreased PAI-1 activity in the heart may contribute to fibrosis ascribing a cardioprotective role to PAI-1. Indeed, PAI-1 deficiency was sufficient to cause spontaneous cardiac fibrosis in mice as a result of increased microvascular permeability, inflammation, and ECM remodeling (Moriwaki *et al.*, 2004; Xu *et al.*, 2010). In agreement with these findings, Angiotensin (Ang)II/salt-induced cardiac fibrosis was more outspoken in PAI-1$^{-/-}$ versus WT mice (Weisberg *et al.*, 2005). Further, after MI induced by coronary ligation, PAI-1$^{-/-}$ mice had more leukocyte infiltration and intramyocardial hemorrhage (Askari *et al.*, 2003). The increase in number of myoendothelial junctions (MEJ) in coronary arterioles of PAI-1$^{-/-}$ hearts after transplantation in WT mice, in addition to the higher number of MEJ in coronary arterioles of WT versus PAI-1$^{-/-}$ mice, supports a role for PAI-1 in increased cardiovascular integrity (Heberlein *et al.*, 2010).

5.1.3. Atherosclerosis—(Re)stenosis

Atherosclerosis is a complex disease involving (repetitive) vascular injury besides lipid accumulation, platelet and fibrin deposition, and cellular migration and proliferation. Thrombus formation, fibrin deposition, and neointimal formation are often closely interrelated. Thus, impact of PAI-1 on thrombus formation and fibrin deposition can be extrapolated to its potential role in atherosclerosis. However, findings are conflicting, and the impact of PAI-1 remains elusive.

Using a model of mechanical and electrical injury-induced intima formation in PAI-1 deficient and WT mice revealed that PAI-1 blocks intimal thickening by inhibiting the migration rather than the proliferation of smooth muscle cells (SMCs; Carmeliet et al., 1997). While carotid artery ligation in PAI-1 deficient mice confirmed the protective effect of PAI-1 against excessive intima formation, the effect was ascribed to a role for PAI-1 as a thrombin inhibitor suppressing the mitogenic activity of thrombin on SMCs (de Waard et al., 2002). Bone marrow transplantation in combination with FeCl$_3$-induced carotid artery injury confirmed that overall lack of PAI-1 was associated with enhanced neointimal formation, while interestingly merely transplanting PAI-1$^{-/-}$ mice with PAI-1$^{+/+}$ marrow resulted in reduced neointimal area and luminal stenosis (Schafer et al., 2006).

In contrast, a series of studies indicated a detrimental effect of PAI-1 on neointima formation. When using a carotid artery ligation, copper- or FeCl$_3$-induced oxidative vascular injury model, PAI-1 deficiency attenuated lesion formation, related to a significant reduction in fibrin deposition, thus removing the provisional matrix for SMC invasion (Peng et al., 2002; Ploplis et al., 2001). PAI-1 deficiency has specifically been shown to protect against TGF-β1 and Ang II-induced vascular remodeling. Whereas TGF-β 1 overexpression induced the formation of a cellular and matrix-rich intima in uninjured carotid arteries primarily by increasing cell migration and matrix accumulation, this effect is not observed in PAI-1$^{-/-}$ mice (Otsuka et al., 2006). In addition, PAI-1$^{-/-}$ mice were protected against Ang II-induced aortic remodeling through a blood pressure-independent mechanism (Weisberg et al., 2005).

Finally, some studies do not support a role for PAI-1 in neointima formation. No difference was seen in PAI-1$^{-/-}$ mice using a carotid artery ligation model (Kawasaki et al., 2001) or in mice overexpressing PAI-1 under the aP2 promoter after severe electrical injury (Lijnen et al., 2004) when comparing with WT animals.

Conflicting observations were also made in models of hypercholesterolemia. Deletion of PAI-1 in apoE^d mice had no significant influence on plaque growth in the proximal aorta but reduced it in the turbulent area of the carotid bifurcation, presumably by reducing fibrin deposition (Eitzman et al., 2000b). This was confirmed by a protective effect of PAI-1 deficiency when analyzing the midportion of the carotid artery using the same genetic model and similar carotid artery injury (Zhu et al., 2001). Further, the importance of the vascular injury site was apparent from the lack of a significant difference when considering spontaneous atherosclerotic lesion development or histological appearance at the base of the aorta in PAI-1$^{-/-}$, WT, or PAI-1 overexpressing animals on either an apoE$^{-/-}$ or an LDLR$^{-/-}$ background (Sjoland et al., 2000). In contrast, Luttun et al. (2002) reported that the loss of PAI-1 in an apoE$^{-/-}$ background did not reduce but stimulate plaque growth at advanced stages of atherosclerosis because of increased matrix deposition, thus revealing an atheroprotective role for PAI-1. In addition, the study indicated that PAI-1 may suppress advanced atherosclerosis by affecting cellular infiltration and matrix accumulation. Bone marrow transplantation studies revealed that SMCs, not macrophages, were the main source of PAI-1 in plaques (Luttun et al., 2002). An inhibitory role of intramural PAI-1 on vascular SMC migration and thus neointimal cellularity, but not lesion size, was confirmed in ApoE$^{-/-}$ mice overexpressing PAI-1 in SMC (Schneider et al., 2004). Accordingly, inhibition of PAI-1 may facilitate development of more plaques that are less prone to rupture.

5.2. Cancer

PAI-1 is elevated in many solid tumors and is often associated with a poor prognosis of cancer (Andreasen, 2007). In contrast, several studies suggest a role for PAI-1 as a cancer inhibitor (Soff et al., 1995; Stefansson et al., 2001).

The growth and metastasis of B16 mouse melanoma tumors in transgenic mice overexpressing PAI-1 or in PAI-1$^{-/-}$ mice was found to be comparable to that of WT mice (Eitzman et al., 1996a). In agreement with this study, no significant difference was seen when comparing the development of primary breast tumors and their metastasis in PAI-1$^{+/+}$ and PAI-1$^{-/-}$ backgrounds (Almholt et al., 2003). Recently, it was shown that PAI-1 deficiency neither did influence tumoral lymphangiogenesis, of importance for metastasis, in two different models of breast cancer (Bruyere et al., 2010), nor did it influence tumor growth in a transgenic mouse model of multistage epithelial carcinogenesis (Masset et al., 2011). Whereas tumor growth and angiogenesis were not influenced, brain metastasis in a transgenic mouse model of ocular tumors was reduced in PAI-1$^{-/-}$ mice (Maillard et al., 2008). While these studies, investigating the impact of PAI-1 on metastasis, have led to conflicting findings, a more consistent story has developed during the past decade with regard to PAI-1 and tumor angiogenesis. It was suggested that PAI-1 deficiency prevents the invasion of transplanted malignant keratinocytes into host tissue and vascularization of the tumor (Bajou et al., 1998). In addition, systemic adenoviral overexpression of PAI-1 lacking the vitronectin-binding site restored the WT phenotype, whereas overexpression of PAI-1 lacking PA inhibitory capacity did not (Bajou et al., 1998, 2001), thus supporting a role for stromal antiproteolytic PAI-1 for tumor invasion and angiogenesis. In support of the latter finding, PAI-1 deficiency protected against angiogenesis and tumor growth/invasion after surface transplantation of two malignant human skin keratinocyte cell lines (Maillard et al., 2005) and implantation of human neuroblastoma cells (Bajou et al., 2008). Further, the growth of transplanted T241 fibrosarcomas was markedly reduced in PAI-1$^{-/-}$ mice relative to WT mice, and this correlated with an inhibition of neovascularization and proliferation in the PAI-1$^{-/-}$ mice (Gutierrez et al., 2000). Interestingly, using the same T241 model system, it was found that PAI-1 did not affect tumor growth or angiogenesis (Curino et al., 2002). The reason for these apparent discrepancies is not clear but may be linked to the potential different amounts of PAI-1 produced by the tumor cells, which may vary greatly between tumors in different animals (Gutierrez et al., 2000). The importance of PAI-1 concentration for angiogenesis was confirmed by different groups (Devy et al., 2002; Lambert et al., 2003; McMahon et al., 2001). Subsequently, using a model in which malignant keratinocytes, either overexpressing PAI-1 or not, were transplanted in mice, either lacking PAI-1 or expressing PAI-1 at various levels, it was shown that host PAI-1 at physiological levels promotes in vivo tumor invasion and angiogenesis (Bajou et al., 2004). In sharp contrast, inhibition of tumor vascularization was observed when PAI-1 was produced at supraphysiologic levels either by host cells or by tumor cells. In addition, use of this model indicated that PAI-1 produced by tumor cells, even at high concentrations, did not overcome the absence of PAI-1 in the host, emphasizing the importance of the cellular source of PAI-1 (Bajou et al., 2004).

Taken together, findings obtained using mice with genetically altered PAI-1 expression indicate that the role of PAI-1 as a determinant of tumoral angiogenesis might vary with experimental setting (tumor-type or injection-site dependent) and might depend on its cellular origin (tumor cells vs. host cells). In addition, PAI-1 concentrations at or near to the normal physiological range appear to promote angiogenesis, while pharmacological levels of PAI-1 appear to inhibit it (Devy et al., 2002; Lambert et al., 2003; McMahon et al., 2001).

6. Conclusions

Subtle structural differences between human and mouse PAI-1, as well as differences between the human and the mouse plasminogen/plasmin system, should be taken into account. Overall small differences in the use of similar models in addition to the use of different models such as the nature of the injury, hyperlipidemic conditions, background strain, among others can often explain or be anticipated to explain the sometimes controversial effects of PAI-1. The different outcomes of the various studies using genetically modified mice do indicate the potential beneficial but also adverse effects that might be associated with the use of PAI-1 inhibitors in humans. New mouse models that mimic human pathology in a more accurate way are continuously being developed. Deleting or overexpressing PAI-1 with or without altered functional regions in these models is anticipated to allow for a better understanding of conflicting results and/or assist in elaborating on current findings. In addition, generation of a currently unavailable PAI-1 conditional knockout mouse will allow for a more systematic evaluation of the contribution of different cell types.

REFERENCES

Alessi, M. C., Peiretti, F., Morange, P., Henry, M., Nalbone, G., and Juhan-Vague, I. (1997). Production of plasminogen activator inhibitor 1 by human adipose tissue: Possible link between visceral fat accumulation and vascular disease. *Diabetes* **46,** 860–867.

Almholt, K., Nielsen, B. S., Frandsen, T. L., Brunner, N., Dano, K., and Johnsen, M. (2003). Metastasis of transgenic breast cancer in plasminogen activator inhibitor-1 gene-deficient mice. *Oncogene* **22,** 4389–4397.

Andreasen, P. A. (2007). PAI-1 - a potential therapeutic target in cancer. *Curr. Drug Targets* **8,** 1030–1041.

Andreasen, P. A., Riccio, A., Welinder, K. G., Douglas, R., Sartorio, R., Nielsen, L. S., Oppenheimer, C., Blasi, F., and Dano, K. (1986). Plasminogen activator inhibitor type-1: Reactive center and amino-terminal heterogeneity determined by protein and cDNA sequencing. *FEBS Lett.* **209,** 213–218.

Andreasen, P. A., Georg, B., Lund, L. R., Riccio, A., and Stacey, S. N. (1990). Plasminogen activator inhibitors: Hormonally regulated serpins. *Mol. Cell. Endocrinol.* **68,** 1–19.

Arnman, V., Nilsson, A., Stemme, S., Risberg, B., and Rymo, L. (1994). Expression of plasminogen activator inhibitor-1 mRNA in healthy, atherosclerotic and thrombotic human arteries and veins. *Thromb. Res.* **76,** 487–499.

Askari, A. T., Brennan, M. L., Zhou, X., Drinko, J., Morehead, A., Thomas, J. D., Topol, E. J., Hazen, S. L., and Penn, M. S. (2003). Myeloperoxidase and plasminogen activator inhibitor 1 play a central role in ventricular remodeling after myocardial infarction. *J. Exp. Med.* **197,** 615–624.

Bajou, K., Noel, A., Gerard, R. D., Masson, V., Brunner, N., Holst-Hansen, C., Skobe, M., Fusenig, N. E., Carmeliet, P., Collen, D., and Foidart, J. M. (1998). Absence of host plasminogen activator inhibitor 1 prevents cancer invasion and vascularization. *Nat. Med.* **4,** 923–928.

Bajou, K., Masson, V., Gerard, R. D., Schmitt, P. M., Albert, V., Praus, M., Lund, L. R., Frandsen, T. L., Brunner, N., Dano, K., Fusenig, N. E., Weidle, U., et al. (2001). The plasminogen activator inhibitor PAI-1 controls in vivo tumor vascularization by interaction with proteases, not vitronectin. Implications for antiangiogenic strategies. *J. Cell Biol.* **152,** 777–784.

Bajou, K., Maillard, C., Jost, M., Lijnen, R. H., Gils, A., Declerck, P., Carmeliet, P., Foidart, J. M., and Noel, A. (2004). Host-derived plasminogen activator inhibitor-1 (PAI-1) concentration is critical for in vivo tumoral angiogenesis and growth. *Oncogene* **23,** 6986–6990.

Bajou, K., Peng, H., Laug, W. E., Maillard, C., Noel, A., Foidart, J. M., Martial, J. A., and DeClerck, Y. A. (2008). Plasminogen activator inhibitor-1 protects endothelial cells from FasL-mediated apoptosis. *Cancer Cell* **14,** 324–334.

Becker, M., Szarvas, T., Wittschier, M., Vom Dorp, F., Totsch, M., Schmid, K. W., Rubben, H., and Ergun, S. (2010). Prognostic impact of plasminogen activator inhibitor type 1 expression in bladder cancer. *Cancer* **116**(19), 4502–4512.

Bijnens, A. P., Knockaert, I., Cousin, E., Kruithof, E. K., and Declerck, P. J. (1997). Expression and characterization of recombinant porcine plasminogen activator inhibitor-1. *Thromb. Haemost.* **77,** 350–356.

Binder, B. R., and Mihaly, J. (2008). The plasminogen activator inhibitor "paradox" in cancer. *Immunol. Lett.* **118,** 116–124.

Binder, B. R., Mihaly, J., and Prager, G. W. (2007). uPAR-uPA-PAI-1 interactions and signaling: A vascular biologist's view. *Thromb. Haemost.* **97,** 336–342.

Booth, N. A., Simpson, A. J., Croll, A., Bennett, B., and MacGregor, I. R. (1988). Plasminogen activator inhibitor (PAI-1) in plasma and platelets. *Br. J. Haematol.* **70,** 327–333.

Brodsky, S., Chen, J., Lee, A., Akassoglou, K., Norman, J., and Goligorsky, M. S. (2001). Plasmin-dependent and -independent effects of plasminogen activators and inhibitor-1 on ex vivo angiogenesis. *Am. J. Physiol. Heart Circ. Physiol.* **281,** H1784–H1792.

Bruyere, F., Melen-Lamalle, L., Blacher, S., Detry, B., Masset, A., Lecomte, J., Lambert, V., Maillard, C., Hoyer-Hansen, G., Lund, L. R., Foidart, J. M., and Noel, A. (2010). Does plasminogen activator inhibitor-1 drive lymphangiogenesis? *PLoS ONE* **5,** e9653.

Carmeliet, P., Kieckens, L., Schoonjans, L., Ream, B., van Nuffelen, A., Prendergast, G., Cole, M., Bronson, R., Collen, D., and Mulligan, R. C. (1993a). Plasminogen activator inhibitor-1 gene-deficient mice. I. Generation by homologous recombination and characterization. *J. Clin. Invest.* **92,** 2746–2755.

Carmeliet, P., Stassen, J. M., Schoonjans, L., Ream, B., van den Oord, J. J., De Mol, M., Mulligan, R. C., and Collen, D. (1993b). Plasminogen activator inhibitor-1 gene-deficient mice. II. Effects on hemostasis, thrombosis, and thrombolysis. *J. Clin. Invest.* **92,** 2756–2760.

Carmeliet, P., Moons, L., Lijnen, R., Janssens, S., Lupu, F., Collen, D., and Gerard, R. D. (1997). Inhibitory role of plasminogen activator inhibitor-1 in arterial wound healing and neointima formation: A gene targeting and gene transfer study in mice. *Circulation* **96,** 3180–3191.

Chmielewska, J., Ranby, M., and Wiman, B. (1983). Evidence for a rapid inhibitor to tissue plasminogen activator in plasma. *Thromb. Res.* **31,** 427–436.

Cleuren, A. C. A., van Hoef, B., Hoylaerts, M. F., van Vlijmen, B. J. M., and Lijnen, H. R. (2009). Short-term ethinylestradiol treatment suppresses inferior caval vein thrombosis in obese mice. *Thromb. Haemost.* **102,** 993–1000.

Cooley, B. C., Szema, L., Chen, C. Y., Schwab, J. P., and Schmeling, G. (2005). A murine model of deep vein thrombosis—Characterization and validation in transgenic mice. *Thromb. Haemost.* **94,** 498–503.

Curino, A., Mitola, D. J., Aaronson, H., McMahon, G. A., Raja, K., Keegan, A. D., Lawrence, D. A., and Bugge, T. H. (2002). Plasminogen promotes sarcoma growth and suppresses the accumulation of tumor-infiltrating macrophages. *Oncogene* **21,** 8830–8842.

Czekay, R.-P., and Loskutoff, D. J. (2004). Unexpected role of plasminogen activator inhibitor 1 in cell adhesion and detachment. *Exp. Biol. Med.* **229,** 1090–1096.

de Waard, V., Arkenbout, E. K., Carmeliet, P., Lindner, V., and Pannekoek, H. (2002). Plasminogen activator inhibitor 1 and vitronectin protect against stenosis in a murine carotid artery ligation model. *Arterioscler. Thromb. Vasc. Biol.* **22,** 1978–1983.

Debrock, S., and Declerck, P. J. (1997). Neutralization of plasminogen activator inhibitor-1 inhibitory properties: Identification of two different mechanisms. *Biochim. Biophys. Acta Protein Struct. Mol. Enzymol.* **1337,** 257–266.

Declerck, P. J., Alessi, M. C., Verstreken, M., Kruithof, E. K., Juhan-Vague, I., and Collen, D. (1988a). Measurement of plasminogen activator inhibitor 1 in biologic fluids with a murine monoclonal antibody-based enzyme-linked immunosorbent assay. *Blood* **71,** 220–225.

Declerck, P. J., De Mol, M., Alessi, M. C., Baudner, S., Paques, E. P., Preissner, K. T., Muller-Berghaus, G., and Collen, D. (1988b). Purification and characterization of a plasminogen activator inhibitor 1 binding protein from human plasma. Identification as a multimeric form of S protein (vitronectin). *J. Biol. Chem.* **263,** 15454–15461.

Declerck, P. J., Juhan-Vague, I., Felez, J., and Wiman, B. (1994). Pathophysiology of fibrinolysis. *J. Intern. Med.* **236,** 425–432.

Declerck, P. J., Verstreken, M., and Collen, D. (1995). Immunoassay of murine t-PA, u-PA and PAI-1 using monoclonal antibodies raised in gene-inactivated mice. *Thromb. Haemost.* **74,** 1305–1309.

Devy, L., Blacher, S., Grignet-Debrus, C., Bajou, K., Masson, V., Gerard, R. D., Gils, A., Carmeliet, G., Carmeliet, P., Declerck, P. J., Noel, A., and Foidart, J. M. (2002). The pro- or antiangiogenic effect of plasminogen activator inhibitor 1 is dose dependent. *FASEB J.* **16,** 147–154.

Dewilde, M., Van De Craen, B., Compernolle, G., Madsen, J. B., Strelkov, S., Gils, A., and Declerck, P. J. (2010). Subtle structural differences between human and mouse PAI-1 reveal the basis for biochemical differences. *J. Struct. Biol.* **171,** 95–101.

Diaz, J. A., Hawley, A. E., Alvarado, C. M., Berguer, A. M., Baker, N. K., Wrobleski, S. K., Wakefield, T. W., Lucchesi, B. R., and Myers, D. D. (2010). Thrombogenesis with continuous blood flow in the inferior vena cava A novel mouse model. *Thromb. Haemost.* **104,** 366–375.

Dieval, J., Nguyen, G., Gross, S., Delobel, J., and Kruithof, E. K. (1991). A lifelong bleeding disorder associated with a deficiency of plasminogen activator inhibitor type 1. *Blood* **77,** 528–532.

Duffy, M. J., McGowan, P. M., and Gallagher, W. M. (2008). Cancer invasion and metastasis: Changing views. *J. Pathol.* **214,** 283–293.

Durand, M. K., Bodker, J. S., Christensen, A., Dupont, D. M., Hansen, M., Jensen, J. K., Kjelgaard, S., Mathiasen, L., Pedersen, K. E., Skeldal, S., Wind, T., and Andreasen, P. A. (2004). Plasminogen activator inhibitor-I and tumour growth, invasion, and metastasis. *Thromb. Haemost.* **91,** 438–449.

Eitzman, D. T., Krauss, J. C., Shen, T., Cui, J., and Ginsburg, J. (1996a). Lack of plasminogen activator inhibitor-1 effect in a transgenic mouse model of metastatic melanoma. *Blood* **87,** 4718–4722.

Eitzman, D. T., McCoy, R. D., Zheng, X., Fay, W. P., Shen, T., Ginsburg, D., and Simon, R. H. (1996b). Bleomycin-induced pulmonary fibrosis in transgenic mice that either lack or overexpress the murine plasminogen activator inhibitor-1 gene. *J. Clin. Invest.* **97,** 232–237.

Eitzman, D. T., Westrick, R. J., Nabel, E. G., and Ginsburg, D. (2000a). Plasminogen activator inhibitor-1 and vitronectin promote vascular thrombosis in mice. *Blood* **95,** 577–580.

Eitzman, D. T., Westrick, R. J., Xu, Z., Tyson, J., and Ginsburg, D. (2000b). Plasminogen activator inhibitor-1 deficiency protects against atherosclerosis progression in the mouse carotid artery. *Blood* **96,** 4212–4215.

Eren, M., Painter, C. A., Atkinson, J. B., Declerck, P. J., and Vaughan, D. E. (2002). Age-dependent spontaneous coronary arterial thrombosis in transgenic mice that express a stable form of human plasminogen activator inhibitor-1. *Circulation* **106,** 491–496.

Eren, M., Gleaves, L. A., Atkinson, J. B., King, L. E., Declerck, P. J., and Vaughan, D. E. (2007). Reactive site-dependent phenotypic alterations in plasminogen activator inhibitor-1 transgenic mice. *J. Thromb. Haemost.* **5,** 1500–1508.

Erickson, L. A., Ginsberg, M. H., and Loskutoff, D. J. (1984). Detection and partial characterization of an inhibitor of plasminogen activator in human platelets. *J. Clin. Invest.* **74,** 1465–1472.

Erickson, L. A., Fici, G. J., Lund, J. E., Boyle, T. P., Polites, H. G., and Marotti, K. R. (1990). Development of venous occlusions in mice transgenic for the plasminogen activator inhibitor-1 gene. *Nature* **346,** 74–76.

Fahim, A. T., Wang, H., Feng, J., and Ginsburg, D. (2009). Transgenic overexpression of a stable plasminogen activator inhibitor-1 variant. *Thromb. Res.* **123,** 785–792.

Farrehi, P. M., Ozaki, C. K., Carmeliet, P., and Fay, W. P. (1998). Regulation of arterial thrombolysis by plasminogen activator inhibitor-1 in mice. *Circulation* **97,** 1002–1008.

Fay, W. P., Shapiro, A. D., Shih, J. L., Schleef, R. R., and Ginsburg, D. (1992). Brief report: Complete deficiency of plasminogen-activator inhibitor type 1 due to a frame-shift mutation. *N. Engl. J. Med.* **327,** 1729–1733.

Fay, W. P., Parker, A. C., Condrey, L. R., and Shapiro, A. D. (1997). Human plasminogen activator inhibitor-1 (PAI-1) deficiency: Characterization of a large kindred with a null mutation in the PAI-1 gene. *Blood* **90,** 204–208.

Fay, W. P., Parker, A. C., Ansari, M. N., Zheng, X., and Ginsburg, D. (1999). Vitronectin inhibits the thrombotic response to arterial injury in mice. *Blood* **93,** 1825–1830.

Fay, W. P., Garg, N., and Sunkar, M. (2007). Vascular functions of the plasminogen activation system. *Arterioscler. Thromb. Vasc. Biol.* **27,** 1231–1237.

Folsom, A. R., Aleksic, N., Park, E., Salomaa, V., Juneja, H., and Wu, K. K. (2001). Prospective study of fibrinolytic factors and incident coronary heart disease: The Atherosclerosis Risk in Communities (ARIC) Study. *Arterioscler. Thromb. Vasc. Biol.* **21,** 611–617.

Francis, C. W. (2002). Plasminogen activator inhibitor-1 levels and polymorphisms. *Arch. Pathol. Lab. Med.* **126,** 1401–1404.

Gao, S., Nielsen, B. S., Krogdahl, A., Sorensen, J. A., Tagesen, J., Dabelsteen, S., Dabelsteen, E., and Andreasen, P. A. (2010). Epigenetic alterations of the SERPINE1 gene in oral squamous cell carcinomas and normal oral mucosa. *Genes Chromosomes Cancer* **49,** 526–538.

Gils, A., Pedersen, K. E., Skottrup, P., Christensen, A., Naessens, D., Deinum, J., Enghild, J. J., Declerck, P. J., and Andreasen, P. A. (2003). Biochemical importance of glycosylation of plasminogen activator inhibitor-1. *Thromb. Haemost.* **90,** 206–217.

Gils, A., Meissenheimer, L. M., Compernolle, G., and Declerck, P. J. (2009). Species-dependent molecular drug targets in plasminogen activator inhibitor-1 (PAI-1). *Thromb. Haemost.* **102,** 609–610.

Ginsburg, D., Zeheb, R., Yang, A. Y., Rafferty, U. M., Andreasen, P. A., Nielsen, L., Dano, K., Lebo, R. V., and Gelehrter, T. D. (1986). cDNA cloning of human plasminogen activator-inhibitor from endothelial cells. *J. Clin. Invest.* **78,** 1673–1680.

Gramling, M. W., and Church, F. C. (2010). Plasminogen activator inhibitor-1 is an aggregate response factor with pleiotropic effects on cell signaling in vascular disease and the tumor microenvironment. *Thromb. Res.* **125,** 377–381.

Gutierrez, L. S., Schulman, A., Brito-Robinson, T., Noria, F., Ploplis, V. A., and Castellino, F. J. (2000). Tumor development is retarded in mice lacking the gene for urokinase-type plasminogen activator or its inhibitor, plasminogen activator inhibitor-1. *Cancer Res.* **60,** 5839–5847.

Hamsten, A., Wiman, B., de Faire, U., and Blomback, M. (1985). Increased plasma levels of a rapid inhibitor of tissue plasminogen activator in young survivors of myocardial infarction. *N. Engl. J. Med.* **313,** 1557–1563.

Heberlein, K. R., Straub, A. C., Best, A. K., Greyson, M. A., Looft-Wilson, R. C., Sharma, P. R., Meher, A., Leitinger, N., and Isakson, B. E. (2010). Plasminogen activator inhibitor-1 regulates myoendothelial junction formation. *Circ. Res.* **106,** 1092–1102.

Hekman, C. M., and Loskutoff, D. J. (1985). Endothelial cells produce a latent inhibitor of plasminogen activators that can be activated by denaturants. *J. Biol. Chem.* **260,** 11581–11587.

Hofmann, K. J., Mayer, E. J., Schultz, L. D., Socher, S. H., and Reilly, C. F. (1992). Purification and characterization of recombinant rabbit plasminogen activator inhibitor-1 expressed in *Saccharomyces cerevisae. Fibrinolysis* **6,** 263–272.

Irigoyen, J. P., Munoz-Canoves, P., Montero, L., Koziczak, M., and Nagamine, Y. (1999). The plasminogen activator system: Biology and regulation. *Cell. Mol. Life Sci.* **56,** 104–132.

Izuhara, Y., Takahashi, S., Nangaku, M., Takizawa, S. Y., Ishida, H., Kurokawa, K., de Strihou, C. V., Hirayama, N., and Miyata, T. (2008). Inhibition of plasminogen activator inhibitor-1 - Its mechanism and effectiveness on coagulation and fibrosis. *Arterioscler. Thromb. Vasc. Biol.* **28,** 672–677.

Janicke, F., Schmitt, M., Pache, L., Ulm, K., Harbeck, N., Hofler, H., and Graeff, H. (1993). Urokinase (uPA) and its inhibitor PAI-1 are strong and independent prognostic factors in node-negative breast cancer. *Breast Cancer Res. Treat.* **24,** 195–208.

Juhan-Vague, I., Alessi, M. C., and Declerck, P. J. (1995). Pathophysiology of fibrinolysis. *Baillières Clin. Haematol.* **8,** 329–343.

Juhan-Vague, I., Pyke, S. D., Alessi, M. C., Jespersen, J., Haverkate, F., and Thompson, S. G. (1996). Fibrinolytic factors and the risk of myocardial infarction or sudden death in patients with angina pectoris. ECAT Study Group. European Concerted Action on Thrombosis and Disabilities. *Circulation* **94,** 2057–2063.

Kaikita, K., Fogo, A. B., Ma, L., Schoenhard, J. A., Brown, N. J., and Vaughan, D. E. (2001). Plasminogen activator inhibitor-1 deficiency prevents hypertension and vascular fibrosis in response to long-term nitric oxide synthase inhibition. *Circulation* **104,** 839–844.

Kaikita, K., Schoenhard, J. A., Painter, C. A., Ripley, R. T., Brown, N. J., Fogo, A. B., and Vaughan, D. E. (2002). Potential roles of plasminogen activator system in coronary vascular remodeling induced by long-term nitric oxide synthase inhibition. *J. Mol. Cell. Cardiol.* **34,** 617–627.

Kawasaki, T., Dewerchin, M., Lijnen, H. R., Vermylen, J., and Hoylaerts, M. F. (2000). Vascular release of plasminogen activator inhibitor-1 impairs fibrinolysis during acute arterial thrombosis in mice. *Blood* **96,** 153–160.

Kawasaki, T., Dewerchin, M., Lijnen, H. R., Vreys, I., Vermylen, J., and Hoylaerts, M. F. (2001). Mouse carotid artery ligation induces platelet-leukocyte-dependent luminal fibrin, required for neointima development. *Circ. Res.* **88,** 159–166.

Kikuchi, S., Umemura, K., Kondo, K., Saniabadi, A. R., and Nakashima, M. (1998). Photochemically induced endothelial injury in the mouse as a screening model for inhibitors of vascular intimal thickening. *Arterioscler. Thromb. Vasc. Biol.* **18,** 1069–1078.

Knoop, A., Andreasen, P. A., Andersen, J. A., Hansen, S., Laenkholm, A. V., Simonsen, A. C., Andersen, J., Overgaard, J., and Rose, C. (1998). Prognostic significance of urokinase-type plasminogen activator and plasminogen activator inhibitor-1 in primary breast cancer. *Br. J. Cancer* **77,** 932–940.

Konecny, G., Untch, M., Pihan, A., Kimmig, R., Gropp, M., Stieber, P., Hepp, H., Slamon, D., and Pegram, M. (2001). Association of urokinase-type plasminogen activator and its inhibitor with disease progression and prognosis in ovarian cancer. *Clin. Cancer Res.* **7,** 1743–1749.

Konstantinides, S., Schafer, K., Thinnes, T., and Loskutoff, D. J. (2001). Plasminogen activator inhibitor-1 and its cofactor vitronectin stabilize arterial thrombi after vascular injury in mice. *Circulation* **103,** 576–583.

Kornelisse, R. F., Hazelzet, J. A., Savelkoul, H. F., Hop, W. C., Suur, M. H., Borsboom, A. N., Risseeuw-Appel, I. M., van der Voort, E., and de Groot, R. (1996). The relationship between plasminogen activator inhibitor-1 and proinflammatory and counterinflammatory mediators in children with meningococcal septic shock. *J. Infect. Dis.* **173,** 1148–1156.

Koschnick, S., Konstantinides, S., Schafer, K., Crain, K., and Loskutoff, D. J. (2005). Thrombotic phenotype of mice with a combined deficiency in plasminogen activator inhibitor 1 and vitronectin. *J. Thromb. Haemost.* **3,** 2290–2295.

Kruithof, E. K., Nicolosa, G., and Bachmann, F. (1987). Plasminogen activator inhibitor 1: Development of a radioimmunoassay and observations on its plasma concentration during venous occlusion and after platelet aggregation. *Blood* **70,** 1645–1653.

Kuhn, W., Schmalfeldt, B., Reuning, U., Pache, L., Berger, U., Ulm, K., Harbeck, N., Spathe, K., Dettmar, P., Hofler, H., Janicke, F., Schmitt, M., *et al.* (1999). Prognostic significance of urokinase (uPA) and its inhibitor PAI-1 for survival in advanced ovarian carcinoma stage FIGO IIIc. *Br. J. Cancer* **79,** 1746–1751.

Kurz, K. D., Main, B. W., and Sandusky, G. E. (1990). Rat model of arterial thrombosis induced by ferric chloride. *Thromb. Res.* **60,** 269–280.

Kwaan, H. C., Wang, J., Svoboda, K., and Declerck, P. J. (2000). Plasminogen activator inhibitor 1 may promote tumour growth through inhibition of apoptosis. *Br. J. Cancer* **82,** 1702–1708.

Lambert, V., Munaut, C., Carmeliet, P., Gerard, R. D., Declerck, P. J., Gils, A., Claes, C., Foidart, J. M., Noel, A., and Rakic, J. M. (2003). Dose-dependent modulation of choroidal neovascularization by plasminogen activator inhibitor type I: Implications for clinical trials. *Invest. Ophthalmol. Vis. Sci.* **44,** 2791–2797.

Lawrence, D., Strandberg, L., Grundstrom, T., and Ny, T. (1989). Purification of active human plasminogen activator inhibitor 1 from *Escherichia coli*. Comparison with natural and recombinant forms purified from eucaryotic cells. *Eur. J. Biochem.* **186,** 523–533.

Lee, M. H., Vosburgh, E., Anderson, K., and McDonagh, J. (1993). Deficiency of plasma plasminogen activator inhibitor 1 results in hyperfibrinolytic bleeding. *Blood* **81,** 2357–2362.

Leik, C. E., Su, E. J., Nambi, P., Crandall, D. L., and Lawrence, D. A. (2006). Effect of pharmacologic plasminogen activator inhibitor-1 inhibition on cell motility and tumor angiogenesis. *J. Thromb. Haemost.* **4,** 2710–2715.

Li, S. H., and Lawrence, D. A. Development of inhibitors of plasminogen activator inhibitors-1. *Methods in Enzymol*. Manuscript submitted for publication.

Lijnen, H. (2005). Pleiotropic functions of plasminogen activator inhibitor-1. *J. Thromb. Haemost.* **3,** 35–45.

Lijnen, H. R., van Hoef, B., Beelen, V., and Collen, D. (1994). Characterization of the murine plasma fibrinolytic system. *Eur. J. Biochem.* **224,** 863–871.

Lijnen, H. R., Maquoi, E., Morange, P., Voros, G., Van Hoef, B., Kopp, F., Collen, D., Juhan-Vague, I., and Alessi, M. C. (2003). Nutritionally induced obesity is attenuated in transgenic mice overexpressing plasminogen activator inhibitor-1. *Arterioscler. Thromb. Vasc. Biol.* **23,** 78–84.

Lijnen, H. R., Van Hoef, B., Umans, K., and Collen, D. (2004). Neointima formation and thrombosis after vascular injury in transgenic mice overexpressing plasminogen activator inhibitor-1 (PAI-1). *J. Thromb. Haemost.* **2,** 16–22.

Loskutoff, D. J., van Mourik, J. A., Erickson, L. A., and Lawrence, D. (1983). Detection of an unusually stable fibrinolytic inhibitor produced by bovine endothelial cells. *Proc. Natl. Acad. Sci. USA* **80,** 2956–2960.

Loskutoff, D. J., Sawdey, M., and Mimuro, J. (1989). Type 1 plasminogen activator inhibitor. *Prog. Hemost. Thromb.* **9,** 87–115.

Lupu, F., Bergonzelli, G. E., Heim, D. A., Cousin, E., Genton, C. Y., Bachmann, F., and Kruithof, E. K. (1993). Localization and production of plasminogen activator inhibitor-1 in human healthy and atherosclerotic arteries. *Arterioscler. Thromb.* **13,** 1090–1100.

Luttun, A., Lupu, F., Storkebaum, E., Hoylaerts, M. F., Moons, L., Crawley, J., Bono, F., Poole, A. R., Tipping, P., Herbert, J. M., Collen, D., and Carmeliet, P. (2002). Lack of plasminogen activator inhibitor-1 promotes growth and abnormal matrix remodeling of advanced atherosclerotic plaques in apolipoprotein E-deficient mice. *Arterioscler. Thromb. Vasc. Biol.* **22,** 499–505.

Maillard, C., Jost, M., Romer, M. U., Brunner, N., Houard, X., Lejeune, A., Munaut, C., Bajou, K., Melen, L., Dano, K., Carmeliet, P., Fusenig, N. E., *et al.* (2005). Host plasminogen activator inhibitor-1 promotes human skin carcinoma progression in a stage-dependent manner. *Neoplasia* **7,** 57–66.

Maillard, C. M., Bouquet, C., Petitjean, M. M., Mestdagt, M., Frau, E., Jost, M., Masset, A. M., Opolon, P. H., Beermann, F., Abitbol, M. M., Foidart, J. M., Perricaudet, M. J., *et al.* (2008). Reduction of brain metastases in plasminogen activator inhibitor-1-deficient mice with transgenic ocular tumors. *Carcinogenesis* **29,** 2236–2242.

Masset, A., Maillard, C., Sounni, N., Jacobs, N., Bruyere, F., Delvenne, P., Tacke, M., Reinheckel, T., Foidart, J. M., Coussens, L., and Noel, A. (2011). Unimpeded skin carcinogenesis in K14-HPV16 transgenic mice deficient for plasminogen activator inhibitor. *Int. J. Cancer* **128,** 283–293.

Matsuno, H., Uematsu, T., Nagashima, S., and Nakashima, M. (1991). Photochemically induced thrombosis model in rat femoral artery and evaluation of effects of heparin and tissue-type plasminogen activator with use of this model. *J. Pharmacol. Methods* **25,** 303–317.

Matsuno, H., Kozawa, O., Niwa, M., Ueshima, S., Matsuo, O., Collen, D., and Uematsu, T. (1999). Differential role of components of the fibrinolytic system in the formation and removal of thrombus induced by endothelial injury. *Thromb. Haemost.* **81,** 601–604.

Matsuo, O., Lijnen, H. R., Ueshima, S., Kojima, S., and Smyth, S. S. (2007). A guide to murine fibrinolytic factor structure, function, assays, and genetic alterations. *J. Thromb. Haemost.* **5,** 680–689.

McDonald, A. P., Meier, T. R., Hawley, A. E., Thibert, J. N., Farris, D. M., Wrobleski, S. K., Henke, P. K., Wakefield, T. W., and Myers, D. D. (2010). Aging is associated with impaired thrombus resolution in a mouse model of stasis induced thrombosis. *Thromb. Res.* **125,** 72–78.

McMahon, B., and Kwaan, H. C. (2008). The plasminogen activator system and cancer. *Pathophysiol. Haemost. Thromb.* **36,** 184–194.

McMahon, G. A., Petitclerc, E., Stefansson, S., Smith, E., Wong, M. K., Westrick, R. J., Ginsburg, D., Brooks, P. C., and Lawrence, D. A. (2001). Plasminogen activator inhibitor-1 regulates tumor growth and angiogenesis. *J. Biol. Chem.* **276,** 33964–33968.

Meade, T. W., Ruddock, V., Stirling, Y., Chakrabarti, R., and Miller, G. J. (1993). Fibrinolytic activity, clotting factors, and long-term incidence of ischaemic heart disease in the Northwick Park Heart Study. *Lancet* **342,** 1076–1079.

Medcalf, R. L. (2007). Fibrinolysis, inflammation, and regulation of the plasminogen activating system. *J. Thromb. Haemost.* **5**(Suppl. 1), 132–142.

Meissenheimer, L. M., Verbeke, K., Declerck, P. J., and Gils, A. (2006). Quantitation of Vervet monkey (Chlorocebus aethiops) plasminogen activator inhibitor-1 in plasma and platelets. *Thromb. Haemost.* **95,** 902–903.

Mesters, R. M., Florke, N., Ostermann, H., and Kienast, J. (1996). Increase of plasminogen activator inhibitor levels predicts outcome of leukocytopenic patients with sepsis. *Thromb. Haemost.* **75,** 902–907.

Mimuro, J., Sawdey, M., Hattori, M., and Loskutoff, D. (1989). cDNA for bovine type 1 plasminogen activator inhibitor (PAI-1). *Nucleic Acid Res.* **17,** 8872.

Moriwaki, H., Stempien-Otero, A., Kremen, M., Cozen, A. E., and Dichek, D. A. (2004). Overexpression of urokinase by macrophages or deficiency of plasminogen activator inhibitor type 1 causes cardiac fibrosis in mice. *Circ. Res.* **95,** 637–644.

Myers, D., Farris, D., Hawley, A., Wrobleski, S., Chapman, A., Stoolman, L., Knibbs, R., Strieter, R., and Wakefield, T. (2002). Selectins influence thrombosis in a mouse model of experimental deep venous thrombosis. *J. Surg. Res.* **108,** 212–221.

Nagai, N., De Mol, M., Lijnen, H. R., Carmeliet, P., and Collen, D. (1999). Role of plasminogen system components in focal cerebral ischemic infarction: A gene targeting and gene transfer study in mice. *Circulation* **99,** 2440–2444.

Nagai, N., Suzuki, Y., Van Hoef, B., Lijnen, H. R., and Collen, D. (2005). Effects of plasminogen activator inhibitor-1 on ischemic brain injury in permanent and thrombotic middle cerebral artery occlusion models in mice. *J. Thromb. Haemost.* **3,** 1379–1384.

Nagai, N., Van Hoef, B., and Lijnen, H. R. (2007). Plasminogen activator inhibitor-1 contributes to the deleterious effect of obesity on the outcome of thrombotic ischemic stroke in mice. *J. Thromb. Haemost.* **5,** 1726–1731.

Nakashima, Y., Plump, A. S., Raines, E. W., Breslow, J. L., and Ross, R. (1994). ApoE-deficient mice develop lesions of all phases of atherosclerosis throughout the arterial tree. *Arterioscler. Thromb.* **14,** 133–140.

Narayanaswamy, M., Wright, K. C., and Kandarpa, K. (2000). Animal models for atherosclerosis, restenosis, and endovascular graft research. *J. Vasc. Interv. Radiol.* **11,** 5–17.

Nekarda, H., Siewert, J. R., Schmitt, M., and Ulm, K. (1994). Tumour-associated proteolytic factors uPA and PAI-1 and survival in totally resected gastric cancer. *Lancet* **343,** 117.

Nilsson, I. M., Ljungner, H., and Tengborn, L. (1985). Two different mechanisms in patients with venous thrombosis and defective fibrinolysis: Low concentration of plasminogen activator or increased concentration of plasminogen activator inhibitor. *Br. Med. J. Clin. Res. Ed.* **290,** 1453–1456.

Nordt, T. K., Sawa, H., Fujii, S., Bode, C., and Sobel, B. E. (1998). Augmentation of arterial endothelial cell expression of the plasminogen activator inhibitor type-1 (PAI-1) gene by proinsulin and insulin in vivo. *J. Mol. Cell. Cardiol.* **30,** 1535–1543.

Numaguchi, Y., Ishii, M., Kubota, R., Morita, Y., Yamamoto, K., Matsushita, T., Okumura, K., and Murohara, T. (2009). Ablation of angiotensin IV receptor attenuates hypofibrinolysis via PAI-1 downregulation and reduces occlusive arterial thrombosis. *Arterioscler. Thromb. Vasc. Biol.* **29,** 2102–2108.

Ny, T., Sawdey, M., Lawrence, D., Millan, J. L., and Loskutoff, D. J. (1986). Cloning and sequence of a cDNA coding for the human beta-migrating endothelial-cell-type plasminogen activator inhibitor. *Proc. Natl. Acad. Sci. USA* **83,** 6776–6780.

Otsuka, G., Agah, R., Frutkin, A. D., Wight, T. N., and Dichek, D. A. (2006). Transforming growth factor beta 1 induces neointima formation through plasminogen activator inhibitor-1-dependent pathways. *Arterioscler. Thromb. Vasc. Biol.* **26,** 737–743.

Owensby, D. A., Morton, P. A., Wun, T. C., and Schwartz, A. L. (1991). Binding of plasminogen activator inhibitor type-1 to extracellular matrix of Hep G2 cells. Evidence that the binding protein is vitronectin. *J. Biol. Chem.* **266,** 4334–4340.

Pannekoek, H., Veerman, H., Lambers, H., Diergaarde, P., Verweij, C. L., van Zonneveld, A. J., and van Mourik, J. A. (1986). Endothelial plasminogen activator inhibitor (PAI): A new member of the Serpin gene family. *EMBO J.* **5,** 2539–2544.

Peng, L., Bhatia, N., Parker, A. C., Zhu, Y., and Fay, W. P. (2002). Endogenous vitronectin and plasminogen activator inhibitor-1 promote neointima formation in murine carotid arteries. *Arterioscler. Thromb. Vasc. Biol.* **22,** 934–939.

Pierangeli, S. S., and Harris, E. N. (1994). Antiphospholipid antibodies in an in vivo thrombosis model in mice. *Lupus* **3,** 247–251.

Pierangeli, S. S., Liu, X. W., Barker, J. H., Anderson, G., and Harris, E. N. (1995). Induction of thrombosis in a mouse model by IgG. IgM and IgA immunoglobulins from patients with the antiphospholipid syndrome. *Thromb. Haemost.* **74,** 1361–1367.

Ploplis, V. A., Cornelissen, I., Sandoval-Cooper, M. J., Weeks, L., Noria, F. A., and Castellino, F. J. (2001). Remodeling of the vessel wall after copper-induced injury is highly attenuated in mice with a total deficiency of plasminogen activator inhibitor-1. *Am. J. Pathol.* **158,** 107–117.

Prendergast, G. C., Diamond, L. E., Dahl, D., and Cole, M. D. (1990). The c-myc-regulated gene mr1 encodes plasminogen activator inhibitor 1. *Mol. Cell. Biol.* **10,** 1265–1269.

Prins, M. H., and Hirsh, J. (1991). A critical review of the evidence supporting a relationship between impaired fibrinolytic activity and venous thromboembolism. *Arch. Intern. Med.* **151,** 1721–1731.

Rijken, D. C., and Lijnen, H. R. (2009). New insights into the molecular mechanisms of the fibrinolytic system. *J. Thromb. Haemost.* **7,** 4–13.

Robbie, L. A., Booth, N. A., Brown, A. J., and Bennett, B. (1996). Inhibitors of fibrinolysis are elevated in atherosclerotic plaque. *Arterioscler. Thromb. Vasc. Biol.* **16,** 539–545.

Sandford, R. M., Bown, M. J., London, N. J., and Sayers, R. D. (2007). The genetic basis of abdominal aortic aneurysms: A review. *Eur. J. Vasc. Endovasc. Surg.* **33,** 381–390.

Schafer, K., Muller, K., Hecke, A., Mounier, E., Goebel, J., Loskutoff, D. J., and Konstantinides, S. (2003). Enhanced thrombosis in atherosclerosis-prone mice is associated with increased arterial expression of plasminogen activator inhibitor-1. *Arterioscler. Thromb. Vasc. Biol.* **23,** 2097–2103.

Schafer, K., Schroeter, M. R., Dellas, C., Puls, M., Nitsche, M., Weiss, E., Hasenfuss, G., and Konstantinides, S. V. (2006). Plasminogen activator inhibitor-1 from bone marrow-derived cells suppresses neointimal formation after vascular injury in mice. *Arterioscler. Thromb. Vasc. Biol.* **26,** 1254–1259.

Schneider, D. J., Hayes, M., Wadsworth, M., Taatjes, H., Rincon, M., Taatjes, D. J., and Sobel, B. E. (2004). Attenuation of neointimal vascular smooth muscle cellularity in atheroma by plasminogen activator inhibitor type 1 (PAI-1). *J. Histochem. Cytochem.* **52,** 1091–1099.

Schneiderman, J., Sawdey, M. S., Keeton, M. R., Bordin, G. M., Bernstein, E. F., Dilley, R. B., and Loskutoff, D. J. (1992). Increased type 1 plasminogen activator inhibitor gene expression in atherosclerotic human arteries. *Proc. Natl. Acad. Sci. USA* **89,** 6998–7002.

Schulman, S., and Wiman, B. (1996). The significance of hypofibrinolysis for the risk of recurrence of venous thromboembolism. Duration of Anticoagulation (DURAC) Trial Study Group. *Thromb. Haemost.* **75,** 607–611.

Seiffert, D., Mimuro, J., Schleef, R. R., and Loskutoff, D. J. (1990). Interactions between type 1 plasminogen activator inhibitor, extracellular matrix and vitronectin. *Cell Differ. Dev.* **32,** 287–292.

Senoo, T., Hattori, N., Tanimoto, T., Furonaka, M., Ishikawa, N., Fujitaka, K., Haruta, Y., Murai, H., Yokoyama, A., and Kohno, N. (2010). Suppression of plasminogen activator inhibitor-1 by RNA interference attenuates pulmonary fibrosis. *Thorax* **65,** 334–340.

Siemens, H. J., Brueckner, S., Hagelberg, S., Wagner, T., and Schmucker, P. (1999). Course of molecular hemostatic markers during and after different surgical procedures. *J. Clin. Anesth.* **11,** 622–629.

Simpson, A. J., Booth, N. A., Moore, N. R., and Bennett, B. (1991). Distribution of plasminogen activator inhibitor (PAI-1) in tissues. *J. Clin. Pathol.* **44,** 139–143.

Singh, I., Smith, A., Vanzieleghem, B., Collen, D., Burnand, K., St-Remy, J. M., and Jacquemin, M. (2002). Antithrombotic effects of controlled inhibition of factor VIII with a partially inhibitory human monoclonal antibody in a murine vena cava thrombosis model. *Blood* **99,** 3235–3240.

Sjoland, H., Eitzman, D. T., Gordon, D., Westrick, R., Nabel, E. G., and Ginsburg, D. (2000). Atherosclerosis progression in LDL receptor-deficient and apolipoprotein

E-deficient mice is independent of genetic alterations in plasminogen activator inhibitor-1. *Arterioscler. Thromb. Vasc. Biol.* **20,** 846–852.

Smith, L. H., Dixon, J. D., Stringham, J. R., Eren, M., Elokdah, H., Crandall, D. L., Washington, K., and Vaughan, D. E. (2006). Pivotal role of PAI-1 in a murine model of hepatic vein thrombosis. *Blood* **107,** 132–134.

Soff, G. A., Sanderowitz, J., Gately, S., Verrusio, E., Weiss, I., Brem, S., and Kwaan, H. C. (1995). Expression of plasminogen activator inhibitor type 1 by human prostate carcinoma cells inhibits primary tumor growth, tumor-associated angiogenesis, and metastasis to lung and liver in an athymic mouse model. *J. Clin. Invest.* **96,** 2593–2600.

Stefansson, S., Petitclerc, E., Wong, M. K., McMahon, G. A., Brooks, P. C., and Lawrence, D. A. (2001). Inhibition of angiogenesis in vivo by plasminogen activator inhibitor-1. *J. Biol. Chem.* **276,** 8135–8141.

Takeshita, K., Hayashi, M., Iino, S., Kondo, T., Inden, Y., Iwase, M., Kojima, T., Hirai, M., Ito, M., Loskutoff, D. J., Saito, H., Murohara, T., *et al.* (2004). Increased expression of plasminogen activator inhibitor-1 in cardiomyocytes contributes to cardiac fibrosis after myocardial infarction. *Am. J. Pathol.* **164,** 449–456.

Thogersen, A. M., Jansson, J. H., Boman, K., Nilsson, T. K., Weinehall, L., Huhtasaari, F., and Hallmans, G. (1998). High plasminogen activator inhibitor and tissue plasminogen activator levels in plasma precede a first acute myocardial infarction in both men and women: Evidence for the fibrinolytic system as an independent primary risk factor. *Circulation* **98,** 2241–2247.

Van de Craen, B., Scroyen, I., Abdelnabi, R., Brouwers, E., Lijnen, H. R., Declerck, P. J., and Gils, A. Characterization of a panel of monoclonal antibodies toward mouse PAI-1 that exert a significant profibrinolytic effect in vivo. *Thromb. Res.* Mar 8. [Epub ahead of print].

Vaughan, D. E. (2005). PAI-1 and atherothrombosis. *J. Thromb. Haemost.* **3,** 1879–1883.

Vaughan, D. E., Declerck, P. J., Van Houtte, E., De Mol, M., and Collen, D. (1990). Studies of recombinant plasminogen activator inhibitor-1 in rabbits. Pharmacokinetics and evidence for reactivation of latent plasminogen activator inhibitor-1 in vivo. *Circ. Res.* **67,** 1281–1286.

Weisberg, A. D., Albornoz, F., Griffin, J. P., Crandall, D. L., Elokdah, H., Fogo, A. B., Vaughan, D. E., and Brown, N. J. (2005). Pharmacological inhibition and genetic deficiency of plasminogen activator inhibitor-1 attenuates angiotensin II/salt-induced aortic remodeling. *Arterioscler. Thromb. Vasc. Biol.* **25,** 365–371.

Westrick, R. J., Winn, M. E., and Eitzman, D. T. (2007). Murine models of vascular thrombosis (Eitzman series). *Arterioscler. Thromb. Vasc. Biol.* **27,** 2079–2093.

Whinna, H. C. (2008). Overview of murine thrombosis models. *Thromb. Res.* **122,** S64–S69.

Wiman, B., Ljungberg, B., Chmielewska, J., Urden, G., Blomback, M., and Johnsson, H. (1985). The role of the fibrinolytic system in deep vein thrombosis. *J. Lab. Clin. Med.* **105,** 265–270.

Xu, Z., Castellino, F. J., and Ploplis, V. A. (2010). Plasminogen activator inhibitor-1 (PAI-1) is cardioprotective in mice by maintaining microvascular integrity and cardiac architecture. *Blood* **115,** 2038–2047.

Xue, Y., Bjorquist, P., Inghardt, T., Linschoten, M., Musil, D., Sjolin, L., and Deinum, J. (1998). Interfering with the inhibitory mechanism of serpins: Crystal structure of a complex formed between cleaved plasminogen activator inhibitor type 1 and a reactive-centre loop peptide. *Structure* **6,** 627–636.

Zaman, A. K., French, C. J., Schneider, D. J., and Sobel, B. E. (2009). A profibrotic effect of plasminogen activator inhibitor type-1 (PAI-1) in the heart. *Exp. Biol. Med. (Maywood)* **234,** 246–254.

Zeheb, R., and Gelehrter, T. D. (1988). Cloning and sequencing of cDNA for the rat plasminogen activator inhibitor-1. *Gene* **73,** 459–468.

Zhu, Y., Carmeliet, P., and Fay, W. P. (1999). Plasminogen activator inhibitor-1 is a major determinant of arterial thrombolysis resistance. *Circulation* **99,** 3050–3055.

Zhu, Y., Farrehi, P. M., and Fay, W. P. (2001). Plasminogen activator inhibitor type 1 enhances neointima formation after oxidative vascular injury in atherosclerosis-prone mice. *Circulation* **103,** 3105–3110.

CHAPTER SIX

Plasminogen Activator Inhibitor Type 2: Still an Enigmatic Serpin but a Model for Gene Regulation

Robert L. Medcalf

Contents

1. Introduction	106
2. PAI-2 and the Plasminogen-Activating System	107
3. General Features of PAI-2	108
3.1. Structural considerations	108
3.2. Clearance receptors	109
3.3. Expression pattern of PAI-2	110
3.4. Role of PAI-2 in skin	110
3.5. Role of PAI-2 in monocyte biology	111
3.6. The role of PAI-2 in metastatic cancer	111
3.7. Association of PAI-2 with retinoblastoma protein	112
3.8. Apoptosis and the innate immune response	113
3.9. PAI-2 expression in the brain and its role as a neuroprotective agent	114
4. PAI-2 Gene Expression and Regulation	115
4.1. Cellular regulation of PAI-2 expression	116
4.2. Transcriptional regulation of PAI-2	116
4.3. Epigenetics	117
4.4. mRNA stability: General principals	118
4.5. Posttranscriptional regulation of PAI-2 expression	119
4.6. Assessment of PAI-2 mRNA decay using tetracycline-regulated expression systems	120
4.7. The role of AU-rich instability elements in the 3′-UTR of PAI-2	121
5. Conclusions	122
6. Methodology: Rapid Run-On Transcription Assay Protocol	123
Acknowledgments	126
References	126

Australian Centre for Blood Diseases, Monash University, Melbourne, Victoria, Australia

Abstract

Plasminogen activator inhibitor type-2 (PAI-2; SERPINB2) is an atypical member of the Ov-serpin family of serine protease inhibitors. While it is an undisputed inhibitor of urokinase and tissue-type plasminogen activator in the extracellular space and on the cell surface, the weight of circumstantial evidence suggests that PAI-2 also fulfills an intracellular role which is independent of plasminogen activator inhibition and indeed may not even involve protease inhibition at all. More and more data continue to implicate a role for PAI-2 in many settings, the most recent associating it as a modulator of the innate immune response. Further to the debates concerning its physiological role, there are few genes, if any, that display the regulation profile of the *PAI-2* gene: PAI-2 protein and mRNA levels can be induced in the order of, not hundred-, but thousand-folds in a process that is controlled at many levels including gene transcription and mRNA stability while an epigenetic component is also likely. The ability of some cells, including monocytes, fibroblasts, and neurons to have the capacity to increase PAI-2 synthesis to such high levels is intriguing enough. So why do these cells have the capacity to synthesize so much of this protein? While tantalizing clues continue to be revealed to the field, an understanding of how this gene is regulated so profoundly has provided insights into the broader mechanics of gene expression and regulation.

1. Introduction

Plasminogen activator inhibitor type 2 (PAI-2; SERPINB2) is a member of the Clade B subgroup of the serine protease inhibitor (serpin) superfamily which includes serpinB3 and B4 (SCCA1 and SCCA2), serpinB5 (maspin), serpinB6 (placental thrombin inhibitor), serpinB8 and B9 (cytoplasmic antiproteinases 2 and 3), and others (Silverman *et al.*, 2004). These so-called Ov-serpins lack a classical secretory signal and as such are localized to the cytoplasmic and even the nucleocytoplasmic compartments (Silverman *et al.*, 2004). Ov-serpin members display marked diversity in the profiles of their target proteases and topological distribution. PAI-2 and maspin are the only Ov-serpins that exist in both the intra- and extracellular compartments. Despite having an expectation of regulating proteolysis, some members of the Ov-serpin family have no known protease inhibitory activity. Indeed, no protease target, either intra- or extracellular, has been identified for maspin (Bass *et al.*, 2002; Pemberton *et al.*, 1995). Yet, despite possessing no recognized antiprotease activity, maspin can modulate angiogenesis, inhibit tumor cell migration and invasion, induce tumor cell apoptosis (Cella *et al.*, 2006; Zhang *et al.*, 1997), and regulate endothelial cell adhesion and migration (Qin and Zhang, 2010).

PAI-2, however, has two known extracellular protease targets, namely the plasminogen activators, urokinase and tissue-type plasminogen activator (u-PA and t-PA). However, the topological distribution of PAI-2 can vary from almost all of PAI-2 being secreted (Ye et al., 1988) to the majority remaining intracellular (Genton et al., 1987; Kruithof et al., 1986; Wohlwend et al., 1987). The predominant intracellular location of PAI-2 raised much speculation for an intracellular protease target and function for this particular Ov-serpin at the very outset. Nonetheless, the established interaction between PAI-2 and both plasminogen activators, albeit mostly derived from *in vitro* studies, guided most research efforts to explore the biology of PAI-2 in this context. While this direction has continued to be explored by a number of groups, a series of unexpected associations of PAI-2 mostly in the innate immune response and in neurobiology has been reported in recent years. This chapter will overview these recent findings, although a major focus will be the regulation of the *PAI-2* gene and the approaches used to unravel the mechanisms underlying its impressive gene expression profile.

2. PAI-2 AND THE PLASMINOGEN-ACTIVATING SYSTEM

The controlled generation of plasmin by the plasminogen activator system is classically associated with two events *in vivo*: t-PA-mediated fibrinolysis in the circulation and u-PA-mediated plasmin generation in the extravascular compartment (Cesarman-Maus and Hajjar, 2005). For the latter, many reports associated extravascular u-PA-mediated plasmin generation with wound closure, cell migration, and tumor metastasis (Medcalf, 2007). Indeed, u-PA was long pursued as a propagator of metastatic spread, and it stood to reason that PAI-2 would be well placed to inhibit these u-PA-mediated pathological events. However, evidence to support the u-PA inhibitory capacity of endogenous PAI-2 as its *major* physiological function has not been compelling. Indeed, the related plasminogen activator inhibitor type-1, namely PAI-1 (SERPINE1), inhibits u-PA $\sim$10-fold more effectively than PAI-2 and also effectively inhibits t-PA, whereas PAI-2 had only minimal activity against single-chain t-PA. PAI-1 is also a fully secreted serpin and hence more available to inhibit u-PA. Mice deficient in PAI-2 (PAI-2$^{-/-}$), produced over a decade ago, display no evidence of a defect in the plasminogen-activating system (Dougherty et al., 1999), although these mice were not subjected to models of metastatic cancer where the extent of u-PA-driven metastasis could be compared. Curiously, PAI-2$^{-/-}$ mice were shown to have an impairment in nutritionally induced adipose tissue development (Lijnen et al., 2007) yet adipose tissue–associated fibrinolytic activity was not affected. While some may argue that PAI-2 may have little influence on u-PA activity *in vivo* (see later), a plethora of new functions, mostly intracellular, has now been attributed to PAI-2. These functions are seemingly diverse and include its ability to influence apoptosis, cell differentiation, the innate immune response, and more recently to act as a neuroprotective agent.

3. General Features of PAI-2

Originally, PAI-2 was defined as a placental-derived u-PA inhibitor by Kawano et al. (1970) and subsequently confirmed by others (Kruithof et al., 1987; Wun and Reich, 1987). Human PAI-2 consists of a single-chain protein of 415 amino acids encoded by a 1900 bp PAI-2 transcript (Schleuning et al., 1987). The majority of PAI-2 is synthesized as a nonglycosylated intracellular (and intranuclear) protein, while as mentioned above a variable proportion (depending on the cell type) is glycosylated and secreted. Since the u-PA inhibitory capacity of both forms of PAI-2 is similar (Mikus et al., 1993), the release of high local concentrations of intracellular PAI-2 from dead or dying cells at sites of inflammation was suggested to provide a source of enriched u-PA inhibitory activity (Medcalf et al., 1988).

3.1. Structural considerations

Ov-serpins, like serpins in general, share a ternary structure consisting of nine alpha helices (A–I), three beta sheets (A–C), and a reactive site loop (RSL) of which the latter engages the target protease (Izuhara et al., 2008; Silverman et al., 2001). Most members of the serpin superfamily have the capability of undergoing dramatic structural changes from a native, stressed state to a relaxed form when engaged with their target protease. The structural basis of this transition became evident when the crystal structure of the trypsin–antitrypsin complex was resolved (Buechler et al., 2001; Stratikos and Gettins, 1999). Following engagement of the protease by the serpin, the RSL is cleaved and the RSL bound to the protease is then rapidly translocated and inserted into beta sheet A creating an additional strand. This conformational change, referred to as the "stressed to relaxed transition," also occurs in PAI-2 (Harrop et al., 1999; Jankova et al., 2001).

Some members of the Ov-serpin family possess an additional feature, being an extension of a domain that bridges helices C and D of the protein. This so-called C–D interhelical domain (Jensen et al., 1994b), or the "CD-loop," has been implicated as a key regulatory domain of PAI-2. Glutamine residues in the CD-loop can be cross-linked to structures in trophoblasts and to fibrin (Jensen et al., 1993, 1994b; Ritchie et al., 2000). The CD-loop has also been shown to bind noncovalently to annexins and to a number of unidentified proteins (Jensen et al., 1996) and more recently to the beta 1 subunit of the proteosome (Fan *et al.*, 2004). The proteosome is a structure that aids in the proteolytic removal of unwanted or misfolded proteins from the cell. This interaction between PAI-2 and the beta 1 subunit of the proteosome has also been confirmed by another group (Major *et al.*, 2011). The biological consequence of this association is unknown. Also, whether PAI-2 is utilizing its protease inhibitory potential or acting in some capacity related to proteosome assembly (Matias *et al.*, 2010) is unknown.

PAI-2 has the capacity to spontaneously polymerize under physiological conditions (Mikus *et al.*, 1993; Mikus and Ny, 1996) in a process that depends to a large extent on the redox status of the cell: PAI-2 can exist as either a stable monomer or a polymer, the latter stabilized by disulphide bonds that connect a cysteine residue within the CD-loop to another cysteine residue at the bottom of the molecule (Wilczynska *et al.*, 2003). The monomeric form is also stabilized by binding to vitronectin while retaining its inhibitory activity. Surprisingly, polymeric PAI-2 also retains its protease inhibitory activity. Unlike a number of other polymerized serpins including neuroserpin (Davis *et al.*, 1999), alpha-2 antitrypsin, and others (as reviewed by Kaiserman *et al.*, 2006), no disease phenotype has been linked with the polymerized form of PAI-2.

3.2. Clearance receptors

One of the hallmarks of plasminogen activator inhibitor–protease interactions is that the complexes formed are rapidly cleared from the cell surface via receptors belonging to the low-density lipoprotein receptor (LDLR) family (Narita *et al.*, 1995; Orth *et al.*, 1992) or by scavenging receptors such as the mannose receptor (Otter *et al.*, 1991). Clearance of PAI-1/t-PA or PAI-1–u-PA complexes occurs via specific members of the LDLR family, notably LRP-1 and vLDLR (Argraves *et al.*, 1995; Herz *et al.*, 1992; Narita *et al.*, 1995; Nguyen *et al.*, 1992; Wing *et al.*, 1991). More recent reports have suggested a role for PAI-2 on the cell surface that is distinct from PAI-1. The implications of this are potentially very important. Although PAI-2 can effectively inhibit cell-surface-bound u-PA and t-PA activity, the complexes formed between PAI-2 and either protease have a distinct biological outcome with regard to internalization and intracellular signaling. First of all, PAI-2, unlike PAI-1, cannot bind to LDLRs directly while PAI-2-containing complexes engage LDLRs via the protease. PAI-1-containing complexes, however, engage LDLRs via a cryptic binding site exposed in PAI-1 itself following complex formation. Moreover, PAI-2-containing complexes display differential endocytosis and signaling properties. While PAI-2 enhances the rate of internalization of u-PA (bound to uPAR) via LDLRs (as seen for PAI-1; Croucher *et al.*, 2007), PAI-2 does not enhance the rate of internalization of t-PA, which is a key feature of PAI-1–t-PA complexes (Lee *et al.*, 2010). In other words, t-PA–PAI-2 complexes are only internalized via LDLRs at the same rate as t-PA alone. This distinguishing feature of PAI-2 compared to PAI-1 was also suggested to enable a more targeted approach to inhibit t-PA-mediated cell-surface plasminogen activation while minimizing intracellular signaling capabilities that are a feature of the high-affinity LDLR binding PAI-1-containing complexes (Lee et al., 2010). Such a notion has potential ramifications in the context of metastatic cancer where protease inhibition, and not cell activation, is preferred.

3.3. Expression pattern of PAI-2

PAI-2 has a restricted tissue-distribution pattern with expression detected at high levels in keratinocytes, activated monocytes, the placenta (Kruithof et al., 1995), and adipocytes (Lijnen et al., 2007). Plasma levels of PAI-2 are usually low or undetectable but high levels can be found in crevicular fluid (Olofsson et al., 2002), saliva (Virtanen et al., 2006), and human tears (Csutak et al., 2008). Plasma levels rise significantly in some forms of monocytic leukemia (Scherrer et al., 1991) and in periodontal disease (Kardesler et al., 2008). Plasma levels of PAI-2 also rise enormously during the third trimester of pregnancy (up to 250 ng/ml) before declining within a week postpartum (Kruithof et al., 1987). The tissue source of plasma PAI-2 is the placenta. Since PAI-2 is highly expressed in trophoblasts (Black et al., 2001; Hofmann et al., 1994), it was proposed that PAI-2 protects the placenta from proteolytic degradation; however, the concentration of PAI-2 is far in excess of that needed to inhibit any suspected protease (i.e., u-PA). PAI-2 forms complexes with other placental proteins, including vitronectin (Radtke et al., 1990), but the functional significance of this is unknown. Lower plasma levels of PAI-2 have been correlated with an increased incidence of some obstetric complications (Roes et al., 2002), including preeclampsia and hydatidiform mole (Reith et al., 1993). While confirmatory reports also link PAI-2 with maternal health, the protective role undertaken by PAI-2 during pregnancy is still a mystery. Mice deficient in PAI-2 develop normally and have normal litter sizes (Dougherty et al., 1999); however, PAI-2 does not appear to be expressed in the mouse placenta (Belin, 1993) which is in stark contrast to its expression in human placental tissue.

3.4. Role of PAI-2 in skin

Another hot-spot for PAI-2 expression is the skin where it is expressed at high levels in the upper layers of the dermis (Lyons-Giordano et al., 1994). During the terminal differentiation of keratinocytes, PAI-2 is cross-linked to the cell membrane via transglutaminase (Oji et al., 2006) and inhibits proliferation and keratinocyte differentiation (Hibino et al., 1999). Despite this expression pattern, PAI-2$^{-/-}$ mice seemingly have normal skin; however, it remains to be seen if the absence of PAI-2 in the skin would have any impact in skin pathology.

Transgenic mice overexpressing PAI-2 in the proliferating layers of mouse epidermis and hair follicle cells are highly susceptible to chemically induced papilloma formation (Zhou *et al.*, 2001). This may be due to the reported antiapoptotic effect of PAI-2 (Section 3.8) since cessation of tumor promoting treatment in control mice resulted in extensive apoptosis of the papilloma but not in the PAI-2 transgenic mouse.

3.5. Role of PAI-2 in monocyte biology

Novel insights into the role of PAI-2 in monocytes came from studies using THP-1 cells. Unlike primary monocytes and essentially all other widely used monocyte-like cell lines (e.g., U-937, K562, HL-60) that express endogenous PAI-2, the THP-1 monocytic cells do not express a functional PAI-2 protein (Gross and Sitrin, 1990). Indeed, the PAI-2 transcript in these cells is missing the first six exons, most likely the consequence of a chromosomal translocation anomaly (Katsikis *et al.*, 2000) and is hence inactive. THP-1 cells became an interesting PAI-2-negative monocytic cell resource to explore the biology of this protein. Taking advantage of these cells, Yu *et al.* (2002) produced stable THP-1 cell lines that expressed either a wild-type PAI-2 or a PAI-2 mutant containing an alanine substitution at the key arginine residue at position 380 that is essential for its interaction with u-PA (i.e., the P1 position). This PAI-2 mutant (PAI-2_{ala380}) was incapable of performing its u-PA inhibitory function. The presence of wild-type PAI-2 in THP-1 cells caused a significant decrease in cell proliferation, reduction in DNA synthesis, and a phenotypic change following phorbol ester-induced differentiation. The ability of PAI-2 to alter differentiation was dependent on its active form since cells expressing PAI-2_{ala380} did not display these changes. This study demonstrated for the first time a role for active PAI-2 in monocytic behavior via protease inhibition. While an undefined intracellular PAI-2 sensitive protease(s) may be involved, additional evidence suggested that the cellular effects were in fact due to disruption of a u-PA/uPAR signaling pathway, since addition of exogenous u-PA negated the antiproliferative effects of PAI-2 (Yu *et al.*, 2002).

3.6. The role of PAI-2 in metastatic cancer

A number of *in vivo* studies have assessed the prognostic relevance of cancer and stromal-derived PAI-2 in the metastatic spread of cancer of the neck, lung, and breast (Borstnar *et al.*, 2002; Duggan *et al.*, 1997; Foekens *et al.*,

1995; Hasina et al., 2003; Yoshino et al., 1998). The only known target for PAI-2, namely u-PA, is strongly implicated in facilitating tumor metastasis, and it is likely that the beneficial effect of PAI-2 seen in these studies is via u-PA inhibition.

PAI-2 was also shown to be downregulated in squamous cell carcinoma cell lines (Hasina et al., 2003). These authors further suggested that the expression level of PAI-2 could be used as a biomarker for squamous cell carcinomas. Whether these cells displayed a concomitant increase in u-PA activity was not determined. A more recent study has suggested that downregulation of the cluster of serpin genes located on chromosome 18q 21.3, including PAI-2, was also associated with the occurrence of oral squamous cell carcinomas (Shiiba et al., 2010). Overexpression of PAI-2 in melanoma cells prevented spontaneous metastasis of transplanted cells (Mueller et al., 1995), while its overexpression in HT-1080 cells reduced u-PA-dependent cell migration *in vitro* and metastatic development *in vivo* (Laug et al., 1993).

The weight of evidence clearly supports some relationship of PAI-2 with cancer. While probable, it remains to be formally proven if this protective effect of PAI-2 is causally linked to its ability to inhibit u-PA, or if this reflects some other role for PAI-2. Evidence needed to support a protective role for PAI-2 via u-PA inhibition would be strengthened with data showing the presence of u-PA–PAI-2 complexes in any of these clinical conditions. Typically, serpins form covalent bonds with their target protease that can be revealed under reduced conditions in SDS-PAGE gels. However, PAI-2–u-PA complexes that are readily produced *in vitro* are rarely seen *in vivo* but this could be a consequence of rapid clearance of these complexes.

3.7. Association of PAI-2 with retinoblastoma protein

Retinoblastoma protein (Rb) is a tumor suppressor gene and critical cell cycle regulator that targets the E2F family of transcription factors (Harbour and Dean, 2000). PAI-2 was shown to colocalize with Rb in the nucleus and to inhibit Rb turnover by protecting it from proteolysis (Darnell et al., 2003). This in turn led to increases in Rb-mediated transcriptional repression of oncogenes. However, the fact that cells from PAI-2$^{-/-}$ mice do not appear to have any alteration in either cell number or proliferation rate (Dougherty et al., 1999) raised questions as to the significance of these findings. Other reports have found no linkage between PAI-2 and Rb (Fish and Kruithof, 2006), while recent findings from the same group studying PAI-2 regulation in HPV-transformed CaSKI cells (a high PAI-2 expressing cell line) have been unable to reproduce these initial findings (Major et al., 2011). Hence, the role of PAI-2 as a modulator of Rb is doubtful.

3.8. Apoptosis and the innate immune response

A landmark publication in 1991 provided *in vitro* evidence to suggest that PAI-2 could inhibit tumor necrosis factor (TNF)-induced apoptosis in HT-1080 fibrosarcoma cells (Kumar and Baglioni, 1991). PAI-2 was also shown to promote an antiapoptotic phenotype when overexpressed in HeLa cells (Dickinson *et al.*, 1995) although other reports have attributed this antiapoptotic effect as a HeLa cell clonal artifact (Fish and Kruithof, 2006). A cleaved form of intracellular PAI-2 has been found in ND4 monocytes undergoing apoptosis (Jensen *et al.*, 1994a). However, some studies have provided contradictory data (Ritchie *et al.*, 2000). Despite this controversy, the *in vivo* study of Zhou *et al.* (2001) where PAI-2 was overexpressed in skin is arguably the most convincing example of an antiapoptotic role for PAI-2 during papilloma formation.

Primary macrophages infected with the bacterium *Bacillus anthracis* are known to trigger an apoptotic response due to the inhibition of p38 MAP kinase signaling pathway (Park *et al.*, 2005). A search for survival genes activated in *B. anthracis*-infected macrophages revealed a critical requirement for the transcription factor CREB (cAMP-responsive element binding protein). CREB is an essential regulator of many signaling pathways, notably cAMP and NF-κβ. However, of all the downstream genes modulated by CREB, induction of PAI-2 was shown to be essential in the survival response of macrophages to this bacterium (Park *et al.*, 2005). This study also used THP-1-derived macrophages to explore their response to lipopolysaccharide (LPS)-mediated signaling and apoptosis. These particular cells, which are devoid of PAI-2, were also shown to be highly sensitive to apoptosis induced by LPS. Overexpression of wild-type PAI-2 rescued THP-1 cells from lethality following *B. anthracis* infection, thereby substantiating *PAI-2* as a relevant survival gene in activated human macrophages. Consistent with these findings, PAI-2 was also shown to inhibit macrophage apoptosis induced by LPS and completely blocked the secretion of the cytokine IL-1β (Greten *et al.*, 2007) which is essential for the initiation of the inflammatory response.

Additional evidence for a role for PAI-2 in the innate immune response came from studies using aryl hydrocarbon receptor (AhR) knockout mice (AhR$^{-/-}$ mice). AhR is a ligand-activated transcription factor that is activated by polycyclic aromatic hydrocarbons such as 2′,3′,7′,8′-tetrachlorodibenzo-*p*-dioxin (TCDD; also known as dioxin), a potent environmental pollutant. Indeed, TCDD has been shown to induce PAI-2 gene expression in a variety of human cell lines including monocytes, keratinocytes, and hepatocytes (Gohl *et al.*, 1996). AhR$^{-/-}$ mice were hypersensitive to LPS-induced septic shock, producing high levels of IL-1β. This was a consequence of dysfunctional macrophages (Sekine *et al.*, 2009). Curiously, macrophages from AhR$^{-/-}$ mice were also shown to express very low levels of PAI-2. Transfection of PAI-2 into macrophages from AhR$^{-/-}$ mice using adenovirus delivery normalized the sensitivity of these cells to LPS with a concomitant decrease in IL-1β secretion. AhR was subsequently shown to increase expression of PAI-2 via NF-κβ, which in turn determined the rate of IL-1β secretion and subsequently the immune response (Sekine *et al.*, 2009).

Exploring further this role of PAI-2 in the innate immune response, Schroder *et al.* (2010a) immunized PAI-2$^{-/-}$ and wild-type mice with ovalbumin (OVA). Surprisingly, IgG antibody titers to OVA were fivefold greater in PAI-2$^{-/-}$ mice than that in littermate controls. The PAI-2$^{-/-}$ mice were also shown to have a greater number of OVA-specific interferon-γ-secreting T-cells in the spleen. Hence, the absence of PAI-2 increases cytokine production in T-cells. In other words, endogenous PAI-2 in antigen presenting cells inhibits the production of key cytokines from T-cells (Th1 cytokines) thereby suppressing Th1 immunity. This observation is consistent with clinical associations seen between PAI-2 dysregulation and polymorphisms and a number of inflammatory diseases including asthma, lupus, scleroderma, and others (Schroder *et al.*, 2010a). The mechanism by which PAI-2 was inhibiting Th1 immunity remains elusive, but it is seemingly unrelated to either u-PA activity or plasmin generation in this system (Schroder *et al.*, 2010b). Taken together, the consistent findings from three independent laboratories support the view that PAI-2 influences the innate immune response although the intracellular mechanism remains to be elucidated.

3.9. PAI-2 expression in the brain and its role as a neuroprotective agent

While many studies on the molecular and cellular biology of PAI-1 have been extrapolated from studies on human or mouse monocytes or monocyte-like cell lines, results of more recent findings implicate *PAI-2* as a major stress response gene in cells of the central nervous system (CNS; Zhang *et al.*, 2009). That PAI-2 is expressed in neurons is not a new finding, as robust expression of PAI-2 in the rodent brain, together with plasminogen, was reported over a decade ago, and both were shown to be potently and rapidly increased following kainate (a glutamate analogue) treatment *in vivo* (Sharon *et al.*, 2002). u-PA expression was also increased in the mouse brain following kainate treatment (Masos and Miskin, 1997) although the expression pattern did not fully overlap with that of PAI-2. Nonetheless, it was proposed that PAI-2, despite its intracellular localization, could act to limit plasmin-induced toxicity in the CNS.

In this recent report, *PAI-2* was identified as one of nine genes for acquired neuroprotection (Zhang *et al.*, 2009). These genes were collectively referred to as "Activity-regulated inhibitor of Death (*AID*)" genes and were identified from a microarray screen as highly induced nuclear calcium-dependent genes in hippocampal neurons during synaptic firing. While the magnitude of the PAI-2 increase (1800-fold increase, the greatest of all the *AID* genes) and short time frame (3 h) for this response is unprecedented, all nine *AID* genes, including *PAI-2*, were shown to individually promote neuroprotection against glutamate analogues *in vivo* when introduced by adenoviral transfer into the brains of wild-type mice (Zhang *et al.*, 2009). Hence, this study uncovered a previously unsuspected protective role for PAI-2 in the brain. Again, like all of the aforementioned associations of PAI-2, the mechanism of neuroprotection afforded by PAI-2 is not known but it is plausible that this mechanism may have some features in common with the protective effect of PAI-2 in immune cells.

4. PAI-2 Gene Expression and Regulation

While subsequent sections will overview the means by which the *PAI-2* gene is regulated by some specific agents, there is one unifying stimulus that seems to be pertinent to the regulation of this gene, that being cellular stress. One could make a strong argument that the *PAI-2* gene is in fact a general stress response gene as its expression is invariably increased in most cells, particularly immune cells, following stress, regardless of the type of stress.

PAI-2 was cloned by groups that had an intent focus on the cell and molecular biology of PAI-2 (Antalis *et al.*, 1988; Schleuning *et al.*, 1987; Ye *et al.*, 1989), and by others serendipitously. For the latter, *PAI-2* was cloned as a TNF-responsive gene in monocytes and fibroblasts (Pytel *et al.*, 1990; Webb *et al.*, 1987) and as a dioxin (TCDD)-responsive gene in keratinocytes (Sutter *et al.*, 1991). *PAI-2* has proven to be an exquisitely regulated gene. The following examples further exemplify this point and its mode of induction being likened to that of a spring-release. Microarray studies identified *PAI-2* as an inducible gene in response to IL-5 (Bystrom *et al.*, 2004), Factor 7/Tissue factor (Camerer *et al.*, 2000), Lp(a) (Buechler *et al.*, 2001), LPS (Suzuki *et al.*, 2000), and again by TNF (Jang *et al.*, 2004).

With hindsight, it is not at all surprising that the *PAI-2* gene is induced by a wide range of growth factors, hormones, cytokines, vasoactive peptides, toxins, and tumor promoters (Dear and Medcalf, 1995; Kruithof *et al.*, 1995; Medcalf and Stasinopoulos, 2005)). However, what is still viewed as a remarkable feature of PAI-2 induction is the sheer magnitude of the effect. For example, the level of PAI-2 mRNA and protein can increase over 1000-fold in monocytes (Medcalf, 1992), fibrosarcoma cells (Maurer and Medcalf, 1996), and neurons (Zhang *et al.*, 2009) and the total synthesized protein, at least in fibroscarcoma cells, constitutes $\sim 0.27\%$

of total protein (Maurer and Medcalf, 1996). Hence, some cells have evolved a capacity to be mini-factories for PAI-2 production.

4.1. Cellular regulation of PAI-2 expression

Regulation of gene expression can occur at many levels, and the most commonly studied parameters have been at the level of transcription and posttranscriptional control. Transcriptional regulation of PAI-2 is arguably the most prominent component underpinning PAI-2 biosynthesis. However, the PAI-2 transcript is also inherently unstable, and many laboratories have provided evidence to support alterations in PAI-2 mRNA stability as critical components in the overall regulation of this serpin.

4.2. Transcriptional regulation of PAI-2

Direct evidence that the *PAI-2* gene was transcriptionally regulated was provided over 20 years ago using nuclear "run-on" transcription assays. This is a very powerful and essentially the *only* method that quantitates the relative changes in the level of a primary transcript within the nucleus of a cell, that is, before it is fully processed and exported to the cytoplasm for translation. Quantitation of gene transcription rates in isolated nuclei using the run-on transcription assay is still the method of choice to assess modulation of gene expression in the context of native chromatin. Nowadays, many groups rely on microarray profiling, but these approaches do not allow one to determine whether changes in transcript levels are a consequence of an alteration in transcription rate or a change in mRNA stability. The original run-on assay (Derman *et al.*, 1981; Greenberg and Ziff, 1984) is a reliable but a labor intensive method and recent protocols have been published based on the original procedure (Smale, 2009). A modification of this procedure is described at the end of this chapter.

Using the run-on procedure, initial studies provided direct evidence that the induction of PAI-2 expression in U-937 cells following phorbol ester treatment involved dramatic increases ($\sim$50-fold) in the rate of PAI-2 transcription (Schleuning *et al.*, 1987). Similar studies in HT-1080 fibrosarcoma cells demonstrated a significant transcriptional component of the *PAI-2* gene in response to TNF (Medcalf *et al.*, 1988). On this particular point, the *PAI-2* and *PAI-1* genes were the *first* genes described that were shown to be transcriptionally regulated by TNF.

The transcriptional responsiveness of the *PAI-2* gene subsequently led to detailed analyses of its gene promoter (Cousin *et al.*, 1991; Kruithof and Cousin, 1988). Gene promoters are the ignition system of genes and harbor regions of DNA (*cis*-acting elements) that provide binding sites for mostly nuclear proteins (transcription factors) that enhance or suppress transcriptional activation. The location of these protein binding sites was revealed using the DNas-1 protection assay, more commonly referred to as "DNA footprinting." In this approach, nuclear protein extracts are incubated with ^{32}P-dATP end-labeled DNA of the promoter of interest. The mixture is then subjected to limited digestion with the enzyme, DNase-1. Regions that harbor protein binding sites are protected from digestion and cleavage fragments are not formed. These regions are readily revealed on a DNA sequencing gel. DNase-1 protection studies revealed that the PAI-2 promoter possessed a congested arrangement of protein binding sites (cis-acting elements). Comparative studies revealed that gene promoters that displayed similar responses to certain agents also harbored similar regulatory elements in their gene promoters and consensus sequences were generated. Regulatory elements revealed in the PAI-2 promoter included AP-1-like consensus elements (AP1a: TGAATCA located between -103 and -97; AP1b: TGAGTAA located between -114 and -108; and a cAMP-responsive element (CRE)-like element: TGACCTCA located between -187 and -182; Cousin et al., 1991; Dear et al., 1997). These sites were shown to have functional activity during transcriptional regulation. For these experiments, constructs harboring the wild-type or mutated PAI-2 gene promoter fused to a reporter gene (usually luciferase) are assessed for reporter gene expression following cell transfection and subsequent treatment with the agonist of interest. The transcription factor CREB itself was shown to be a major player in the transcriptional regulation of PAI-2 as phorbol ester induction of PAI-2 in HT-1080 cells was largely inhibited in cells transfected with a plasmid expressing a dominant negative mutant of the CREB protein (Costa et al., 2000). CREB was also shown to control expression of the *PAI-2* gene in activated monocytes (Park et al., 2005).

A repressor element located between -219 and -1100 of the PAI-2 promoter was suggested to play a role during TNF induction (Dear et al., 1996) as deletion of this region enhanced the transcriptional response to TNF. The identification of the sequence within this region and trans-acting factors responsible for this activity were not reported. Antalis et al. (1996) characterized 5.1 kb of the PAI-2 promoter region in U937 cells by deletion analysis and found an additional repressive region at an upstream location. This "silencer" activity was localized to a 28-bp sequence containing a 12-bp palindrome at position -1832, CTCTCTAGAGAG, which was termed PAI-2-upstream silencer element-1 (PAUSE-1). Although more details on the sequence binding requirements of this element were defined (Ogbourne and Antalis, 2001), the mechanism by which this silencer functioned and the identification of associated binding proteins was not explored.

4.3. Epigenetics

Epigenetics is the study of inherited changes in phenotype or gene expression caused by mechanisms other than changes in the underlying DNA sequence. Indeed, epigenetics has proven to be of paramount importance in the control of gene expression. Notwithstanding the importance of regulatory DNA elements and their associated binding factors in gene transcription, chemical modification to DNA (that can be transient or maintained through cell division) can modulate transcriptional activity of genes. For example, methylation of cytosine residues in DNA to 5-methylcytosine can depress transcriptional activation of gene promoters. Hence, highly methylated genes are generally less active. Histones provide another target for epigenetic influence. Histones are positively charged proteins that interact with DNA to form nucleosomes. Histone interaction can influence the positioning of individual nucleosomes relative to regulatory sequence elements and the folding of nucleosomes into higher-order structures (Eberharter and Becker, 2002). Transcription occurs more favorably when nucleosomes are in a more open configuration. This can occur when histones–DNA interactions are weakened through histone modification via acetylation, methylation, and sumoylation among others. Of these, acetylation is the most commonly studied modification.

Evidence for an epigenetic component in PAI-2 gene expression is accumulating. PAI-2 gene expression is modulated by agents that alter the acetylation status of histones. Histone deacetylase inhibitors (i.e., trichostatin A, TSA), for example, modulate PAI-2 gene expression, and such changes have been implicated in the alteration of PAI-2 expression in some malignancies (Foltz *et al.*, 2006) and have been implicated in the changes in PAI-2 gene expression and other serpins (notably SERPINA3) in normal pregnancy and in preeclampsia (Chelbi and Vaiman, 2008; Chelbi *et al.*, 2007). Epigenetic control of SERPINE1 (PAI-1; Gao *et al.*, 2010) and some of the Ov-serpins, including Maspin (Bellido *et al.*, 2010; Ogasawara *et al.*, 2004), have also been reported, and this aspect of gene regulation of serpins is likely to attract more interest as epigenetics gains more prominence in human pathology.

4.4. mRNA stability: General principals

Posttranscriptional control of gene expression is critically important for controlling the levels of transiently induced transcripts. It is a particularly valuable process to allow for a near-immediate increase in mRNA abundance with minimal support of the transcriptional machinery. By necessity, these transiently induced transcripts have extremely short half-lives and are rapidly removed from the cell. The ability of a transcript to alter its longevity in a cell is most commonly mediated by regulatory mRNA elements located in the $3'$-untranslated region ($3'$-UTR). These elements are usually rich in adenylate (A) and uridylate (U) residues and are commonly referred to as "AU-rich elements" or AREs. The best characterized of these AREs are found in highly unstable mRNAs (e.g., cytokines, oncogenes; Garneau *et al.*, 2007). The $3'$-UTR can harbor multiple-AREs that can either interact with each other or act independently to define the fate of a transcript under constitutive conditions and in response to specific physiological states (Chen et al., 1995; Winzen et al., 2004, 2007). AREs are usually 50–100 nucleotide (nt) in length and contain single or multiple copies of a core consensus motif AUUUA, UUAUUUA (U/A)(U/A), or UUAUUUAUU embedded within a U-rich sequence (Chen and Shyu, 1995; Lagnado et al., 1994), and have been classed in three groups (groups I, II, III), depending on their particular AU-rich sequence content (Chen and Shyu, 1995). A database compiling ARE-containing transcripts predicted that ∼8% of human genes code for transcripts that contain AREs (Bakheet et al., 2006; Khabar, 2005).

AREs function as destabilizing elements by recruiting ARE binding proteins that, in turn, interact with deadenylases (that remove adenine residues from the polyA tail of mRNA) and with components of the exosome (promoting $3'$–$5'$ decay) or various components of the decapping proteins (promoting $5'$–$3'$ decay; Garneau et al., 2007). The best characterized ARE binding mRNA-destabilizing proteins include tristetraprolin (TTP), KH-splicing regulatory protein (KSRP), and AU-rich binding factor-1 (AUF-1), while the embryonic lethal abnormal vision (ELAV) proteins such as HuR (Eberhardt et al., 2007; Garneau et al., 2007) are known to stabilize a range of ARE-containing transcripts.

4.5. Posttranscriptional regulation of PAI-2 expression

Although PAI-2 induction involves substantial changes at the level of transcription, posttranscriptional events are also important in modulating its expression. Evidence for this has been provided from a number of independent laboratories dating back over 18 years. For example, phorbol ester-mediated induction of PAI-2 mRNA in PL-21 myeloid leukemia cells was coincident with an increase in the half-life of the PAI-2 transcript from 2–5 h. Similarly, in HL-60 cells, induction of PAI-2 mRNA following treatment with cycloheximide was associated with a fourfold increase in PAI-2 mRNA stability (Antalis and Dickinson, 1992; Niiya et al., 1994). Conversely, *suppression* of PAI-2 mRNA by the glucocorticoid, dexamethasone, was also shown to be associated with an acceleration in the decay rate of the PAI-2 transcript (Pytel et al., 1990). Finally, the marked increase in PAI-2 mRNA expression in response to the phosphatase inhibitor, okadaic acid (OA), was shown to be blocked by addition of the cAMP analogue, 8-bromo-cAMP. This inhibitory effect on PAI-2 mRNA expression levels was presumed to be occurring posttranscriptionally, since OA-induced increase in PAI-2 transcription was unaffected by cAMP (Medcalf, 1992).

Most of the earlier approaches used to assess changes in mRNA stability have relied on the use of general transcription inhibitors, that is, actinomycin D (Act-D) or 5,6-dichloro-β-D-ribofuranosylbenzimidazole

(DRB). However, the problem with the use of such compounds is that it assumes that the mRNA decay process itself is not dependent upon on-going transcription. For example, labile proteins that mediate mRNA decay may require on-going transcription to produce sufficient protein to implement mRNA turnover. The use of Act-D or DRB would block the synthesis of the protein(s) in question and the ensuing alteration in the mRNA decay rate. Hence, some results may be misleading. Despite the limitations in the use of general transcription inhibitors, these early findings laid the groundwork that supported the view that PAI-2 mRNA stability is a regulated process involved in both induction and downregulation of the *PAI-2* gene.

The PAI-2 transcript is inherently unstable (Maurer *et al.*, 1999; Maurer and Medcalf, 1996; Stasinopoulos *et al.*, 2010), and the vast majority of research reported to date on PAI-2 mRNA stability has been derived from studies exploring its decay rate under constitutive conditions. Whether mRNA stability *per se* is indeed modulated during PAI-2 induction or whether the posttranscriptional component of PAI-2 gene regulation was reflected elsewhere, for example, at the level of translation is still an open question. The inherent instability of the PAI-2 transcript was initially shown to require a classical nonameric ARE (UUAUUUAUU) in its $3'$-UTR (Maurer and Medcalf, 1996). This element was subsequently shown to possess a binding site for a number of intracellular binding proteins associated with mRNA metabolism including the mRNA-stabilizing protein HuR (Maurer *et al.*, 1999) and the mRNA-destabilizing protein, TTP (Yu *et al.*, 2003). In addition to the instability element in the $3'$-UTR, an mRNA instability region was also localized to a short stretch within exon 4 of the coding region. This stretch of mRNA provided a binding site for proteins of $\sim$52 kDa and other unidentified proteins as determined using RNA electrophoretic mobility shift and UV-cross-linking assays (Tierney and Medcalf, 2001), (see Fig. 6.1). Curiously, this "coding region mRNA instability determinant" was found to be present within the coding region of a number of other unstable transcripts, including c-myc, the u-PA receptor, and VEGF (Tierney and Medcalf, 2001) implying a more general feature of posttranscriptional gene expression.

4.6. Assessment of PAI-2 mRNA decay using tetracycline-regulated expression systems

To overcome the limitations of using general transcription inhibitors (i.e., Act-D or DRB), tetracycline (TET)-regulated expression vectors are now widely used to study mRNA decay rates. The TET-regulated expression system (TET-ON or TET-OFF) is a powerful genetic tool that permits the expression of any gene construct introduced into either cultured cells or transgenic animals to be precisely controlled. It requires a regulatory

Gene Regulation of PAI-2

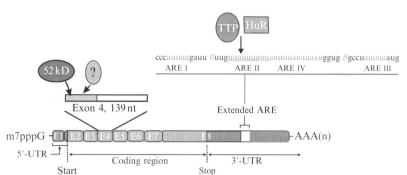

Figure 6.1 *Schematic representation of regulatory domains in the PAI-2 transcript that influence steady state mRNA levels.* The PAI-2 mRNA is encoded by eight exons (E). mRNA instability elements located in exon 4 of the PAI-2 transcript are indicated. An unidentified 52 kDa protein interacts with the first 50 nt of exon 4. Another unidentified proteins ("?") are also likely to interact with this region. The critical regulatory region in the 3′-UTR harboring the "extended ARE" involved in posttranscriptional regulation of PAI-2 is shown. A series of AREs within this region are presented in the order ARE I, II, IV, and III. Of these, ARE II containing a nonameric ARE (underlined text) is the most important ARE and provides a binding site for the mRNA-binding proteins HuR and TTP. m7pppG refers to the mRNA cap with AAA(n) denoting the adenine residues in the poly(A) tail.

component based on the prokaryotic tetracycline repressor (TetR) and a response plasmid that expresses the gene of interest under control of the TET-response element (Schonig *et al.*, 2010). Depending on the TetR binding moiety of the TET-controlled transactivator, the reporter gene can be switched on (TET-ON) or off (TET-OFF) by the addition of TET or more commonly by its derivative, doxycycline. For mRNA decay studies, the TET-OFF approach is desired as addition of doxycycline represses transcription allowing quantitation of the subsequent rate of mRNA decay over time. Using the TET-OFF approach, recent reports have indicated that the regulation of PAI-2 mRNA decay is more complex than previously thought.

4.7. The role of AU-rich instability elements in the 3′-UTR of PAI-2

Although the nonameric ARE within the 3′-UTR of PAI-2 is certainly a major player in the control of constitutive PAI-2 mRNA decay, this ARE alone was not fully responsible for destabilizing the PAI-2 transcript, since deletion or mutagenesis of this sequence only partially reversed the destabilizing capacity of the PAI-2 3′-UTR (Maurer and Medcalf, 1996;

Stasinopoulos *et al.*, 2010). Hence, other important instability elements were likely to reside in the 3′-UTR that had not been detected in earlier studies. Sequence analysis of the 3′-UTR revealed the existence of a number of potential ARE or ARE-like instability elements and were denoted as ARE I, II, III, and IV (with the ARE II being the nonamer). ARE I and III represent classical AU-rich sequences, while ARE IV was considered on sequence grounds to be atypical (Stasinopoulos *et al.*, 2010). All four AREs were located within a short 74 nt U-rich stretch of the 3′-UTR that flanked the nonameric motif.

The relative importance of these four AREs was evaluated in a systematic mutagenesis screening approach in which the wild-type full-length PAI-2 3′-UTR or mutant full-length constructs containing one or multiple ARE mutations were fused to the TET-regulated (TET-OFF) beta-globin reporter transcript. Comparisons of the decay rates of these chimeric transcripts in stably transfected HT-1080 fibrosarcoma cells following docycycline treatment revealed some redundancy in ARE usage (i.e., ARE III was dispensable); however, evidence for a cooperative interplay was revealed between ARE I, II, and IV (Stasinopoulos *et al.*, 2010). These additional elements were shown to be conserved between species and to optimize the destabilizing capacity with the nonameric element to ensure complete mRNA instability. A schematic representation of the posttranscriptional regulation of PAI-2 is provided in Fig. 6.1.

What remains to be elucidated is precisely how these mRNA decay determinants cooperate and the identification and role of associated mRNA binding proteins. Although HuR and TTP associate with the nonameric element *in vitro*, the influence of these proteins on the newly identified instability elements remains to be seen. Although much has been learned about the regulatory elements that underpin the constitutive expression of PAI-2 mRNA, another major question is the importance of these regions in the 3′-UTR during induction of PAI-2 expression (i.e., by TNF, LPS, etc.) in monocytes and other immune cells and the extent of their involvement during PAI-2 induction in neurons which remains to be determined.

5. Conclusions

PAI-2 continues to be implicated in a plethora of biological activities; the most recent additions to this growing list include its role in the innate immune response and as a neuroprotective agent. From a classical serpin viewpoint, the only genuinely convincing role to date has been its ability to inhibit u-PA, yet for the most part, this is restricted to *in vitro* observations. A genuine biological role for PAI-2 as a regulator of the proteolytic activity of the plasminogen activators is presumed, but has not been rigorously proven. The growing body of PAI-2 associated intracellular events is pointing PAI-2 into many directions creating ambiguity. Evolutionary biologists may argue against the principle of any one protein having more than one basic function, and a unifying mechanism is now needed that can account for these apparently unrelated events. The unprecedented magnitude of response of the *PAI-2* gene to toxins and cytokines has provided strong circumstantial evidence to link PAI-2 with inflammation and tissue repair and as a general stress-responsive gene. This feature is also consistent with a role for PAI-2 in the innate immune response. The impressive scale of regulation of this Ov-serpin has also provided a relevant model gene/ transcript to explore basic mechanisms and concepts of gene regulation, at the level of both transcription and posttranscription, which has, up until now, been immensely rewarding.

6. METHODOLOGY: RAPID RUN-ON TRANSCRIPTION ASSAY PROTOCOL

In this assay, primary transcripts that are in the process of elongation are captured in isolated nuclei due to the depletion of ribonucleotide substrates. Hence, transcripts are "stalled" at a certain point in time on DNA but transcriptional elongation can be continued and allowed to finish by supplying fresh ribonucleotides in the presence of a radioactive tracer. Initiation of transcription however cannot occur in isolated nuclei.

The method outlined below, referred to as the "rapid run-on" method, uses a similar approach as the original protocol (Greenberg and Ziff, 1984) to label nascent RNA in isolated nuclei, but incorporates the acid phenol (nonbuffered phenol) procedure (Chomczynski and Sacchi, 1987) to separate RNA from DNA.

1. Nuclei from $\sim 10^7$ cells are isolated by disrupting the cell membrane with 1 ml of 0.5% NP-40 lysis solution (10 mM Tris–HCl, pH 7.4; 10 mM NaCl; 3 mM MgCl$_2$; 1 mM EDTA; 1 mM PMSF; 1 mM DTT; and 0.5% NP-40). The concentration of NP-40 needs to be determined empirically, particularly if using more fragile nonadherent cells. NP-40 (0.2%) is recommended for monocytic cells (i.e., U937 cells).
2. Cells are mixed, left on ice for 5 min, and then centrifuged for 10 s at full speed at 4 °C in a standard microcentrifuge. After removing the supernatant, the nuclear pellet is resuspended in 1 ml of the same NP-40 lysis buffer and again centrifuged to remove contaminating cytoplasmic material. The supernatant is *completely* removed and the nuclear pellet resuspended in 120 μl per 10^7 cells of nuclear storage buffer (50 mM Tris–HCl, pH 8, 3; 5 mM MgCl$_2$; 0.1 mM EDTA; 40%

glycerol) and kept as 100 μl aliquots. Isolated nuclei can be used immediately or snap frozen in liquid nitrogen and kept at $-80\,^\circ$C.

3. *In vitro* transcription in isolated nuclei is performed by adding to each aliquot of nuclei, 100 μl of 2× reaction buffer (10 mM Tris–HCl, pH 8.0; 5 mM MgCl$_2$; 300 mM KCl; 1 mM DTT; 1 mM each of ATP, CTP, GTP; 6 U RNase inhibitor) containing 10 μl of ^{32}P-UTP (3000 Ci/mmol) then placed in a 30 $^\circ$C water bath for 30 min to allow elongation of initiated nascent transcripts to continue.

4. To disrupt chromatin and release most of the RNA, 600 μl of high salt buffer (0.5 M NaCl; 50 mM MgCl$_2$; 2 mM CaCl$_2$; 10 mM Tris–HCl, pH 7.4 containing 40 μg/ml DNase-1) is added, and the samples are incubated at 30 $^\circ$C for 5 min. Two hundred microliters of SDS buffer (5% SDS; 0.5 M Tris–HCl, pH 7.4; 0.125 M EDTA containing 10 μl of proteinase K [20 mg/ml]) is added, and the samples are incubated at 37 $^\circ$C for 30 min.

5. The next step is to isolate RNA from the nuclei in the most efficient manner. In the original description of the procedure, this also involved additional DNase 1 and proteinase K digestions. These steps are no longer required as they have been replaced with the nonbuffered phenol extraction procedure. Samples (∼1 ml) are transferred to 10 ml sterile plastic tubes. One hundred microliters of 2 M sodium acetate and 1 ml of nonbuffered phenol "acid phenol" are then added, and the samples vortexed. Two hundred microliters of chloroform–isoamylalcohol (25:1) is added, and samples vortexed hard for at least 15 s, then centrifuged at 3000 rpm for 5 min and the supernatant transferred to fresh 10 ml plastic tubes. The phenol phase of the first 10 ml tube is "back extracted" by adding 2 ml of TE, vortexed, and centrifuged for 5 min at 3000 rpm. The upper phase is removed and added to the first supernatant solution to give a total volume of ∼3 ml extracted material.

At this stage, the labeled RNA can be precipitated with ethanol (add 300 μl of 3 M sodium acetate and 8 ml 100% ethanol), and the protocol continued at step 9 below. However, if background signal proves to be a concern, then the offending unincorporated material can be removed as described in steps 6–9 below:

6. To remove unincorporated ^{32}P-UTP, 3 ml of 10% TCA containing 60 mM sodium pyrophosphate is added and the samples left on ice for 30 min. TCA precipitates the RNA, and in doing so, the samples may appear slightly cloudy. Unincorporated ^{32}P-UTP is removed by passing the samples through a Millipore type HA (0.45 μm) filter disk using a Millipore "Swinnex" housing apparatus (cat no: SX002500) and a 10-ml disposable syringe. RNA will be retained on the top of the filter, while the flow through is collected into the same tube from which it was removed. Filters are then washed with 10 ml of 5% TCA and 30 mM

sodium pyrophosphate. The assembly can now be unscrewed and the filter containing the labeled RNA removed and placed RNA-side facing up into a sterile scintillation vial.

7. To remove the bound RNA from the filter, 1.5 ml of RNA elution buffer (1 mM Tris–HCl, pH 7.4; 5 mM EDTA) is added to the scintillation vial and the samples incubated at 65 °C for 5 min. The eluted material is added to a sterile 10 ml plastic tube. Approximately 80% of the RNA is obtained during this step. The RNA remaining on the filter is recovered by adding 750 µl of the RNA elution buffer and incubated at 65 °C for 5 min. The solution is removed and added to the first collected material. This step is repeated one more time giving a total volume of 3 ml of eluted material.

8. Samples are subjected to chloroform extraction (3 ml) and centrifuged for 5 min at 3000 rpm. The supernatant is removed and labeled RNA precipitated by adding 300 µl 3 M sodium acetate and 8 ml 100% ethanol. Samples can be left overnight at $-20°$ or placed at $-80°$ for 1 h.

9. The 10-ml RNA/ethanol solution is transferred to 30 ml siliconized Corex glass tubes and centrifuged at 9000 rpm for 20 min at 4 °C. The supernatant is removed and the radioactive pellet dissolved in 100 µl of TE and transferred to an eppendorf tube and placed on ice. RNA remaining in the Corex tube is recovered by adding another 100 µl of TE to the tube and rotating the TE around the inner surface. This washing step is repeated to give a total volume of recovered RNA of 300 µl. Samples are then precipitated with ethanol (800 µl 100% ethanol plus 30 µl 3 M sodium acetate) and placed at -80 °C for 30 min.

Hybridization procedure

10. After centrifugation of the RNA samples, the supernatant is removed and the pellet washed with 70% ethanol and dissolved in 300 µl hybridization buffer (50 mM HEPES, pH 7.4; 300 mM NaCl; 0.2% SDS; 200 µg/ml denatured salmon sperm DNA; 1× Denhardts [without BSA]). It is important to make sure that all of the RNA is dissolved. To facilitate this process, the RNA can be initially dissolved in $\sim$20 µl of sterile water before adding the hybridization buffer. If required, the radioactivity in multiple samples can be quantitated and normalized (highly recommended).

Preparation of immobilized and hybridization

11. Plasmid or genomic DNA- or single-stranded sequences (preferred) containing the genes of interest are linearized (if plasmid or DNA) with the appropriate restriction enzyme or denatured by boiling in the presence of 100 mM NaOH (if plasmid or DNA) and neutralized in 5.0 ml of 6× SSC. DNA is immobilized onto nitrocellulose (2 µg/slot)

following standard procedures using a slot blot apparatus (i.e., HYBRI-SLOT™ manifold Bethesda Research Laboratories, USA or equivalent) then baked at 80 °C for 2 h. Individual filter strips containing the immobilized DNA are cut out[1] and incubated in prehybridization buffer (50 mM HEPES, pH 7.4; 300 mM NaCl; 10 mM EDTA; 0.2% SDS; 1000 μg/ml denatured salmon sperm DNA; 5× Denhardts [without BSA]) for at least 1 h at 65 °C. Strips are then removed and added to the labeled RNA in hybridization buffer. Hybridization is performed at 65 °C in a water bath for up to 36 h.
12. After hybridization, filter strips are washed at 65 °C with 2× SSC/0.1% SDS for 30 min, then with 2× SSC alone for up to 2 h with at least three changes of the washing buffer. Filters are then treated with RNase A (10 μg/ml in 2× SSC) for 30 min at 37 °C, then washed at this temperature for another 30 min in 2× SSC. Filter strips are finally placed DNA side facing up on paper using tape, covered with plastic wrapping and exposed to X-ray film with an intensifying screen.

ACKNOWLEDGMENTS

The author would like to thank colleagues at the Australian Centre for Blood Diseases at Monash University for critical reading of this chapter. The authors' laboratory is supported with grants obtained from the National Health and Medical Research Council (NHMRC) of Australia.

REFERENCES

Antalis, T. M., and Dickinson, J. L. (1992). Control of plasminogen-activator inhibitor type 2 gene expression in the differentiation of monocytic cells. *Eur. J. Biochem.* **205,** 203–209.

Antalis, T. M., Clark, M. A., Barnes, T., Lehrbach, P. R., Devine, P. L., Schevzov, G., Goss, N. H., Stephens, R. W., and Tolstoshev, P. (1988). Cloning and expression of a cDNA coding for a human monocyte-derived plasminogen activator inhibitor. *Proc. Natl. Acad. Sci. USA* **85,** 985–989.

Antalis, T. M., Costelloe, E., Muddiman, J., Ogbourne, S., and Donnan, K. (1996). Regulation of the plasminogen activator inhibitor type-2 gene in monocytes: Localization of an upstream transcriptional silencer. *Blood* **88,** 3686–3697.

Argraves, K. M., Battey, F. D., MacCalman, C. D., McCrae, K. R., Gafvels, M., Kozarsky, K. F., Chappell, D. A., Strauss, J. F., 3rd, and Strickland, D. K. (1995). The very low density lipoprotein receptor mediates the cellular catabolism of lipoprotein lipase and urokinase-plasminogen activator inhibitor type I complexes. *J. Biol. Chem.* **270,** 26550–26557.

---

[1] Strips are usually cut 0.8 × 0.3 cm, depending on the position of the immobilized DNA. It is also advisable to mark the side of the filter that contains the DNA. If multiple strips are to be cohybridized (this can be done with approximately five different strips), it is recommended to appropriately label each strip. A common means is to cut each strip in a different manner.

Bakheet, T., Williams, B. R., and Khabar, K. S. (2006). ARED 3.0: The large and diverse AU-rich transcriptome. *Nucleic Acids Res.* **34**, D111–D114.
Bass, R., Fernandez, A. M., and Ellis, V. (2002). Maspin inhibits cell migration in the absence of protease inhibitory activity. *J. Biol. Chem.* **277**, 46845–46848.
Belin, D. (1993). Biology and facultative secretion of plasminogen activator inhibitor-2. *Thromb. Haemost.* **70**, 144–147.
Bellido, M. L., Radpour, R., Lapaire, O., De Bie, I., Hosli, I., Bitzer, J., Hmadcha, A., Zhong, X. Y., and Holzgreve, W. (2010). MALDI-TOF mass array analysis of RASSF1A and SERPINB5 methylation patterns in human placenta and plasma. *Biol. Reprod.* **82**, 745–750.
Black, S., Yu, H., Lee, J., Sachchithananthan, M., and Medcalf, R. L. (2001). Physiologic concentrations of magnesium and placental apoptosis: Prevention by antioxidants. *Obstet. Gynecol.* **98**, 319–324.
Borstnar, S., Vrhovec, I., Svetic, B., and Cufer, T. (2002). Prognostic value of the urokinase-type plasminogen activator, and its inhibitors and receptor in breast cancer patients. *Clin. Breast Cancer* **3**, 138–146.
Buechler, C., Ullrich, H., Ritter, M., Porsch-Oezcueruemez, M., Lackner, K. J., Barlage, S., Friedrich, S. O., Kostner, G. M., and Schmitz, G. (2001). Lipoprotein (a) up-regulates the expression of the plasminogen activator inhibitor 2 in human blood monocytes. *Blood* **97**, 981–986.
Bystrom, J., Wynn, T. A., Domachowske, J. B., and Rosenberg, H. F. (2004). Gene microarray analysis reveals interleukin-5-dependent transcriptional targets in mouse bone marrow. *Blood* **103**, 868–877.
Camerer, E., Gjernes, E., Wiiger, M., Pringle, S., and Prydz, H. (2000). Binding of factor VIIa to tissue factor on keratinocytes induces gene expression. *J. Biol. Chem.* **275**, 6580–6585.
Cella, N., Contreras, A., Latha, K., Rosen, J. M., and Zhang, M. (2006). Maspin is physically associated with [beta]1 integrin regulating cell adhesion in mammary epithelial cells. *FASEB J.* **20**, 1510–1512.
Cesarman-Maus, G., and Hajjar, K. A. (2005). Molecular mechanisms of fibrinolysis. *Br. J. Haematol.* **129**, 307–321.
Chelbi, S. T., and Vaiman, D. (2008). Genetic and epigenetic factors contribute to the onset of preeclampsia. *Mol. Cell. Endocrinol.* **282**, 120–129.
Chelbi, S. T., Mondon, F., Jammes, H., Buffat, C., Mignot, T. M., Tost, J., Busato, F., Gut, I., Rebourcet, R., Laissue, P., Tsatsaris, V., Goffinet, F., *et al.* (2007). Expressional and epigenetic alterations of placental serine protease inhibitors: SERPINA3 is a potential marker of preeclampsia. *Hypertension* **49**, 76–83.
Chen, C. Y., and Shyu, A. B. (1995). AU-rich elements: Characterization and importance in mRNA degradation. *Trends Biochem. Sci.* **20**, 465–470.
Chen, C. Y., Xu, N., and Shyu, A. B. (1995). mRNA decay mediated by two distinct AU-rich elements from c-fos and granulocyte-macrophage colony-stimulating factor transcripts: Different deadenylation kinetics and uncoupling from translation. *Mol. Cell. Biol.* **15**, 5777–5788.
Chomczynski, P., and Sacchi, N. (1987). Single-step method of RNA isolation by acid guanidinium thiocyanate-phenol-chloroform extraction. *Anal. Biochem.* **162**, 156–159.
Costa, M., Shen, Y., and Medcalf, R. L. (2000). Overexpression of a dominant negative CREB protein in HT-1080 cells selectively disrupts plasminogen activator inhibitor type 2 but not tissue-type plasminogen activator gene expression. *FEBS Lett.* **482**, 75–80.
Cousin, E., Medcalf, R. L., Bergonzelli, G. E., and Kruithof, E. K. (1991). Regulatory elements involved in constitutive and phorbol ester-inducible expression of the plasminogen activator inhibitor type 2 gene promoter. *Nucleic Acids Res.* **19**, 3881–3886.

Croucher, D. R., Saunders, D. N., Stillfried, G. E., and Ranson, M. (2007). A structural basis for differential cell signalling by PAI-1 and PAI-2 in breast cancer cells. *Biochem. J.* **408,** 203–210.

Csutak, A., Silver, D. M., Tozser, J., Steiber, Z., Bagossi, P., Hassan, Z., and Berta, A. (2008). Plasminogen activator inhibitor in human tears after laser refractive surgery. *J. Cataract Refract. Surg.* **34,** 897–901.

Darnell, G. A., Antalis, T. M., Johnstone, R. W., Stringer, B. W., Ogbourne, S. M., Harrich, D., and Suhrbier, A. (2003). Inhibition of retinoblastoma protein degradation by interaction with the serpin plasminogen activator inhibitor 2 via a novel consensus motif. *Mol. Cell. Biol.* **23,** 6520–6532.

Davis, R. L., Shrimpton, A. E., Holohan, P. D., Bradshaw, C., Feiglin, D., Collins, G. H., Sonderegger, P., Kinter, J., Becker, L. M., Lacbawan, F., Krasnewich, D., Muenke, M., *et al.* (1999). Familial dementia caused by polymerization of mutant neuroserpin. *Nature* **401,** 376–379.

Dear, A. E., and Medcalf, R. L. (1995). The cellular and molecular of plasminogen activator inhibitor type-2. *Fibrinolysis* **9,** 321–330.

Dear, A. E., Shen, Y., Ruegg, M., and Medcalf, R. L. (1996). Molecular mechanisms governing tumor-necrosis-factor-mediated regulation of plasminogen-activator inhibitor type-2 gene expression. *Eur. J. Biochem.* **241,** 93–100.

Dear, A. E., Costa, M., and Medcalf, R. L. (1997). Urokinase-mediated transactivation of the plasminogen activator inhibitor type 2 (PAI-2) gene promoter in HT-1080 cells utilises AP-1 binding sites and potentiates phorbol ester-mediated induction of endogenous PAI-2 mRNA. *FEBS Lett.* **402,** 265–272.

Derman, E., Krauter, K., Walling, L., Weinberger, C., Ray, M., and Darnell, J. E., Jr. (1981). Transcriptional control in the production of liver-specific mRNAs. *Cell* **23,** 731–739.

Dickinson, J. L., Bates, E. J., Ferrante, A., and Antalis, T. M. (1995). Plasminogen activator inhibitor type 2 inhibits tumor necrosis factor alpha-induced apoptosis. Evidence for an alternate biological function. *J. Biol. Chem.* **270,** 27894–27904.

Dougherty, K. M., Pearson, J. M., Yang, A. Y., Westrick, R. J., Baker, M. S., and Ginsburg, D. (1999). The plasminogen activator inhibitor-2 gene is not required for normal murine development or survival. *Proc. Natl. Acad. Sci. USA* **96,** 686–691.

Duggan, C., Kennedy, S., Kramer, M. D., Barnes, C., Elvin, P., McDermott, E., O'Higgins, N., and Duffy, M. J. (1997). Plasminogen activator inhibitor type 2 in breast cancer. *Br. J. Cancer* **76,** 622–627.

Eberhardt, W., Doller, A., Akool, el-S., and Pfeilschifter, J. (2007). Modulation of mRNA stability as a novel therapeutic approach. *Pharmacol. Ther.* **114,** 56–73.

Eberharter, A., and Becker, P. B. (2002). Histone acetylation: A switch between repressive and permissive chromatin. Second in review series on chromatin dynamics. *EMBO Rep.* **3,** 224–229.

Fan, J., Zhang, Y. Q., Li, P., Hou, M., Tan, L., Wang, X., and Zhu, Y. S. (2004). Interaction of plasminogen activator inhibitor-2 and proteasome subunit, beta type 1. *Acta Biochim. Biophys. Sin. (Shanghai)* **36,** 42–46.

Fish, R. J., and Kruithof, E. K. (2006). Evidence for serpinB2-independent protection from TNF-alpha-induced apoptosis. *Exp. Cell Res.* **312,** 350–361.

Foekens, J. A., Buessecker, F., Peters, H. A., Krainick, U., van Putten, W. L., Look, M. P., Klijn, J. G., and Kramer, M. D. (1995). Plasminogen activator inhibitor-2: Prognostic relevance in 1012 patients with primary breast cancer. *Cancer Res.* **55,** 1423–1427.

Foltz, G., Ryu, G. Y., Yoon, J. G., Nelson, T., Fahey, J., Frakes, A., Lee, H., Field, L., Zander, K., Sibenaller, Z., Ryken, T. C., Vibhakar, R., *et al.* (2006). Genome-wide analysis of epigenetic silencing identifies BEX1 and BEX2 as candidate tumor suppressor genes in malignant glioma. *Cancer Res.* **66,** 6665–6674.

Gao, S., Nielsen, B. S., Krogdahl, A., Sorensen, J. A., Tagesen, J., Dabelsteen, S., Dabelsteen, E., and Andreasen, P. A. (2010). Epigenetic alterations of the SERPINE1 gene in oral squamous cell carcinomas and normal oral mucosa. *Genes Chromosom. Cancer* **49**, 526–538.

Garneau, N. L., Wilusz, J., and Wilusz, C. J. (2007). The highways and byways of mRNA decay. *Nat. Rev. Mol. Cell Biol.* **8**, 113–126.

Genton, C., Kruithof, E. K., and Schleuning, W. D. (1987). Phorbol ester induces the biosynthesis of glycosylated and nonglycosylated plasminogen activator inhibitor 2 in high excess over urokinase-type plasminogen activator in human U-937 lymphoma cells. *J. Cell Biol.* **104**, 705–712.

Gohl, G., Lehmkoster, T., Munzel, P. A., Schrenk, D., Viebahn, R., and Bock, K. W. (1996). TCDD-inducible plasminogen activator inhibitor type 2 (PAI-2) in human hepatocytes, HepG2 and monocytic U937 cells. *Carcinogenesis* **17**, 443–449.

Greenberg, M. E., and Ziff, E. B. (1984). Stimulation of 3T3 cells induces transcription of the c-fos proto-oncogene. *Nature* **311**, 433–438.

Greten, F. R., Arkan, M. C., Bollrath, J., Hsu, L. C., Goode, J., Miething, C., Goktuna, S. I., Neuenhahn, M., Fierer, J., Paxian, S., Van Rooijen, N., Xu, Y., *et al.* (2007). NF-kappaB is a negative regulator of IL-1beta secretion as revealed by genetic and pharmacological inhibition of IKKbeta. *Cell* **130**, 918–931.

Gross, T. J., and Sitrin, R. G. (1990). The THP-1 cell line is a urokinase-secreting mononuclear phagocyte with a novel defect in the production of plasminogen activator inhibitor-2. *J. Immunol.* **144**, 1873–1879.

Harbour, J. W., and Dean, D. C. (2000). Rb function in cell-cycle regulation and apoptosis. *Nat. Cell Biol.* **2**, E65–E67.

Harrop, S. J., Jankova, L., Coles, M., Jardine, D., Whittaker, J. S., Gould, A. R., Meister, A., King, G. C., Mabbutt, B. C., and Curmi, P. M. (1999). The crystal structure of plasminogen activator inhibitor 2 at 2.0 Å resolution: Implications for serpin function. *Structure* **7**, 43–54.

Hasina, R., Hulett, K., Bicciato, S., Di Bello, C., Petruzzelli, G. J., and Lingen, M. W. (2003). Plasminogen activator inhibitor-2: A molecular biomarker for head and neck cancer progression. *Cancer Res.* **63**, 555–559.

Herz, J., Clouthier, D. E., and Hammer, R. E. (1992). LDL receptor-related protein internalizes and degrades uPA-PAI-1 complexes and is essential for embryo implantation. *Cell* **71**, 411–421.

Hibino, T., Matsuda, Y., Takahashi, T., and Goetinck, P. F. (1999). Suppression of keratinocyte proliferation by plasminogen activator inhibitor-2. *J. Invest. Dermatol.* **112**, 85–90.

Hofmann, G. E., Glatstein, I., Schatz, F., Heller, D., and Deligdisch, L. (1994). Immunohistochemical localization of urokinase-type plasminogen activator and the plasminogen activator inhibitors 1 and 2 in early human implantation sites. *Am. J. Obstet. Gynecol.* **170**, 671–676.

Izuhara, K., Ohta, S., Kanaji, S., Shiraishi, H., and Arima, K. (2008). Recent progress in understanding the diversity of the human ov-serpin/clade B serpin family. *Cell. Mol. Life Sci.* **65**, 2541–2553.

Jang, W. G., Kim, H. S., Park, K. G., Park, Y. B., Yoon, K. H., Han, S. W., Hur, S. H., Park, K. S., and Lee, I. K. (2004). Analysis of proteome and transcriptome of tumor necrosis factor alpha stimulated vascular smooth muscle cells with or without alpha lipoic acid. *Proteomics* **4**, 3383–3393.

Jankova, L., Harrop, S. J., Saunders, D. N., Andrews, J. L., Bertram, K. C., Gould, A. R., Baker, M. S., and Curmi, P. M. (2001). Crystal structure of the complex of plasminogen activator inhibitor 2 with a peptide mimicking the reactive center loop. *J. Biol. Chem.* **276**, 43374–43382.

Jensen, P. H., Lorand, L., Ebbesen, P., and Gliemann, J. (1993). Type-2 plasminogen-activator inhibitor is a substrate for trophoblast transglutaminase and factor XIIIa. Transglutaminase-catalyzed cross-linking to cellular and extracellular structures. *Eur. J. Biochem.* **214,** 141–146.

Jensen, P. H., Cressey, L. I., Gjertsen, B. T., Madsen, P., Mellgren, G., Hokland, P., Gliemann, J., Doskeland, S. O., Lanotte, M., and Vintermyr, O. K. (1994a). Cleaved intracellular plasminogen activator inhibitor 2 in human myeloleukaemia cells is a marker of apoptosis. *Br. J. Cancer* **70,** 834–840.

Jensen, P. H., Schuler, E., Woodrow, G., Richardson, M., Goss, N., Hojrup, P., Petersen, T. E., and Rasmussen, L. K. (1994b). A unique interhelical insertion in plasminogen activator inhibitor-2 contains three glutamines, Gln83, Gln84, Gln86, essential for transglutaminase-mediated cross-linking. *J. Biol. Chem.* **269,** 15394–15398.

Jensen, P. H., Jensen, T. G., Laug, W. E., Hager, H., Gliemann, J., and Pepinsky, B. (1996). The exon 3 encoded sequence of the intracellular serine proteinase inhibitor plasminogen activator inhibitor 2 is a protein binding domain. *J. Biol. Chem.* **271,** 26892–26899.

Kaiserman, D., Whisstock, J. C., and Bird, P. I. (2006). Mechanisms of serpin dysfunction in disease. *Expert Rev. Mol. Med.* **8,** 1–19.

Kardesler, L., Buduneli, N., Biyikoglu, B., Cetinkalp, S., and Kutukculer, N. (2008). Gingival crevicular fluid PGE2, IL-1beta, t-PA, PAI-2 levels in type 2 diabetes and relationship with periodontal disease. *Clin. Biochem.* **41,** 863–868.

Katsikis, J., Yu, H., Maurer, F., and Medcalf, R. (2000). The molecular basis for the aberrant production of plasminogen activator inhibitor type 2 in THP-1 monocytes. *Thromb. Haemost.* **84,** 468–473.

Kawano, T., Morimoto, K., and Uemura, Y. (1970). Partial purification and properties of urokinase inhibitor from human placenta. *J. Biochem. (Tokyo)* **67,** 333–342.

Khabar, K. S. (2005). The AU-rich transcriptome: More than interferons and cytokines, and its role in disease. *J. Interferon Cytokine Res.* **25,** 1–10.

Kruithof, E. K., and Cousin, E. (1988). Plasminogen activator inhibitor 2. Isolation and characterization of the promoter region of the gene. *Biochem. Biophys. Res. Commun.* **156,** 383–388.

Kruithof, E. K., Vassalli, J. D., Schleuning, W. D., Mattaliano, R. J., and Bachmann, F. (1986). Purification and characterization of a plasminogen activator inhibitor from the histiocytic lymphoma cell line U-937. *J. Biol. Chem.* **261,** 11207–11213.

Kruithof, E. K., Tran-Thang, C., Gudinchet, A., Hauert, J., Nicoloso, G., Genton, C., Welti, H., and Bachmann, F. (1987). Fibrinolysis in pregnancy: A study of plasminogen activator inhibitors. *Blood* **69,** 460–466.

Kruithof, E. K., Baker, M. S., and Bunn, C. L. (1995). Biological and clinical aspects of plasminogen activator inhibitor type 2. *Blood* **86,** 4007–4024.

Kumar, S., and Baglioni, C. (1991). Protection from tumor necrosis factor-mediated cytolysis by overexpression of plasminogen activator inhibitor type-2. *J. Biol. Chem.* **266,** 20960–20964.

Lagnado, C. A., Brown, C. Y., and Goodall, G. J. (1994). AUUUA is not sufficient to promote poly(A) shortening and degradation of an mRNA: The functional sequence within AU-rich elements may be UUAUUUA(U/A)(U/A). *Mol. Cell. Biol.* **14,** 7984–7995.

Laug, W. E., Cao, X. R., Yu, Y. B., Shimada, H., and Kruithof, E. K. (1993). Inhibition of invasion of HT1080 sarcoma cells expressing recombinant plasminogen activator inhibitor 2. *Cancer Res.* **53,** 6051–6057.

Lee, J. A., Croucher, D. R., and Ranson, M. (2010). Differential endocytosis of tissue plasminogen activator by serpins PAI-1 and PAI-2 on human peripheral blood monocytes. *Thromb. Haemost.* **104,** 1133–1142.

Lijnen, H. R., Frederix, L., and Scroyen, I. (2007). Deficiency of plasminogen activator inhibitor-2 impairs nutritionally induced murine adipose tissue development. *J. Thromb. Haemost.* **5,** 2259–2265.

Lyons-Giordano, B., Loskutoff, D., Chen, C. S., Lazarus, G., Keeton, M., and Jensen, P. J. (1994). Expression of plasminogen activator inhibitor type 2 in normal and psoriatic epidermis. *Histochemistry* **101,** 105–112.

Major, L., Schroder, W. A., Gardner, J., Fish, R. J., and Suhrbier, A. (2011). Human papilloma virus transformed CaSki cells constitutively express high levels of functional SerpinB2. *Exp. Cell. Res.* **317,** 338–347.

Masos, T., and Miskin, R. (1997). mRNAs encoding urokinase-type plasminogen activator and plasminogen activator inhibitor-1 are elevated in the mouse brain following kainate-mediated excitation. *Brain Res. Mol. Brain Res.* **47,** 157–169.

Matias, A. C., Ramos, P. C., and Dohmen, R. J. (2010). Chaperone-assisted assembly of the proteasome core particle. *Biochem. Soc. Trans.* **38,** 29–33.

Maurer, F., and Medcalf, R. L. (1996). Plasminogen activator inhibitor type 2 gene induction by tumor necrosis factor and phorbol ester involves transcriptional and post-transcriptional events. Identification of a functional nonameric AU-rich motif in the 3'-untranslated region. *J. Biol. Chem.* **271,** 26074–26080.

Maurer, F., Tierney, M., and Medcalf, R. L. (1999). An AU-rich sequence in the 3'-UTR of plasminogen activator inhibitor type 2 (PAI-2) mRNA promotes PAI-2 mRNA decay and provides a binding site for nuclear HuR. *Nucleic Acids Res.* **27,** 1664–1673.

Medcalf, R. L. (1992). Cell- and gene-specific interactions between signal transduction pathways revealed by okadaic acid. Studies on the plasminogen activating system. *J. Biol. Chem.* **267,** 12220–12226.

Medcalf, R. L. (2007). Fibrinolysis, inflammation, and regulation of the plasminogen activating system. *J. Thromb. Haemost.* **5**(Suppl 1), 132–142.

Medcalf, R. L., and Stasinopoulos, S. J. (2005). The undecided serpin. The ins and outs of plasminogen activator inhibitor type 2. *FEBS J.* **272,** 4858–4867.

Medcalf, R. L., Kruithof, E. K., and Schleuning, W. D. (1988). Plasminogen activator inhibitor 1 and 2 are tumor necrosis factor/cachectin-responsive genes. *J. Exp. Med.* **168,** 751–759.

Mikus, P., and Ny, T. (1996). Intracellular polymerization of the serpin plasminogen activator inhibitor type 2. *J. Biol. Chem.* **271,** 10048–10053.

Mikus, P., Urano, T., Liljestrom, P., and Ny, T. (1993). Plasminogen-activator inhibitor type 2 (PAI-2) is a spontaneously polymerising SERPIN. Biochemical characterisation of the recombinant intracellular and extracellular forms. *Eur. J. Biochem.* **218,** 1071–1082.

Mueller, B. M., Yu, Y. B., and Laug, W. E. (1995). Overexpression of plasminogen activator inhibitor 2 in human melanoma cells inhibits spontaneous metastasis in scid/scid mice. *Proc. Natl. Acad. Sci. USA* **92,** 205–209.

Narita, M., Bu, G., Herz, J., and Schwartz, A. L. (1995). Two receptor systems are involved in the plasma clearance of tissue-type plasminogen activator (t-PA) in vivo. *J. Clin. Invest.* **96,** 1164–1168.

Nguyen, G., Self, S. J., Camani, C., and Kruithof, E. K. (1992). Demonstration of a specific clearance receptor for tissue-type plasminogen activator on rat Novikoff hepatoma cells. *J. Biol. Chem.* **267,** 6249–6256.

Niiya, K., Taniguchi, T., Shinbo, M., Ishikawa, T., Tazawa, S., Hayakawa, Y., and Sakuragawa, N. (1994). Different regulation of plasminogen activator inhibitor 2 gene expression by phorbol ester and cAMP in human myeloid leukemia cell line PL-21. *Thromb. Haemost.* **72,** 92–97.

Ogasawara, S., Maesawa, C., Yamamoto, M., Akiyama, Y., Wada, K., Fujisawa, K., Higuchi, T., Tomisawa, Y., Sato, N., Endo, S., Saito, K., and Masuda, T. (2004).

Disruption of cell-type-specific methylation at the Maspin gene promoter is frequently involved in undifferentiated thyroid cancers. *Oncogene* **23**, 1117–1124.

Ogbourne, S. M., and Antalis, T. M. (2001). Characterisation of PAUSE-1, a powerful silencer in the human plasminogen activator inhibitor type 2 gene promoter. *Nucleic Acids Res.* **29**, 3919–3927.

Oji, V., Oji, M. E., Adamini, N., Walker, T., Aufenvenne, K., Raghunath, M., and Traupe, H. (2006). Plasminogen activator inhibitor-2 is expressed in different types of congenital ichthyosis: In vivo evidence for its cross-linking into the cornified cell envelope by transglutaminase-1. *Br. J. Dermatol.* **154**, 860–867.

Olofsson, A., Matsson, L., and Kinnby, B. (2002). Plasminogen activating capacity in gingival fluid from deteriorating and stable periodontal pockets. *J. Periodontal Res.* **37**, 60–65.

Orth, K., Madison, E. L., Gething, M. J., Sambrook, J. F., and Herz, J. (1992). Complexes of tissue-type plasminogen activator and its serpin inhibitor plasminogen-activator inhibitor type 1 are internalized by means of the low density lipoprotein receptor-related protein/alpha 2-macroglobulin receptor. *Proc. Natl. Acad. Sci. USA* **89**, 7422–7426.

Otter, M., Barrett-Bergshoeff, M. M., and Rijken, D. C. (1991). Binding of tissue-type plasminogen activator by the mannose receptor. *J. Biol. Chem.* **266**, 13931–13935.

Park, J. M., Greten, F. R., Wong, A., Westrick, R. J., Arthur, J. S., Otsu, K., Hoffmann, A., Montminy, M., and Karin, M. (2005). Signaling pathways and genes that inhibit pathogen-induced macrophage apoptosis—CREB and NF-kappaB as key regulators. *Immunity* **23**, 319–329.

Pemberton, P. A., Wong, D. T., Gibson, H. L., Kiefer, M. C., Fitzpatrick, P. A., Sager, R., and Barr, P. J. (1995). The tumor suppressor maspin does not undergo the stressed to relaxed transition or inhibit trypsin-like serine proteases. Evidence that maspin is not a protease inhibitory serpin. *J. Biol. Chem.* **270**, 15832–15837.

Pytel, B. A., Peppel, K., and Baglioni, C. (1990). Plasminogen activator inhibitor type-2 is a major protein induced in human fibroblasts and SK-MEL-109 melanoma cells by tumor necrosis factor. *J. Cell. Physiol.* **144**, 416–422.

Qin, L., and Zhang, M. (2010). Maspin regulates endothelial cell adhesion and migration through an integrin signaling pathway. *J. Biol. Chem.* **285**, 32360–32369.

Radtke, K. P., Wenz, K. H., and Heimburger, N. (1990). Isolation of plasminogen activator inhibitor-2 (PAI-2) from human placenta. Evidence for vitronectin/PAI-2 complexes in human placenta extract. *Biol. Chem. Hoppe Seyler* **371**, 1119–1127.

Reith, A., Booth, N. A., Moore, N. R., Cruickshank, D. J., and Bennett, B. (1993). Plasminogen activator inhibitors (PAI-1 and PAI-2) in normal pregnancies, pre-eclampsia and hydatidiform mole. *Br. J. Obstet. Gynaecol.* **100**, 370–374.

Ritchie, H., Lawrie, L. C., Crombie, P. W., Mosesson, M. W., and Booth, N. A. (2000). Cross-linking of plasminogen activator inhibitor 2 and alpha 2-antiplasmin to fibrin (ogen). *J. Biol. Chem.* **275**, 24915–24920.

Roes, E. M., Sweep, C. G., Thomas, C. M., Zusterzeel, P. L., Geurts-Moespot, A., Peters, W. H., and Steegers, E. A. (2002). Levels of plasminogen activators and their inhibitors in maternal and umbilical cord plasma in severe preeclampsia. *Am. J. Obstet. Gynecol.* **187**, 1019–1025.

Scherrer, A., Kruithof, E. K., and Grob, J. P. (1991). Plasminogen activator inhibitor-2 in patients with monocytic leukemia. *Leukemia* **5**, 479–486.

Schleuning, W. D., Medcalf, R. L., Hession, C., Rothenbuhler, R., Shaw, A., and Kruithof, E. K. (1987). Plasminogen activator inhibitor 2: Regulation of gene transcription during phorbol ester-mediated differentiation of U-937 human histiocytic lymphoma cells. *Mol. Cell. Biol.* **7**, 4564–4567.

Schonig, K., Bujard, H., and Gossen, M. (2010). The power of reversibility regulating gene activities via tetracycline-controlled transcription. *Methods Enzymol.* **477**, 429–453.

Schroder, W. A., Le, T. T., Major, L., Street, S., Gardner, J., Lambley, E., Markey, K., Macdonald, K. P., Fish, R. J., Thomas, R., and Suhrbier, A. (2010a). A physiological function of inflammation-associated SerpinB2 is regulation of adaptive immunity. *J. Immunol.* **184,** 2663–2670.

Schroder, W. A., Le, T. T., Major, L., Street, S., Gardner, J., Lambley, E., Markey, K., MacDonald, K. P., Fish, R. J., Thomas, R., and Suhrbier, A. (2010b). A physiological function of inflammation-associated SerpinB2 is regulation of adaptive immunity. *J. Immunol.* **184,** 2663–2670.

Sekine, H., Mimura, J., Oshima, M., Okawa, H., Kanno, J., Igarashi, K., Gonzalez, F. J., Ikuta, T., Kawajiri, K., and Fujii-Kuriyama, Y. (2009). Hypersensitivity of aryl hydrocarbon receptor-deficient mice to lipopolysaccharide-induced septic shock. *Mol. Cell. Biol.* **29,** 6391–6400.

Sharon, R., Abramovitz, R., and Miskin, R. (2002). Plasminogen mRNA induction in the mouse brain after kainate excitation: Codistribution with plasminogen activator inhibitor-2 (PAI-2) mRNA. *Brain Res. Mol. Brain Res.* **104,** 170–175.

Shiiba, M., Nomura, H., Shinozuka, K., Saito, K., Kouzu, Y., Kasamatsu, A., Sakamoto, Y., Murano, A., Ono, K., Ogawara, K., Uzawa, K., and Tanzawa, H. (2010). Downregulated expression of SERPIN genes located on chromosome 18q21 in oral squamous cell carcinomas. *Oncol. Rep.* **24,** 241–249.

Silverman, G. A., Bird, P. I., Carrell, R. W., Church, F. C., Coughlin, P. B., Gettins, P. G., Irving, J. A., Lomas, D. A., Luke, C. J., Moyer, R. W., Pemberton, P. A., Remold-O'Donnell, E., *et al.* (2001). The serpins are an expanding superfamily of structurally similar but functionally diverse proteins. Evolution, mechanism of inhibition, novel functions, and a revised nomenclature. *J. Biol. Chem.* **276,** 33293–33296.

Silverman, G. A., Whisstock, J. C., Askew, D. J., Pak, S. C., Luke, C. J., Cataltepe, S., Irving, J. A., and Bird, P. I. (2004). Human clade B serpins (ov-serpins) belong to a cohort of evolutionarily dispersed intracellular proteinase inhibitor clades that protect cells from promiscuous proteolysis. *Cell. Mol. Life Sci.* **61,** 301–325.

Smale, S. T. (2009). Nuclear run-on assay. *Cold Spring Harb. Protoc.* doi:10.1101/pdb.prot5329.

Stasinopoulos, S., Mariasegaram, M., Gafforini, C., Nagamine, Y., and Medcalf, R. L. (2010). The plasminogen activator inhibitor 2 transcript is destabilized via a multicomponent 3′ UTR localized adenylate and uridylate-rich instability element in an analogous manner to cytokines and oncogenes. *FEBS J.* **277,** 1331–1344.

Stratikos, E., and Gettins, P. G. (1999). Formation of the covalent serpin-proteinase complex involves translocation of the proteinase by more than 70 A and full insertion of the reactive center loop into beta-sheet A. *Proc. Natl. Acad. Sci. USA* **96,** 4808–4813.

Sutter, T. R., Guzman, K., Dold, K. M., and Greenlee, W. F. (1991). Targets for dioxin: Genes for plasminogen activator inhibitor-2 and interleukin-1 beta. *Science* **254,** 415–418.

Suzuki, T., Hashimoto, S., Toyoda, N., Nagai, S., Yamazaki, N., Dong, H. Y., Sakai, J., Yamashita, T., Nukiwa, T., and Matsushima, K. (2000). Comprehensive gene expression profile of LPS-stimulated human monocytes by SAGE. *Blood* **96,** 2584–2591.

Tierney, M. J., and Medcalf, R. L. (2001). Plasminogen activator inhibitor type 2 contains mRNA instability elements within exon 4 of the coding region. Sequence homology to coding region instability determinants in other mRNAs. *J. Biol. Chem.* **276,** 13675–13684.

Virtanen, O. J., Siren, V., Multanen, J., Farkkila, M., Leivo, I., Vaheri, A., and Koskiniemi, M. (2006). Plasminogen activators and their inhibitors in human saliva and salivary gland tissue. *Eur. J. Oral Sci.* **114,** 22–26.

Webb, A. C., Collins, K. L., Snyder, S. E., Alexander, S. J., Rosenwasser, L. J., Eddy, R. L., Shows, T. B., and Auron, P. E. (1987). Human monocyte Arg-Serpin cDNA. Sequence, chromosomal assignment, and homology to plasminogen activator-inhibitor. *J. Exp. Med.* **166,** 77–94.

Wilczynska, M., Lobov, S., and Ny, T. (2003). The spontaneous polymerization of plasminogen activator inhibitor type-2 and Z-antitrypsin are due to different molecular aberrations. *FEBS Lett.* **537,** 11–16.

Wing, L. R., Hawksworth, G. M., Bennett, B., and Booth, N. A. (1991). Clearance of t-PA, PAI-1, and t-PA-PAI-1 complex in an isolated perfused rat liver system. *J. Lab. Clin. Med.* **117,** 109–114.

Winzen, R., Gowrishankar, G., Bollig, F., Redich, N., Resch, K., and Holtmann, H. (2004). Distinct domains of AU-rich elements exert different functions in mRNA destabilization and stabilization by p38 mitogen-activated protein kinase or HuR. *Mol. Cell. Biol.* **24,** 4835–4847.

Winzen, R., Thakur, B. K., Dittrich-Breiholz, O., Shah, M., Redich, N., Dhamija, S., Kracht, M., and Holtmann, H. (2007). Functional analysis of KSRP interaction with the AU-rich element of interleukin-8 and identification of inflammatory mRNA targets. *Mol. Cell. Biol.* **27,** 8388–8400.

Wohlwend, A., Belin, D., and Vassalli, J. D. (1987). Plasminogen activator-specific inhibitors in mouse macrophages: In vivo and in vitro modulation of their synthesis and secretion. *J. Immunol.* **139,** 1278–1284.

Wun, T. C., and Reich, E. (1987). An inhibitor of plasminogen activation from human placenta. Purification and characterization. *J. Biol. Chem.* **262,** 3646–3653.

Ye, R. D., Wun, T. C., and Sadler, J. E. (1988). Mammalian protein secretion without signal peptide removal. Biosynthesis of plasminogen activator inhibitor-2 in U-937 cells. *J. Biol. Chem.* **263,** 4869–4875.

Ye, R. D., Ahern, S. M., Le Beau, M. M., Lebo, R. V., and Sadler, J. E. (1989). Structure of the gene for human plasminogen activator inhibitor-2. The nearest mammalian homologue of chicken ovalbumin. *J. Biol. Chem.* **264,** 5495–5502.

Yoshino, H., Endo, Y., Watanabe, Y., and Sasaki, T. (1998). Significance of plasminogen activator inhibitor 2 as a prognostic marker in primary lung cancer: Association of decreased plasminogen activator inhibitor 2 with lymph node metastasis. *Br. J. Cancer* **78,** 833–839.

Yu, H., Maurer, F., and Medcalf, R. L. (2002). Plasminogen activator inhibitor type 2: A regulator of monocyte proliferation and differentiation. *Blood* **99,** 2810–2818.

Yu, H., Stasinopoulos, S., Leedman, P., and Medcalf, R. L. (2003). Inherent instability of plasminogen activator inhibitor type 2 mRNA is regulated by tristetraprolin. *J. Biol. Chem.* **278,** 13912–13918.

Zhang, M., Maass, N., Magit, D., and Sager, R. (1997). Transactivation through Ets and Ap1 transcription sites determines the expression of the tumor-suppressing gene maspin. *Cell Growth Differ.* **8,** 179–186.

Zhang, S. J., Zou, M., Lu, L., Lau, D., Ditzel, D. A., Delucinge-Vivier, C., Aso, Y., Descombes, P., and Bading, H. (2009). Nuclear calcium signaling controls expression of a large gene pool: Identification of a gene program for acquired neuroprotection induced by synaptic activity. *PLoS Genet.* **5,** e1000604.

Zhou, H. M., Bolon, I., Nichols, A., Wohlwend, A., and Vassalli, J. D. (2001). Overexpression of plasminogen activator inhibitor type 2 in basal keratinocytes enhances papilloma formation in transgenic mice. *Cancer Res.* **61,** 970–976.

CHAPTER SEVEN

The SerpinB1 Knockout Mouse: A Model for Studying Neutrophil Protease Regulation in Homeostasis and Inflammation

Charaf Benarafa

Contents

1. Generation of *SerpinB1* Knockout Mice — 136
   1.1. Background — 136
   1.2. Characterization of the mouse homologues of SERPINB1 — 137
   1.3. Plasmid construct design and embryonic stem cell targeting — 137
2. Models of Lung Infection and Inflammation — 139
   2.1. Neutrophil proteases and inhibitors in the lungs — 139
   2.2. Acute lung infection model — 140
   2.3. MPO activity in tissues and cell-free bronchoalveolar lavage — 141
3. Neutrophil Homeostasis — 142
   3.1. Role of serpinB1 in neutrophil homeostasis in the bone marrow — 142
   3.2. Immunophenotyping of mouse neutrophils in bone marrow and blood — 143
   3.3. Measuring elastase activity in mouse bone marrow fluid — 145
Acknowledgments — 146
References — 146

Abstract

SerpinB1 is a clade B serpin, or ov-serpin, found at high levels in the cytoplasm of neutrophils. SerpinB1 inhibits neutrophil serine proteases, which are important in killing microbes. When released from granules, these potent enzymes also destroy host proteins and contribute to morbidity and mortality in inflammatory diseases including emphysema, chronic obstructive pulmonary disease, cystic fibrosis, arthritis, and sepsis. Studies of *serpinB1*-deficient mice have established a crucial role for this serpin in *Pseudomonas aeruginosa* infection by preserving lung antimicrobial proteins from proteolysis and by protecting

Theodor Kocher Institute, University of Bern, Bern, Switzerland lung-recruited neutrophils from a premature death. SerpinB1$^{-/-}$ mice also have a severe defect in the bone marrow reserve of mature neutrophils demonstrating a key role for serpinB1 in cellular homeostasis. Here, key methods used to generate and characterize *serpinB1*$^{-/-}$ mice are described including intranasal inoculation, myeloperoxidase activity, flow cytometry analysis of bone marrow myeloid cells, and elastase activity. *SerpinB1*-knockout mice provide a model to dissect the pathogenesis of inflammatory disease characterized by protease: antiprotease imbalance and may be used to assess the efficacy of therapeutic compounds.

1. GENERATION OF *SERPINB1* KNOCKOUT MICE

1.1. Background

The 2007 Nobel Prize in Physiology or Medicine awarded to Mario R. Capecchi, Martin J. Evans, and Oliver Smithies "for their discoveries of principles for introducing specific gene modifications in mice by the use of embryonic stem cells" is a prominent acknowledgment of the biomedical advances made from the studies of knockout mice. Correspondingly, the understanding of human serpin functions has been strongly substantiated by studies of serpin mutant mice. The relevance of using gene targeting in mice and the potential strength of such models is directly correlated with the degree of functional homology between the genes in the two species. Differences in the pattern of tissue expression and mouse gene duplication as observed in mouse clade A and clade B serpin loci are significant challenges in the design and generation of relevant models (Askew *et al.*, 2004; Horvath *et al.*, 2004; Kaiserman *et al.*, 2002).

SERPINB1, also named monocyte neutrophil elastase inhibitor (MNEI) and leukocyte elastase inhibitor (LEI), was first identified as a product of monocytes and neutrophils with a high inhibitory activity against neutrophil elastase (Dubin *et al.*, 1992; Remold-O'Donnell *et al.*, 1992). SERPINB1 also efficiently inhibits proteinase-3 and cathepsin G, two serine proteases stored together with elastase in neutrophil primary granules (Cooley *et al.*, 2001). Like other clade B serpins, SERPINB1 lacks a cleavable hydrophobic signal sequence and has a largely cytoplasmic localization. However, SERPINB1 is also found in airway fluids during lung inflammatory disease (Cooley *et al.*, 2010; Davies *et al.*, 2010; Yasumatsu *et al.*, 2006). Indeed, SERPINB1 can be secreted by an alternative mechanism used by other leaderless cytoplasmic proteins such as IL-1 family members (Keller *et al.*, 2008). However, the relative importance of this secreting pathway and the release of SERPINB1 in the extracellular milieu after cell death remains to be established. The rapid and efficient inhibition of all three neutrophil granule proteases by SERPINB1 is only equaled by α1-antitrypsin (α1-AT, encoded by *SERPINA1*), the well-studied plasma inhibitor of neutrophil proteases. Here, the generation and relevance of the *serpinB1* knockout mouse model are reviewed and key methodological features used to characterize the *serpinB1* knockout mice and the role of SERPINB1 in lung infection and neutrophil homeostasis are described.

1.2. Characterization of the mouse homologues of SERPINB1

SERPINB1 is an ancestral clade B serpin or ov-serpin (Benarafa and Remold-O'Donnell, 2005; Kaiserman and Bird, 2005). The specificity determining residues of the reactive center loop (RCL) of SERPINB1 are strikingly conserved in vertebrates, which indicates a deeply rooted inhibitory function. One mouse orthologue, named *serpinB1a*, has a conserved RCL, and inhibits all three neutrophil proteases (Benarafa *et al.*, 2002). The human and mouse genes share the same broad pattern of tissue expression, including high levels in myeloid cells. Three additional SERPINB1 homologues are found close to *serpinB1a* on an expanded serpin locus on mouse chromosome 13: *serpinB1b*, *serpinB1c*, and *serpinB1*-ps1 (Kaiserman *et al.*, 2002). Of these, serpinB1b can form inhibitory complexes with cathepsin G but is degraded by elastase and is not expressed in leukocytes. *Serpinb1c* and *serpinB1*-ps1 are pseudogenes and do not encode functional serpins due to early stop codons and missing exons (Benarafa *et al.*, 2002). Thus, *serpinB1a* can be considered as the only physiologically relevant and functional homologue of human SERPINB1 in mice and a targeting construct was generated to delete *serpinB1a*.

1.3. Plasmid construct design and embryonic stem cell targeting

Because both human and mouse SERPINB1 are ubiquitously expressed, a conditional knockout strategy using Cre-loxP technology was chosen to evaluate the effects of serpinB1 deletion in specific tissues/cells and to anticipate potential setbacks caused by unexpected severe phenotype or embryonic lethality, as observed for other serpins (Gao *et al.*, 2004; Ishiguro *et al.*, 2000; Nagai *et al.*, 2000). The generation and initial characterization of *serpinB1*-deficient mice has been described previously (Benarafa *et al.*, 2007). The official nomenclature for this allele is *serpinb1a$^{tm1.1Cben}$* but, for simplicity, they will be referred to as *serpinB1$^{-/-}$* in this chapter. Key features that were considered in the generation of *serpinB1$^{-/-}$* mice and additional recommendations in the design of a new gene-targeting project include the removal of the selection cassette and the choice of ES cells. To generate *serpinB1$^{-/-}$* mice, we used a strategy with three loxP sites with a floxed exon 7 and floxed hygromycin-thymidine kinase (CMV-HYG/TK) double selection cassette (Fig. 7.1), a targeting strategy

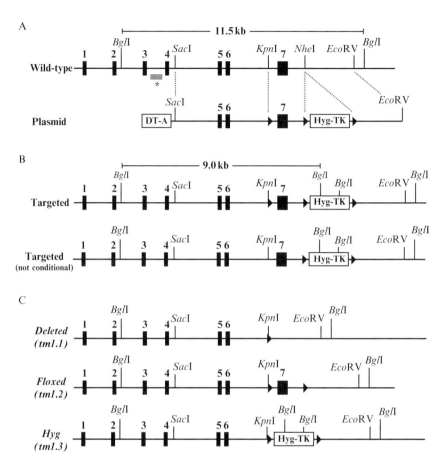

Figure 7.1 Targeting vector for *serpinB1*-knockout mouse. (A) The vector contained a loxP site in intron 6 and a floxed CMV-HYG/TK double selection cassette in the 3′ flanking region of *serpinB1a* downstream of exon 7. The vector was integrated into the *serpinB1a* locus by homologous recombination and clones selected with hygromycin. Selection was enhanced by the use of a PGK-DTA negative selection cassette to eliminate the majority of clones with nonhomologous recombination. (B) The homologous recombination event generated two types of targeted ES clones with (conditional) or without (nonconditional) integration of the loxP site in intron 6, which were identified by PCR. (C) Targeted conditional ES clones were then transfected to transiently express Cre recombinase and to excise the floxed CMV-HYG/TK cassette via site-specific recombination. Clones were selected with gancyclovir to eliminate clones where the deletion of the cassette did not occur (untransfected targeted clones or clones with the HYG cassette only). ES clones with a deleted allele (*serpinb1a*$^{Cbentm1.1}$) as well as a floxed allele (*serpinb1a*$^{Cbentm1.2}$) were generated simultaneously. LoxP sites are indicated by black triangles and position of Southern blot probe is shown with an asterisk.

previously shown to efficiently generate both deleted and floxed alleles for conditional deletion (Yu *et al.*, 2000). The removal of the selection cassette is a crucial feature since cassettes remaining within the targeting locus can directly affect neighboring gene expression levels, including in loci containing serpins and proteases (Revell *et al.*, 2005; Scarff *et al.*, 2003). 129S6/W4 ES cells were used to generate *serpinB1*$^{-/-}$ mice because this ES cell line had a low passage number and was isogenic to 129S6/SvEvTac wild-type mice commercially available (Taconic), which allowed direct study of the knock-out mice without the immediate need for a lengthy backcrossing. The availability of high-quality ES cell lines in the C57BL/6 background as well as targeted ES cells through the International Knockout Mouse Consortium (http://www.knockoutmouse.org/) now provides further alternatives for generating serpin knockout mice. Additional challenges remain for generating mice lacking more than one serpin gene found in a single locus. This is desired when a single human gene has several homologues in mice such as α1-antitrypsin (*serpinA1*) or to study genes with overlapping functions such as some subsets of clade B serpins.

2. Models of Lung Infection and Inflammation

2.1. Neutrophil proteases and inhibitors in the lungs

Neutrophils are the first leukocytes recruited to injured tissues in response to danger signals. Neutrophil elastase, cathepsin G, and proteinase-3 are carried at high concentration in granules of neutrophils and are major elements of the innate immune defense armamentarium against invading microbes (Belaaouaj *et al.*, 2000; Pham, 2006; Weinrauch *et al.*, 2002). Mice deficient for elastase, cathepsin G, or both are defective in killing pathogenic bacteria and fungi (Belaaouaj *et al.*, 1998; Reeves *et al.*, 2002; Tkalcevic *et al.*, 2000). These neutrophil proteases are, however, a double-edged sword in infection and are detrimental when released in excess of the antiprotease shield, such as in cystic fibrosis for which development of effective protease inhibitors and well-designed clinical trials are needed (Griese *et al.*, 2008; Kelly *et al.*, 2008). High levels of free neutrophil proteases in the lungs are inversely correlated with pulmonary function in cystic fibrosis (Mayer-Hamblett *et al.*, 2007). While the role of plasma inhibitor α1-AT, as well as nonserpin family secreted leukocyte proteinase inhibitor (SLPI) and elafin have been long established as part of the antiprotease shield of the lungs (Greene and McElvaney, 2009), recent evidence increasingly support a role for SERPINB1 in regulating neutrophil proteases in human pulmonary disease (Cooley *et al.*, 2010; Davies *et al.*, 2010).

High levels of free neutrophil proteases in human pulmonary disease and their correlation with negative outcome indicate that the natural antiprotease shield is insufficient. In this respect, $serpinB1^{-/-}$ mice reproduce an important feature of cystic fibrosis and other pulmonary diseases characterized by high levels of endogenous proteases, destruction of antimicrobial lung proteins, and failure to clear both mucoid and nonmucoid strains of *P. aeruginosa* (Benarafa *et al.*, 2007). Further, treatment with recombinant SERPINB1 downregulated the excessive inflammatory cytokine response, reduced proteolysis of surfactant protein-D and dramatically rescued bacterial clearance (Benarafa *et al.*, 2007). $SerpinB1^{-/-}$ mice thus provide an *in vivo* model that may be a useful to evaluate protease inhibitors *in vivo* in the context of infection and inflammation.

2.2. Acute lung infection model

2.2.1. Experimental design

Groups of mice should be matched for age, sex, and body weight as these factors can significantly influence the response to infection. A pilot study should be designed to identify the inoculum dose and end points based on the goals of the experiment and in accordance with local animal ethics committees.

2.2.2. Intranasal inoculation

The depth of the anesthesia is crucial to the correct delivery of the inoculum to the lungs. If the anesthesia is too shallow, the mice may swallow part or all the inoculum resulting in highly variable results. In our hands, a single intraperitoneal injection of 100 mg/kg ketamine and 10 mg/kg xylazine has shown to be a safe and reliable anesthetic protocol for intranasal inoculation. For best results, mice should be kept for 10–15 min with a minimum of light and sound stimuli once the ketamine/xylazine has been injected. The depth of the anesthesia is best assessed by the loss of the pedal reflex, which is usually attained after 10–20 min.

The inoculum is then applied by rapidly pipetting 10–20 µl onto each naris. Note that some mice may stop breathing 10–30 s after application of the inoculum, especially if slightly overdosed with ketamine/xylazine. This transient apnea is especially observed in C57BL/6 background but rarely in 129S6 mice, which can receive a 10% overdose and not present postinoculation apnea. Therefore, the breathing pattern should be carefully monitored after inoculation in sensitive strains, and 2–3 pressure pinches of the mouse thorax between index and thumb are usually sufficient to stimulate the mouse to breathe again if it has stopped breathing for a few seconds. Adequate anesthetic dosage is crucial to obtaining reproducible inoculation.

2.2.3. Tissue harvest

Blood, bronchoalveolar lavage (BAL) fluid, and organs are harvested to evaluate microbial clearance, inflammatory markers, and tissue injury. Different groups of mice may be required for different processing of the lungs such as histopathology, homogenate, or BAL.

At time of harvest, mice are anesthetized with ketamine/xylazine, as above, for blood collection from the retro-orbital plexus. Blood from one heparinized capillary tube (60–70 μl) is sufficient for total blood-cell counts and for flow cytometry analysis. More uncoagulated blood can then be collected for preparation of serum for cytokine analysis. Spleen and right lung are dissected using sterile technique and homogenized in 1 ml buffer (1% proteose peptone or PBS) in autoclaved glass tubes (75 × 12 mm). Between samples, the probe of the tissue homogenizer is wiped, cleaned, flamed with ethanol, and cooled in sterile ice-cold water. Several aliquots of tissue homogenates can be used immediately for determination of bacterial load, or mixed with Trizol LS for RNA extraction and RT-PCR analysis, or directly frozen and used later for further analysis, such as myeloperoxidase (MPO) activity assay (see below), cytokine ELISA, and Western blots. Left lungs are fixed and processed for immunohistochemistry.

2.3. MPO activity in tissues and cell-free bronchoalveolar lavage

MPO is specifically expressed in granulocytes and is stored in primary granules together with serine proteases. MPO activity assay has been widely used as a semi-quantitative marker of neutrophil recruitment in tissues (Bradley et al., 1982). We have also used this assay to quantify MPO activity in cell-free BAL as a measure of neutrophil necrosis (Benarafa et al., 2007). In this assay, MPO activity is measured using a spectrophotometric assay in 50 mM potassium phosphate (pH 6) containing 0.167 mg/ml o-dianisidine dihydrochloride and 0.0005% hydrogen peroxide.

2.3.1. Materials and sample preparation

2.3.1.1. MPO standard Prepare 10 μl aliquots of MPO (M6908; Sigma) to be used as standard by resuspending 5 units in 250 μl of distilled H_2O and desiccate under vacuum (Speedvac) for 20 min and store at $-20\ °C$. Do not freeze in solution, as the activity will be lost.

2.3.1.2. Assay buffers Make a stock solution of 1 M potassium phosphate buffer by preparing 250 ml of 1 M KH_2PO_4 and gradually adding 1 M $K_2HPO_4 \cdot H_2O$ until pH 6.0. Autoclave and store at room temperature. For a 96-well plate assay, prepare 100 ml of 50 mM potassium phosphate buffer from the 1 M stock. Prepare a solution (5 ml) of 1% hexadecyltrimethylammonium bromide (HDTA) in 50 mM potassium phosphate.

2.3.1.3. Sample preparation Lungs are dissected and excess blood is blotted on sterile gauze and weighed. Lung lobes (50–200 mg) are then homogenized in 1 ml of ice-cold buffer (PBS, 1% peptone or 50 mM potassium phosphate have shown equal results) using a tissue homogenizer (Ultra-Turrax, IKA) and kept on ice. Aliquots of homogenized tissues can be frozen as MPO contained in granules will remain active. If measuring free MPO in BAL or other cell-free fluid, no special sample preparation is required but samples cannot be frozen before MPO measurements.

2.3.2. MPO assay

Mix 100 µl of tissue homogenate with an equal volume of 50 mM phosphate buffer 1% HDTA (final conc. 0.5% HDTA). HDTA is required to disrupt cellular membranes and release MPO from azurophil granules. Vortex well and incubate for 5 min at room temperature. Centrifuge at $10,000 \times g$ for 10 min at 4 °C. Harvest supernatant and keep on ice until plating. Prepare MPO standard dilutions by resuspending an aliquot of vacuum-dried MPO (0.2 unit) in 50 µl of 50 mM potassium phosphate 0.5% HDTA. Vortex and make serial 1:2 dilutions in the same buffer to a total of six further concentrations. Aliquot 7 µl of MPO standard dilutions, blank, and each sample in triplicate at the center of wells of a flat-bottom microplate. Resuspend 10 mg o-dianisidine substrate tablet (D9154; Sigma) in 1 ml distilled H_2O and vortex well. For a 96-well microplate, prepare 24 ml of assay buffer at room temperature with 23.6 ml of 50 mM potassium phosphate, 12 µl of 1% H_2O_2, and 400 µl of 10 mg/ml o-dianisidine solution (added last). Promptly add 200 µl of assay buffer with freshly added o-dianisidine to each well of the microplate using a multichannel pipette. Incubate at room temperature and read absorbance at 450 nm every 10–15 min for up to 45 min as the reaction develops a yellow-orange color. MPO activity is expressed in unit per milligram of tissue or milliliter of fluid using a standard curve.

3. Neutrophil Homeostasis

3.1. Role of serpinB1 in neutrophil homeostasis in the bone marrow

We recently reported that the bone marrow reserve of mature neutrophils is considerably reduced in *serpinB1$^{-/-}$* mice and that this bone marrow reserve pool is completely exhausted during acute lung injury (Benarafa *et al.*, 2011). In steady state, the number of mature neutrophils in the bone marrow is 20-

fold higher than in the blood circulation (Boxio *et al.*, 2004; Chervenick *et al.*, 1968), where neutrophils remain only for a few hours before being disposed of by macrophages in spleen, liver, and bone marrow. In response to inflammatory signals in the periphery, the bone marrow neutrophils are rapidly called up to the blood for extravasation in injured tissues. Therefore, the constitution and maintenance of a neutrophil reserve in the bone marrow is essential for a timely and vast mobilization in response to infection.

Neutrophils originate from hematopoietic stem cells via common myeloid and granulocyte/macrophage progenitors. They then differentiate into either neutrophil precursors (myeloblasts) or monocyte/macrophage precursors (promonocytes). Myeloblasts, which are committed to the neutrophil lineage, differentiate into promyelocytes and myelocytes, which actively proliferate. The serine proteases elastase, cathepsin G, and proteinase-3 are synthesized only at the promyelocyte stage and stored in an active form in primary granules (Borregaard, 2010). Interestingly, we found that serpinB1 is expressed at intermediate levels in stem and progenitors cells and then massively increased at the promyelocytes stage, coinciding with the target protease synthesis. As neutrophils develop and mature, protease production ceases but high protease concentration remains in granules. SerpinB1 mRNA expression and protein levels also remain high in the cytoplasm of myelocytes to mature neutrophils as well as in circulating neutrophils (Benarafa *et al.*, 2011). Myeloid and neutrophil progenitors, which do not yet carry protease targets of serpinB1, are normal in *serpinB1$^{-/-}$* bone marrow and the neutrophil reserve defect due to serpinB1 deletion is largely restricted to postmitotic mature neutrophils (Benarafa *et al.*, 2011). Thus, serpinB1 plays a central role in neutrophil homeostasis in sustaining a healthy reserve of bone marrow neutrophils, which are required in acute inflammatory responses.

3.2. Immunophenotyping of mouse neutrophils in bone marrow and blood

3.2.1. Cell surface markers for mouse neutrophils and monocytes

Multicolor flow cytometry analysis using subset-specific cell surface markers is an efficient way to evaluate cells committed to the neutrophil and monocyte lineages in the bone marrow. The evaluation of proliferative and differentiation potential of early progenitors is best performed in methylcellulose clonogenic assays (StemCell Technologies) and are not described here.

Cells committed to the monocytic lineage express CD115 (c-fms), the M-CSF receptor, and can thus be distinguished from cells of other lineages or from noncommitted cells. Approximately half of the bone marrow CD115$^+$ monocytic cells do not express CD11b, which is induced before release into the circulation. No single marker has been validated to identify all the cells in the mouse neutrophil lineage. However, a combination of cell

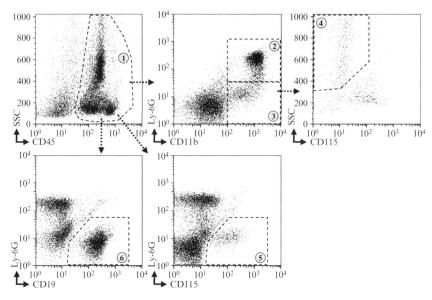

Figure 7.2 Gating strategy for flow cytometry analysis of mouse bone marrow myeloid leukocyte subsets. Flow cytometry dotplots of mouse bone marrow cells after erythrocyte lysis to identify CD45$^+$ leukocytes (gate 1). Within gate 1, three subsets are directly identified as mature neutrophils (gate 2; CD11b$^+$Ly-6G$^+$), promonocytes/monocytes (gate 5; CD115$^+$), and B cells (gate 6; CD19$^+$). Within gate 3 (CD11b$^+$Ly-6G^{neg}), myelocytes are identified as SSC^{hi}CD115^{neg} (gate 4). In $serpinB1^{-/-}$ bone marrow, the relative number of neutrophils (gate 2) is significantly reduced (Benarafa et al., 2011).

surface markers and cell granularity can be used to identify these cells from the myelocytes stage (Fig. 7.2). A high quantity of granules are produced at the promyelocyte/myelocyte stages and this correlates with high side scatter (SSC^{hi}) in flow cytometry. Bone marrow myelocytes start expressing CD11b before expressing Ly-6G, a neutrophil-specific GPI-anchored protein. Therefore, myelocytes can easily be distinguished from monocytic cells because of their high side scatter (SSC^{hi}) within the CD11b$^+$Ly-6G^{neg} subset.

Circulating neutrophils can be unequivocally identified by high levels of both CD11b and Ly-6G, whereas blood monocytes constitute all the CD11b$^+$Ly-6G^{neg} subset. A subset of blood monocytes and all blood neutrophils express Ly-6C and Ly-6B.2 (7/4 antigen) (Rosas et al., 2010), two glycosylphosphatidylinositol (GPI)-anchored proteins related to neutrophil-specific Ly-6G. The Gr-1 antigen defined by the binding of the monoclonal antibody RB6-8C5 recognizes both Ly-6G and Ly-6C (Fleming et al., 1993).

3.2.2. Sample collection and staining

Blood is collected from the retro-orbital sinus of sedated mice using heparinized microhematocrit tubes. Cell counts in blood are measured on an automatic blood cell counter. To isolate mouse bone marrow cells, dissect femurs from euthanized mice, cut the extremities of the bones and flush each femur with a syringe filled with 5 ml PBS or IMDM using a fine (25–31-gauge) needle until the inside of the bone appears white. Obtain a single cell suspension by gently pipeting the cells and passing through a 70 μm strainer. Count the cells using a hemocytometer and Türks solution. Wash bone marrow cells with PBS containing 1% FCS and 0.09% sodium azide and resuspend at $5-10 \times 10^7$ cells/ml.

For staining, aliquot in FACS tubes 50 μl of whole blood or 50–100 μl of bone marrow cells at $5-10 \times 10^7$ cells/ml. Block low-affinity Fc receptors by adding anti-CD16/CD32 at 1 μg/ml (final concentration) and incubate 10 min on ice. Stain the cells by adding premixed fluorochrome-labeled monoclonal antibody clones 1A8 (Ly-6G), 30F11 (CD45), M1/70 (CD11b), and AFS98 (CD115) at 1 μg/ml (except AFS98, 2 μg/ml). Incubate on ice and protect from light for at least 30 min, wash twice, and analyze on a four-color flow cytometer.

3.3. Measuring elastase activity in mouse bone marrow fluid

Neutrophil elastase released within the bone marrow degrades adhesion molecules and inactivates chemokine-receptor pairs in the bone marrow (Levesque *et al.*, 2001, 2003a,b). Initial findings suggested that neutrophil elastase regulates mobilization of hematopoietic stem cells following treatment with G-CSF or cyclophosphamide. The increased transient elastase activity has been attributed to downregulation of serpina3 (Winkler *et al.*, 2005), but studies of protease-deficient mice have shown that stem cell mobilization does not require elastase activity (Levesque *et al.*, 2004). After cyclophosphamide treatment, serpinB1−/− mice have a comparable increase in elastase activity in the bone marrow fluid as wild-type mice (Benarafa *et al.*, 2011). Elastase activity is measured in bone marrow fluid using a colorimetric assay.

3.3.1. Reagents and solutions

Human sputum elastase (Elastin Products, SE563) is reconstituted in PBS (pH 7.1) at 1 mg/ml and used as standard. Assay buffer 20 mM TRIS (pH 7.4), 500 mM NaCl, 0.05% Tween-20, 4 mM DTT. Substrate N-Methoxysuccinyl-Ala-Ala-Pro-Val p-nitroanilide (Sigma; M4765) stock solution (80 mM) is prepared in dimethylformamide and stored at $-20\ °C$.

3.3.2. Sample preparation and measurements

Bone marrow fluid is collected by flushing each femur with 0.5 ml PBS. Cells are centrifuged and the supernatant is aliquoted and stored frozen.

Add 100 μl of assay buffer to all wells, and add 50 μl of samples, standard, and blank, each in duplicate. Incubate for 3 min at 37 °C. Prepare N-Methoxysuccinyl-Ala-Ala-Pro-Val p-nitroanilide (Sigma; M4765) substrate solution by mixing 200 μl of 80 mM stock in 8 ml of assay buffer at 37 °C. Add 50 μl of substrate solution and measure absorbance at 405 nm.

ACKNOWLEDGMENTS

This work was supported by grants from the Flight Attendant Medical Research Institute (FAMRI), the Swiss National Science Foundation (grant 310030-127464), and the EU (FP7 IRG Marie Curie Actions).

REFERENCES

Askew, D. J., Askew, Y. S., Kato, Y., Turner, R. F., Dewar, K., Lehoczky, J., and Silverman, G. A. (2004). Comparative genomic analysis of the clade B serpin cluster at human chromosome 18q21: Amplification within the mouse squamous cell carcinoma antigen gene locus. *Genomics* **84**, 176–184.

Belaaouaj, A., McCarthy, R., Baumann, M., Gao, Z., Ley, T. J., Abraham, S. N., and Shapiro, S. D. (1998). Mice lacking neutrophil elastase reveal impaired host defense against gram negative bacterial sepsis. *Nat. Med.* **4**, 615–618.

Belaaouaj, A., Kim, K. S., and Shapiro, S. D. (2000). Degradation of outer membrane protein A in *Escherichia coli* killing by neutrophil elastase. *Science* **289**, 1185–1188.

Benarafa, C., and Remold-O'Donnell, E. (2005). The ovalbumin serpins revisited: Perspective from the chicken genome of clade B serpin evolution in vertebrates. *Proc. Natl. Acad. Sci. USA* **102**, 11367–11372.

Benarafa, C., Cooley, J., Zeng, W., Bird, P. I., and Remold-O'Donnell, E. (2002). Characterization of four murine homologs of the human ov-serpin monocyte neutrophil elastase inhibitor MNEI (SERPINB1). *J. Biol. Chem.* **277**, 42028–42033.

Benarafa, C., Priebe, G. P., and Remold-O'Donnell, E. (2007). The neutrophil serine protease inhibitor serpinb1 preserves lung defense functions in *Pseudomonas aeruginosa* infection. *J. Exp. Med.* **204**, 1901–1909.

Benarafa, C., LeCuyer, T. E., Baumann, M., Stolley, J. M., Cremona, T. P., and Remold-O'Donnell, E. (2011). SerpinB1 protects the mature neutrophil reserve in the bone marrow. *J. Leukoc. Biol.* doi:10.1189/jlb.0810461.

Borregaard, N. (2010). Neutrophils, from marrow to microbes. *Immunity* **33**, 657–670.

Boxio, R., Bossenmeyer-Pourie, C., Steinckwich, N., Dournon, C., and Nusse, O. (2004). Mouse bone marrow contains large numbers of functionally competent neutrophils. *J. Leukoc. Biol.* **75**, 604–611.

Bradley, P. P., Priebat, D. A., Christensen, R. D., and Rothstein, G. (1982). Measurement of cutaneous inflammation: Estimation of neutrophil content with an enzyme marker. *J. Invest. Dermatol.* **78**, 206–209.

Chervenick, P. A., Boggs, D. R., Marsh, J. C., Cartwright, G. E., and Wintrobe, M. M. (1968). Quantitative studies of blood and bone marrow neutrophils in normal mice. *Am. J. Physiol.* **215**, 353–360.

Cooley, J., Takayama, T. K., Shapiro, S. D., Schechter, N. M., and Remold-O'Donnell, E. (2001). The serpin MNEI inhibits elastase-like and chymotrypsin-like serine proteases through efficient reactions at two active sites. *Biochemistry* **40**, 15762–15770.

Cooley, J., Sontag, M. K., Accurso, F. J., and Remold-O'Donnell, E. (2010). SerpinB1 in CF airway fluids: Quantity, molecular form and mechanism of elastase inhibition. *Eur. Respir. J.* doi: 10.1183/09031936.00073710.

Davies, P. L., Spiller, O. B., Beeton, M. L., Maxwell, N. C., Remold-O'Donnell, E., and Kotecha, S. (2010). Relationship of proteinases and proteinase inhibitors with microbial presence in chronic lung disease of prematurity. *Thorax* **65,** 246–251.

Dubin, A., Travis, J., Enghild, J. J., and Potempa, J. (1992). Equine leukocyte elastase inhibitor. Primary structure and identification as a thymosin-binding protein. *J. Biol. Chem.* **267,** 6576–6583.

Fleming, T. J., Fleming, M. L., and Malek, T. R. (1993). Selective expression of Ly-6G on myeloid lineage cells in mouse bone marrow. RB6-8C5 mAb to granulocyte-differentiation antigen (Gr-1) detects members of the Ly-6 family. *J. Immunol.* **151,** 2399–2408.

Gao, F., Shi, H. Y., Daughty, C., Cella, N., and Zhang, M. (2004). Maspin plays an essential role in early embryonic development. *Development* **131,** 1479–1489.

Greene, C. M., and McElvaney, N. G. (2009). Proteases and antiproteases in chronic neutrophilic lung disease—Relevance to drug discovery. *Br. J. Pharmacol.* **158,** 1048–1058.

Griese, M., Kappler, M., Gaggar, A., and Hartl, D. (2008). Inhibition of airway proteases in cystic fibrosis lung disease. *Eur. Respir. J.* **32,** 783–795.

Horvath, A. J., Forsyth, S. L., and Coughlin, P. B. (2004). Expression patterns of murine antichymotrypsin-like genes reflect evolutionary divergence at the Serpina3 locus. *J. Mol. Evol.* **59,** 488–497.

Ishiguro, K., Kojima, T., Kadomatsu, K., Nakayama, Y., Takagi, A., Suzuki, M., Takeda, N., Ito, M., Yamamoto, K., Matsushita, T., Kusugami, K., Muramatsu, T., et al. (2000). Complete antithrombin deficiency in mice results in embryonic lethality. *J. Clin. Invest.* **106,** 873–878.

Kaiserman, D., and Bird, P. I. (2005). Analysis of vertebrate genomes suggests a new model for clade B serpin evolution. *BMC Genomics* **6,** 167.

Kaiserman, D., Knaggs, S., Scarff, K. L., Gillard, A., Mirza, G., Cadman, M., McKeone, R., Denny, P., Cooley, J., Benarafa, C., Remold-O'Donnell, E., Ragoussis, J., et al. (2002). Comparison of human chromosome 6p25 with mouse chromosome 13 reveals a greatly expanded ov-serpin gene repertoire in the mouse. *Genomics* **79,** 349–362.

Keller, M., Ruegg, A., Werner, S., and Beer, H. D. (2008). Active caspase-1 is a regulator of unconventional protein secretion. *Cell* **132,** 818–831.

Kelly, E., Greene, C. M., and McElvaney, N. G. (2008). Targeting neutrophil elastase in cystic fibrosis. *Expert Opin. Ther. Targets* **12,** 145–157.

Levesque, J. P., Takamatsu, Y., Nilsson, S. K., Haylock, D. N., and Simmons, P. J. (2001). Vascular cell adhesion molecule-1 (CD106) is cleaved by neutrophil proteases in the bone marrow following hematopoietic progenitor cell mobilization by granulocyte colony-stimulating factor. *Blood* **98,** 1289–1297.

Levesque, J. P., Hendy, J., Takamatsu, Y., Simmons, P. J., and Bendall, L. J. (2003a). Disruption of the CXCR4/CXCL12 chemotactic interaction during hematopoietic stem cell mobilization induced by GCSF or cyclophosphamide. *J. Clin. Invest.* **111,** 187–196.

Levesque, J. P., Hendy, J., Winkler, I. G., Takamatsu, Y., and Simmons, P. J. (2003b). Granulocyte colony-stimulating factor induces the release in the bone marrow of proteases that cleave c-KIT receptor (CD117) from the surface of hematopoietic progenitor cells. *Exp. Hematol.* **31,** 109–117.

Levesque, J. P., Liu, F., Simmons, P. J., Betsuyaku, T., Senior, R. M., Pham, C., and Link, D. C. (2004). Characterization of hematopoietic progenitor mobilization in protease-deficient mice. *Blood* **104,** 65–72.

Mayer-Hamblett, N., Aitken, M. L., Accurso, F. J., Kronmal, R. A., Konstan, M. W., Burns, J. L., Sagel, S. D., and Ramsey, B. W. (2007). Association between pulmonary function and sputum biomarkers in cystic fibrosis. *Am. J. Respir. Crit. Care Med.* **175,** 822–828.

Nagai, N., Hosokawa, M., Itohara, S., Adachi, E., Matsushita, T., Hosokawa, N., and Nagata, K. (2000). Embryonic lethality of molecular chaperone hsp47 knockout mice is associated with defects in collagen biosynthesis. *J. Cell Biol.* **150,** 1499–1506.

Pham, C. T. (2006). Neutrophil serine proteases: Specific regulators of inflammation. *Nat. Rev. Immunol.* **6,** 541–550.

Reeves, E. P., Lu, H., Jacobs, H. L., Messina, C. G., Bolsover, S., Gabella, G., Potma, E. O., Warley, A., Roes, J., and Segal, A. W. (2002). Killing activity of neutrophils is mediated through activation of proteases by K+ flux. *Nature* **416,** 291–297.

Remold-O'Donnell, E., Chin, J., and Alberts, M. (1992). Sequence and molecular characterization of human monocyte/neutrophil elastase inhibitor. *Proc. Natl. Acad. Sci. USA* **89,** 5635–5639.

Revell, P. A., Grossman, W. J., Thomas, D. A., Cao, X., Behl, R., Ratner, J. A., Lu, Z. H., and Ley, T. J. (2005). Granzyme B and the downstream granzymes C and/or F are important for cytotoxic lymphocyte functions. *J. Immunol.* **174,** 2124–2131.

Rosas, M., Thomas, B., Stacey, M., Gordon, S., and Taylor, P. R. (2010). The myeloid 7/4-antigen defines recently generated inflammatory macrophages and is synonymous with Ly-6B. *J. Leukoc. Biol.* **88,** 169–180.

Scarff, K. L., Ung, K. S., Sun, J., and Bird, P. I. (2003). A retained selection cassette increases reporter gene expression without affecting tissue distribution in SPI3 knockout/GFP knock-in mice. *Genesis* **36,** 149–157.

Tkalcevic, J., Novelli, M., Phylactides, M., Iredale, J. P., Segal, A. W., and Roes, J. (2000). Impaired immunity and enhanced resistance to endotoxin in the absence of neutrophil elastase and cathepsin G. *Immunity* **12,** 201–210.

Weinrauch, Y., Drujan, D., Shapiro, S. D., Weiss, J., and Zychlinsky, A. (2002). Neutrophil elastase targets virulence factors of enterobacteria. *Nature* **417,** 91–94.

Winkler, I. G., Hendy, J., Coughlin, P., Horvath, A., and Levesque, J. P. (2005). Serine protease inhibitors serpina1 and serpina3 are down-regulated in bone marrow during hematopoietic progenitor mobilization. *J. Exp. Med.* **201,** 1077–1088.

Yasumatsu, R., Altiok, O., Benarafa, C., Yasumatsu, C., Bingol-Karakoc, G., Remold-O'Donnell, E., and Cataltepe, S. (2006). SERPINB1 upregulation is associated with in vivo complex formation with neutrophil elastase and cathepsin G in a baboon model of bronchopulmonary dysplasia. *Am. J. Physiol. Lung Cell. Mol. Physiol.* **291,** L619–L627.

Yu, H., Kessler, J., and Shen, J. (2000). Heterogeneous populations of ES cells in the generation of a floxed Presenilin-1 allele. *Genesis* **26,** 5–8.

CHAPTER EIGHT

INVESTIGATING MASPIN IN BREAST CANCER PROGRESSION USING MOUSE MODELS

Michael P. Endsley *and* Ming Zhang

Contents

1. Introduction	150
2. Transgenic Mouse Models	151
2.1. WAP-*maspin* transgenic mice	151
2.2. Bitransgenic mice	154
2.3. Limitations for transgenic mouse models	156
3. Syngeneic Tumor Model	157
3.1. Development of syngeneic breast cancer model	157
3.2. *Maspin*-mediated gene therapy in mouse model of breast cancer	159
4. Conclusion	161
References	162

Abstract

Clade B serpin family of proteins regulate a variety of cellular functions including cell adhesion and motility. One key member of the clade B serpin family is maspin (SERPINB5). Maspin is classified as a type II tumor suppressor that regulates cell adhesion and invasion. It is expressed in normal mammary epithelial cells but is reduced in benign breast tumors and absent in invasive breast carcinomas. Although maspin regulates cell apoptosis, cell adhesion, migration, and invasion in breast cancer cell culture systems, mouse models are necessary to verify this *in vivo*. In this chapter, we review the development of transgenic and syngeneic mouse models to study the role of maspin in mammary tumorigenesis and in normal mammary development.

Robert H. Lurie Comprehensive Cancer Center and Center for Genetic Medicine, Department of Molecular Pharmacology and Biological Chemistry, Northwestern University, Feinberg School of Medicine, Chicago, Illinois, USA

1. Introduction

Enzymatic breakdown of bioactive macromolecules is an important component involved in maintaining cellular homeostasis. However, excessive enzymatic activity is often detrimental to cellular processes and can be associated with many pathological states such as cancer. In conjunction with the evolutionary development of cellular enzymes, regulators of enzymatic functions have also developed. A unique family of proteins, termed serpins (an acronym derived from *se*rine *p*rotease *in*hibito*rs*), have evolved as negative regulators of cellular serine proteases.

Serpins have evolved with living organisms and are found in all domains of life from bacteria to humans (Irving *et al.*, 2000, 2002). Specifically, human DNA minimally encodes for 35 different members that are separated into nine different clades (labeled A–I). The characterization of these different clade serpins is determined by their amino acid homology, tertiary protein structure, and mechanism of protease inhibition (Irving *et al.*, 2000).

In 1993, the clade B serpins (originally termed ov-serpins due to their similarity to chicken ovalbumin) were identified by amino acid similarities and other molecular factors, with the most significant being the absence of cleavable N-terminal signal peptide (Remold-O'Donnell, 1993). There are two different types of clade B serpins; inhibitory and noninhibitory. Inhibitory clade B serpins, such as plasminogen activator inhibitor 1 (PAI-1 or *SERPINB2*), utilize a reactive site loop (RSL) functional domain to inhibit their target serine proteases (i.e., PAI-1 inhibits tPA and uPA) by an irreversible suicide substrate mechanism (Potempa *et al.*, 1994; Silverman *et al.*, 2001). Whereas, noninhibitory clade B serpins like ovalbumin and pigment epithelium-derived factor (PEDF or *SERPINF1*), lack protease inhibitor function but are still able to function as a storage protein and neural differentiation factor, respectively (Gettins *et al.*, 2002; Tombran-Tink *et al.*, 1992). There have been many studies implicating the clade B serpin family of proteins having multiple functions in addition to enzymatic inhibition but none more apparent than maspin (*SERPINB5*).

Maspin stands for *ma*mmary homologue to *serpin*s since it was originally identified by differential expression screening from normal mammary epithelial cells and breast tumors. *Maspin* is classified as a type II tumor suppressor gene since it is transcriptionally repressed in neoplastic tissues (Zou *et al.*, 1994). In cell culture models, maspin acts as a potent tumor suppressor through decreased cell invasion and motility (Sheng *et al.*, 1994, 1996). Of particular interest to our laboratory, maspin is implicated as a key regulator of breast cancer metastasis.

Although studies in cell culture systems are good for investigating the cancer cell specifically, the results do not always correlate with what is observed *in vivo*. Increasing reports implicate the tumor microenvironment in regulating and/or assisting tumor progression. Therefore, developing mouse models can provide a more comprehensive understanding of the tumor microenvironment complexity and its role in tumor progression. To verify maspin tumor suppressor functions *in vivo*, we have developed a variety of transgene mouse *maspin* and breast cancer models to investigate the functions of maspin in tumor progression and normal mouse mammary gland development. In this chapter, we will review the development of mouse models to study the role of the unique clade B serpin, maspin.

2. Transgenic Mouse Models

2.1. WAP-*maspin* transgenic mice

2.1.1. Rationale of the experimental approach

Throughout development and pregnancy, the female mammary gland undergoes cycles of growth, morphogenesis, and involution. During these cycles, the interaction between epithelial cells and the extracellular matrices (ECM) are extensively regulated (Alexander *et al.*, 1996; Sympson *et al.*, 1994; Talhouk *et al.*, 1991). However, upon transformation and tumor progression, the regulation of cell–ECM interactions is lost. Since maspin is a key regulator of cell–ECM interactions and thus cell migration and invasion, we wanted to study the effect of maspin in both normal mammary gland development and in mammary tumorigenesis.

In order to investigate maspin during development and tumor progression, we first needed to develop a mouse model where *maspin* was transcriptionally overexpressed in mammary luminal epithelial cells. This was achieved by developing a maspin transgene that was under control by the whey acidic protein (WAP) promoter, a mammary luminal cell-specific protein (Hennighausen and Sippel, 1982; Zhang *et al.*, 1999). Since WAP is exclusively expressed from midpregnancy through lactation, these WAP-*maspin* transgenic mice can assess the role of selective *maspin* overexpression in the mammary gland, specifically during pregnancy (Campbell *et al.*, 1984; Hennighausen and Sippel, 1982; Pittius *et al.*, 1988).

2.1.2. Generation of WAP-*maspin* transgene

Mouse *maspin* was originally isolated using recombinant phages from a mouse mammary cDNA library obtained from Dr. Robert Friis (University of Bern). Today, *maspin* cDNA can simply be purchased from many commercial vendors. Amplify mouse *maspin* cDNA by PCR primers flanked by *Kpn*I and *Sal*I sites (sense, 5′-CGGTACCGGATCCATG-GATGCCCTGAGACTGGCA; antisense, 5′-TCCCCCGGGTCGAC-TACAGACAAGTTCCCTGAGA; Zhang *et al.*, 1997). Then, digest the *maspin* cDNA and WAP genomic sequence (originally obtained from an

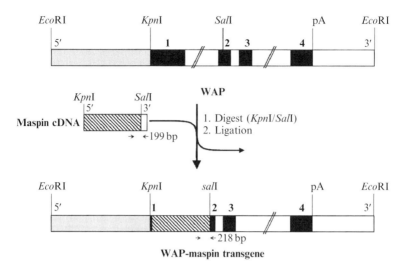

Figure 8.1 Generation of WAP-*maspin* transgene. Schematic involving the insertion of maspin cDNA sequence into WAP genomic DNA through *Kpn*I and *Sal*I digestion and subsequent ligation. Gray, WAP promoter; black, WAP exons; arrows, PCR primer locations to determine difference between endogenous (199 bp) and transgene (218 bp) *maspin* expression in founder mice.

*Eco*RI digest containing all exons, introns, poly-adenylate tail sequences) by *Kpn*I and *Sal*I (Andres *et al.*, 1987). The *Kpn*I cleavage site of WAP is located 25 bp downstream of the WAP start codon sequence and the *Sal*I site cleaves through the middle of exon 2 (Fig. 8.1). Therefore, ligation of the *maspin* cDNA into the digested WAP gene produces mouse maspin cDNA expression under the regulation of the WAP promoter sequence. Transform the plasmid containing WAP-*maspin* into XL-1 Blue *Escherichia coli* cells by heat shock and select clones with ampicillin (100 μg/1 ml LB broth). Purify the procured WAP-*maspin* plasmid using the Qiagen Max-iPrep kit, according to manufacturer's instructions. After propagation, sequence the maspin cDNA portion of the WAP-*maspin* transgene to verify the absence of any mutations.

2.1.3. Isolation and purification of WAP-*mapsin* construct

The successful generation of transgenic mice begins immediately with preparation of a clean, pure DNA construct, as a trace amount of the vector construct is toxic to mouse zygotes. The isolation and purification of the WAP-*maspin* transgene involves multiple steps. First, separate the WAP-*maspin* transgene from the vector by digesting ~100 μg of vector DNA with *Not*I and *Hin*dIII restriction enzymes. Next, isolate the WAP-maspin transgene from the vector by running the completed digest on a 1% (w/v) low-melting point agarose gel in TAE buffer (40 mM Tris–acetate, 1 mM EDTA, 0.1% (v/v) acetic acid, pH 8.5). Normally, DNA is visualized by ethidium bromide (0.5 μg/ml) stain under ultraviolet light but this process can cause DNA nicking thus damaging the transgene. To circumvent this problem, run a small amount of sample with the molecular weight markers on the other side of the gel that can be separated, ethidium bromide stained, and visualized for the transgene location. Realign the ethidium bromide stained section of the gel with the unstained section and excise the area corresponding to the WAP-maspin transgene, then weigh the excised gel section and place into an eppendorf tube. There are many similar commercially available kits for purifying DNA from agarose gels; we purified using QIAEX II gel extraction kit (Qiagen), according to the manufacturer's instructions. Briefly, add QX1 buffer and QIAEX II to the excise gel fragment and incubate at 50 °C for 10 min, with intermittent vortexing. Centrifuge the samples (16,000×g for 30 s) and wash the pellet once with QX1 buffer followed by two washes with PE buffer. Allow the pellet to air-dry (∼30 min) and elute the DNA by adding 20 μl of microinjection buffer (10 mM Tris–HCl, pH 7.5, and 0.1 mM EDTA) followed by vortexing. Incubate the mixture at 50 °C for 5 min, centrifuge, and carefully transfer DNA containing supernatant to a fresh eppendorf tube. To enhance recovery, an additional 20 μl of microinjection buffer can be added to QIAEX II pellet and repeat extraction. Quantify the DNA concentration and dilute WAP-*maspin* DNA to a final concentration of 2 ng/μl. Run 2.5 μl of recovered sample on a 0.7% (w/v) agarose gel to assess the recovery and purification of the WAP-*maspin* transgene.

2.1.4. Generation of WAP-*maspin* transgenic mice

In this section, we will briefly describe the procedure used to develop WAP-*maspin* mice. However, there are many comprehensive, step-by-step protocols published explaining the complex nature of this section that can be used to obtain more in-depth experimental details (Cho et al., 2009; Conner, 2004; Ittner and Gotz, 2007; Nagy et al., 2010). There are three key steps necessary to successfully develop transgenic founder mice; isolation of fertilized eggs, injection of WAP-*maspin* transgene construct into zygotes, and implantation of modified zygotes into pseudopregnant mice.

In order to isolate the fertilized eggs necessary for microinjection, C57BL/6 female mice were given an interperitoneal (i.p.) injection of pregnant mare serum gonadotropin (PMSG) on day 1 and an injection (i.p.) of human chorionic gonadotropin (HCG) on day 3 to induce superovulation. The night of day 3, superovulating C57BL/6 mice were mated with either BALB/c stud or vasectomized male mice to produce pregnant and pseudopregnant female mice, respectively. On the following day, inspect for the presence of a vaginal plug, signifying pregnancy or pseudopregnancy. Place pseudopregnant females aside, they will be used later. Sacrifice all pregnant female mice, remove the reproductive organs, isolate the oviducts, and harvest the fertilized eggs. Conceptually, microinjection is simple but technically very challenging (detailed technical procedure are reviewed in Cho et al., 2009; Conner, 2004; Ittner and Gotz, 2007; Nagy et al., 2010). Briefly, fertilized eggs are grasped by pipette aspiration and injected with the WAP-maspin DNA construct into the pronucleus by positive pressure. Surgically implant 10–20 microinjected zygotes into the oviduct of a pseudopregnant C57BL/6 female mouse and should deliver pups 20 days later. Using DNA from tail biopsies, determine transgenic founders by either Southern blot or PCR methodologies.

Using PCR, decipher the transgene founders using specific primers for either endogenous *maspin* (sense, 5′-GATGGTGGTGAGTCCATC; antisense, 5′-TCCCCCGGGTCGACTACAGACAAGTTCCCTGAGA) or WAP-*maspin* (sense, 5′-GATGGTGGTGAGTCCATC; antisense, 5′-GCTCTAGAGGTGTACATGTCATGACACAGTCGAC). Transgene and endogenous *maspin* products were PCR amplified using 35 cycles of the following reaction conditions: 95 °C for 1 min, 52 °C for 1 min, and 72 °C for 1.5 min.

After founders are identified, backcross them with C57BL/6 mice to generate F1 lines and screen litters for transgene by PCR analysis as stated above. These F1 lines were first used to investigate the effect of maspin overexpression on mouse mammary gland development. This study demonstrated that maspin induced epithelial cells apoptosis during pregnancy and disrupted milk gene production and lactation (Zhang et al., 1997). Since the primary goal of generating WAP-*maspin* transgenic mice was to test the protective role of maspin during tumor progression, these WAP-*maspin* transgenic mice could be crossed with a tumorigenic mouse model.

2.2. Bitransgenic mice

2.2.1. Rationale of the experimental approach

Maspin has been shown to inhibit invasion and motility of mammary carcinoma cells in culture but the role of maspin during metastasis needed to be investigated (Pemberton et al., 1997; Sheng et al., 1996; Sternlicht and Barsky, 1997; Zhang et al., 1997; Zou et al., 1994). Metastasis is a complicated, multifaceted process that is best studied in animal models. In the pioneering maspin study, Zou et al. (1994) demonstrated that orthotopic implantation of human breast carcinoma cells overexpressing *maspin* into nude mice exhibited reduced metastasis compared to *neo* transfection control cells. However, recent investigations have begun to clarify the important influence of the tumor microenvironment on tumor progression, in particular, immune cells (which are lacking in nude mice; DeNardo et al.,

2008). Although orthotopic xenograft is excellent for predicting drug response in human tumors, the transgenic mouse model may be better for examining the role of specific genes (like *maspin*) in tumor development and progression (Richmond and Su, 2008). Therefore, to investigate the protective role of maspin on tumor progression, we crossed our WAP-*maspin* transgenic mice with the well-established mammary tumorigenic mouse strain, WAP-Simian Virus 40 T antigen (TAg; Li et al., 1996; Tzeng et al., 1993; Zhang et al., 1999).

WAP-TAg mice develop mammary tumors with 100% frequency that is initiated by pregnancy (due to WAP promoter; Li et al., 1996; Tzeng et al., 1993). Specifically, TAg induces tumorigenesis via inactivation of both the tumor-suppressing p53 and retinoblastoma (pRb) family of proteins (Dyson et al., 1989; Li et al., 2000; Mietz et al., 1992). p53 is the most commonly mutated gene in many human cancers, with mutations estimated to occur in 50% of all cancers. Additionally, p53 gene mutations are found in 20–40% of invasive breast cancers (de Cremoux et al., 1999; Gasco et al., 2002). Therefore, the WAP-TAg background will serve as an adequate model of breast cancer, where we can investigate the tumor-suppressing role of maspin. The powerful WAP-*maspin*/WAP-TAg bitransgenic mouse model verified maspin as a bona fine tumor suppressor (Zhang et al., 2000).

2.2.2. Mice
Cross WAP-TAg male mice (in a NMRI background) with WAP-maspin female mice in a C57BL/6 background. Analyze progeny tail DNA by PCR, as previously described, and classify them into four distinct genotypes groups in a mixed C57BL/6/NMRI background: bitransgenic WAP-*maspin*/WAP-TAg, WAP-TAg, WAP-maspin, and nontransgenic mice (Li et al., 1996; Zhang et al., 1999). Since the WAP promoter is activated during pregnancy, cage 8-week-old female mice with stud male mice to induce breeding, thus transgene expression. Determine pregnancies by plug appearance and confirm by delivery.

2.2.3. Tumor progression
Breed mice from the four groups (listed above) and examine biweekly for first palpable tumor. Measure tumor growth rate biweekly using caliper measurements of tumor volume until the primary palpable tumor reaches 2.5 cm in diameter then euthanize the mice (tumors should only be present in mice expressing the WAP-TAg transgene). Calculate tumor volume using a formula: (length $\times$ width^2)/2 (Tsujii et al., 1998). Continuously breed female mice so that they undergo multiple cycles of pregnancy and parturition. At the time of euthanasia, determine and collect lung tissues from mice for future analysis.

2.2.4. Mammary tissue biopsies and histology

Remove either the left or right #4 inguinal mammary gland at day 17 after first pregnancy for mammary gland biopsies. Fix mammary tissues and tumors in 10% neutral formalin buffer; embed in paraffin, and section at 5 μm. At this point, tissue sections can be evaluated by immunostaining. Briefly, treat tissue sections with proteinase K (10 μg/ml) for 10 min at 37 °C and quench with 0.03% (v/v) H_2O_2 in PBS for 30 min at 25 °C. Block tissue sections with 10% normal horse serum in PBS for 1 h before incubation with the primary antibody. To determine if tumor cells were derived from the WAP-TAg transgene, probe tissue sections with anti-TAg (1:1000 in blocking solution) monoclonal antibody (Santa Cruz Biotechnologies) for 2 h at 25 °C. Wash slides three times for 5 min with PBS and incubate with goat anti-mouse (1:400 in blocking solution) secondary antibody (Santa Cruz Biotechnologies) for 1 h at 25 °C. Wash slides as stated above and incubate with an avidin–biotin-peroxidase complex (Vector Laboratories) for 30 min. Wash slides and incubate with 3,3'-diaminobenzidine (DAB, Sigma Aldrich) for 1–5 min depending on color development. Rinse slide, dehydrate, and counterstain nucleus using hematoxylin. Visualize using an inverted microscope with a 40× objective and process utilizing NIH Image software.

To analyze lung metastatic lesions, cut formalin-fixed lung tissues into 5 μm sections with each section separated from the next one by 50 μm. Stain three sections separated by 300 μm (sections 1, 6, and 11) with hematoxylin and eosin (H&E) and examine for the presence of metastases. The other sections can be used for TAg immunostaining, as stated above. Quantify lung metastases by capturing microscopic images of H&E-stained sections with a digital camera and process the images by NIH Image software. To determine the volume of pulmonary metastases (foci/lung area), calculate the image size (pixels) for each lung sample (equivalent to lung area) and then calculate the number of tumor foci per area of 10^4 pixels.

2.3. Limitations for transgenic mouse models

The SV40 TAg model is a strong tool for studying cancer progression; however, there are also limitations to this model (Shi et al., 2003). In this model, *maspin* expression is driven by the WAP promoter and is only strongly activated during pregnancy. Therefore, the WAP-TAg and bitransgenic mice require mating throughout the study to maintain transgene expression. It has also been shown that TAg-expressing mammary cells can continue to be tumorigenic-independent from the WAP promoter, which will influence experimental results (Li et al., 1996; Tzeng et al., 1993). Since TAg expression inactivates p53 *and* transcription of endogenous *maspin* is regulated by p53, the WAP-TAg and bitransgenic mice exhibit decreases in endogenous *maspin* expression (Zhang *et al.*, 2000; Zou *et al.*, 2000). Therefore, these results present compounding factors that can alter the balance between tumor promotion and suppressive factors. In order to counteract this potential tumorigenic effect, *maspin* expression must be specifically increased in tumors either by systemic delivery or cloning the maspin transgene under the control of a constitutive promoter. Subsequently, we developed a new syngeneic breast tumor mouse model to accomplish this goal.

3. Syngeneic Tumor Model

3.1. Development of syngeneic breast cancer model

3.1.1. Rationale of the experimental approach

Although the transgenic mouse model is appropriate for investigating the role of maspin in tumor progression, developing these mice is expensive, time-consuming, and technically challenging. Moreover, transcription of *maspin* and other tumor suppressive genes are regulated by p53 and pRb, an unwanted side-effect from WAP-TAg mice. Therefore, we wanted to develop another animal model that did not take as long as transgenic mice but could also specifically increase *maspin* expression in tumor cells without other compounding factors (like those seen in WAP-TAg transgenic mice).

We have known for decades that the mammary gland is a natural site for the implantation of both normal and neoplastic cells (Deome *et al.*, 1959; Medina, 1996; Medina and Daniel, 1996). In conjunction, a developed serial transplantation model selects for murine mammary cells that exhibit alveolar hyperplastic outgrowth and eventually produces transformed mammary cell populations. TM40D, an example of a mammary outgrowth cell line, was characterized as tumorigenic and metastatic (Kittrell *et al.*, 1992; Stickeler *et al.*, 1999). Therefore, we developed a syngeneic mammary tumor model where TM40D cells are implanted into the #4 mammary gland of BALB/c female mice to develop into invasive, metastatic tumors.

3.1.2. Establishing TM40D maspin transfectants

There are two ways to overexpress *maspin* in TM40D cells, stable clones from antibiotic selection (group 1) or stable retroviral transfection (group 2). Therefore, we will briefly review the development of these *maspin* transfectants. For group 1, insert human maspin cDNA (1.3 kb full length) into pEF1 expression vector (Invitrogen) by restriction enzyme digestion of *Eco*RI and *Xba*I (*note*: human and mouse *maspin* act similarly in this model so either cDNA can be used). In the pEF1 vector, *maspin* is regulated by the human elongation factor 1α promoter, which allows for general transcription. Mix 5.0–10.0 × 10^6 TM40D cells with 3 μg of either the pEF1-vector or pEF1-*maspin* constructs into an electroporation cuvette (0.4 cm path length). Using a Bio-Rad Cell Porator, apply an electrical pulse (200 V, 980 pF) to the cell–plasmid mixture then leave electroporated cells for 10 min. After incubation, transfer cells to a 100 mm dish containing growth media (DMEM/F12 supplemented with 2% adult bovine serum, 10 mg/ml insulin, 5 ng/ml epidermal growth factor, 5 μg/ml linoleic acid, 5 mg/ml bovine serum albumin, 200 units/ml nystatin, and 50 μg/ml gentamicin). Three days after electroporation, select TM40D clones with 300 μg/ml G418 for 2 weeks. Isolate total RNA and whole cell lysates to identify *maspin* clones by RT-PCR (using *maspin* specific primers: sense, 5′-AATTTAAGGTGGAAAAGATG; antisense, 5′-TCTATGGAATCC CCATCTTC) and Western blot (anti-maspin antibody obtained from BD Biosciences) analysis, respectively.

For group 2, human *maspin* cDNA were cloned into pS2-GFP (a retroviral vector derived from the pS2 family of retroviral vectors) wherein maspin cDNA and GFP were expressed independently from 5′ long terminal repeat and pECM promoters, respectively. Produce infective viral particles by transfecting the retroviral packaging HEK293T cells with pS2-*maspin* GFP or pS2-GFP plasmid constructs along with the pECO plasmid (Clontech) in Fugene reagent (Roche). After 72 h, remove the media containing viral supernatants and add 8 μg/ml polybrene. Place the transfection media onto subconfluent TM40D cells and incubate for 72 h at 37 °C. After transfection, split the transduced TM40D cells and place them into selection media (growth media supplemented with 100 μg/ml zeocin) for about 2 weeks to establish stable clones. Again, identify stable retroviral transfectants by RT-PCR (see above for primer sequences) and Western blot (anti-maspin antibody obtained from BD Biosciences) analysis.

3.1.3. Implantation of modified TM40D cells into BALB/c mice

For implantation studies, use 8-week-old female BALB/c mice (Harlan Sprague Dawley, Inc.). For group 1, evenly divide sisters for implantation of either TM40D-GFP or TM40D-*Maspin* clones. Grow tumor cells to about 70–85% of confluence before harvesting for implantation. Inject 5.0×10^5 (in 10 μl PBS) of either maspin or control TM40D cells into each #4 mammary gland. Make sure to inject the same cell type into both #4 mammary glands. Use a total of 10 mice per cell type, which corresponds to 20 total injection sites (cells injected into left and right mammary gland per mouse). All of the mice were sacrificed when the primary tumors grew to ~ 1 cm in diameter. For group 2 mice, inject 2.0×10^5 cells (in 10 μl PBS) of pS2-GFP or pS2-*maspin* into #4 mammary glands. Allow mice to grow for ~ 35 days after the appearance of primary tumor. Monitor tumor initiation every 2 days by palpation, tumor growth rate, and volume every-other day by caliper measurement (Shi *et al.*, 2001, 2002). Histology and immunostaining of proteins can be performed as described earlier (Section 2.2.4).

Our study demonstrated that maspin inhibited breast tumor growth and tumor metastasis. When maspin was integrated to chromosome by retrovirus, its inhibition was more effective compared to pEF plasmid-mediated maspin overexpression. Overall, this study strongly demonstrated that maspin might serve as an important antitumor and antimetastasis factor in cancer therapy (Shi et al., 2002).

3.2. *Maspin*-mediated gene therapy in mouse model of breast cancer

3.2.1. Rationale of the experimental approach

Gene therapy presents a novel, potent therapeutic approach for cancer. Gene therapy can help develop new strategies for selectively killing tumorigenic cells or arresting cell growth. The key gene therapy targets are tumor suppressor genes that are typically mutated or inactivated during tumorigenesis. Experimentally, gene therapy requires both a good animal model and an effective delivery system that exhibits low toxicity (McCormick, 2001). Therefore, we established this model to facilitate studies investigating the therapeutic value of the maspin gene in breast cancer therapy.

3.2.2. Development and implantation of high invasive MMTV-PyV (PyV MT-high) cells into FVB mice

The PyV MT parental tumor cell line was isolated from tumors of MMTV-PyV MT transgenic mice in the laboratory of William Muller (University of Toronto). To select for high invasiveness, pass cells through a Boyden chamber assay as previously described (Hendrix et al., 1987). Briefly, cover 10 μm polycarbonate membrane transwells with Matrigel (Becton Dickinson, Inc.) and plate 1×10^5 cells into the upper chamber and allow cells to migrate for 24 h to the lower chamber. Collect and culture the invasive cells (cells that migrated to the lower chamber). Subject the invasive cells to another round of Boyden chamber selection. The cells that result from double section are named PyV MT-high for high invasiveness and are ready for experimentation.

Grow PyV MT-high cells in DMEM with 10% FBS to 70–85% confluence before injections. Harvest cells and normalize to a concentration of 1.0×10^4 cells/μl PBS. Inject 7-week-old female FVB mice (Harlan, Inc.) into each #4 mammary gland with 5.0×10^5 cells in 10 μl PBS. Calculate tumor volume and growth rate as stated earlier (Section 3.2.1)

3.2.3. Liposome preparation

A number of methods have been developed to transfer DNA into eukaryotic cells; however, most of these methods suffer from problems related to cellular toxicity, poor reproducibility, and/or inefficiency of DNA delivery. Improved formulations of cationic lipids have increased DNA delivery in tissue culture systems; however, in animals, the intravenous (i.v.) delivery of DNA by cationic lipids has been less efficient (Templeton et al., 1997). Improved efficiency of DNA:liposome complexes have been identified using an equimolar concentration of cationic lipid:neutral lipid, namely 1,2-bis(oleoyloxy)-3-(trimethylammonio)propane:cholesterol (DOTAP: Chol) liposomes. Develop these DOTAP:Chol liposomes as previously described (Templeton et al., 1997). Briefly, mix equimolar concentrations of the lipids and dissolve in HPLC-grade chloroform in a 1-l round bottom flask. Remove excess chloroform on a Buchi rotary evaporator at 30 °C for 30 min. Remnants of the mixture are a clear thin film that can be dried under vacuum for 15 min. Hydrate the film with 5% (w/v) dextrose in water to a final concentration of 20 mM. Heat the hydrated lipid film by rotating in a water bath at 50 °C for 45 min then 35 °C for 10 min. Allow the mixture to stand overnight in the parafilm covered flask at room temperature. The next day, sonicate at low frequency for 5 min at 50 °C then transfer to a tube and heat for 10 min at 50 °C. Last, sequentially filter through Whatman filters 1.0, 0.45, 0.2, and 0.1 μm. The DOTAP: Chol liposomes are now ready for use.

3.2.4. Determine efficiency of liposome:gene construct

Prepare DNA:liposome complexes the day before injection. First, monitor the DOTAP:Chol liposomes delivery efficiency by transfecting a chloramphenicol acetyltransferase (CAT) DNA reporter construct (p4119) into mice (Templeton et al., 1997). To do this, inject 40 μg p4119 CAT plasmid DNA in 80 μl of 4 mM DOTAP:Chol into the tail veins of 6-week-old (~20 g) BALB/c mice. In addition, inject one control group of mice with DOTAP:Chol only, to ascertain liposome toxicity. Sacrifice mice 24 h after injection, harvest organs, and snap freeze them in liquid nitrogen. Extract tissue proteins as described (Stribling et al., 1992). Briefly, homogenize lung and spleen tissues in 250 mM Tris–HCl (pH 7.5) supplemented with 5 mM EDTA or for liver, heart, kidney, and mammary gland tissues in 250 mM Tris–HCl (pH 7.5) plus 5 mM EDTA and protease inhibitors aprotinin, E-64, and leupeptin (Roche). Lyse cells by three freeze/thaw cycles, heat lysate (65 °C for 10 min), and centrifuge (16,000×g for 2 min). Normalize protein concentrations with the BCA protein assay kit (Pierce). Assay CAT activities in tissue extracts using the CAT ELISA kit (Roche) per manufacturer's instructions and then correct all CAT protein determinations from no CAT plasmid extracts controls.

3.2.5. Maspin gene delivery and analysis of tumor

By injecting the PyV MT-high cells, all mice should develop palpable tumors about 6 weeks after implantation. Deliver pEF1-maspin (human *maspin* cDNA) and pEF1-vector control using an optimized concentration of 150 mg of DNA in 200 μl of 4 mM DOTAP:Chol, amount based on previous findings (Templeton *et al.*, 1997). During the first week (when all tumors have a volume smaller than 0.5 cm^3), administer intratumoral injections (50 μl/site) of the DNA:liposome complexes twice to the mice. During weeks 2–5, administer 200 μl (per mouse) of the DNA:liposome mixture into the tail vein (i.v.), twice a week. Sacrifice mice 6 days after the last treatment. In addition, assay for any potential toxicity of DNA:liposome complex by treating two FVB female mice without tumor transplantation with DNA:liposome complex at the dosage mentioned above through tail vein injection twice a week for 8 weeks. Isolate total RNA from samples and conduct RT-PCR analysis using specific human maspin primers (sense, 5′-AATTTAAGGTGGAAAAGATG; antisense, 5′-TCTATGGAATCC CCATCTTC). Normalize *maspin* expression to the control L19 ribosomal gene using specific primers: sense, 5′-CTGAAGGTCAAAGGGAATGT; antisense, 5′- GGACAGAGTCTTGATGATCTC.

Additionally, examine metastasis by running PCR analysis for polyoma middle T antigen gene, a marker for PyV MT-high tumor cells. First, digest liver, spleen, and lymph nodes (containing pooled inguinal lymph nodes) in buffer containing 10 mM Tris–HCl (pH 7.5), 10 mM EDTA, 10 mM NaCl, 0.5% Sarcosyl, and 0.5 mg/ml proteinase K. Then, isolate DNA by ethanol precipitation and run PCR amplification using primers: sense, 5′-GGAAGCAAGTACTTCACAAGGG; antisense, 5′-GGAAAGTCAC-TAGGAGCA. Use the following PCR reaction conditions: 95 °C for 1 min, 52 °C for 1 min, 72 °C for 1 min and 30 s, for 35 cycles. Run immunohistochemical analysis and determine tumor volume, growth rate, and rate of metastasis as described earlier (Section 2.2.4).

Using the above gene therapy approach, we showed systemic delivery of a *maspin* DNA:liposome complex inhibits breast tumor growth. We observed that both the tumor size and overall tumor growth rate were significantly decreased for maspin-treated tumors relative to the controls. In addition, a significant decrease in lung metastasis was observed in the maspin-treated mice, as evidenced by a decrease in both the size and number of lung tumor foci. We further demonstrated that such inhibition of breast cancer progression is mediated by increased apoptosis in the maspin-treated tumors.

4. Conclusion

When developing new experimental hypotheses, it is important to choose the appropriate model system that adequately tests the experimental hypothesis. Metastasis is a complex, multistep process involving the detachment of neoplastic cells from the primary tumor, degradation of the basement membrane, tumor cell migration, intravasation and extravasation of the tumor cells, and establishment in distant organs. Cell culture systems allow for direct human cell analysis in an artificially derived system, which are appropriate for answering some basic science hypotheses but needs some *in vivo* confirmation prior to use as therapy in humans. Yet, the keystone animal models are necessary to recapitulate the complex networks that are involved in tumorigenesis. In particular, many cells that comprise the tumor microenvironment are present in animal models. Currently, there are few animal models available for the analysis of breast tumor metastasis. In this chapter, we describe *maspin* transgenic and bitransgenic mice consisting of maspin transgenic and oncogenic mice targeting mammary gland-specific expression of oncogenes (Sections 1 and 2). These transgenic mice allow for proper examination of maspin in tumor development and progression but not without some limitations. Some include long-term observation (months), variation in tumor development, and not all of these tumors are metastatic. To examine future therapeutic options regarding maspin, we developed a better metastasis model using a syngeneic model that can also be used for gene therapy (Section 3). Together, this chapter demonstrates the ability to generate mouse models to investigate the role of maspin in tumor progression. Similar strategy can be applied to study other tumor suppressor genes in their control of cancer progression *in vivo*.

REFERENCES

Alexander, C. M., Howard, E. W., Bissell, M. J., and Werb, Z. (1996). Rescue of mammary epithelial cell apoptosis and entactin degradation by a tissue inhibitor of metalloproteinases-1 transgene. *J. Cell Biol.* **135**, 1669–1677.

Andres, A. C., Schonenberger, C. A., Groner, B., Hennighausen, L., LeMeur, M., and Gerlinger, P. (1987). Ha-ras oncogene expression directed by a milk protein gene promoter: Tissue specificity, hormonal regulation, and tumor induction in transgenic mice. *Proc. Natl. Acad. Sci. USA* **84,** 1299–1303.

Campbell, S. M., Rosen, J. M., Hennighausen, L. G., Strech-Jurk, U., and Sippel, A. E. (1984). Comparison of the whey acidic protein genes of the rat and mouse. *Nucleic Acids Res.* **12**, 8685–8697.

Cho, A., Haruyama, N., and Kulkarni, A. B. (2009). Generation of transgenic mice. *Curr. Protoc. Cell Biol.* **42,** 19.11.1–19.11.22.

Conner, D. A. (2004). Transgenic mouse production by zygote injection. *Curr. Protoc. Mol. Biol.* **68,** 23.9.1–23.9.29.

de Cremoux, P., Vincent Salomon, A., Liva, S., Dendale, R. M., Bouchind'homme, B., Martin, E., Sastre-Garau, X., Magdelenat, H., Fourquet, A., and Soussi, T. (1999). p53 mutation as a genetic trait of typical medullary breast carcinoma. *J. Natl. Cancer Inst.* **91,** 641–643.

DeNardo, D., Johansson, M., and Coussens, L. (2008). Immune cells as mediators of solid tumor metastasis. *Cancer Metastasis Rev.* **27,** 11–18.

Deome, K. B., Faulkin, L. J., Jr., Bern, H. A., and Blair, P. B. (1959). Development of mammary tumors from hyperplastic alveolar nodules transplanted into gland-free mammary fat pads of female C3H mice. *Cancer Res.* **19,** 515–520.

Dyson, N., Buchkovich, K., Whyte, P., and Harlow, E. (1989). The cellular 107K protein that binds to adenovirus E1A also associates with the large T antigens of SV40 and JC virus. *Cell* **58,** 249–255.

Gasco, M., Shami, S., and Crook, T. (2002). The p53 pathway in breast cancer. *Breast Cancer Res.* **4,** 70–76.

Gettins, P. G., Simonovic, M., and Volz, K. (2002). Pigment epithelium-derived factor (PEDF), a serpin with potent anti-angiogenic and neurite outgrowth-promoting properties. *J. Biol. Chem.* **383,** 1677–1682.

Hendrix, M. J., Seftor, E. A., Seftor, R. E., and Fidler, I. J. (1987). A simple quantitative assay for studying the invasive potential of high and low human metastatic variants. *Cancer Lett.* **38,** 137–147.

Hennighausen, L. G., and Sippel, A. E. (1982). Characterization and cloning of the mRNAs specific for the lactating mouse mammary gland. *Eur. J. Biochem.* **125,** 131–141.

Irving, J. A., Pike, R. N., Lesk, A. M., and Whisstock, J. C. (2000). Phylogeny of the serpin superfamily: Implications of patterns of amino acid conservation for structure and function. *Genome Res.* **10,** 1845–1864.

Irving, J. A., Steenbakkers, P. J., Lesk, A. M., Op den Camp, H. J., Pike, R. N., and Whisstock, J. C. (2002). Serpins in prokaryotes. *Mol. Biol. Evol.* **19,** 1881–1890.

Ittner, L. M., and Gotz, J. (2007). Pronuclear injection for the production of transgenic mice. *Nat. Protoc.* **2,** 1206–1215.

Kittrell, F. S., Oborn, C. J., and Medina, D. (1992). Development of mammary preoplasias in vivo from mouse mammary epithelial cell lines in vitro. *Cancer Res.* **52,** 1924–1932.

Li, M., Hu, J., Heermeier, K., Hennighausen, L., and Furth, P. A. (1996). Expression of a viral oncoprotein during mammary gland development alters cell fate and function: Induction of p53-independent apoptosis is followed by impaired milk protein production in surviving cells. *Cell Growth Differ.* **7,** 3–11.

Li, M., Lewis, B., Capuco, A. V., Laucirica, R., and Furth, P. A. (2000). WAP-TAg transgenic mice and the study of dysregulated cell survival, proliferation, and mutation during breast carcinogenesis. *Oncogene* **19,** 1010–1019.

McCormick, F. (2001). Cancer gene therapy: Fringe or cutting edge? *Nat. Rev. Cancer* **2,** 130–141.

Medina, D. (1996). The mammary gland: A unique organ for the study of development and tumorigenesis. *J. Mammary Gland Biol. Neoplasia* **1,** 5–19.

Medina, D., and Daniel, C. (1996). Experimental models of development, function, and neoplasia. *J. Mammary Gland Biol. Neoplasia* **1,** 3–4.

Mietz, J. A., Unger, T., Huibregtse, J. M., and Howley, P. M. (1992). The transcriptional transactivation function of wild-type p53 is inhibited by SV40 large T-antigen and by HPV-16 E6 oncoprotein. *EMBO J.* **11,** 5013–5020.

Nagy, A., Gertsenstein, M., Vintersten, K., and Behringer, R. (2010). In vitro screen to obtain widespread, transgenic expression in the mouse. *Cold Spring Harb. Protoc.* **8,** 1101.

Pemberton, P. A., Tipton, A. R., Pavloff, N., Smith, J., Erickson, J. R., Mouchabeck, Z. M., and Kiefer, M. C. (1997). Maspin is an intracellular serpin that partitions into secretory vesicles and is present at the cell surface. *J. Histochem. Cytochem.* **45,** 1697–1706.

Pittius, C. W., Sankaran, L., Topper, Y. J., and Hennighausen, L. (1988). Comparison of the regulation of the whey acidic protein gene with that of a hybrid gene containing the whey acidic protein gene promoter in transgenic mice. *Mol. Endocrinol.* **2,** 1027–1032.

Potempa, J., Korzus, E., and Travis, J. (1994). The serpin superfamily of proteinase inhibitors: Structure, function, and regulation. *J. Biol. Chem.* **269,** 15957–15960.

Remold-O'Donnell, E. (1993). The ovalbumin family of serpin proteins. *FEBS Lett.* **315,** 105–108.

Richmond, A., and Su, Y. (2008). Mouse xenograft models vs GEM models for human cancer therapeutics. *Dis. Model Mech.* **1,** 78–82.

Sheng, S., Pemberton, P. A., and Sager, R. (1994). Production, purification, and characterization of recombinant maspin proteins. *J. Biol. Chem.* **269,** 30988–30993.

Sheng, S., Carey, J., Seftor, E. A., Dias, L., Hendrix, M. J., and Sager, R. (1996). Maspin acts at the cell membrane to inhibit invasion and motility of mammary and prostatic cancer cells. *Proc Natl. Acad. Sci. USA* **93,** 11669–11674.

Shi, H. Y., Zhang, W., Liang, R., Abraham, S., Kittrell, F. S., Medina, D., and Zhang, M. (2001). Blocking tumor growth, invasion, and metastasis by maspin in a syngeneic breast cancer model. *Cancer Res.* **61,** 6945–6951.

Shi, Y., Liang, R., Templeton, N., and Zhang, M. (2002). Characterization and systemic treatment of maspin in a breast metastasis model. *Mol. therapy* **5**(6), 755–761.

Shi, H. Y., Zhang, W., Liang, R., Kittrell, F., Templeton, N. S., Medina, D., and Zhang, M. (2003). Modeling human breast cancer metastasis in mice: Maspin as a paradigm. *Histol. Histopathol.* **18,** 201–206.

Silverman, G. A., Bird, P. I., Carrell, R. W., Church, F. C., Coughlin, P. B., Gettins, P. G., Irving, J. A., Lomas, D. A., Luke, C. J., Moyer, R. W., Pemberton, P. A., Remold-O'Donnell, E., *et al.* (2001). The serpins are an expanding superfamily of structurally similar but functionally diverse proteins. Evolution, mechanism of inhibition, novel functions, and a revised nomenclature. *J. Biol. Chem.* **276,** 33293–33296.

Sternlicht, M. D., and Barsky, S. H. (1997). The myoepithelial defense: A host defense against cancer. *Med. Hypotheses* **48,** 37–46.

Stickeler, E., Kittrell, F., Medina, D., and Berget, S. M. (1999). Stage-specific changes in SR splicing factors and alternative splicing in mammary tumorigenesis. *Oncogene* **18,** 3574–3582.

Stribling, R., Brunette, E., Liggitt, D., Gaensler, K., and Debs, R. (1992). Aerosol gene delivery in vivo. *Proc. Natl. Acad. Sci. USA* **89,** 11277–11281.

Sympson, C. J., Talhouk, R. S., Alexander, C. M., Chin, J. R., Clift, S. M., Bissell, M. J., and Werb, Z. (1994). Targeted expression of stromelysin-1 in mammary gland provides evidence for a role of proteinases in branching morphogenesis and the requirement for an intact basement membrane for tissue-specific gene expression. *J. Cell Biol.* **125,** 681–693.

Talhouk, R. S., Chin, J. R., Unemori, E. N., Werb, Z., and Bissell, M. J. (1991). Proteinases of the mammary gland: Developmental regulation in vivo and vectorial secretion in culture. *Development* **112,** 439–449.

Templeton, N. S., Lasic, D. D., Frederik, P. M., Strey, H. H., Roberts, D. D., and Pavlakis, G. N. (1997). Improved DNA: Liposome complexes for increased systemic delivery and gene expression. *Nat. Biotechnol.* **15,** 647–652.

Tombran-Tink, J., Li, A., Johnson, M. A., Johnson, L. V., and Chader, G. J. (1992). Neurotrophic activity of interphotoreceptor matrix on human Y79 retinoblastoma cells. *J. Comp. Neurol.* **317,** 175–186.

Tsujii, M., Kawano, S., Tsuji, S., Sawaoka, H., Hori, M., and DuBois, R. N. (1998). Cyclooxygenase regulates angiogenesis induced by colon cancer cells. *Cell* **93,** 705–716.

Tzeng, Y. J., Guhl, E., Graessmann, M., and Graessmann, A. (1993). Breast cancer formation in transgenic animals induced by the whey acidic protein SV40 T antigen (WAP-SV-T) hybrid gene. *Oncogene* **8,** 1965–1971.

Zhang, M., Sheng, S., Maass, N., and Sager, R. (1997). mMaspin: The mouse homolog of a human tumor suppressor gene inhibits mammary tumor invasion and motility. *Mol. Med.* **3,** 49–59.

Zhang, M., Magit, D., Botteri, F., Shi, H. Y., He, K., Li, M., Furth, P., and Sager, R. (1999). Maspin plays an important role in mammary gland development. *Dev. Biol.* **215,** 278–287.

Zhang, M., Shi, Y., Magit, D., Furth, P. A., and Sager, R. (2000). Reduced mammary tumor progression in WAP-TAg/WAP-maspin bitransgenic mice. *Oncogene* **19**, 6053–6058.

Zou, Z., Anisowicz, A., Hendrix, M. J., Thor, A., Neveu, M., Sheng, S., Rafidi, K., Seftor, E., and Sager, R. (1994). Maspin, a serpin with tumor-suppressing activity in human mammary epithelial cells. *Science* **263**, 526–529.

Zou, Z., Gao, C., Nagaich, A. K., Connell, T., Saito, S., Moul, J. W., Seth, P., Appella, E., and Srivastava, S. (2000). p53 regulates the expression of the tumor suppressor gene maspin. *J. Biol. Chem.* **275**, 6051–6054.

CHAPTER NINE

Hsp47 as a Collagen-Specific Molecular Chaperone

Yoshihito Ishida *and* Kazuhiro Nagata

Contents

1. Introduction	168
2. Hsp47 as a Collagen-Binding Protein in the ER	169
3. Interaction and Recognition of Collagen by Hsp47	170
4. A Phenotype and Abnormal Collagen Maturation in Hsp47 Knockout Mice	171
5. Possible Roles of Hsp47 in Procollagen Maturation in the ER	172
6. Hsp47 Null Cells: A Tool for Studying the Fate of Misfolded Collagen	174
7. Regulation of Hsp47 Expression and Its Clinical Importance	176
References	178

Abstract

Heat shock protein (HSP) 47 is a 47 kDa collagen-binding glycoprotein localized in the endoplasmic reticulum (ER). It belongs to the serpin family and contains a serpin loop, although it does not have serine protease inhibitory activity. The induction of Hsp47 by heat shock is regulated by a heat shock element in its promoter region, while the constitutive and tissue-specific expression of Hsp47 correlates with that of collagen and is regulated via enhancer elements located in the promoter and intron regions. Hsp47 transiently binds to procollagen in the ER and dissociates in the *cis*-Golgi or ER-Golgi intermediate compartment region (ERGIC). Gene ablation studies indicated that Hsp47 is essential for embryonic development and the maturation of several types of collagen. The requirement for Hsp47 in collagen maturation may reflect its ability to inhibit collagen aggregation by binding procollagen in the ER and facilitate triple helix formation. In Hsp47-deficient cells, misfolded procollagen aggregates in the ER are degraded by the autophagy–lysosome pathway but not through the ubiquitin proteasome pathway. Hsp47 may be a therapeutic target for collagen-related disorders such as fibrosis, which feature abnormal accumulations of collagen and increased expression of Hsp47. This is supported by mouse models of

Laboratory of Molecular and Cellular Biology, Department of Molecular Biosciences, Faculty of Life Sciences, Kyoto Sangyo University, Kyoto, Japan fibrosis in which knockdown of Hsp47 clearly decreased the accumulation of collagen in fibrotic tissues and prevented the promotion of fibrosis. On the other hand, mutations in Hsp47 cause collagen-related genetic diseases such as osteogenesis imperfecta. Thus, Hsp47 is an indispensible molecular chaperone specific for collagen that is important in several major human diseases.

1. INTRODUCTION

In response to a range of stresses such as heat shock, cells respond by inducing a group of proteins called heat shock proteins (HSPs) (Richter et al., 2010). Most HSPs, many of which are well conserved from bacteria to mammals, function as molecular chaperones which facilitate the correct folding of nascent polypeptides into their native conformations and prevent aggregate formation by misfolded proteins in response to various stress conditions (Bukau et al., 2006). Thus, molecular chaperones maintain protein homeostasis under normal and stress conditions.

Molecular chaperones are located in various organelles, although many kinds are located at a high concentration in the endoplasmic reticulum (ER), a major organelle in which approximately 30% of total cellular proteins are synthesized, particularly secretory and membrane proteins (Ghaemmaghami et al., 2003; Huh et al., 2003). In addition to molecular chaperones such as glucose-regulated protein 78 (BiP), productive folding of nascent polypeptides in the ER also requires various enzymes involved in N-glycosylation and disulfide bond formation between cysteine residues (Anelli and Sitia, 2008). Only correctly folded proteins are allowed to be transported from the ER to their final destinations. In response to various stresses in the ER ("ER stress"), once protein folding is impaired and misfolded proteins are accumulated in the ER, which triggers the ER stress, various ER-resident molecular chaperones are induced. Molecular chaperones thus induced by ER stress are involved in refolding of misfolded proteins as well as degradation of potentially toxic misfolded or aggregated proteins, the mechanism of which is called "ER quality control."

Hsp47 is an ER-resident stress protein with a unique character as a molecular chaperone that specifically binds to procollagen. Interestingly, Hsp47 is the only stress protein in the ER induced by heat shock; the other ER stress proteins are induced by ER stress. While other molecular chaperones have broad substrate specificity, Hsp47 specifically binds to procollagens in the ER and in addition, Hsp47 has a characteristic expression pattern in cells and tissues that correlates closely with collagen expression (Nagata, 1996, 1998). Hsp47 belongs to the serpin (serine protease inhibitor) superfamily but does not have inhibitory activity for serine proteases (Nagata, 2003a).

In this chapter, we review the molecular function of Hsp47 in collagen biogenesis, the importance of Hsp47 in mouse development, and the clinical importance of Hsp47 in various collagen-related diseases including fibrosis.

2. Hsp47 as a Collagen-Binding Protein in the ER

Hsp47 was first identified in chick embryos as a 47 kDa collagen-binding protein resident in the ER, with a basic isoelectric point ($pI = 9.0$; Nagata and Yamada, 1986). Hsp47 was shown to be a homologue of colligin in mice (Kurkinen et al., 1984) and gp46 in rats (Cates et al., 1987). The molecular function of Hsp47 is one of the most well characterized among serpin family proteins located in the ER (Ragg, 2007).

The expression of Hsp47 is dramatically enhanced by heat shock (Nagata and Yamada, 1986), but not by ER stress. This is a unique feature of Hsp47 because all other mammalian stress proteins in the ER are induced by ER stress, not by heat shock. In addition to the heat shock induction, the constitutive expression of Hsp47 always correlates with that of collagen (see later). For example, the expression of Hsp47 in chick embryo fibroblasts (CEFs) decreased after malignant transformation with Rous sarcoma virus (Nagata et al., 1986). Analysis of the nucleotide sequence of Hsp47 cDNA revealed that it belongs to the serpin superfamily (Hsp47 is also known as serpin H1) and has a signal sequence at the N-terminus, two N-glycosylation sites, and an ER-retention signal consisting of Arg-Asp-Glu-Leu (RDEL) at the C-terminus (Hirayoshi et al., 1991; Hughes et al., 1987). Localization of Hsp47 in the ER was confirmed by immunofluorescence and immunoelectron microscopic analysis (Saga et al., 1987), and deletion of the RDEL sequence from its C-terminus caused the rapid secretion of Hsp47 from the cells (Satoh et al., 1996).

Hsp47 binds to collagen in a pH-dependent manner: it binds to collagen at a neutral pH and dissociates from collagen below pH 6.3 (Saga et al., 1987). By taking advantage of this property, Hsp47 can be easily purified using collagen-coupled Sepharose beads after elution with a low pH buffer (around pH 6.0).

Hsp47 transiently associates with procollagen in the pH-neutral ER, but dissociates from it in the lower pH environment of the cis-Golgi or ER-Golgi intermediate compartment (ERGIC) (Nakai et al., 1992). Hsp47 has been shown to prevent the aggregation of procollagen in vitro (Thomson and Ananthanarayanan, 2000). Therefore, the association of Hsp47 with procollagen in the ER may function to prevent the aggregation and/or bundle formation of procollagen in the ER, thereby facilitating the transport of procollagen from the ER to Golgi.

3. INTERACTION AND RECOGNITION OF COLLAGEN BY HSP47

In vitro analysis using the surface plasmon resonance revealed that recombinant mouse Hsp47 can bind to collagen types I–V with dissociation constants of 10^{-6}–10^{-7} M (Natsume *et al.*, 1994). Estimates of dissociation and association rate constants suggested a rapid association and dissociation, which is likely to contribute to the transient nature of the interaction between Hsp47 and procollagen in the secretory pathway (Nakai *et al.*, 1992). Based on pulse-chase experiments, Hsp47 can rapidly bind to newly synthesized procollagen within the ER, followed by the rapid secretion of procollagen within a further 10–15 min. Treatment of cells with brefeldin A1, which inhibits the secretory pathway between the ER and *cis*-Golgi, blocked the dissociation of Hsp47 from procollagen, while treatment with bafilomycin A1, which inhibits the transport of secretory proteins at the post-*cis*-Golgi or post-*trans*-Golgi, did not affect the interaction (Satoh *et al.*, 1996). Thus, these authors suggested that Hsp47 transiently binds to procollagen in the ER and rapidly dissociates at the *cis*-Golgi or ERGIC due to the low pH at these sites. After dissociation from procollagen, it is thought that Hsp47 is recycled back to the ER in a process mediated by the KDEL receptor (Fig. 9.1) (Satoh *et al.*, 1996).

The consensus sequence on procollagen that was required for the binding of Hsp47 was determined using a synthetic peptide approach. Multiple types of collagen share a common Gly-X-Y repeat in their primary sequence in which the Y position is often hydroxylated Proline (Pro) (Berg and Prockop, 1973). Synthetic peptides containing more than

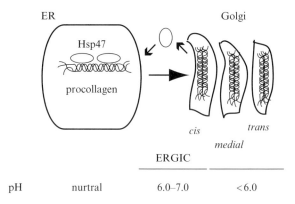

Figure 9.1 The interaction of Hsp47 and collagen. Hsp47 transiently binds to procollagen in the ER and dissociates in the *cis*-Golgi or ERGIC, whose pH is around 6.0. The binding fashion of Hsp47 and collagen is totally dependent on pH.

seven Gly-X-Y repeats were able to bind to Hsp47 (Koide *et al.*, 1999). The binding of Hsp47 with these synthetic peptides was stronger at 24 °C than at 30 °C. Considering that synthetic peptides containing more than seven Gly-X-Y repeats can form a triple helix at 24 °C, but not at 30 °C, it was suggested that Hsp47 preferentially binds to the triple helix form of collagen-like peptides (Koide *et al.*, 2000). Interestingly, replacement of Pro with Arg at the Y position markedly increased the binding with Hsp47, suggesting that a Gly-X-Arg sequence contained in the Gly-X-Y repeats serves as a binding motif for Hsp47 when these polypeptides form a triple helix (Koide *et al.*, 2002). When the C-termini of the synthetic model peptides of collagen linked to each other by the disulfide bond formation via the cysteine residues introduced at the C-terminus, the binding of Hsp47 was markedly enhanced, further suggesting that triple helix formation is a prerequisite for the interaction of Hsp47 with collagen model peptides (Koide *et al.*, 2006). In addition to these *in vitro* analyses, this conclusion was also supported by two independent studies using semipermeabilized cells (Hendershot and Bulleid, 2000; Tasab *et al.*, 2000, 2002).

4. A Phenotype and Abnormal Collagen Maturation in Hsp47 Knockout Mice

To further study the biological function of Hsp47 *in vivo*, *hsp47* gene was disrupted in mice using the homologous recombination technique (Nagai *et al.*, 2000). Mice lacking *hsp47* did not survive beyond 11.5 days postcoitus (dpc), indicating that Hsp47 is essential for mouse development. At 10.5 dpc, the embryos of hsp47$^{-/-}$ mice were still viable but were about one size third of those of the wild type and showed developmental retardation and cardiac hypertrophy.

Collagen fibril accumulation in the tissues was markedly decreased in *hsp47*$^{-/-}$ mice. In *hsp47*$^{-/-}$ cells, the secretion of type I and IV collagens was delayed (Ishida *et al.*, 2006; Matsuoka *et al.*, 2004) leading to the intracellular accumulation of procollagen. Immunohistological analysis of *hsp47*$^{-/-}$ embryos and fibroblasts showed that type I and IV procollagens accumulated in the ER, causing dilation of the ER (Ishida *et al.*, 2006; Marutani *et al.*, 2004). The secretion and accumulation of other extracellular matrix (ECM) proteins such as fibronectin and laminin were unaffected, indicating that *hsp47* deficiency specifically altered procollagen maturation.

N-propeptides were not processed in type I procollagen secreted from *hsp47*$^{-/-}$ cells and were retained within the collagen sequence in the extracellular spaces (Ishida *et al.*, 2006). Type I and IV collagens secreted from *hsp47*$^{-/-}$ cells were sensitive to trypsin/chymotrypsin digestion, unlike the collagens from wild-type cells (Matsuoka *et al.*, 2004; Nagai

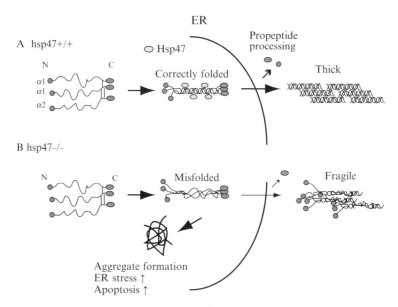

Figure 9.2 Abnormal maturation in $hsp47^{-/-}$ cells. Collagen forms trimer in the ER, and triple helix formation proceeds from C-terminus to the N-terminus in a zipper-like manner. (A) Correctly folded collagen is secreted to outer cell, and rigid collagen fibrils are formed by assisting of Hsp47. (B) Without Hsp47, collagen folding is impaired, and it is retained in the ER. Aggregate formation and induction of ER stress are observed, finally triggering ER stress induced apoptosis. N-propeptides of procollagen secreted from $hsp47^{-/-}$ cells are not processed owing to improper folding of triple helix.

et al., 2000). These data suggested that procollagens were not correctly folded into the triple helix in the ER of $hsp47^{-/-}$ embryos and fibroblasts (Fig. 9.2).

It is likely that ER stress was induced in $hsp47^{-/-}$ embryo due to the accumulation of procollagen in the ER. The induction of ER stress in $hsp47^{-/-}$ cells was demonstrated by the induction of CHOP (Marutani et al., 2004) and by the splicing of XBP-1 mRNA. Apoptosis caused by ER stress was also observed in $hsp47^{-/-}$ cells (Ishida et al., 2009).

5. Possible Roles of Hsp47 in Procollagen Maturation in the ER

After incorporating the possible function of Hsp47, we can draw the following brief sketch of procollagen maturation in the ER. Newly synthesized polypeptides of procollagen, which are translated as procollagen chains, are cotranslationally inserted into the ER (Lamande and Bateman,

1999; Nagata, 1996). In the case of type I collagen, two α1 chains and one α2 chain form a trimer linked by disulfide bonds at the C-terminal domain. Intramolecular folding of polypeptides inserted into the ER is prevented by molecular chaperones such as protein disulfide isomerase (PDI) until the three α chains form a triple helix from the C-terminus. Various folding enzymes, oxidoreductases, and molecular chaperones are involved in this process (Bachinger et al., 1980; Bulleid et al., 1997; Engel and Prockop, 1991; Lamande and Bateman, 1999).

Collagen has a unique characteristic that its molecular surface is still hydrophobic after it completes folding and triple helix formation. In the ER, Hsp47 binds to the hydrophobic surface of the triple helix region of procollagen ER and is thought to facilitate the transport of procollagen to *cis*-Golgi by preventing procollagen aggregate formation (Bonfanti et al., 1998). Procollagen molecules are unstable even after triple helix formation, and unfolding can occur at body temperature (Leikina et al., 2002; Makareeva and Leikin, 2007). Hsp47 is presumed to prevent local unfolding or reverse unfolding of triple helix back to the C-terminus, thereby facilitating procollagen triple helix formation (Fig. 9.3).

After completion of triple helix formation, procollagen is transported from the ER to the Golgi apparatus. During this process, Hsp47 dissociates

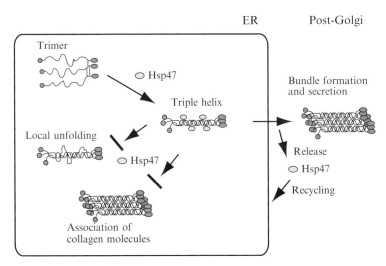

Figure 9.3 A possible role in Hsp47 for collagen maturation. Newly synthesized collagen is inserted into the ER, and trimer and triple helix formation occurs. Hsp47 preferentially binds to triple helical procollagen. One hypothesis is that Hsp47 binds to triple helical procollagen to prevent association of collagen molecule and formation of aggregation in the ER since collagens have a characteristic to associate and form aggregate under the neutral condition. Another hypothesis is that Hsp47 hampers the local unfolding of triple helical procollagen. (See Color Insert.)

from procollagen presumably in the *cis*-Golgi in response to the decreased pH of this compartment. Dissociated Hsp47 would be trapped by KDEL receptors in the Golgi and retrogradely transported from the Golgi to the ER (Nakai *et al.*, 1992; Satoh *et al.*, 1996). Procollagen is then secreted from the cells, where the *N*- and *C*-propeptides are cleaved by specific proteases, followed by the formation of collagen bundles in the ECM (Dombrowski and Prockop, 1988; Li *et al.*, 1996). In $hsp47^{-/-}$ cells, collagen forms aggregates (Ishida *et al.*, 2006), and collagen fibrillogenesis is repaired by the addition of Hsp47 into the $hsp47^{-/-}$ cells.

6. HSP47 NULL CELLS: A TOOL FOR STUDYING THE FATE OF MISFOLDED COLLAGEN

In $hsp47^{-/-}$ fibroblasts, triple helix formation of procollagen was totally impaired and its secretion was markedly inhibited or delayed, leading to the accumulation of procollagen in the ER (Ishida *et al.*, 2006). In general, cells are equipped with two strategies to dispose of intracellular misfolded proteins: the autophagy–lysosome and ubiquitin–proteasome systems (Vembar and Brodsky, 2008). We investigated which of these pathways contributed to the disposal of accumulated procollagen in the $hsp47^{-/-}$ cells.

The accumulation of misfolded proteins in the ER causes ER stress, which activates the unfolded protein response (UPR). The UPR was activated in $hsp47^{-/-}$ cells due to the accumulation of misfolded procollagen in the ER. Once the UPR is activated, the misfolded proteins are removed by ER-associated degradation (ERAD) (Hoseki *et al.*, 2010; Yoshida, 2007), in which misfolded substrates are retrogradely transported from the lumen of the ER to the cytosol and then degraded by proteasomes after polyubiquitination. However, we found that the misfolded procollagen that accumulated in $hsp47^{-/-}$ cells was not degraded by ERAD (Ishida *et al.*, 2009).

Another major mechanism for elimination of damaged proteins is the autophagy–lysosome pathway. Autophagy is a bulk degradation system that is involved in the stress response to nutrient starvation or organelle turnover (Mizushima, 2009; Mizushima *et al.*, 2008). The disruption of autophagic activity caused the severe accumulation of ubiquitinated proteins and damaged mitochondria, suggesting that autophagy is a dominant clearance system for maintaining cellular and protein homeostasis (Iwata *et al.*, 2006; Komatsu *et al.*, 2005, 2006, 2007a,b).

We found that the procollagen that accumulated in $hsp47^{-/-}$ cells was eliminated by autophagy–lysosome system (Ishida *et al.*, 2009). The amount of intracellular procollagen in $hsp47^{-/-}$ cells was not affected by treatment with the proteasome inhibitors MG132 and lactacystin. In contrast, the treatment of $hsp47^{-/-}$ cells with the lysosome inhibitors bafilomycin A1 and E64d/pepA increased the amount of accumulated procollagen in the ER, indicating that the lysosome pathway was active in the disposal of these collagens. Further, the knockdown and overexpression of autophagy-related genes, including ATG5 (Kuma et al., 2004) and ATG4 mutant (Fujita et al., 2009), caused an increase in the amount of accumulated procollagen in $hsp47^{-/-}$ cells (Ishida et al., 2009). Interestingly, the down-regulation of autophagic activity in these siRNA experiments resulted in apoptosis of the $hsp47^{-/-}$ cells, suggesting that the autophagic disposal of misfolded procollagen contributes to cell survival.

Many mutations have been identified in collagen genes, particularly type I collagen, which cause severe genetic diseases such as osteogenesis imperfecta (OI) (Marini et al., 2007). In many of these diseases, procollagen fails to form a correctly folded triple helix and its secretion is hampered, leading to a failure to construct normal collagen fibers in the ECM and to the accumulation of misfolded procollagen in the ER. The fate of the misfolded procollagen that accumulates in the ER in these genetic diseases is not known (Bateman et al., 2009).

We explored the mechanism for the disposal of the mutant procollagen in the ER by using Mov13 cells, which do not synthesize α1 chains of type I collagen because of a genetic mutation (Ishida et al., 2009). In Mov13 cells, the type I procollagen α2 chain is degraded after transportation to lysosomes (Gotkin et al., 2004). When the type I collagen α1 chain was introduced into Mov13 cells, the α1 and α2 chains were able to form a triple helix that was secreted into the ECM. However, if an α1 chain harboring a mutation at the triple helix region was introduced, it formed a trimer with the α2 chain but failed to correctly form a triple helix. These misfolded trimers of procollagen were degraded via the autophagy–lysosome pathway, similar to the $hsp47^{-/-}$ cells. Interestingly, if an α1 chain harboring a mutation in the trimer-forming region at the C-terminus was introduced into the Mov13 cells, the α1 chain could not make a trimer and was degraded by ERAD, which is similar mechanism to that reported for OI in which there was a mutation in the type I collagen α1 chain (Fitzgerald et al., 1999). Taken together, it appears that the trimeric form of misfolded procollagen is eliminated by autophagy in the case of specific mutations in collagen or genetic deficiency of Hsp47. However, misfolded α1 chains that are unable to form a trimer are removed via the ERAD pathway (Fig. 9.4) (Ishida and Nagata, 2009).

Since Hsp47 is essential for mouse development, mutations in human hsp47 would be expected to be lethal. However, point mutations in Hsp47 cause OI in humans (Christiansen et al., 2010) and Dachshunds (Drogemuller et al., 2009). These results again emphasize the importance of Hsp47 as a collagen-specific molecular chaperone required for the formation of procollagen triple helices in the ER.

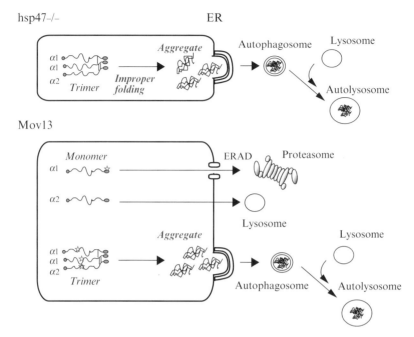

Figure 9.4 A degradation pathway of misfoled procollagen. Type I collagen trimers in $hsp47^{-/-}$ cells accumulate as aggregates in the ER. These misfolded procollagens are eliminated via autophagy–lysosome pathway, but not ERAD. Mutant collagen trimers due to genetic mutation form aggregates in the ER, and these are degraded via autophagy–lysosome pathway. In contrast, the monomeric α1 propeptide species is removed from the ER by the ERAD pathway, while monomeric α2 peptide is eliminated via lysosome, not autophagy independent pathway. Red stars indicate mutations of collagen. (For interpretation of the references to color in this figure legend, the reader is referred to the Web version of this chapter.)

7. Regulation of Hsp47 Expression and Its Clinical Importance

As described above, Hsp47 expression is induced by heat shock but not by ER stresses. The constitutive expression of Hsp47 correlates tightly with collagen expression in various tissues and cell types. Indeed, Hsp47 expression is observed in collagen-producing cells such as fibroblasts, cartilages, and adipocytes, but not in collagen nonproducing cells such as neurocytes and myelocytes (Nagata, 2003b). Hsp47 possesses a Heat Shock Element (HSE) in its promoter region, and its induction by heat shock is mediated by the binding of the Heat Shock Factor (HSF) transcription factor to the HSE (Hirayoshi et al., 1991; Takechi et al., 1992). A Sp-1-binding site 280 base pairs upstream from the transcription initiation site is necessary for basal expression. Two domains in the intron region are necessary for tissue-specific expression of Hsp47: one is BS5-B in the first intron providing a binding site for Sp2/Sp3 and Zf9, a Kruppel-like factor (KLF), and the other is EP7-D in the second intron, the binding factor (s) to which has not yet been identified (Hirata *et al.*, 1999). Interestingly, the expression of Hsp47 is highly responsive to changes in gravity; hypergravity induced the upregulation of Hsp47 expression, while low- or zero-gravity caused a marked downregulation in Hsp47 expression. These responses to gravity occurred both *in vitro* and *in vivo* (Oguro *et al.*, 2006). Maintaining the expression of Hsp47 may be a strategy for preventing the osteoporosis that results from zero-gravity; for example, in the space shuttle (Fig. 9.5).

Fibrosis is an intractable disease characterized by the abnormal accumulation of collagen, most often type I collagen, in a wide range of organs including liver, lung, and kidney (Naitoh *et al.*, 2005; Razzaque and Taguchi, 1997; Razzaque *et al.*, 1998). There are no effective therapeutic strategies for the treatment of fibrosis. Hsp47 might be a promising target for the therapy of the collagen-related diseases such as liver cirrhosis. As described above, the expression of Hsp47 correlates tightly with that of collagen in normal tissues as well as various fibrotic tissues, including liver and kidney (Brown *et al.*, 2005; Masuda *et al.*, 1994; Sunamoto *et al.*, 1998a). In an experimental glomerulosclerosis model induced by antithymocyte serum, knockdown of Hsp47 using antisense RNA resulted in reduced collagen accumulation in mouse kidney (Ohashi *et al.*, 2004; Sunamoto *et al.*, 1998b). Sato *et al.* recently reported that collagen secretion and accumulation in the liver during the progression of cirrhosis were dramatically suppressed by Hsp47 knockdown using siRNA packaged in

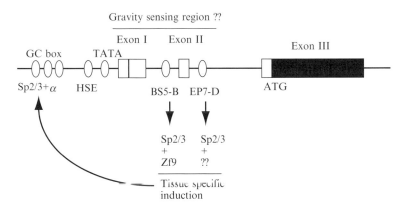

Figure 9.5 A schematic in promoter and regulatory element of Hsp47. Basal and tissue specific expression is controlled by the GC box, BS5-B, and EP7-D. Sp2/3 and Zf9 bind to these elements and regulates expression in Hsp47. A region which senses gravity condition may be around first or second intron.

vitamin A-coupled liposomes, which preferentially target hepatic stellate cells (Sato *et al.*, 2008). We propose that small molecule inhibitors of the interaction between Hsp47 and procollagen in the ER may be effective in the treatment of fibrotic diseases by preventing collagen secretion into the extracellular spaces. Hsp47 may be a novel therapeutic target for the treatment of fibrotic diseases.

REFERENCES

Anelli, T., and Sitia, R. (2008). Protein quality control in the early secretory pathway. *EMBO J.* **27,** 315–327.

Bachinger, H. P., Bruckner, P., Timpl, R., Prockop, D. J., and Engel, J. (1980). Folding mechanism of the triple helix in type-III collagen and type-III pN-collagen. Role of disulfide bridges and peptide bond isomerization. *Eur. J. Biochem.* **106,** 619–632.

Bateman, J. F., Boot-Handford, R. P., and Lamande, S. R. (2009). Genetic diseases of connective tissues: Cellular and extracellular effects of ECM mutations. *Nat. Rev. Genet.* **10,** 173–183.

Berg, R. A., and Prockop, D. J. (1973). The thermal transition of a non-hydroxylated form of collagen. Evidence for a role for hydroxyproline in stabilizing the triple-helix of collagen. *Biochem. Biophys. Res. Commun.* **52,** 115–120.

Bonfanti, L., Mironov, A. A., Jr., Martinez-Menarguez, J. A., Martella, O., Fusella, A., Baldassarre, M., Buccione, R., Geuze, H. J., Mironov, A. A., and Luini, A. (1998). Procollagen traverses the Golgi stack without leaving the lumen of cisternae: Evidence for cisternal maturation. *Cell* **95,** 993–1003.

Brown, K. E., Broadhurst, K. A., Mathahs, M. M., Brunt, E. M., and Schmidt, W. N. (2005). Expression of HSP47, a collagen-specific chaperone, in normal and diseased human liver. *Lab. Invest.* **85,** 789–797.

Bukau, B., Weissman, J., and Horwich, A. (2006). Molecular chaperones and protein quality control. *Cell* **125,** 443–451.

Bulleid, N. J., Dalley, J. A., and Lees, J. F. (1997). The C-propeptide domain of procollagen can be replaced with a transmembrane domain without affecting trimer formation or collagen triple helix folding during biosynthesis. *EMBO J.* **16,** 6694–6701.

Cates, G. A., Nandan, D., Brickenden, A. M., and Sanwal, B. D. (1987). Differentiation defective mutants of skeletal myoblasts altered in a gelatin-binding glycoprotein. *Biochem. Cell Biol.* **65,** 767–775.

Christiansen, H. E., Schwarze, U., Pyott, S. M., AlSwaid, A., Al Balwi, M., Alrasheed, S., Pepin, M. G., Weis, M. A., Eyre, D. R., and Byers, P. H. (2010). Homozygosity for a missense mutation in SERPINH1, which encodes the collagen chaperone protein HSP47, results in severe recessive osteogenesis imperfecta. *Am. J. Hum. Genet.* **86,** 389–398.

Dombrowski, K. E., and Prockop, D. J. (1988). Cleavage of type I and type II procollagens by type I/II procollagen N-proteinase. Correlation of kinetic constants with the predicted conformations of procollagen substrates. *J. Biol. Chem.* **263,** 16545–16552.

Drogemuller, C., Becker, D., Brunner, A., Haase, B., Kircher, P., Seeliger, F., Fehr, M., Baumann, U., Lindblad-Toh, K., and Leeb, T. (2009). A missense mutation in the SERPINH1 gene in Dachshunds with osteogenesis imperfecta. *PLoS Genet.* **5,** e1000579.

Engel, J., and Prockop, D. J. (1991). The zipper-like folding of collagen triple helices and the effects of mutations that disrupt the zipper. *Annu. Rev. Biophys. Biophys. Chem.* **20,** 137–152.

Fitzgerald, J., Lamande, S. R., and Bateman, J. F. (1999). Proteasomal degradation of unassembled mutant type I collagen pro-alpha1(I) chains. *J. Biol. Chem.* **274**, 27392–27398.

Fujita, N., Noda, T., and Yoshimori, T. (2009). Atg4B(C74A) hampers autophagosome closure: A useful protein for inhibiting autophagy. *Autophagy* **5**, 88–89.

Ghaemmaghami, S., Huh, W. K., Bower, K., Howson, R. W., Belle, A., Dephoure, N., O'Shea, E. K., and Weissman, J. S. (2003). Global analysis of protein expression in yeast. *Nature* **425**, 737–741.

Gotkin, M. G., Ripley, C. R., Lamande, S. R., Bateman, J. F., and Bienkowski, R. S. (2004). Intracellular trafficking and degradation of unassociated proalpha2 chains of collagen type I. *Exp. Cell Res.* **296**, 307–316.

Hendershot, L. M., and Bulleid, N. J. (2000). Protein-specific chaperones: The role of hsp47 begins to gel. *Curr. Biol.* **10**, R912–R915.

Hirata, H., Yamamura, I., Yasuda, K., Kobayashi, A., Tada, N., Suzuki, M., Hirayoshi, K., Hosokawa, N., and Nagata, K. (1999). Separate cis-acting DNA elements control cell type- and tissue-specific expression of collagen binding molecular chaperone HSP47. *J. Biol. Chem.* **274**, 35703–35710.

Hirayoshi, K., Kudo, H., Takechi, H., Nakai, A., Iwamatsu, A., Yamada, K. M., and Nagata, K. (1991). HSP47: A tissue-specific, transformation-sensitive, collagen-binding heat shock protein of chicken embryo fibroblasts. *Mol. Cell. Biol.* **11**, 4036–4044.

Hoseki, J., Ushioda, R., and Nagata, K. (2010). Mechanism and components of endoplasmic reticulum-associated degradation. *J. Biochem.* **147**, 19–25.

Hughes, R. C., Taylor, A., Sage, H., and Hogan, B. L. (1987). Distinct patterns of glycosylation of colligin, a collagen-binding glycoprotein, and SPARC (osteonectin), a secreted Ca^{2+}-binding glycoprotein. Evidence for the localisation of colligin in the endoplasmic reticulum. *Eur. J. Biochem.* **163**, 57–65.

Huh, W. K., Falvo, J. V., Gerke, L. C., Carroll, A. S., Howson, R. W., Weissman, J. S., and O'Shea, E. K. (2003). Global analysis of protein localization in budding yeast. *Nature* **425**, 686–691.

Ishida, Y., and Nagata, K. (2009). Autophagy eliminates a specific species of misfolded procollagen and plays a protective role in cell survival against ER stress. *Autophagy* **5**, 1217–1219.

Ishida, Y., Kubota, H., Yamamoto, A., Kitamura, A., Bachinger, H. P., and Nagata, K. (2006). Type I collagen in Hsp47-null cells is aggregated in endoplasmic reticulum and deficient in N-propeptide processing and fibrillogenesis. *Mol. Biol. Cell* **17**, 2346–2355.

Ishida, Y., Yamamoto, A., Kitamura, A., Lamande, S. R., Yoshimori, T., Bateman, J. F., Kubota, H., and Nagata, K. (2009). Autophagic elimination of misfolded procollagen aggregates in the endoplasmic reticulum as a means of cell protection. *Mol. Biol. Cell* **20**, 2744–2754.

Iwata, J., Ezaki, J., Komatsu, M., Yokota, S., Ueno, T., Tanida, I., Chiba, T., Tanaka, K., and Kominami, E. (2006). Excess peroxisomes are degraded by autophagic machinery in mammals. *J. Biol. Chem.* **281**, 4035–4041.

Koide, T., Asada, S., and Nagata, K. (1999). Substrate recognition of collagen-specific molecular chaperone HSP47. Structural requirements and binding regulation. *J. Biol. Chem.* **274**, 34523–34526.

Koide, T., Aso, A., Yorihuzi, T., and Nagata, K. (2000). Conformational requirements of collagenous peptides for recognition by the chaperone protein HSP47. *J. Biol. Chem.* **275**, 27957–27963.

Koide, T., Takahara, Y., Asada, S., and Nagata, K. (2002). Xaa-Arg-Gly triplets in the collagen triple helix are dominant binding sites for the molecular chaperone HSP47. *J. Biol. Chem.* **277**, 6178–6182.

Koide, T., Nishikawa, Y., Asada, S., Yamazaki, C. M., Takahara, Y., Homma, D. L., Otaka, A., Ohtani, K., Wakamiya, N., Nagata, K., and Kitagawa, K. (2006). Specific recognition of the collagen triple helix by chaperone HSP47. II. The HSP47-binding structural motif in collagens and related proteins. *J. Biol. Chem.* **281,** 11177–11185.

Komatsu, M., Waguri, S., Ueno, T., Iwata, J., Murata, S., Tanida, I., Ezaki, J., Mizushima, N., Ohsumi, Y., Uchiyama, Y., Kominami, E., Tanaka, K., *et al.* (2005). Impairment of starvation-induced and constitutive autophagy in Atg7-deficient mice. *J. Cell Biol.* **169,** 425–434.

Komatsu, M., Waguri, S., Chiba, T., Murata, S., Iwata, J., Tanida, I., Ueno, T., Koike, M., Uchiyama, Y., Kominami, E., and Tanaka, K. (2006). Loss of autophagy in the central nervous system causes neurodegeneration in mice. *Nature* **441,** 880–884.

Komatsu, M., Ueno, T., Waguri, S., Uchiyama, Y., Kominami, E., and Tanaka, K. (2007a). Constitutive autophagy: Vital role in clearance of unfavorable proteins in neurons. *Cell Death Differ.* **14,** 887–894.

Komatsu, M., Waguri, S., Koike, M., Sou, Y. S., Ueno, T., Hara, T., Mizushima, N., Iwata, J., Ezaki, J., Murata, S., Hamazaki, J., Nishito, Y., *et al.* (2007b). Homeostatic levels of p62 control cytoplasmic inclusion body formation in autophagy-deficient mice. *Cell* **131,** 1149–1163.

Kuma, A., Hatano, M., Matsui, M., Yamamoto, A., Nakaya, H., Yoshimori, T., Ohsumi, Y., Tokuhisa, T., and Mizushima, N. (2004). The role of autophagy during the early neonatal starvation period. *Nature* **432,** 1032–1036.

Kurkinen, M., Taylor, A., Garrels, J. I., and Hogan, B. L. (1984). Cell surface-associated proteins which bind native type IV collagen or gelatin. *J. Biol. Chem.* **259,** 5915–5922.

Lamande, S. R., and Bateman, J. F. (1999). Procollagen folding and assembly: The role of endoplasmic reticulum enzymes and molecular chaperones. *Semin. Cell Dev. Biol.* **10,** 455–464.

Leikina, E., Mertts, M. V., Kuznetsova, N., and Leikin, S. (2002). Type I collagen is thermally unstable at body temperature. *Proc. Natl. Acad. Sci. USA* **99,** 1314–1318.

Li, S. W., Sieron, A. L., Fertala, A., Hojima, Y., Arnold, W. V., and Prockop, D. J. (1996). The C-proteinase that processes procollagens to fibrillar collagens is identical to the protein previously identified as bone morphogenic protein-1. *Proc. Natl. Acad. Sci. USA* **93,** 5127–5130.

Makareeva, E., and Leikin, S. (2007). Procollagen triple helix assembly: An unconventional chaperone-assisted folding paradigm. *PLoS ONE* **2,** e1029.

Marini, J. C., Forlino, A., Cabral, W. A., Barnes, A. M., San Antonio, J. D., Milgrom, S., Hyland, J. C., Korkko, J., Prockop, D. J., De Paepe, A., Coucke, P., Symoens, S., *et al.* (2007). Consortium for osteogenesis imperfecta mutations in the helical domain of type I collagen: Regions rich in lethal mutations align with collagen binding sites for integrins and proteoglycans. *Hum. Mutat.* **28,** 209–221.

Marutani, T., Yamamoto, A., Nagai, N., Kubota, H., and Nagata, K. (2004). Accumulation of type IV collagen in dilated ER leads to apoptosis in Hsp47-knockout mouse embryos via induction of CHOP. *J. Cell Sci.* **117,** 5913–5922.

Masuda, H., Fukumoto, M., Hirayoshi, K., and Nagata, K. (1994). Coexpression of the collagen-binding stress protein HSP47 gene and the alpha 1(I) and alpha 1(III) collagen genes in carbon tetrachloride-induced rat liver fibrosis. *J. Clin. Invest.* **94,** 2481–2488.

Matsuoka, Y., Kubota, H., Adachi, E., Nagai, N., Marutani, T., Hosokawa, N., and Nagata, K. (2004). Insufficient folding of type IV collagen and formation of abnormal basement membrane-like structure in embryoid bodies derived from Hsp47-null embryonic stem cells. *Mol. Biol. Cell* **15,** 4467–4475.

Mizushima, N. (2009). Physiological functions of autophagy. *Curr. Top. Microbiol. Immunol.* **335,** 71–84.

Mizushima, N., Levine, B., Cuervo, A. M., and Klionsky, D. J. (2008). Autophagy fights disease through cellular self-digestion. *Nature* **451,** 1069–1075.

Nagai, N., Hosokawa, M., Itohara, S., Adachi, E., Matsushita, T., Hosokawa, N., and Nagata, K. (2000). Embryonic lethality of molecular chaperone hsp47 knockout mice is associated with defects in collagen biosynthesis. *J. Cell Biol.* **150,** 1499–1506.

Nagata, K. (1996). Hsp47: A collagen-specific molecular chaperone. *Trends Biochem. Sci.* **21,** 22–26.

Nagata, K. (1998). Expression and function of heat shock protein 47: A collagen-specific molecular chaperone in the endoplasmic reticulum. *Matrix Biol.* **16,** 379–386.

Nagata, K. (2003a). HSP47 as a collagen-specific molecular chaperone: Function and expression in normal mouse development. *Semin. Cell Dev. Biol.* **14,** 275–282.

Nagata, K. (2003b). Therapeutic strategy for fibrotic diseases by regulating the expression of collagen-specific molecular chaperone HSP47. *Nippon Yakurigaku Zasshi* **121,** 4–14.

Nagata, K., and Yamada, K. M. (1986). Phosphorylation and transformation sensitivity of a major collagen-binding protein of fibroblasts. *J. Biol. Chem.* **261,** 7531–7536.

Nagata, K., Saga, S., and Yamada, K. M. (1986). A major collagen-binding protein of chick embryo fibroblasts is a novel heat shock protein. *J. Cell Biol.* **103,** 223–229.

Naitoh, M., Kubota, H., Ikeda, M., Tanaka, T., Shirane, H., Suzuki, S., and Nagata, K. (2005). Gene expression in human keloids is altered from dermal to chondrocytic and osteogenic lineage. *Genes Cells* **10,** 1081–1091.

Nakai, A., Satoh, M., Hirayoshi, K., and Nagata, K. (1992). Involvement of the stress protein HSP47 in procollagen processing in the endoplasmic reticulum. *J. Cell Biol.* **117,** 903–914.

Natsume, T., Koide, T., Yokota, S., Hirayoshi, K., and Nagata, K. (1994). Interactions between collagen-binding stress protein HSP47 and collagen. Analysis of kinetic parameters by surface plasmon resonance biosensor. *J. Biol. Chem.* **269,** 31224–31228.

Oguro, A., Sakurai, T., Fujita, Y., Lee, S., Kubota, H., Nagata, K., and Atomi, Y. (2006). The molecular chaperone HSP47 rapidly senses gravitational changes in myoblasts. *Genes Cells* **11,** 1253–1265.

Ohashi, S., Abe, H., Takahashi, T., Yamamoto, Y., Takeuchi, M., Arai, H., Nagata, K., Kita, T., Okamoto, H., Yamamoto, H., and Doi, T. (2004). Advanced glycation end products increase collagen-specific chaperone protein in mouse diabetic nephropathy. *J. Biol. Chem.* **279,** 19816–19823.

Ragg, H. (2007). The role of serpins in the surveillance of the secretory pathway. *Cell. Mol. Life Sci.* **64,** 2763–2770.

Razzaque, M. S., and Taguchi, T. (1997). Collagen-binding heat shock protein (HSP) 47 expression in anti-thymocyte serum (ATS)-induced glomerulonephritis. *J. Pathol.* **183,** 24–29.

Razzaque, M. S., Kumatori, A., Harada, T., and Taguchi, T. (1998). Coexpression of collagens and collagen-binding heat shock protein 47 in human diabetic nephropathy and IgA nephropathy. *Nephron* **80,** 434–443.

Richter, K., Haslbeck, M., and Buchner, J. (2010). The heat shock response: Life on the verge of death. *Mol. Cell* **40,** 253–266.

Saga, S., Nagata, K., Chen, W. T., and Yamada, K. M. (1987). pH-dependent function, purification, and intracellular location of a major collagen-binding glycoprotein. *J. Cell Biol.* **105,** 517–527.

Sato, Y., Murase, K., Kato, J., Kobune, M., Sato, T., Kawano, Y., Takimoto, R., Takada, K., Miyanishi, K., Matsunaga, T., Takayama, T., and Niitsu, Y. (2008). Resolution of liver cirrhosis using vitamin A-coupled liposomes to deliver siRNA against a collagen-specific chaperone. *Nat. Biotechnol.* **26,** 431–442.

Satoh, M., Hirayoshi, K., Yokota, S., Hosokawa, N., and Nagata, K. (1996). Intracellular interaction of collagen-specific stress protein HSP47 with newly synthesized procollagen. *J. Cell Biol.* **133,** 469–483.

Sunamoto, M., Kuze, K., Iehara, N., Takeoka, H., Nagata, K., Kita, T., and Doi, T. (1998a). Expression of heat shock protein 47 is increased in remnant kidney and correlates with disease progression. *Int. J. Exp. Pathol.* **79,** 133–140.

Sunamoto, M., Kuze, K., Tsuji, H., Ohishi, N., Yagi, K., Nagata, K., Kita, T., and Doi, T. (1998b). Antisense oligonucleotides against collagen-binding stress protein HSP47 suppress collagen accumulation in experimental glomerulonephritis. *Lab. Invest.* **78,** 967–972.

Takechi, H., Hirayoshi, K., Nakai, A., Kudo, H., Saga, S., and Nagata, K. (1992). Molecular cloning of a mouse 47-kDa heat-shock protein (HSP47), a collagen-binding stress protein, and its expression during the differentiation of F9 teratocarcinoma cells. *Eur. J. Biochem.* **206,** 323–329.

Tasab, M., Batten, M. R., and Bulleid, N. J. (2000). Hsp47: A molecular chaperone that interacts with and stabilizes correctly-folded procollagen. *EMBO J.* **19,** 2204–2211.

Tasab, M., Jenkinson, L., and Bulleid, N. J. (2002). Sequence-specific recognition of collagen triple helices by the collagen-specific molecular chaperone HSP47. *J. Biol. Chem.* **277,** 35007–35012.

Thomson, C. A., and Ananthanarayanan, V. S. (2000). Structure–function studies on hsp47: pH-dependent inhibition of collagen fibril formation in vitro. *Biochem. J.* **349**(Pt 3), 877–883.

Vembar, S. S., and Brodsky, J. L. (2008). One step at a time: Endoplasmic reticulum-associated degradation. *Nat. Rev. Mol. Cell Biol.* **9,** 944–957.

Yoshida, H. (2007). ER stress and diseases. *FEBS J.* **274,** 630–658.

CHAPTER TEN

Assays for the Antiangiogenic and Neurotrophic Serpin Pigment Epithelium-Derived Factor

Preeti Subramanian,* Susan E. Crawford,[†] *and*
S. Patricia Becerra*

Contents

1. Introduction	184
2. Purification of PEDF Protein	185
2.1. Interphotoreceptor matrix	185
2.2. Vitreous humor	186
2.3. Aqueous humor	186
2.4. Plasma	186
2.5. Recombinant PEDF	187
2.6. Biochemical fractionation	187
3. Techniques to Assay PEDF	188
3.1. Immunochemical assays	188
3.2. Binding activities	190
4. Neurotrophic Assays	194
4.1. Neurite-outgrowth analyses	195
4.2. Retina cell survival assays	196
4.3. Protection against oxidative damage	196
5. Antiangiogenic Assays	197
5.1. Chick embryo aortic arch assay	197
5.2. Directed *in vivo* angiogenesis assay	198
5.3. Cell migration	198
5.4. Choroidal neovascularization	199
5.5. Corneal pocket	201
Acknowledgments	202
References	202

* Section of Protein Structure and Function, National Eye Institute, NIH, Bethesda, Maryland, USA
[†] Department of Surgery and Pathology, NorthShore University Research Institute, Evanston, Illinois, USA

Methods in Enzymology, Volume 499
ISSN 0076-6879, DOI: 10.1016/B978-0-12-386471-0.00010-9

Abstract

Pigment epithelium-derived factor (PEDF) is a secreted serpin that exhibits a variety of interesting biological activities. The multifunctional PEDF has neurotrophic and antiangiogenic properties, and acts in retinal differentiation, survival, and maintenance. It is also antitumorigenic and antimetastatic, and has stem cell self-renewal properties. It is widely distributed in the human body and exists in abundance in the eye as a soluble extracellular glycoprotein. Its levels are altered in diseases characterized by retinopathies and angiogenesis. Its mechanisms of neuroprotection and angiogenesis are associated with receptor interactions at cell-surface interfaces and changes in protein expression. This serpin lacks demonstrable serine protease inhibitory activity, but has binding affinity to extracellular matrix components and cell-surface receptors. Here we describe purification protocols, methods to quantify PEDF, and determine interactions with specific molecules, as well as neurotrophic and angiogenesis assays for this multifunctional protein.

1. INTRODUCTION

Pigment epithelium-derived factor (PEDF) is an extracellular serpin protein that lacks demonstrable serine protease inhibitory activity, but exhibits a variety of interesting biological properties (Becerra, 1997, 2006; Filleur et al., 2009). It is broadly distributed in the human body. It exists in abundance in the human eye, but its levels are altered in diseases characterized by retinopathies, such as age-related macular degeneration, diabetic retinopathy (Barnstable and Tombran-Tink, 2004; Bouck, 2002). The multifunctional PEDF has neurotrophic and antiangiogenic properties and acts in retinal differentiation, survival, and maintenance. Efficacy was demonstrated (i) in retinoblastoma cells and primary developing motor neurons by promoting neurite-outgrowth; (ii) in primary retinal cells, primary cerebellar granule cell neurons, primary hippocampal neurons, and primary motor neurons by protecting against apoptotic cell death that is associated with toxins and oxidative stress; as well as (iii) in a variety of endothelial cells by promoting proapoptotic mechanisms and preventing cell migration, and by inhibiting endothelial tube formation and vessel sprouting (Tombran-Tink and Barnstable, 2003). PEDF also has antitumorigenic and antimetastatic activities (Broadhead et al., 2009) and it has self-renewal properties on neural stem cell and human embryonic stem cell (Gonzalez et al., 2010; Ramirez-Castillejo et al., 2006).

The importance of PEDF in the development, maintenance, and function of the retina and CNS is evident in animal models for inherited and light-induced retinal degeneration, as well as for degeneration of spinal cord motor neurons. Ocular neovascularization- and retinal degeneration-related animal models have prompted clinical development. Clinical trials to assess the safety of a viral expression vector for PEDF in the context of age-related macular degeneration have been performed. Moreover, PEDF is a potential diagnostic tool for several ocular diseases triggered by pathological neovascularization, retinal degenerations, or tumors. Given the above and that studies on the mechanisms of PEDF action are associated with receptor interactions at cell-surface interfaces and changes in protein expression, there are great interests in methodologies to measure PEDF levels and the interactions of PEDF with components of its natural milieu. Here we describe purification protocols, quantification and binding assays, and bioassays for this multifunctional protein.

2. Purification of PEDF Protein

PEDF is a soluble, extracellular, monomeric glycoprotein of an apparent molecular weight of ∼50,000 (Wu *et al.*, 1995). In mammalian eyes, it is abundant in the vitreous (Wu and Becerra, 1996), and is highly concentrated in the interphotoreceptor matrix (IPM; Wu *et al.*, 1995). It is also present in aqueous humor (Ortego *et al.*, 1996), blood (Petersen *et al.*, 2003), cerebrospinal fluid (Kuncl *et al.*, 2002), bone and cartilage (Quan *et al.*, 2005). PEDF amounts ≤1% of total soluble protein of these sources and can be purified from them successfully. Bovine eyes are a good source of PEDF because of their large size, that is, the volume of a bovine vitreous cavity is about three to four times larger than that of human and monkey vitreous and it dislodges easily than that of primate eyes.

2.1. Interphotoreceptor matrix

The soluble components of the IPM can be obtained by the "no-cut" method to assure that the extracellular fluid is free of significant cellular contamination (Adler, 1989). The following is a description of an extraction procedure from bovine eyes that can be adapted to samples from other species:

1. All procedures are to be performed at 4 °C and fresh adult bovine eyes are kept on ice during dissection.
2. 30–50 eyes are dissected at a time. Each eye is dissected as follows: the periocular tissue is trimmed, and the anterior segment and the vitreous are removed leaving an eyecup.
3. A solution of phosphate buffered saline (PBS) is gently introduced between the neural retina and the RPE with a needle at 0.5 ml per eye. The eye is rocked to allow the PBS solution to run through the entire surface.

4. The PBS solution is extracted by aspiration, avoiding breakage of the retina membrane, and is transferred to a 50-ml tube.
5. The pooled washes are subjected to centrifugation at $1500 \times g$ for 15 min to remove cellular debris.
6. The supernatant is filtered through a 0.45-μm syringe filter and stored at $-80\ °C$. This filtrate is termed IPM wash.

2.2. Vitreous humor

1. After trimming the periocular tissue, remove the anterior segment and carefully peel the vitreous body from the neural retina.
2. The vitreous is collected at approximately 10–12 ml per eye and is placed in a 50-ml tube.
3. Homogenize vitreous gel using a Brinkman Polytron, model PT-MR 3000 (Kinematica AG, Littau, Switzerland) three times at 9500 rpm for 3 s.
4. The liquefied homogenates are subjected to centrifugation at $1300 \times g$ for 15 min to remove cellular debris. The supernatant is ready to use or store at $-80\ °C$ until use.

2.3. Aqueous humor

1. Extract aqueous humor from eyes by keratocentesis and by aspirating the humor with a syringe connected to a 25-gauge needle.
2. The pooled aqueous humor extracts are subjected to centrifugation at $1300 \times g$ for 15 min to remove cellular debris. The volume of a humor per bovine eye is approximately 1.5 ml. The supernatant is ready to use or store at $-80\ °C$ until use.

2.4. Plasma

1. Blood is collected using blood collection tubes with heparin (BD Vacutainer, BD Diagnostics, Oxford, UK), and centrifuged at $1000 \times g$ for 15 min at room temperature. The serum can be used immediately or stored at $-80\ °C$.
2. Serum samples are albumin-depleted using the Qproteome Murine Albumin Depletion Kit following manufacturer's instructions (Qiagen, Valencia, CA, USA) before use. This step removes the most abundant protein in serum and allows a more precise analysis of low-abundant proteins, such as PEDF.

2.5. Recombinant PEDF

Heterologous expression of PEDF from several species can be achieved using prokaryotic or mammalian cells. *Escherichia coli* cells expressing PEDF can be used as starting material for purification (Becerra *et al.*, 1993). Stably transfected mammalian cells (e.g., BHK, HEK-293 cells) with expression vectors containing full-length PEDF cDNA produce and secrete PEDF to the conditioned media (Duh *et al.*, 2002; Perez-Mediavilla *et al.*, 1998; Sanchez-Sanchez *et al.*, 2008; Stratikos *et al.*, 1996). Transfected cells are cultured to confluency in roller bottles with the complete culturing media for 24 h and then without serum for 24 h in repetitive cycles. After each cycle, media without serum is harvested, filtered, and stored at $-80\ °C$ or used immediately as starting material for biochemical fractionation.

2.6. Biochemical fractionation

Starting with an IPM wash or vitreous sample, PEDF protein can be purified >150-fold to near homogeneity by ammonium sulfate fractionation and cation-exchange chromatography, with a recovery of >40% (Wu *et al.*, 1995). The PEDF from bovine extracts remains in suspension at 45% ammonium sulfate saturation, and can be used as a first purification step. For PEDF from other sources, 80% ammonium sulfate precipitation can be used as a protein concentration step. Highly purified recombinant PEDF protein has been obtained at milligram amount per liter of cell culture (Stratikos *et al.*, 1996).

2.6.1. Ammonium sulfate fractionation

1. A total of 258 mg ammonium sulfate is added per 1 ml extract (above).
2. The suspension is stirred for 2 h and then centrifuged at $40,000 \times g$ for 2 h.
3. The supernatant fraction (S_{45}) is mixed with an additional 226 mg ammonium sulfate per milliliter to achieve 80% saturation, and stirred and fractionated as in step 2.
4. The precipitated fraction (P_{80}) is resuspended thoroughly in PBS solution; for example, use 1 ml PBS to resuspend P_{80} from IPM from every 25 eyes or from vitreous from 1.6 eyes.
5. The suspension is dialyzed against buffer S (50 mM sodium phosphate, pH 6.4, 1 mM DTT, 10% glycerol) containing 50 mM NaCl for 2 h each of three changes of buffer.
6. The dialyzate is centrifuged and passed through a filter (0.45 μm) to remove particulate material.

2.6.2. Cation-exchange column chromatography

A variety of cation-exchange resins and column sizes can be used, for example, S-Sepharose Fast Flow (Pharmacia) at 1 ml-bed volume for IPM dialyzate from 80 eyes, or vitreal dialyzate from 16 eyes; Mono-S HR5/5 column (10 cm × 1 cm, Pharmacia) for IPM from >500 eyes, or vitreous from more than 100 eyes; or 1.67 ml-bed volume POROS S (Applied BioSystems). The columns can be attached to standard automated fast protein liquid chromatography (FPLC) or perfusion chromatography systems. After equilibration with buffer S containing 50 mM NaCl, the samples are loaded on the columns. The unbound material is washed with the same buffer and the bound material is eluted with a linear gradient from 50 to 500 mM NaCl in buffer S at a flow rate 0.8–3 ml/min. PEDF elutes at about 200 ± 50 mM NaCl.

2.6.3. Anion-exchange column chromatography

The PEDF-containing fractions from the cation-exchange column chromatography are pooled and concentrated, for example, by ultrafiltration with Centricon-30 (Amicon, Beverly, MA). The filtrate is desalted, for example, by ultrafiltration or dialysis against buffer Q (50 mM Tris, pH 8.2) containing 50 mM NaCl. A variety of resins and column sizes may be used, for example, Q-Sepharose Fast Flow (Pharmacia), POROS Q (Applied BioSystems). The columns can be attached to standard automated FPLC or perfusion chromatography systems. After equilibration with buffer Q containing 50 mM NaCl, the samples are loaded on the columns. The unbound material is washed with the same buffer and the bound material is eluted with a linear gradient from 50 to 500 mM NaCl in buffer Q at a flow rate 0.8–3 ml/min. PEDF elutes at about 200 ± 50 mM NaCl. The PEDF-containing fractions are pooled and concentrated and stored at −80 °C.

3. Techniques to Assay PEDF

3.1. Immunochemical assays

Polyclonal and monoclonal antibodies to PEDF are available from commercial sources. Purified native and recombinant full-length PEDF can be detected as a ∼50,000-MW protein in Western blots. Sensitivity of detection of the PEDF-immunoreactive signal is very similar between the colorimetric and chemiluminescent detection methods after Western blotting.

3.1.1. Immunoblot reaction with polyclonal antibody Ab-rPEDF

Solutions and reagents:

a. TBST (20 mM Tris–Cl, pH 7.5, 150 mM NaCl, 0.05% Tween 20)
b. 1% BSA in TBST (filter after dissolving, store at 4°C)

c. Rabbit polyclonal AbPEDF antibody (e.g., BioProducts MD, LLC) or monoclonal anti-PEDF (e.g., Millipore, MAB1059, clone 10F12.2)
d. Biotinylated antirabbit IgG (H + L) [affinity purified antibody made in goat, human serum absorbed KPL cat. No. 16-15-16] (in 50% glycerol, store at 4°C); or biotinylated anti-mouse IgG (H + L) (affinity purified antibody made in goat, human serum absorbed KPL cat. No. 16-15-16) in 50% glycerol, store at 4 °C
d. ABC: Vectastin ABC elite kit (Vector labs, Inc cat. No. PK-6100) store at 4 °C
e. TBS (20 mM Tris–Cl, pH 7.5, 150 mM NaCl)
f. HRP color development reagent (Bio-Rad cat. No. 170-6534) stored at -20 °C

1. Incubate blot in 1% BSA/TBST for 1 h at room temperature—or 16 h at 4 °C—with gentle rocking (about 15–20 ml per membrane of 8 cm × 7 cm)
2. Incubate blot in polyclonal or monoclonal AbPEDF 1:1000–1:4000 in 1% BSA/TBST for 1 h at room temperature or more diluted than 1:4000 at 4 °C for 16–24 h with gentle rocking (about 10 ml per membrane of 8 cm × 7 cm). (*Note*: the diluted antiserum may be reused later, save if necessary, store at 4 °C. Its titer might decrease with time and use, whatever comes first.)
3. Wash membrane with 25 ml of TBST, rock at room temperature for 5 min, three times
4. Incubate blot in biotinylated anti-rabbit IgG (H + L) or anti-mouse IgG (H + L) 1:1000 in 1% BSA/TBST, with gentle rocking for 1 h at room temperature (about 10 ml per membrane 8 cm × 7 cm). (Prepare ABC, see step 6.)
5. Wash membrane with 25 ml of TBST, rocking at room temperature for 5 min each wash, three times
6. Incubate blot in ABC for 30 min at room temperature
    Note: ABC must be prepared in advanced!
    ABC: Vectastin ABC elite kit:
    to 10 ml TBST add three drops Solution A, mix by inversion
    then add three drops Solution B, mix by inversion
    let ABC stand 30 min at room temperature before using
7. Wash membrane with 25 ml of TBS, rock at room temperature for 5 min each wash, three times
8. Develop color with freshly prepared solution D

Solution D:

– dissolve 12 mg of HRP color development reagent in 4 ml methanol
– add 20 ml TBS, mix
– mix in 12:1 30% H_2O_2
– let color develop up to 30 min, if necessary

9. Rinse blot with water to stop development reaction and air dry. Store membrane between pieces of 3MM paper and cover with aluminum wrap

3.1.2. ELISA
The best way to quantify PEDF in heterologous samples is by ELISA. Several ELISA kits are available in the market. We have used one by BioProducts MD because of its high sensitivity and it specifically detects PEDF. The following link is for step-by-step instructions on its use http://www.bioproductsmd.com/Documents/PEDF_ELISA_Manual.pdf

3.2. Binding activities
PEDF has affinity for extracellular matrix components such as glycosaminoglycans and collagens (Alberdi et al., 1998; Becerra et al., 2008; Meyer et al., 2002). The following procedures are for assaying binding to heparin, heparin sulfate, hyaluronan, and collagen.

3.2.1. Glycosaminoglycans
3.2.1.1. Glycosaminoglycan-affinity column chromatography Heparin- or hyaluronan-affinity resins can be prepared as described earlier (Alberdi et al., 1998; Becerra et al., 2008) or from commercial sources (e.g., Sigma). Optimum PEDF binding is obtained in phosphate buffers containing NaCl concentrations below 100 mM and at pH values between 6 and 7.

1. Affinity resins are packed in Polyprep chromatography columns (Bio-Rad) to yield 0.5 ml settled bed volume and equilibrated with buffer H (20 mM NaCl, 20 mM sodium phosphate, pH 6.5, and 10% glycerol).
2. A solution of protein (PEDF up to 40 μg) in buffer H is applied to the appropriate affinity column and incubated with the resin at 4 °C for 30 min.
3. The glycosaminoglycan-affinity columns are washed with more than 10 column volumes of the incubation buffer.
4. Elution is with an NaCl step-gradient in buffer H at 1.5 column volumes per fraction.

3.2.1.2. Cetylpyridinium chloride precipitation Hyaluronan binding in solution can be assayed by precipitation with cetylpyridinium chloride (CPC). Optimum PEDF binding to hyaluronan occurs in buffers with pH 7.5–8.0 and with NaCl concentrations ≤ 300 mM (Becerra et al., 2008).

1. Solutions of 10–100 μg/ml PEDF and 10–1000 μg/ml hyaluronan are mixed in PBS containing BSA as carrier (e.g., 250 μg/ml) and incubated at 4 °C for 60 min.

2. Add 1× volume of 2.5% CPC in PBS and incubate at 37 °C for 1 h.
3. Separate the precipitate by centrifugation in an Eppendorf centrifuge at 16,000×g for 15 min at 4 °C.
4. Discard the supernatant and wash the precipitate with 200 μl of 1% CPC in PBS.
5. Resuspend the precipitate in 25–30 μl of sample buffer for SDS-PAGE.
6. Protein detection in gels is performed with Coomassie Blue staining or immunostaining after Western blotting with antibodies to PEDF.

3.2.2. Heparan sulfate proteoglycan

These assays were based on the separation of complexes formed between PEDF and other proteins by ultrafiltration in which PEDF complexes >100-kDa are retained by a membrane of M_r 100,000 exclusion limit, while free PEDF molecules of 50 kDa are filtered through.

3.2.2.1. Radioactive free assay

1. PEDF (120 μg/ml) is mixed with Heparan sulfate proteoglycan (HSPG; 50–300 μg/ml) in buffer H in a final volume of 100 μl.
2. The mixtures are incubated at 4 °C for 30 min and then ultrafiltered through membranes with a molecular cut-off >100,000 (e.g., Centricon-100 devices, Millipore).
3. The concentrated material is diluted 20-fold with incubation buffer and ultrafiltered. This is repeated four times.
4. Aliquots of the concentrated samples are analyzed by SDS-PAGE.

3.2.2.2. Radioactivity binding assay

1. Radiolabeled [^{125}I]PEDF at 90 ng/ml (6.25 μCi/ml) and increasing concentrations of HSPG are mixed in buffer H (10 μl) and incubated at 4 °C for 30 min.
2. Free and bound PEDF are separated by ultrafiltration through Microcon-100 (Amicon). The concentrated material is diluted 40-fold with incubation buffer and washed as described above.
3. Each Microcon retenate cup is transferred to a scintillation vial and mixed with 5 ml of Bio-Safe II liquid scintillation solution (Research Products International, Corp.) by extensive vortexing. Radioactivity is determined using a β-counter (Beckman, model LS 3801).

3.2.3. Collagen

3.2.3.1. Solution binding assays
PEDF proteins (100 μg/ml) are mixed with collagen (100 μg/ml) in PBS, pH 7.4, containing 10% glycerol and incubated at 4 °C for 1 h. If stock solutions of collagen contain acetic acid,

NaOH can be added to neutralize the pH of the reaction mixtures. PEDF complexes are separated from free PEDF by ultrafiltration as described above for HSPG using Centricon-100 devices. Bound PEDF can be analyzed in the retained material by SDS-PAGE and visualized by Coomassie Blue staining or immunostaining, and it can be quantified by ELISA.

3.2.3.2. Solid-phase binding assays

1. Binding reactions are performed with ^{125}I-PEDF (e.g., 2 nM) and increasing amounts of unlabeled PEDF in 0.1% BSA/PBS, pH 7.4, to collagen I immobilized on plastic of 24-well plates.
2. After incubations at 4 °C for 90 min with gentle rocking, the binding solution is removed, and the wells are washed three times with 0.1% BSA/PBS.
3. Then 1 N NaOH is added to the wells, incubated at room temperature for 30 min, and transferred to scintillation vials to determine the amount of radioactivity using a β-scintillation counter.
4. Nonspecific binding is determined from fractions with > 100-fold molar excess of unlabeled PEDF over radioligand.
5. Binding data can be analyzed by nonlinear regression using GraphPad Prism software.

3.2.3.3. Surface plasmon resonance assays

Assays for PEDF-collagen I interactions can be performed immobilizing either 4 ng of collagen I or PEDF on a CM5 sensor chip, by N-hydroxysuccinimide (NHS)/EDC activation, followed by covalent amine coupling of the proteins to the surface using Biacore 3000. Treat the carboxymethyl-dextran surface of the sensor chip with NHS/EDC to activate it in preparation for amine coupling. The NHS/EDC creates reactive ester groups where the carboxyl groups were on the dextran (only about 40% of the COO$^-$ groups are derivatized). The protein designed to be bound to the activated CM5 chip is exposed to the dextran with the reactive esters. Primary amines on the protein (e.g., N-terminal and possibly lysine groups) then attach to the dextran by nucleophilic substitution of the reactive ester. The remaining free surface (about 60%) is then blocked with 0.1 M Tris, pH 8.0, and the matrix washed with 0.5 M NaCl solution and then reequilibrated with binding buffer (PBS, 10% glycerol). Eight different dilutions of PEDF or collagen I are prepared in binding buffer with concentrations ranging from 0 to 1.0 μm and injected from low to high concentration, and then the series is repeated, to study the interaction of both free PEDF on a collagen I matrix and the inverse orientation. Each injection is followed by a 0.5 M NaCl regeneration step. The data are then fitted to several binding models for a kinetic analysis. The best fittings are obtained with a simple 1:1 Langmuir model for the collagen surface binding assay and with a bivalent analyte model for the opposite orientation.

3.2.4. PEDF receptor proteins
3.2.4.1. Radiolabeled ^{125}I-PEDF binding assays
3.2.4.1.1. Cells in suspension

1. Cells in suspension (6×10^5 cells/ml) are incubated at 4 °C for 15 min before the addition of ^{125}I-PEDF (0.1–2 nM). *Note*: binding is enhanced when binding buffer is media conditioned overnight without serum (Alberdi *et al.*, 2003).
2. The binding reaction mixture is incubated at 4 °C for a period of time between 15 and 90 min.
3. The reaction is terminated by the addition of 10 ml of ice-cold PBS supplemented with 0.1% BSA.
4. Immediately the reaction is subjected to filtration under vacuum through Whatman GF/C filters presoaked in 0.3% polyethylenimine.
5. Finally, the filters are washed with 10 ml of the ice-cold 1% BSA in PBS.

3.2.4.1.2. Attached cells

1. Attached cells are cultured in 24-well plates, to 90% confluency or containing 5×10^5 cells/well.
2. Cells are washed with 0.5 ml of 0.1% BSA in culturing medium (binding buffer) three times before the addition of ^{125}I-PEDF in binding buffer.
3. After incubation at 4 °C for 15–90 min, the unbound PEDF is washed with 0.5 ml of binding buffer three times.
4. Then the cells are lysed by incubation with 0.5 ml of 1 M NaOH at room temperature for 30 min.

3.2.4.1.3. Determination of bound PEDF Filters and cell lysates are placed in scintillation vials, mixed with 5 ml of Bio-Safe II liquid scintillation solution (Research Products International, Corp.), incubated at room temperature overnight, and then mixed by extensive vortexing before determining the radioactivity using a β-counter (Beckman, model LS 3801). Alternatively, bound and free radioligand are separated by centrifugation of cell suspensions followed by three washes with 1% BSA in PBS, and the bound radioactivity is determined in the cell pellets using a gamma counter (Wallac) or after SDS-PAGE and autoradiography. Nonspecific binding is defined as the amount of bound radioactivity in the presence of saturating concentrations of unlabeled ligand, and specific binding as bound radioactivity minus nonspecific binding. Data are analyzed using the Minitab statistical program and Microsoft Excel for linear regression, as well as GraphPad Prism for nonlinear regression and Scatchard analyses.

3.2.4.2. Ligand-affinity column chromatography

1. Fresh detergent-soluble membrane fractions from tissues or cell in culture are prepared as described (Alberdi *et al.*, 1999; Aymerich *et al.*, 2001).
2. Highly purified PEDF protein is coupled to beads of preactivated hydrophilic, cross-linked bis-acrylamide/azlactone copolymers (3M Emphaze Ultralink; Pierce, Rockford, IL).
3. Detergent-soluble membrane proteins (0.1–1 mg) are passed through a column of resin without ligand (2 ml).
4. The unbound material is mixed with PEDF-coupled resin (2 ml; ~6 mg PEDF/ml resin) and gently rotated at 4 °C for 1 h.
5. The material is packed in a column, washed with buffer D (20 column volumes or until absorbance at 280 nm was undetectable), followed by 1 M NaCl in buffer D (10 column volumes).
6. The bound material is eluted with 0.1 M glycine buffer, pH 11, 10% glycerol, 1 mM CaCl$_2$, 0.15 NaCl, and 0.25% CHAPS (10 column volumes).
7. Eluted proteins are concentrated to 100 μl by ultrafiltration with microconcentrators (Centricon-30, Millipore).

3.2.4.3. Ligand blot

1. Detergent-soluble membrane proteins are resolved by SDS-PAGE under nonreducing conditions and transferred to a 0.2-μm nitrocellulose membrane.
2. The membrane is first washed with 1% NP-40 in TBS for 15 min and then twice with TBS at 25 °C for 10 min each.
3. The blot is incubated with blocking solution (1% BSA in TBST, containing TBS with 0.05% Tween 20) at 25 °C for 2 h.
4. The blot is incubated with ^{125}I-PEDF (2 nM) in blocking solution at 4 °C for 16 h.
5. The blot is washed three times with TBST at 25 °C for 15 min to remove the unbound ligand.
6. The blot is air dried, and exposed to X-ray film (BioMax ML, Eastman Kodak Co., Rochester, NY) to detect bound radioligand by autoradiography.

4. Neurotrophic Assays

PEDF acts on the retina *in vivo* and on live cells. It supports normal development of photoreceptor neurons (Jablonski *et al.*, 2000). In addition, it can delay the death of photoreceptors in mouse models of inherited retinal degenerations (Cayouette et al., 1999), and it can protect photorecepotors from light-induced damage (Cao et al., 2001). It can protect the neural retina against ischemic injury (Takita et al., 2003). Moreover, PEDF can induce morphological differentiation of retinoblastoma cells into a neuronal phenotype. It can promote neurite-outgrowth on Y-79 and Weri cells as well as in spinal cord motor neurons (Becerra, 1997; Houenou et al., 1999). It can protect mixed retina cells, retinal ganglion cells, retinal pigment epithelial cells, and developing hippocampal neurons from death by several insults (DeCoster et al., 1999; Notari et al., 2005; Pang et al., 2007; Tsao et al., 2006).

4.1. Neurite-outgrowth analyses

1. Human Y-79 retinoblastoma cells (ATCC) are cultured in suspension in MEM supplemented with 15% fetal bovine serum and antibiotics (100 units/ml penicillin and 100 pg/ml streptomycin) at 37 °C in a humidified incubator under 5% CO_2.
2. Cells are propagated for two passages after receipt and then frozen in the same MEM medium containing 10% dimethyl sulfoxide. Separate aliquots of cells are then used for each differentiation experiment.
3. After thawing, cells are kept in suspension culture without further passaging in serum-containing MEM until the appropriate number of cells is available.
4. Cells are collected by centrifugation, washed twice, and resuspended in PBS and counted.
5. For treatment with PEDF, 2.5×10^5 cells are seeded into each well in 6-well plates (Nunc, Inc.) with 2 ml of serum-free medium consisting of MEM supplemented with 1 mM sodium pyruvate, 10 mM HEPES, 1× nonessential amino acids, 1 mM L-glutamine, and 0.1% ITS mix and antibiotics as above.
6. Approximately 12–16 h later, PEDF (50–200 ng/ml) is added to the medium.
7. The cultures are incubated and kept undisturbed for 7 days. Cells under these conditions remain in suspension.
8. On the 8th day after treatment, cells are transferred to 6-well plates precoated with poly-D-lysine (Collaborative Research); once the cells attach to the substrate (about 6–8 h), the old medium is replaced with 2 ml of fresh serum-free medium. The cultures are maintained under these conditions for up to 11 days.
9. Using an Olympus CK2 phase-contrast microscope, postattachment cultures are examined daily for morphological differentiation and quantification of neurite-outgrowth.

4.2. Retina cell survival assays

PEDF can protect cell death induced by serum starvation in a rat retinal precursor R28 cell line (Notari et al., 2005). The R28 cell line is derived from postnatal day 6 Sprague–Dawley rat retina, immortalized with the 12S E1A gene of adenovirus using incompetent retroviral vector (Seigel et al., 1996). These cells are provided by Dr. Gail Siegel (SUNY, Buffalo, NY). The survival activity assay for PEDF is performed as follows using cells with passage numbers 45–55:

1. R28 cells are cultured in Dulbecco's modified Eagle's medium (DMEM) incomplete media with 3% sodium bicarbonate, 1× MEM nonessential amino acids, 1× MEM vitamins, 2 mM L-glutamine, gentamicin 0.1 mg/ml, 10% of bovine calf serum, and 1% of Penicillin/Streptomycin (P/S) at 37 °C with 5% CO_2. Cells are propagated maintaining them at not more than 70% confluency and are detached from the flask using 5 mM EDTA in 1× PBS.
2. Cells are seeded at a density of 2×10^4 cells/well in 24-well plates in media containing 5% of bovine calf serum, and allowed to attach for 8 h.
3. At the end of 8 h, media is removed, cells are washed once with 1× PBS, and serum-free media or serum-free media containing human recombinant PEDF at desired concentrations is added to each well.
4. Cell viability is measured at the end of 48 h, for example, using the CellTiter-Glo™ viability assay kit (Promega). This kit uses a unique, stable form of luciferase to measure ATP as an indicator of viable cells and the luminescent signal produced is proportional to the number of viable cells present in culture. Briefly, cells are washed once with 1× PBS. This is followed by addition of 100 μl of PBS and 100 μl of Cell-Glo reagent (thawed to room temperature) to each well. The plate is incubated for 10 min at room temperature and luminescence signal can be measured using Envision automated plate reader (Perkin Elmer, MA).

4.3. Protection against oxidative damage

PEDF protects retinal pigment epithelium from apoptosis induced by oxidative stress as demonstrated earlier (Mukherjee et al., 2007). The assay is performed as follows:

1. ARPE-19 cells (ATCC) are cultured in DMEM/F-12 (1:1), penicillin/streptomycin, 0.5 mg/ml Geneticin, and 10% fetal bovine serum at 37 °C, 5% CO_2.
2. Cells are trypsinized and plated at a density of 1×10^5 cells/well in 24-well plates. The cells are allowed to grow to 100% confluency (72 h) which is very critical for the assay.

3. Media is removed and media containing 0.5% serum is added to starve the cells for 8 h. Then the cells are treated with 30% H_2O_2 to achieve the desired final concentrations and TNF α (10 ng/ml). The optimal H_2O_2 concentration for oxidative damage can be determined in preliminary experiments using concentration ranges (400–800 μM). PEDF (10 ng/ml) along with DHA (30 nM) are added along with H_2O_2 and TNF-α.
4. Cells are incubated for 16 h and fixed with methanol for 15 min at room temperature.
5. Cells are then washed with 1× PBS followed by addition of 5 μM Hoechst reagent in PBS for 15 min at room temperature.
6. Cells are washed once with 1× PBS and different fields are imaged using UV fluorescence under a Nikon Eclipse TE2000-U microscope.
7. The percentage of pyknotic cell nuclei as seen by condensed morphology in the field is counted from the digital images.

5. Antiangiogenic Assays

5.1. Chick embryo aortic arch assay

The chick embryo aortic arch assay is an *ex vivo* angiogenesis assay as previously described (Martinez *et al.*, 2004).

1. Aortic rings of approximately 0.8 mm in length are prepared from the five aortic arches of 13-day-old chicken embryos (CBT Farms, Chestertown, MD).
2. The soft connective tissue of the adventitia layer is carefully removed with tweezers.
3. Each aortic ring is placed in the center of a well in a 48-well plate and covered with 10 μl of synthetic matrix (Matrigel; BD Biosciences, San Jose, CA).
4. After the matrix solidified, 300 μl of growth-factor–free human endothelial serum-free basal growth medium (Invitrogen) containing the proper concentration of the test substances is added to each well.
5. The plates are kept in a humid incubator at 37 °C in 5% CO_2 for 24–36 h.
6. Microvessels sprouting from each aortic ring are photographed in an inverted microscope and the area covered by the newly formed capillaries is estimated.
7. Endothelial cell growth supplement (ECGS; Biomedical Collaborative Products, Bedford, MA) is used at 400 μg/ml as an angiogenesis promoter. Six independent rings per treatment are measured as replicates.

5.2. Directed in vivo angiogenesis assay

Analysis and quantitation of angiogenesis is done using a directed *in vivo* angiogenesis assay (DIVAA) as previously described (Martinez *et al.*, 2002).

1. Ten millimeter long, surgical-grade silicone tubes with only one end open (angioreactors) are filled with 20 µl of synthetic matrix alone or mixed with VEGF and/or rhuPEDF.
2. After the matrix solidified, the angioreactors are implanted subcutaneously into the dorsal flanks of anesthetized athymic nude mice (National Cancer Institute [NCI] colony).
3. After 11 days, the mice are injected intravenously (IV) with 25 mg/ml FITC-dextran (100 µl/mouse; Sigma-Aldrich) 20 min before the angioreactors are removed.
4. Quantitation of neovascularization in the angioreactors is determined as the amount of fluorescence trapped in the implants and is measured in a spectrophotometer (HP; Perkin Elmer Life Sciences). Eight implants are used per treatment point as replicates.

This protocol was approved by the internal NIH animal committee and was in compliance with the ARVO Statement for the Use of Animals in Ophthalmic and Vision Research.

5.3. Cell migration

This assay was modified from a protocol previously described (Polverini *et al.*, 1991):

1. Prepare gelatinized 0.5 or 0.8 µm Nuclepore membranes as follows: incubate overnight in 0.5 M glacial acetic acid at room temperature with *gentle* shaking, wash three times for 1 h each in autoclaved milli-Q H$_2$O (mQH$_2$O) at room temperature with gentle shaking, incubate overnight in 0.01% gelatin in sterile PBS (Ca^{2+} and Mg^{2+} free); Difco gelatin (cat. No. 214340) for at least 4 h to O/N at room temperature (I prefer O/N); rinse in sterile mQH$_2$O for 5 min and air dry 1 h. Store membranes between pieces of whatmann paper or between the papers that come with the membranes and use within ~1–2 months.
2. Rinse a confluent T75 flask of microvascular endothelial cells with PBS, and replace with basal media + 0.1% BSA to starve cells. Incubate at least 4 h to O/N (16 h, maximum).
3. Harvest the cells by trypsinization, resuspending at a cell concentration of 1.0–1.5 × 10^6 cells/ml in basal media + 0.1% BSA.
4. Rinse Boyden chamber wells with 1× PBS (Ca^{2+} and Mg^{2+} free), 29 µl/well and flick chamber to remove PBS.

5. Add 29 µl of the cell suspension per well in Boyden chamber, remixing cells after every 3rd to 4th well by pipetting up and down.
6. Using forceps, carefully place membrane over cells, *shiny side up* and avoiding air bubbles between the cells and membrane. Assemble chamber per the manufacturer's instructions.
7. Invert Boyden chamber, wrap loosely in aluminum foil (to prevent evaporation), and incubate 1.5–2 h at 37 °C, 5% CO_2 to allow cells to attach to the lower side of the membrane.
8. Prepare 210 µl of test samples and controls in media + 0.1% BSA. The media + 0.1% BSA serves as the negative control, and VEGF (100–1000 pg/ml) or bFGF2 (10–50 ng/ml) in media + 0.1% BSA, positive control. Test PEDF for inhibition against positive control or test substance at 1–10 n*M*. For other test samples, an initial dilution curve may be required to determine optimal dose.
9. Load 51–52 µl per well in quadruplicate for each control and test sample and incubate as above for 3–4 h, loosely wrapped in foil.
10. Dismantle the chamber as per the manufacturer's instructions and notch membrane to maintain orientation.
11. Fix the cells by 1-min incubation in fixative, then stain the membrane for one to two each in solution 1 (orange) and then solution 2 (purple), rinse briefly in PBS and air dry on paper towels for 1–3 h or O/N, shiny side up (be sure membrane does not stick to paper).
12. Cut the membrane in half, notching the second half in a manner to maintain orientation, and mount on glass slide, shiny side up, using a convenient mounting medium (Cytoseal XYL), then coverslip.
13. Count cells that have migrated to the upper side of the membrane which is the side in which the pores of the membrane are in focus. The majority of cells should be on the side of the membrane in which the pores are not in focus. Count 10 randomly chosen high powered fields for each replicate of each sample well at 100× (oil immersion). Choose a field by scanning the lower side of the membrane (unmigrated cells), then refocus to the upper side to count migrated cells to avoid field bias.

5.4. Choroidal neovascularization

PEDF and PEDF-derived peptides can attenuate choroidal neovascularization (CNV) induced with laser (Amaral and Becerra, 2010). The following method has been described earlier (Campos *et al.*, 2006).

5.4.1. Laser-induced CNV

1. Brown Norway male rats (Charles River Laboratories, Rockville, MD) weighing between 300 and 350 g are used.

2. Rats are anesthetized with an intraperitoneal injection of a 40–80 mg/kg ketamine (Fort Dodge Animal Health, Fort Dodge, IA) and 10–12 mg/kg xylazine (Ben Venue Laboratories, Bedford, OH) mixture.
3. Topical 0.5% proparacaine is applied and pupils are dilated with a mixture of 1% tropicamide and 2.5% phenylephrine (Alcon Fort Worth, TX).
4. Hot pads maintain the body temperature while rats are placed in front of a slit lamp.
5. Four to eight shots surrounding the optic nerve are placed with an ND:YAG 532-nm laser (Alcon) using a 5.4-mm contact fundus laser lens (Ocular Instruments, Bellevue, WA), a spot size of 50 μm, power between 80 and 90 mW, and 0.100 s of exposure time. The end point "bubble formation" assures breakage of Bruch's membrane (Dobi *et al.*, 1989).
6. The animals are euthanatized by CO_2 exposure after specific periods of time after injury.

This protocol was approved by the internal NIH animal committee and was in compliance with the ARVO Statement for the Use of Animals in Ophthalmic and Vision Research.

5.4.2. Flatmount technique

1. With the nictitans membrane (nasal) used for orientation, eyes are enucleated and immediately fixed in 4% paraformaldehyde (EM Grade; Polysciences, Inc. Warrington, PA) in PBS (9 g/l NaCl, 0.232 g/l KH_2PO_4, 0.703 g/l Na_2HPO_4, pH 7.3) for 1 h.
2. Under a dissecting microscope, the anterior segment, and crystalline lens are removed, and the retinas are detached and separated from the optic nerve head with fine curved scissors.
3. The remaining eye cups are washed with cold ICC buffer (0.5% BSA, 0.2% Tween 20, 0.05% sodium azide) in PBS.
4. A 1:1000 dilution of a 10-mg solution of 4′,6-diamidino-2-phenylindole (DAPI), a 1:100 dilution of a 1-μg/μl solution of isolectin IB_4 conjugated with Alexa Fluor 568, and a 1:100 dilution of a 0.2 units/μl solution of phalloidin conjugated with Alexa Fluor 488 (Invitrogen-Molecular Probes, Eugene, OR) are prepared in ICC buffer and centrifuged for 1 min at $2040 \times g$.
5. Alternatively, CD11b (MCA275R; Serotec, Oxford, UK) conjugated with Alexa Fluor 488, an antibody that labels microglia in retina and brain, is used at dilutions of 1:200 to identify retinal microglia.
6. A humidified chamber is prepared, the eye cups are covered with fluorescent dyes prepared as described earlier, incubated at 4 °C with gentle rotation for 4 h, and washed with cold ICC buffer.

7. Radial cuts are made toward the optic nerve head, and the sclera-choroid/RPE complexes are flatmounted (Gel-mount; Biomedia Corp. Foster City, CA), covered, and sealed.

5.4.3. Subconjunctival injections

Protein injections can begin immediately after laser injury and be repeated daily until day 4. Doses of 10, 1, 0.1, and 0.01 pmol PEDF, or PEDF peptides per injection have been used. Lasered eyes with no injection, PBS injection, and angiostatin are used as negative and positive controls. The animals are euthanatized by CO_2 exposure at day 7. The effect of PEDF after CNV has been induced can be tested. In another set of experiments, daily PEDF injections can start at the 7th day after laser and continue until day 11 after laser. The animals are euthanatized by CO_2 exposure at day 14 after laser.

1. Protein or peptide stock solutions are diluted in PBS and filtered sterilized. Dilutions are prepared so that 2 µl contain the desired dosage per animal.
2. Two microliters are injected into the subconjunctiva of each eye while the animal under general anesthesia for restraint. Samples are administered daily.

5.5. Corneal pocket

1. Prepare Hydron pellets in sterile hood. Cut nylon mesh (Spectrum, cat. No. 148391) into rectangles roughly 3 cm × 4 cm and sterilize by soaking in 70% ethanol for 30 min and allow to air dry. Suspend 5 mg Sucralfate (Bukh, MediTec, cat. No. 95092601) in 10 µl test substance in PBS. For positive control, use 5 µl of 250 ng/ml FGF2. Test PEDF at 8 nM. Add 10 µl 12% Hydron (Interferon Sciences, New Brunswick, NJ) prepared at least 10 h in advance in 96% EtOH. Resuspend well.
2. Spread resulting solution on mesh area of approximately 10 × 5 squares of mesh. Repeat several times if necessary. Allow 10–15 min for gel to polymerize. The pellets will be embedded in the mesh. Spread additional 5 µl of Hydron solution on both sides of the embedded pellet area. Allow another 5–10 min to dry. Carefully remove threads of the mesh in one dimension, releasing the pellets (watch for static electricity).
3. Anesthetize female Fischer 344 rats (Harlan Industries, Indianapolis, IN) weighing 120–140 g, and rinse the eye with several drops of proparacaine, blot. Assay can be performed in mice as well.
4. Make an incision across the center of the eye with #15 surgical blade. Using a modified iris spatula, create a pocket. Insert a pellet with fine forceps, and seal the incision with ophthalmic antibiotic ointment.

5. Briefly, Hydron pellets (Interferon Sciences) of <5 μm were prepared containing the test sample CM or bFGF (0.15 μM, positive control). Pellets were implanted into the avascular corneas of anesthetized rats 1.0–1.5 mm from the limbus.
6. At 7 days postimplantation, perfuse animal with colloidal carbon to visualize the vascularity and for a permanent record of the response.

ACKNOWLEDGMENTS

This work was supported in part by the Intramural Research Program of the NIH, NEI.

REFERENCES

Adler, A. J. (1989). Selective presence of acid hydrolases in the interphotoreceptor matrix. *Exp. Eye Res.* **49**, 1067–1077.

Alberdi, E., et al. (1998). Pigment epithelium-derived factor (PEDF) binds to glycosaminoglycans: Analysis of the binding site. *Biochemistry* **37**, 10643–10652.

Alberdi, E., et al. (1999). Binding of pigment epithelium-derived factor (PEDF) to retinoblastoma cells and cerebellar granule neurons. Evidence for a PEDF receptor. *J. Biol. Chem.* **274**, 31605–31612.

Alberdi, E. M., et al. (2003). Glycosaminoglycans in human retinoblastoma cells: Heparan sulfate, a modulator of the pigment epithelium-derived factor–receptor interactions. *BMC Biochem.* **4**, 1.

Amaral, J., and Becerra, S. P. (2010). Effects of human recombinant PEDF protein and PEDF-derived peptide 34-mer on choroidal neovascularization. *Invest. Ophthalmol. Vis. Sci.* **51**, 1318–1326.

Aymerich, M. S., et al. (2001). Evidence for pigment epithelium-derived factor receptors in the neural retina. *Invest. Ophthalmol. Vis. Sci.* **42**, 3287–3293.

Barnstable, C. J., and Tombran-Tink, J. (2004). Neuroprotective and antiangiogenic actions of PEDF in the eye: Molecular targets and therapeutic potential. *Prog. Retin. Eye Res.* **23**, 561–577.

Becerra, S. P. (1997). Structure–function studies on PEDF. A noninhibitory serpin with neurotrophic activity. *Adv. Exp. Med. Biol.* **425**, 223–237.

Becerra, S. P. (2006). Focus on molecules: Pigment epithelium-derived factor (PEDF). *Exp. Eye Res.* **82**, 739–740.

Becerra, S. P., et al. (1993). Overexpression of fetal human pigment epithelium-derived factor in *Escherichia coli*. A functionally active neurotrophic factor. *J. Biol. Chem.* **268**, 23148–23156.

Becerra, S. P., et al. (2008). Pigment epithelium-derived factor binds to hyaluronan. Mapping of a hyaluronan binding site. *J. Biol. Chem.* **283**, 33310–33320.

Bouck, N. (2002). PEDF: Anti-angiogenic guardian of ocular function. *Trends Mol. Med.* **8**, 330–334.

Broadhead, M. L., et al. (2009). In vitro and in vivo biological activity of PEDF against a range of tumors. *Expert Opin. Ther. Targets* **13**, 1429–1438.

Campos, M., et al. (2006). A novel imaging technique for experimental choroidal neovascularization. *Invest. Ophthalmol. Vis. Sci.* **47**, 5163–5170.

Cao, W., et al. (2001). In vivo protection of photoreceptors from light damage by pigment epithelium-derived factor. *Invest. Ophthalmol. Vis. Sci.* **42,** 1646–1652.

Cayouette, M., et al. (1999). Pigment epithelium-derived factor delays the death of photoreceptors in mouse models of inherited retinal degenerations. *Neurobiol. Dis.* **6,** 523–532.

DeCoster, M. A., et al. (1999). Neuroprotection by pigment epithelial-derived factor against glutamate toxicity in developing primary hippocampal neurons. *J. Neurosci. Res.* **56,** 604–610.

Dobi, E. T., et al. (1989). A new model of experimental choroidal neovascularization in the rat. *Arch. Ophthalmol.* **107,** 264–269.

Duh, E. J., et al. (2002). Pigment epithelium-derived factor suppresses ischemia-induced retinal neovascularization and VEGF-induced migration and growth. *Invest. Ophthalmol. Vis. Sci.* **43,** 821–829.

Filleur, S., et al. (2009). Characterization of PEDF: A multi-functional serpin family protein. *J. Cell. Biochem.* **106,** 769–775.

Gonzalez, R., et al. (2010). Screening the mammalian extracellular proteome for regulators of embryonic human stem cell pluripotency. *Proc. Natl. Acad. Sci. USA* **107,** 3552–3557.

Houenou, L. J., et al. (1999). Pigment epithelium-derived factor promotes the survival and differentiation of developing spinal motor neurons. *J. Comp. Neurol.* **412,** 506–514.

Jablonski, M. M., et al. (2000). Pigment epithelium-derived factor supports normal development of photoreceptor neurons and opsin expression after retinal pigment epithelium removal. *J. Neurosci.* **20,** 7149–7157.

Kuncl, R. W., et al. (2002). Pigment epithelium-derived factor is elevated in CSF of patients with amyotrophic lateral sclerosis. *J. Neurochem.* **81,** 178–184.

Martinez, A., et al. (2002). The effects of adrenomedullin overexpression in breast tumor cells. *J. Natl. Cancer Inst.* **94,** 1226–1237.

Martinez, A., et al. (2004). Proadrenomedullin NH2-terminal 20 peptide is a potent angiogenic factor, and its inhibition results in reduction of tumor growth. *Cancer Res.* **64,** 6489–6494.

Meyer, C., et al. (2002). Mapping the type I collagen-binding site on pigment epithelium-derived factor. Implications for its antiangiogenic activity. *J. Biol. Chem.* **277,** 45400–45407.

Mukherjee, P. K., et al. (2007). Neurotrophins enhance retinal pigment epithelial cell survival through neuroprotectin D1 signaling. *Proc. Natl. Acad. Sci. USA* **104,** 13152–13157.

Notari, L., et al. (2005). Pigment epithelium-derived factor is a substrate for matrix metalloproteinase type 2 and type 9: Implications for downregulation in hypoxia. *Invest. Ophthalmol. Vis. Sci.* **46,** 2736–2747.

Ortego, J., et al. (1996). Gene expression of the neurotrophic pigment epithelium-derived factor in the human ciliary epithelium. Synthesis and secretion into the aqueous humor. *Invest. Ophthalmol. Vis. Sci.* **37,** 2759–2767.

Pang, I. H., et al. (2007). Pigment epithelium-derived factor protects retinal ganglion cells. *BMC Neurosci.* **8,** 11.

Perez-Mediavilla, L. A., et al. (1998). Sequence and expression analysis of bovine pigment epithelium-derived factor. *Biochim. Biophys. Acta* **1398,** 203–214.

Petersen, S. V., et al. (2003). Pigment-epithelium-derived factor (PEDF) occurs at a physiologically relevant concentration in human blood: Purification and characterization. *Biochem. J.* **374,** 199–206.

Polverini, P. J., et al. (1991). Assay and purification of naturally occurring inhibitor of angiogenesis. *Methods Enzymol.* **198,** 440–450.

Quan, G. M., et al. (2005). Localization of pigment epithelium-derived factor in growing mouse bone. *Calcified Tissue Int.* **76,** 146–153.

Ramirez-Castillejo, C., *et al.* (2006). Pigment epithelium-derived factor is a niche signal for neural stem cell renewal. *Nat. Neurosci.* **9,** 331–339.

Sanchez-Sanchez, F., *et al.* (2008). Expression and purification of functional recombinant human pigment epithelium-derived factor (PEDF) secreted by the yeast Pichia pastoris. *J. Biotechnol.* **134,** 193–201.

Seigel, G. M., *et al.* (1996). Expression of glial markers in a retinal precursor cell line. *Mol. Vis.* **2,** 2.

Stratikos, E., *et al.* (1996). Recombinant human pigment epithelium-derived factor (PEDF): Characterization of PEDF overexpressed and secreted by eukaryotic cells. *Protein Sci.* **5,** 2575–2582.

Takita, H., *et al.* (2003). Retinal neuroprotection against ischemic injury mediated by intraocular gene transfer of pigment epithelium-derived factor. *Invest. Ophthalmol. Vis. Sci.* **44,** 4497–4504.

Tombran-Tink, J., and Barnstable, C. J. (2003). PEDF: A multifaceted neurotrophic factor. *Nat. Rev. Neurosci.* **4,** 628–636.

Tsao, Y. P., *et al.* (2006). Pigment epithelium-derived factor inhibits oxidative stress-induced cell death by activation of extracellular signal-regulated kinases in cultured retinal pigment epithelial cells. *Life Sci.* **79,** 545–550.

Wu, Y. Q., and Becerra, S. P. (1996). Proteolytic activity directed toward pigment epithelium-derived factor in vitreous of bovine eyes. Implications of proteolytic processing. *Invest. Ophthalmol. Vis. Sci.* **37,** 1984–1993.

Wu, Y. Q., *et al.* (1995). Identification of pigment epithelium-derived factor in the interphotoreceptor matrix of bovine eyes. *Protein Expr. Purif.* **6,** 447–456.

CHAPTER ELEVEN

THE *DROSOPHILA* SERPINS: MULTIPLE FUNCTIONS IN IMMUNITY AND MORPHOGENESIS

Jean Marc Reichhart,* David Gubb,[†] *and* Vincent Leclerc*

Contents

1. Introduction	206
2. The Range of *Drosophila* Serpin Functions	207
2.1. Conservation of serpin functions	207
2.2. Genetic analysis of serpin functions	211
3. Techniques for Analysis of Immune Response	214
3.1. Activation of the Toll or IMD pathways	214
3.2. Methods for immune challenge	216
3.3. Hemolymph collection	217
3.4. Survival assays	218
3.5. Control of endemic infections in laboratory stocks	218
Acknowledgments	220
References	220

Abstract

Members of the serpin superfamily of proteins have been found in all living organisms, although rarely in bacteria or fungi. They have been extensively studied in mammals, where many rapid physiological responses are regulated by inhibitory serpins. In addition to the inhibitory serpins, a large group of noninhibitory proteins with a conserved serpin fold have also been identified in mammals. These noninhibitory proteins have a wide range of functions, from storage proteins to molecular chaperones, hormone transporters, and tumor suppressors. In contrast, until recently, very little was known about insect serpins in general, or *Drosophila* serpins in particular. In the last decade, however, there has been an increasing interest in the serpin biology of insects. It is becoming clear that, like in mammals, a similar wide range of physiological responses are regulated in insects and that noninhibitory serpin-fold proteins also play key roles in insect biology. *Drosophila* is also an important model

* Université de Strasbourg, UPR 9022 CNRS, IBMC, 15 rue Descartes, Strasbourg, France
[†] Unidad de Genómica Funcional, CIC bioGUNE, Parque Tecnológico de Vizcaya, Vizcaya, España, Spain

organism that can be used to study human pathologies (among which serpinopathies or other protein conformational diseases) and mechanisms of regulation of proteolytic cascades in health or to develop strategies for control of insect pests and disease vectors. As most of our knowledge on insect serpins comes from studies on the *Drosophila* immune response, we survey here the *Drosophila* serpin literature and describe the laboratory techniques that have been developed to study serpin-regulated responses in this model genetic organism.

1. Introduction

Members of the serpin superfamily of proteins have been found in all living organisms, although rarely in bacteria or fungi (Irving *et al.*, 2000). They have been extensively studied in mammals, where many rapid physiological responses are regulated by inhibitory serpins (Silverman *et al.*, 2001). For example, the coagulation, inflammatory, and complement pathways are controlled by antithrombin, α_1-antitrypsin and C1-Inhibitor, respectively (Bruce *et al.*, 1994; Cicardi *et al.*, 1998; Lomas *et al.*, 1992); while plasminogen activator inhibitor-1 (PAI-1) modulates angiogenesis, affecting both wound-healing and tumor growth (Providence and Higgins, 2004). In addition to the inhibitory serpins, a large group of noninhibitory proteins such as ovalbumin (Huntington and Stein, 2001) and Hsp47 (Dafforn *et al.*, 2001) with a conserved serpin fold have been identified. These noninhibitory proteins have a wide range of functions, from storage proteins to molecular chaperones, hormone transporters, and tumor suppressors (Silverman *et al.*, 2001). In contrast, until recently, very little was known about insect serpins in general, or *Drosophila* serpins in particular. In the last decade, however, there has been an increasing interest in the serpin biology of insects. It is becoming clear that, like in mammals, a similar wide range of physiological responses are regulated in insects and that noninhibitory serpin-fold proteins also play key roles in insect biology (Gubb *et al.*, 2007; Reichhart, 2005). These studies in insects are important for modeling of human serpinopathies (Lomas and Mahadeva, 2002; Lomas *et al.*, 2005) and protein conformational diseases (Lomas and Carrell, 2002), understanding the normal mechanisms of the regulation of proteolytic cascades in health, and developing strategies for control of insect pests and disease vectors. As most of our knowledge on insect serpins comes from studies on the *Drosophila* immune response, we will survey here the *Drosophila* serpin literature and describe the laboratory techniques that have been developed to study serpin-regulated responses in this model genetic organism.

Studies on mammalian inhibitory serpins have established a "suicide-cleavage" mechanism in which the serpins interact with their target proteinases to form an inactive, covalently linked, serpin/proteinase complex (Gettins, 2002). The serpin fold consists of three β-sheets (A–C) with eight or nine α-helical linkers and represents 350–400 amino acids (Irving *et al.*, 2000). As such, the serpin fold represents a large molecular weight protease inhibitor, bigger than the chymotrypsin protease fold. Sticking out from the serpin core is an exposed reactive center loop (RCL) that acts as bait for the target protease. In the native state, serpins are in a metastable (stressed, S) conformation. Proteinase cleavage within the RCL allows the serpin structure to undergo a transition to the stable (relaxed, R) conformation. During this process, the proteinase is translocated through 70 Å, from the upper to the lower pole of the serpin, and the RCL inserts as an extra strand within β-sheet A. The protease is denatured by crushing against the bottom of the serpin core and the denatured serpin/proteinase complex is targeted for degradation (Huntington *et al.*, 2000). This mechanism imposes several constraints on the serpin fold. In particular, the energy store held in the stressed configuration is critical and appears to be dependent on the specific arrangement of β-sheets and α-helical linkers; at least by the criterion that this structural organization is constant, although the identity of relatively few of the individual amino acids are strongly conserved (Irving *et al.*, 2000). In addition, the S → R transition requires a run of relatively small amino acids in the "hinge region" to allow insertion into β-sheet A of the cut RCL loop (Irving *et al.*, 2000). Serpin-fold proteins that lack this flexible hinge region are unable to undergo the S → R transition and are not active protease inhibitors.

2. The Range of *Drosophila* Serpin Functions

2.1. Conservation of serpin functions

2.1.1. Comparison with mammalian serpins

In mammals, the largest group of serpins consists of humoral protease inhibitors, such as antitrypsin and antithrombin, which regulate rapid physiological responses. Intracellular protease inhibitors, such as PAI-2, are also common. In addition, there are both intracellular and humoral noninhibitory serpins, such as PEDF (Huntington and Stein, 2001) and Hsp47 (Dafforn *et al.*, 2001). All four of these functional groups are represented in the *Drosophila* genome (Table 11.1).

The human complement of 35 serpins has been grouped into nine Clades, reflecting sequence similarity, gene structure, and chromosomal synteny (Irving *et al.*, 2000). The number of serpins per genome is highly

Table 11.1 The *Drosophila* serpins

Serpin	FlyBase identifier	AAs	P1/P1'	Pathway/ function	Synonym
Spn27A	CG11331	447-25	K/F	Morph. and PO	
Spn28B	CG6717	378-15	K/K		
Spn28Da	CG31902	384-15	L/S		
Spn28Db	CG33121	346	–		
Spn28Dc	CG7219	536-16	S/G	Melanization	Spn28D
Spn28F	CG8137	375-18	Y/S		Spn2
Spn31A	CG4804	382	–		
Spn38F	CG9334	372-15	K/S	Immunity	Spn3
Spn42Da	CG9453	A 392	R/A	HDEL signal	Spn4-A
		B 424-28	R/A	HDEL signal	Spn4-B
		D 411-28	T/S		Spn4-D
		E, F 411-28	A/S		Spn4-E, F
		G 379	T/S		Spn4-G
		H,K,L 379	A/S		Spn4-H,K,L
		I 406-28	V/A		Spn4-I
		J 374	V/A		Spn4-J
Spn42Db	CG9454	388	K/G		
Spn42Dc	CG9455	403	M/M		
Spn42Dd	CG9456	372-15	R/A		Spn1
Spn42De	CG9460	404-20	E/S		
Spn43Aa	CG12172	390-22	M/S		
Spn43Ab	CG1865	393-18	–		
Spn43Ac	CG1857	476-23	L/S	Toll, immunity	
Spn43Ad	CG1859	407-22	–		
Spn47C	CG7722	382-?20	–		
Spn53F	CG10956	379-16	–		
Spn55B	CG10913	374	R/M		
Spn75F	CG32203	356-19	–		
Spn76A	CG3801	388-22	–	Immunity	
Spn77Ba	CG6680	450-24	K/A	Melanization	
Spn77Bb	CG6663	362-23	–		
Spn77Bc	CG6289	416-22	–	Seminal fluid	
Spn85F	CG12807	640-20	?	SerpXXin	
Spn88Ea	CG18525	427-18	S/A	PO, wing	Spn5
Spn88Eb	CG6687	426-18	S/S		
Spn100A	CG1342	649-19	?	SerpXXin	
	CG14470	1976	–		

variable within the vertebrate lineage and is increased to 64 in mice, largely as a result of expansion and divergence of inhibitory serpins from Clade A (Askew et al., 2007) and Clade B (Askew et al., 2004; Kaiserman and Bird, 2005). The *Drosophila* serpin genes, however, do not group convincingly within clearly defined Clades. Still less is it possible to match clusters of related *Drosophila* serpins with the mammalian Clade groupings. Further, it is surprisingly difficult to identify the *Drosophila melanogaster* serpin orthologues in distant *Drosophilid* species (Garrett et al., 2009) or within mosquitoes and other insects. For this reason, the *D. melanogaster* serpins are generally designated by their cytological location or FlyBase CG gene identifiers, without assigning them to novel Clades (Reichhart, 2005).

2.1.2. Conservation of RCL sequences

Within the *Drosophilid* species, the RCL sequences of inhibitory serpins are strongly conserved within the orthologues that can be identified, including the putative proteinase cleavage sites. Inhibitory serpins rarely change their proteinase targets except by a duplication/divergence mechanism. Noninhibitory serpins appear to derive from inhibitory serpins, but not the reverse (Garrett et al., 2009). Concerning the inhibitory serpins, *D. melanogaster* has 17 genes that carry the run of small amino acids characteristic of the flexible hinge region of active inhibitors. However, one of these genes, *Spn42Da*, encodes eight protein isoforms with four different RCL sequences (see below). On this basis, the *D. melanogaster* genome encodes 24 inhibitory serpin activities, 17 having a secretion peptide and 7 without (Table 11.1). This number is comparable to about 31 inhibitory serpins in humans, based on data from Irving et al. (2007) and about 58 inhibitory serpins in mice, based on data from Askew et al. (2007). By comparison, the number of putative target proteases in these species are 211 chymotrypsin-fold proteases in *Drosophila*, 176 in humans, and 277 in mice. In principle then, the *Drosophila* genome encodes rather fewer serpins and rather more proteases than the human genome. Serpin-regulated proteolytic signaling cascades could play a central role in regulating a wide range of physiological responses as in mammals.

2.1.3. Macroglobulins and tight-binding protease inhibitors

In addition to serpin-family inhibitors, the *Drosophila* genome encodes macroglobulin-family inhibitors and 10 families of tight-binding inhibitors, such as Kazal and Kunitz (Gubb et al., 2010). Tight-binding inhibitors have a similar bait region to the serpin RCL, but form a precise lock and key fit with the catalytic site of their target protease. The tight-binding inhibitory domains tend to be shorter than 100 amino acids, but these domains are frequently incorporated into larger proteins as dimers, or multimers.

2.1.4. Terminal peptides and size-conservation of the serpin core

In striking contrast to the tight-binding inhibitory domains, the core serpin fold is always within the 350–400 amino acid size range and this fold has not been found incorporated into double, or multiple-domain inhibitors. A number of serpins are known with short terminal extensions to the core fold, but this is rare. For example, the N-terminal extension of Heparin cofactor II interacts with an exosite on thrombin, and cleavage of this extension reduces the serpin's specificity for its substrate (Baglin *et al.*, 2002), Similarly, the thermophilic bacterial serpin, tengpin, carries an N-terminal extension that binds to the serpin core and stabilizes the native conformation at high temperatures (Zhang *et al.*, 2007). Having said this, it is surprising to find that the *Drosophila* genome encodes eight proteins with homology to serpins that are outside the normal size range (Spn27A, Spn28Dc, Spn43Ac, Spn77Ba, Spn85F, Spn88Ea, Spn88Eb, and Spn100A; Table 11.1). Analysis with the NCB conserved domain server indicates that the core serpin folds of Spn27A, Spn28Dc, Spn43Ac, and Spn77Ba each carries a long N-terminal extension of about 45, 92, 77, and 51 amino acids, respectively (after cleavage of putative secretion signal peptides). All four of these serpins have conserved orthologues across the 12 *Drosophilid* species. The Spn88Ea and Spn88Eb serpins carry short N-terminal extensions of about 20 and 18 amino acids, respectively. The Spn85F and Spn100A serpins, however, are well above the normal size range, at 620 and 630 amino acids, respectively. Both proteins have C- and N-terminal homology to serpins, with an internal segment of 220–230 amino acids, which is composed of unrelated sequences. These proteins represent a serpin-related fold, which has apparently arisen by insertion of a novel segment of peptide-coding sequence. It is unclear whether these serpin-related folds might have inhibitory functions; in both cases the putative RCL hinge regions contain a string of small amino acids, although neither matches exactly the *Drosophilid* serpin flexible hinge-region consensus (Garrett *et al.*, 2009). Spn85F and Spn100A have widely diverged from each other (Reichhart, 2005), but both have conserved orthologues in all of the 12 sequenced *Drosophilid* genomes. In addition, the *D. melanogaster* genome also encodes a putative nonsecreted protein of 1976 amino acids (CG14470) that includes a C-terminal block of 350 amino acids with homology to the serpin core. To the best of our knowledge, this is the only example of a serpin-related fold that is incorporated into a large protein. Again *CG14470* has orthologues in the other *Drosophilid* genomes, *Apis melifera* (Zou *et al.*, 2006) and other insects, implying that the gene represents a conserved function, rather than being the result of a recent fusion of peptide-coding sequences that happens to be in-frame, but lacking activity.

Analysis with the conserved domain server also identifies a single serpin, Spn75F, with a long C-terminal extension. Spn75F shows only about 282 amino acids with serpin homology. It is a putative noninhibitory serpin-fold protein that apparently lacks the RCL segment, although the conserved Pro369 and Phe384 residues (α_1-antitrypsin numbering) that form the "gate" at the start of s4B (Irving et al., 2000) are retained.

2.2. Genetic analysis of serpin functions

2.2.1. The innate immune response

In *Drosophila*, the best-documented facet of the immune response involves the activation of two signal transduction cascades, the Toll and IMD (immune deficiency) pathways (Ferrandon et al., 2007). These pathways lead to transcription of genes encoding antimicrobial peptides (AMPs) in the fat body, and the subsequent secretion of these peptides into the hemolymph. This defence mechanism discriminates between Gram-negative bacteria, which activate mostly the IMD pathway, and Gram-positive bacteria and fungi, which stimulate the Toll pathway. Both pathways culminate in the activation of NF-κB-related transcription factors. Direct binding of bacterial peptidoglycan (PGN) to a transmembrane receptor activates the IMD pathway. In contrast, the transmembrane receptor Toll does not recognize microbial determinants directly, but is activated by binding of an endogenous ligand, the proteolytically cleaved form of the cytokine-like protein Spaetzle (Spz). The detection of Gram-positive bacterial or fungal cell wall components requires a combination of different circulating receptors. Binding of microbial ligands to these molecules induces the activation of proteolytic cascades leading to the cleavage of Spz and subsequent Toll activation.

Five of the *Drosophila* serpins have been shown to have functions related to the immune response (*Spn43Ac*, *Spn27A*, *Spn28Dc*, *Spn77Ba*, and *Spn88Ea*). Of these, the best studied is *Spn43Ac* (*necrotic, nec, CG1857*). This serpin regulates the systemic response to fungal and Gram-positive bacterial infections, via the Toll signaling pathway. Nec mutants express high levels of the antifungal peptide Drosomycin even in the absence of immune challenge (Levashina et al., 1999). Nec has a broad inhibitory spectrum for elastase, thrombin, and chymotrypsin-like proteases, *in vitro* (Robertson et al., 2003). The N-terminal extension of the Nec core serpin is cleaved on immune challenge, which increases the substrate specificity for porcine pancreatic elastase (Pelte et al., 2006). The length and sequence of this N-terminal is poorly conserved within the *Drosophilidae*, being about 50 amino acids in *Drosophila sechellia* and 125 amino acids in *Drosophila grimshawi* (Garrett et al., 2009). Interestingly, the other three serpins with long N-terminal extensions (Spn27A, Spn28Dc, and Spn77Ba) and one of those with a short N-terminal (Spn88Ea) also have immune-related functions, although whether these termini are cleaved on immune challenge is not known.

Spn27A (*CG1131*) regulates pathogen melanization, mediated through the phenoloxidase (PO) pathway and mutants lacking this serpin, show high levels of PO in the hemolymph, associated with melanization of internal tissues (De Gregorio *et al.*, 2002; Ligoxygakis *et al.*, 2002). The *Spn27A* transcript is upregulated following either septic injury with Gram-positive or Gram-negative bacteria or natural infection with the fungus *Beauvaria bassiana*. *Spn28Dc* (*CG7219*, *Spn28D*) also regulates the PO pathway, but in response to wounding (Scherfer *et al.*, 2008). The melanization reaction in trachea is regulated by *Spn77Ba* (*CG6680*; Tang *et al.*, 2008). *Spn88Ea* (*CG18525*, *Spn5*; Ahmad *et al.*, 2009) is, like *Spn77Ba*, involved in activation of the Toll pathway and subsequent systemic Drosomycin expression.

In addition to deregulating the Toll pathway, *nec* mutations also cause a cellular necrosis phenotype (Green *et al.*, 2000) and act as suppressors of melanotic tumors (Avet-Rochex *et al.*, 2010), while *Spn27A* regulates the embryonic Toll pathway controlling dorso–ventral axis formation (Hashimoto *et al.*, 2003; Ligoxygakis *et al.*, 2003) and *Spn88Ea* affects wing inflation (Charron *et al.*, 2008). Given the large number of proteases in *Drosophila*, it seems likely that many inhibitory serpins may have multiple targets in different tissues and developmental stages. However, the regulation of the PO cascade appears to involve multiple serpins. This situation has been described in the beetle *Tenebrio molitor*, where *three different serpins* inhibit the proteases of the PO cascade in larval hemolymph (Jiang *et al.*, 2009).

2.2.2. Seminal fluid proteins

Apart from the large number of serpins with roles in the innate immune response, a significant proportion of the remainder form components of seminal fluid and accessory gland secretions. Spn76A (CG3801, Acp76A) is transferred to the female genital tract during mating (Coleman *et al.*, 1995) and is one of the seven putative noninhibitory serpin transcripts (*Spn28Da, Spn28Db, Spn53F, Spn75F, Spn76A, Spn77Ba, Spn77Bc*) and two putative inhibitory serpin transcripts (*Spn28B* and *Spn28F*) that are expressed at high levels in the male accessory gland (http://flyatlas.org/). All but one of these serpins (*Spn28Db*) carry a signal peptide (Table 11.1) and could therefore be components of seminal fluid. Five of these serpins have also been identified as orthologues of ESTs from a *Drosophila simulans* accessory gland cDNA library (*Spn28F, Spn38F, Spn53F, Spn76A,* and *Spn77Bc*; Mueller *et al.*, 2005). It is well established that accessory gland proteins (Acps) in general evolve more rapidly than nonreproductive proteins (Swanson *et al.*, 2001). In particular, orthologues for none of the five *D. simulans* Acp serpins were identified in *Drosophila pseudoobscura* (Mueller *et al.*, 2005). Strickingly, most of *D. melanogaster* serpins, for which orthologues could not be identified in the *Drosophilid* outgroups (Garrett *et al.*, 2009), correspond to the nine putative Acps listed above. In particular, discounting these nine, the remaining serpins are strongly conserved within the *Drosophilid* group: all but three have been identified in *D. pseudoobscura*, and all but five in *D. grishawi* (Garrett *et al.*, 2009). It is unclear what targets most of the Acp serpins interact with, but clearly most of them are not protease inhibitors. Several seminal fluid components modify female reproductive behavior (Mueller *et al.*, 2007; Ram and Wolfner, 2007). It is also known that many noninhibitory serpins play a role in hormone binding. The one well-studied example of an *Acp* protease inhibitor, however (Acp62F) is a tight-binding inhibitor rather than a serpin (Lung *et al.*, 2002).

2.2.3. Alternatively spliced serpin transcripts and multiple protease targets

The splice variants of *Spn42Da* (*CG9453, Spn4*) are particularly interesting in that three alternative 5' exons provide different localization signals, while four alternative 3' exons give alternative RCL segments. Eleven splice variants have been identified, but several of these encode the same protein variants, so that eight distinct protein isoforms are generated (Table 11.1). These isoforms are targeted against subtilisin-, chymotrypsin-like serine proteases and papain-like cystein proteases (Bruning *et al.*, 2007; Richer *et al.*, 2004). Alternative "chimeric" serpin isoforms were first detected in *Bombyx mori* (Sasaki, 1991) and the splicing mechanism elucidated in *Manduca sexta* (Jiang *et al.*, 1994, 1996). Alternatively, spliced serpin transcripts have since been described in nematodes, mosquitoes, and other insects (Danielli *et al.*, 2003; Kruger *et al.*, 2002). In the *Drosophilid* subgroup, exon 5 of *Spn42Da*, which encodes a furin-targeted RCL involved in the protein secretion pathway, is maintained in orthologues across all 12 species. Exons 6–8 are maintained within the *melanogaster* subgroup, but duplicate and diverge widely in the outgroups (Borner and Ragg, 2008) generating a set of RCL sequences with novel putative P1/P1' sites (Garrett *et al.*, 2009).

2.2.4. Identification of target proteases and the degradation of serpin/protease complexes in the hemolymph

Direct biochemical approaches for identifying the target proteases of individual serpins are limited by the small amounts of starting material that can be obtained. *In vitro* experiments with transgenic serpins and proteases have identified the types of protease substrates for individual serpins in several cases. Given the large number of proteases encoded within the *Drosophila* genome and the broad substrate specificity of many serpins, however, the true physiological targets remain uncertain. One of the difficulties with identifying humoral serpin/proteases complexes in *Drosophila* is that these are rapidly taken up from hemolymph and degraded. In mammals, serpin/protease complexes are removed from the blood plasma via receptors of the low-density lipoprotein (LDL) family in the liver (Kasza *et al.*, 1997; Kounnas *et al.*, 1996). In *Drosophila*, humoral serpins are synthesized and exported from the fat body, but endocytosed and degraded in athrocytes (Soukup *et al.*, 2009). The Nec serpin is taken up by the LDL receptor LpR1, probably as the serpin/protease complex. In any case, although the native serpin can be readily detected on Western blots of hemolymph proteins, putative serpin/protease bands are very faint.

An example where a combination of biochemical and genetic approaches have identified serpin/protease complexes is Spn27A in the embryo (Hashimoto *et al.*, 2003). The proteases involved in the embryonic Toll signaling cascade have been identified by genetic screens including Easter that activates Spz. Given that a higher molecular weight form of Easter had been detected biochemically (Misra *et al.*, 1998) and the inhibitory spectrum of transgenic Easter is trypsin-like (Hashimoto *et al.*, 2003), identified Spn27A by looking for a serpin with an RCL that carried a trypsin-like bait and resembled the cleavage site that activates Spz.

The genetic and molecular tools that have been developed for the *Drosophila* model organism are extremely powerful and allow the functions of mutant and genetically engineered proteins to be studied *in vivo*. We hope that future refinements in the sensitivity of biochemical techniques will allow these to be applied to the small volumes of fluids available from *Drosophila*. The combination of genetic and biochemical technologies should lead to new insights into the role of serpins and their target serine proteases.

3. Techniques for Analysis of Immune Response

3.1. Activation of the Toll or IMD pathways

3.1.1. Inducers

Both Gram-positive bacteria and fungi are able to activate the Toll pathway whereas Gram-negative bacteria activate the IMD pathway. The extracellular proteolytic cascade, which results in the cleavage of the Toll ligand is split into two branches, one sensitive to exogenous proteases (danger signaling), and the other responding to fungal and Gram-positive bacterial pathogen associated molecular patterns (PAMP; El Chamy *et al.*, 2008; Gottar *et al.*, 2006). During most infections with microbial pathogens, both danger signaling and PAMP side-branches of the Toll pathway will be activated. It is possible that the IMD pathway may also be activated through both microbial determinants and proteolytic activities (Schmidt *et al.*, 2008).

To investigate the separate contributions of these side-branches, it is useful to be able to activate each branch independently of the other. Activating only the PAMP responses can be achieved by injection of PGN or by immune challenge with dead bacteria or fungal spores. Alternatively, nonpathogenic microorganisms, which do not secrete virulence factors, can be used to selectively activate the PAMP response, *Micrococcus luteus* (for the Toll pathway) and *Escherichia coli* (for the IMD pathway). Activation of the danger-signaling pathway can be achieved by injection of pathogen-derived proteases. To activate both danger-signaling and PAMP pathways together, flies can be challenged with pathogenic microorganisms, such as the bacterium *Enterococcus faecalis* or the fungus *B. bassiana*. When using live pathogenic microorganisms for infection, survival after immune challenge can also be evaluated as a measure of the effectiveness of the immune response. Flies can be naturally infected (i.e., without injuring the cuticle) with filamentous fungi and Gram-negative bacteria. There is, however, no established model for natural infection of flies with Gram-positive bacteria.

3.1.2. Read-outs for immune pathways activation
The classical read-out for Toll and IMD pathways activation consists in measuring target gene expression level (Lemaitre *et al.*, 1995). The most widely used reporter gene for Toll pathway activation is *drosomycin*, which encodes an antimicrobial peptide. Maximal expression of *drosomycin* occurs around 24 h after immune challenge. The activation of the IMD pathway is monitored by measuring expression of the *diptericin* gene 6–9 h after immune challenge. Transcription of these target genes can be monitored by Northern blotting experiments, quantitative-RT-PCR analysis or using transgenic reporter flies (Jung *et al.*, 2001).

3.1.3. Factors to consider when making an immune challenge
We should first point out that the *Drosophila* immune response varies depending on the circadian rhythm (Lee and Edery, 2008). A stronger response and an increased survival are observed when flies are infected in the middle of the night. Most researchers are probably not inclined to infect flies during the small hours of the night, but for the sake of reproducibility, flies should always be challenged at the same time. For similar reasons, flies should always be kept on a regular day/night cycle, either in a culture room or an incubator.

The immune response is strongly affected by the genetic background in fly strains. Considerable variability exists in the ability of different strains of *D. melanogaster* to mount an immune response (e.g., see Lazzaro *et al.*, 2004). For this reason, serious attention should be paid when choosing the correct control fly strains. In particular, when using the *upstream activating sequence (UAS)-Gal4* system (Brand and Perrimon, 1993), we have repeatedly observed that some *Gal4* driver strains show reduced immune challenge dependent activation of IMD pathway. Controls for activation of the IMD pathway should include a strain containing the relevant driver together with an irrelevant UAS-driven gene such as GFP. Initial observations suggest that Toll pathway is not affected (unpublished observations).

There is a low variable basal level of constitutive *drosomycin* expression in some epithelia (Tzou *et al.*, 2000). For this reason, expression levels of target genes should be normalized using a control in which the immune response is strongly activated. When working on the Toll pathway, flies challenged with *M. luteus* are often a suitable control.

3.2. Methods for immune challenge

3.2.1. Septic injury with needles

The simplest way to challenge flies is to wound them with a needle dipped into a bacterial culture (Lemaitre *et al.*, 1995). We prefer tungsten needles sharpened electrolytically, but glass needles can also be used (Schmidt *et al.*, 2008). A piece of 0.25 mm tungsten wire is linked to the cathode of a 6-V generator, dipped into a 1 M NaOH solution (with a carbon electrode as anode) and gently pulled out of the solution several times. This electrolytic sharpening produces a smoothly tapering tip to the tungsten wire.

Having dipped the needle in bacterial culture, flies are pricked on one side of the thorax, just below the wing attachment site. For reproducible results, we find that the diameter of the needle must be constant and the same person should handle the complete series of any set of experiments. For immune challenge with nonpathogenic strains such as *M. luteus*, we use the pellet obtained by centrifugation of an overnight culture. For challenge with pathogenic strains, the bacterial concentration and virulence state are critical. It is important to control both the optical density and the growth phase of the culture. As an example, for *E. faecalis* we use an exponential culture of the *OG1RF* wild-type strain at OD 0.5.

3.2.2. Direct injection of pathogens

The dose of bacteria delivered to the body cavity by septic injury with a contaminated needle cannot be precisely controlled or calibrated. To deliver a defined quantity of bacterial, yeast, or fungal spores, cultures diluted in PBS can be injected into the *Drosophila* body cavity using a glass capillary. We routinely inject 4–50 nL of solution with a Drummond Nanoject II (Drummond Scientific, Broomall, PA) into a single fly. Glass capillary needles are made using 3.5″ glass capillary tube with a P97 Model Suttler needle-puller (as a rough guide, we set the parameters: Heat 347, Pull 225, Velocity 50 and Time 150 on this particular machine). The glass needle is back-filled with mineral oil (Voltalef H10S) before taking up the diluted culture to allow for precise volume ejection. As with septic injuries, injections are made into the thorax.

This same injection setup is also ideal for introducing any aqueous solution into the fly. In particular, the immune-response elicitors PGN, β-(1,3)-glucan and proteases have been injected dissolved in PBS: PGN (9.2 nL at 5–10 mg/mL; Leulier *et al.*, 2003; Stenbak *et al.*, 2004) and β-(1,3)-glucans (alkali-insoluble fraction of the *Aspergillus fumigatus* fungal cell wall at 4.6 nL of a 2.5 mg/mL; Gottar *et al.*, 2006). For protease injection, 18.3 nL of a commercial solution of proteases (diluted 1:2000 in PBS) from *Bacillus subtilis* (over 16 U/g; P5985; Sigma-Aldrich) or *Aspergillus oryzae* (over 500 U/g; P6110; Sigma-Aldrich) is a suitable dose (El Chamy *et al.*, 2008). Under these conditions, *drosomycin* expression is maximal at 12 h after injection.

Dead bacteria, yeast, and fungal spores have also been injected using this method. Dead bacteria are prepared from bacterial cultures (absorbance 0.1 at 600 nm) that are heated for 20 min at 95 °C, cooled for 20 min at 4 °C, then heated again for 20 min at 95 °C (El Chamy *et al.*, 2008). Overnight cultures of the yeast *Candida albicans* are harvested by centrifugation and washed three times with ice-cold PBS. The yeast cells are killed, either by incubation at 95 °C for 5 min or by fixation with 4% paraformaldehyde (PFA) for 20 min at 4 °C, followed by two washes with PBS. Similar treatments can be used to kill *B. bassiana* or *Metarhizium anisopliae* spores. Filamentous fungi are grown on malt-agar plates in order to collect the spores. The spores are then killed by incubation with 1.5 M NaOH solution twice for 30 min at 70 °C, followed by four washes with PBS 0.01% Tween 20. A volume of 4.6 nL of spores or yeast suspension is injected (Gottar *et al.*, 2006).

3.2.3. Natural infection

With entomopathogenic fungi, natural infection is performed simply by shaking anesthetized flies for a few minutes in a Petri dish containing a sporulating culture of fungi. Flies covered by spores are then placed into fresh *Drosophila* culture vials and incubated at 29 °C (Lemaitre *et al.*, 1997).

For Gram-positive bacteria, an oral infection method has been described by Cox and Gilmore (2007), but only used in this single publication. Oral infection has been used several times for Gram-negative bacteria (Basset *et al.*, 2000; Nehme *et al.*, 2007; Vodovar *et al.*, 2005). Two different ways of feeding flies with bacteria have been described: bacteria are either mixed with the feeding medium or diluted in a 50 mM sucrose solution on a filter paper. Flies can be fed continuously on bacteria or transferred after a period of time to normal food or sucrose solution.

3.3. Hemolymph collection

Hemolymph can be collected with the same Nanoject II apparatus used for injection. Rather thinner glass capillaries are used (parameters set to Heat 347, Pull 100, Velocity 150, and Time 150) as this allows hemolymph into the needle by capillary action. As before, the needle is inserted into the mesothorax below the wing hinge and kept just under the cuticle. With most fly strains, around 1 μg of protein per fly can be collected. Hemolymph tends to coagulate rapidly on exposure to air but can be used immediately for rescue experiments, by injection into a single recipient fly (Gobert et al., 2003; Michel et al., 2001). Alternatively, the collected fluid can be diluted by injecting into a 10-μL drop of PBS (or other suitable buffer) supplemented with a cocktail of protease inhibitors (Complete protease inhibitor, Roche). A convenient way to do this is to have the drop sit on a small piece of parafilm on top of an ice bucket.

For experiments on the melanization cascade, prophenol oxydase activity can be tested in hemolymph: 10 μg of protein are suspended in 40 μL of PBS containing protease inhibitors. One hundred and twenty microliters of a saturated solution of L-3,4-dihydroxyphenylalanine (L-DOPA) are added to each sample. OD measurements are taken 5 min later at 492 nm (Ligoxygakis et al., 2002). Alternatively PO cleavage can be analyzed by Western blotting (Leclerc et al., 2006).

3.4. Survival assays

Survival assays are performed on batches of 20–30 flies. After infection, flies are incubated at 25–29 °C depending on the optimal growth temperature of the microorganism used to infect the flies. Infections are performed by septic injury or injection of bacteria, yeast, or fungal spores (Alarco et al., 2004; Gottar et al., 2006; Lemaitre et al., 1995, 1996).

3.5. Control of endemic infections in laboratory stocks

Several natural infections are able to affect flies viability in normal conditions or after a stress like an immune challenge, an injury or a heat shock. Special care should be taken by laboratories working in the field of insect immunity as endemic infections in laboratory stocks affect the fly immune response and resistance to experimental infections. Mites are potential vectors of infections spreading them between fly stocks. Mite infection of fly cultures should therefore be avoided. Most information about fly stock infections and the way how to get rid of them can be found in Ashburner's "*Drosophila*, a Laboratory Handbook" (Ashburner et al., 2005). We would like to add a few comments on three common infections that have a clearly demonstrated adverse effect on immunity experiments.

3.5.1. Wolbachia

Wolbachia pipientis are maternally transmitted, Gram-negative, obligate intracellular bacteria. *D. melanogaster* strains are commonly infected with *Wolbachia*, which are known to confer increased resistance to insect viruses (Hedges et al., 2008; Teixeira et al., 2008). We have no data about the effect of *Wolbachia* infection on fly resistance to other infections, although this is a likely mechanism for driving *Wolbachia* infections through insect populations. The status of most fly stocks with respect to *Wolbachia* is unknown. It is therefore highly recommended to check for the presence of *Wolbachia* before using a new stock. *Wolbachia* can be detected by PCR using Wsp primers (81F TGGTCCAATAAGTGATGAAGAAAC and 691R AAAAATTAAACGCTACTCCA; Zhou *et al.*, 1998). We routinely use a second nested PCR (wolb Fw TTGTAGCCTGCTATGGTA-TAACT and wolb Rv GAATAGGTATGATTTTCATGT; P. Giammarinaro and D. Ferrandon, personal communication). Wolbachia infection is cured by a 0.03% tetracycline treatment in the fly food.

3.5.2. Microsporidia

Microsporidia are obligate intracellular parasites. The infection can stay asymptomatic for a long time in fly stocks. The first sign of microsporidial infections is a reduced adult life span (more extreme at 29 °C than at 25 and 18 °C) and an increased susceptibility to immune challenge, which can be misleading when interpreting survival experiments with other pathogens. While acute infections can wipe out a fly stock, microsporidial infections are often chronic at low level and may go unrecognized. The characteristic features are prolonged development, pupal mortality, reduced fertility, reduced fecundity, and bloated abdomens. A characteristic of sterile eggs is that they tend to start development, but die and turn black, unlike unfertilized eggs, which remain white for several weeks. Most *Drosophila* strains show the occasional black eggs and pupae, but high levels are a warning sign. Microsporidial infections can be detected by PCR (Franzen *et al.*, 2005). Rapid expansion of stocks and dechorionation of embryos eliminate vertical transmission and remove most of the parasites. This treatment will allow amplification of the stock, but addition of Fumagillin-B at 8 g/l in the fly food is required to completely eliminate the infection. The Fumagillin-B treatment is toxic to flies and can therefore only be applied to an expanded stock (Ashburner *et al.*, 2005; G. Boulianne, S. Niehus, and D. Ferrandon, personal communication).

3.5.3. Nora virus

Nora virus is a newly identified picornavirus (Habayeb *et al.*, 2006, 2009a,b), which infects *Drosophila*. The virus is transmitted by feces and is present in a large proportion of laboratory stocks with highly variable viral loads. Nora virus infection can be almost asymptomatic under normal conditions. Infected flies show poor health at 29 °C and are susceptible to heat shock or immune stress (S. Mueller and J. L. Imler, personal communication), which affects the results of viability experiments. No efficient treatment has so far been described, but Nora virus can be detected by Q-RT-PCR (Habayeb *et al.*, 2006).

ACKNOWLEDGMENTS

Work in the laboratory of J. M. R. and V. L. was supported by CNRS, Université de Strasbourg and Grants from ANR (Drosovir ANR-09-MIEN-007), FRM (DEQ 20090515422), and ERC (Immudroso 2009-AdG-20090506). D. G. was supported by a Spanish Ministry of Education and Science Grant with additional support from the Department of Industry, Tourism and Trade of the Government of the Autonomous Community of the Basque Country (Etortek Research Programs) and The Department of Education, Universities and Research of the Basque Government (Basic and Applied Research Program), and from the Innovation Technology Department of Bizkaia County.

REFERENCES

Ahmad, S. T., Sweeney, S. T., Lee, J. A., Sweeney, N. T., and Gao, F. B. (2009). Genetic screen identifies serpin5 as a regulator of the toll pathway and CHMP2B toxicity associated with frontotemporal dementia. *Proc. Natl. Acad. Sci. USA* **106,** 12168–12173.

Alarco, A. M., Marcil, A., Chen, J., Suter, B., Thomas, D., and Whiteway, M. (2004). Immune-deficient *Drosophila melanogaster*: A model for the innate immune response to human fungal pathogens. *J. Immunol.* **172,** 5622–5628.

Ashburner, M., Golic, K. G., and Hawley, R. S. (2005). *Drosophila*: A Laboratory Handbook. 2nd edn. Cold Spring Harbor Laboratory Press, Cold Spring Harbor, New York.

Askew, D. J., Askew, Y. S., Kato, Y., Turner, R. F., Dewar, K., Lehoczky, J., and Silverman, G. A. (2004). Comparative genomic analysis of the clade B serpin cluster at human chromosome 18q21: Amplification within the mouse squamous cell carcinoma antigen gene locus. *Genomics* **84,** 176–184.

Askew, D. J., Coughlin, P., and Bird, P. I. (2007). Mouse serpins and transgenic studies. *In* "Molecular and Cellular Aspects of the Serpinopathies and Disorders in Serpin Activity," (D. A. Lomas and G. A. Silverman, eds.), pp. 101–129. World Scientific Publishing Co. Pte. Ltd, Singapore.

Avet-Rochex, A., Boyer, K., Polesello, C., Gobert, V., Osman, D., Roch, F., Auge, B., Zanet, J., Haenlin, M., and Waltzer, L. (2010). An in vivo RNA interference screen identifies gene networks controlling *Drosophila melanogaster* blood cell homeostasis. *BMC Dev. Biol.* **10,** 65.

Baglin, T. P., Carrell, R. W., Church, F. C., Esmon, C. T., and Huntington, J. A. (2002). Crystal structures of native and thrombin-complexed heparin cofactor II reveal a multistep allosteric mechanism. *Proc. Natl. Acad. Sci. USA* **99,** 11079–11084.

Basset, A., Khush, R. S., Braun, A., Gardan, L., Boccard, F., Hoffmann, J. A., and Lemaitre, B. (2000). The phytopathogenic bacteria *Erwinia carotovora* infects *Drosophila* and activates an immune response. *Proc. Natl. Acad. Sci. USA* **97,** 3376–3381.

Borner, S., and Ragg, H. (2008). Functional diversification of a protease inhibitor gene in the genus *Drosophila* and its molecular basis. *Gene* **415,** 23–31.

Brand, A. H., and Perrimon, N. (1993). Targeted gene expression as a means of altering cell fates and generating dominant phenotypes. *Development* **118,** 401–415.

Bruce, D., Perry, D. J., Borg, J. Y., Carrell, R. W., and Wardell, M. R. (1994). Thromboembolic disease due to thermolabile conformational changes of antithrombin Rouen-VI (187 Asn–>Asp). *J. Clin. Invest.* **94,** 2265–2274.

Bruning, M., Lummer, M., Bentele, C., Smolenaars, M. M., Rodenburg, K. W., and Ragg, H. (2007). The Spn4 gene from *Drosophila melanogaster* is a multipurpose defence tool directed against proteases from three different peptidase families. *Biochem. J.* **401,** 325–331.

Charron, Y., Madani, R., Combepine, C., Gajdosik, V., Hwu, Y., Margaritondo, G., and Vassalli, J. D. (2008). The serpin Spn5 is essential for wing expansion in *Drosophila melanogaster*. *Int. J. Dev. Biol.* **52,** 933–942.

Cicardi, M., Bergamaschini, L., Cugno, M., Beretta, A., Zingale, L. C., Colombo, M., and Agostoni, A. (1998). Pathogenetic and clinical aspects of C1 inhibitor deficiency. *Immunobiology* **199,** 366–376.

Coleman, S., Drahn, B., Petersen, G., Stolorov, J., and Kraus, K. (1995). A *Drosophila* male accessory gland protein that is a member of the serpin superfamily of proteinase inhibitors is transferred to females during mating. *Insect Biochem. Mol. Biol.* **25,** 203–207.

Cox, C. R., and Gilmore, M. S. (2007). Native microbial colonization of *Drosophila melanogaster* and its use as a model of *Enterococcus faecalis* pathogenesis. *Infect. Immun.* **75,** 1565–1576.

Dafforn, T. R., Della, M., and Miller, A. D. (2001). The molecular interactions of heat shock protein 47 (Hsp47) and their implications for collagen biosynthesis. *J. Biol. Chem.* **276,** 49310–49319.

Danielli, A., Kafatos, F. C., and Loukeris, T. G. (2003). Cloning and characterization of four Anopheles gambiae serpin isoforms, differentially induced in the midgut by Plasmodium berghei invasion. *J. Biol. Chem.* **278,** 4184–4193.

De Gregorio, E., Han, S. J., Lee, W. J., Baek, M. J., Osaki, T., Kawabata, S., Lee, B. L., Iwanaga, S., Lemaitre, B., and Brey, P. T. (2002). An immune-responsive Serpin regulates the melanization cascade in *Drosophila*. *Dev. Cell* **3,** 581–592.

El Chamy, L., Leclerc, V., Caldelari, I., and Reichhart, J. M. (2008). Sensing of "danger signals" and pathogen-associated molecular patterns defines binary signaling pathways "upstream" of Toll. *Nat. Immunol.* **9,** 1165–1170.

Ferrandon, D., Imler, J. L., Hetru, C., and Hoffmann, J. A. (2007). The *Drosophila* systemic immune response: Sensing and signalling during bacterial and fungal infections. *Nat. Rev. Immunol.* **7,** 862–874.

Franzen, C., Fischer, S., Schroeder, J., Scholmerich, J., and Schneuwly, S. (2005). Morphological and molecular investigations of *Tubulinosema ratisbonensis* gen. nov., sp. nov. (Microsporidia: Tubulinosematidae fam. nov.), a parasite infecting a laboratory colony of *Drosophila melanogaster* (Diptera: Drosophilidae). *J. Eukaryot. Microbiol.* **52,** 141–152.

Garrett, M., Fullaondo, A., Troxler, L., Micklem, G., and Gubb, D. (2009). Identification and analysis of serpin-family genes by homology and synteny across the 12 sequenced Drosophilid genomes. *BMC Genomics* **10,** 489.

Gettins, P. G. (2002). Serpin structure, mechanism, and function. *Chem. Rev.* **102,** 4751–4804.

Gobert, V., Gottar, M., Matskevich, A. A., Rutschmann, S., Royet, J., Belvin, M., Hoffmann, J. A., and Ferrandon, D. (2003). Dual activation of the *Drosophila* toll pathway by two pattern recognition receptors. *Science* **302,** 2126–2130.

Gottar, M., Gobert, V., Matskevich, A. A., Reichhart, J. M., Wang, C., Butt, T. M., Belvin, M., Hoffmann, J. A., and Ferrandon, D. (2006). Dual detection of fungal infections in *Drosophila* via recognition of glucans and sensing of virulence factors. *Cell* **127,** 1425–1437.

Green, C., Levashina, E., McKimmie, C., Dafforn, T., Reichhart, J. M., and Gubb, D. (2000). The necrotic gene in *Drosophila* corresponds to one of a cluster of three serpin transcripts mapping at 43A1.2. *Genetics* **156,** 1117–1127.

Gubb, D., Robertson, A., Dafforn, T., Troxler, L., and Reichhart, J. M. (2007). *Drosophila* serpins: Regulatory cascades in innate immunity and morphogenesis. *In* "Molecular and Cellular Aspects of the Serpinopathies and Disorders in Serpin Activity," (D. A. Lomas and G. A. Silverman, eds.), pp. 207–227. World Scientific Publishing Co. Pte. Ltd, Singapore.

Gubb, D., Sanz-Parra, A., Barcena, L., Troxler, L., and Fullaondo, A. (2010). Protease inhibitors and proteolytic signalling cascades in insects. *Biochimie* **92,** 1749–1759.

Habayeb, M. S., Ekengren, S. K., and Hultmark, D. (2006). Nora virus, a persistent virus in *Drosophila*, defines a new picorna-like virus family. *J. Gen. Virol.* **87,** 3045–3051.

Habayeb, M. S., Cantera, R., Casanova, G., Ekstrom, J. O., Albright, S., and Hultmark, D. (2009a). The *Drosophila* Nora virus is an enteric virus, transmitted via feces. *J. Invertebr. Pathol.* **101,** 29–33.

Habayeb, M. S., Ekstrom, J. O., and Hultmark, D. (2009b). Nora virus persistent infections are not affected by the RNAi machinery. *PLoS ONE* **4,** e5731.

Hashimoto, C., Kim, D. R., Weiss, L. A., Miller, J. W., and Morisato, D. (2003). Spatial regulation of developmental signaling by a serpin. *Dev. Cell* **5,** 945–950.

Hedges, L. M., Brownlie, J. C., O'Neill, S. L., and Johnson, K. N. (2008). Wolbachia and virus protection in insects. *Science* **322,** 702.

Huntington, J. A., and Stein, P. E. (2001). Structure and properties of ovalbumin. *J. Chromatogr. B Biomed. Sci. Appl.* **756,** 189–198.

Huntington, J. A., Read, R. J., and Carrell, R. W. (2000). Structure of a serpin–protease complex shows inhibition by deformation. *Nature* **407,** 923–926.

Irving, J. A., Pike, R. N., Lesk, A. M., and Whisstock, J. C. (2000). Phylogeny of the serpin superfamily: Implications of patterns of amino acid conservation for structure and function. *Genome Res.* **10,** 1845–1864.

Irving, J. A., Cabrita, L. D., Kaiserman, D., Worrall, M. M., and Whisstock, J. C. (2007). Evolution and classification of the serpin Superfamily. *In* "Molecular and Cellular Aspects of the Serpinopathies and Disorders in Serpin Activity," (D. A. Lomas and G. A. Silverman, eds.), pp. 1–33. World Scientific Publishing Co. Pte. Ltd, Singapore.

Jiang, H., Wang, Y., and Kanost, M. R. (1994). Mutually exclusive exon use and reactive center diversity in insect serpins. *J. Biol. Chem.* **269,** 55–58.

Jiang, H., Wang, Y., Huang, Y., Mulnix, A. B., Kadel, J., Cole, K., and Kanost, M. R. (1996). Organization of serpin gene-1 from *Manduca sexta*. Evolution of a family of alternate exons encoding the reactive site loop. *J. Biol. Chem.* **271,** 28017–28023.

Jiang, R., Kim, E. H., Gong, J. H., Kwon, H. M., Kim, C. H., Ryu, K. H., Park, J. W., Kurokawa, K., Zhang, J., Gubb, D., and Lee, B. L. (2009). Three pairs of protease–serpin complexes cooperatively regulate the insect innate immune responses. *J. Biol. Chem.* **284,** 35652–35658.

Jung, A. C., Criqui, M. C., Rutschmann, S., Hoffmann, J. A., and Ferrandon, D. (2001). Microfluorometer assay to measure the expression of beta-galactosidase and green fluorescent protein reporter genes in single *Drosophila* flies. *Biotechniques* **30**(594–598), 591–600.

Kaiserman, D., and Bird, P. I. (2005). Analysis of vertebrate genomes suggests a new model for clade B serpin evolution. *BMC Genomics* **6,** 167.

Kasza, A., Petersen, H. H., Heegaard, C. W., Oka, K., Christensen, A., Dubin, A., Chan, L., and Andreasen, P. A. (1997). Specificity of serine proteinase/serpin complex binding to very-low-density lipoprotein receptor and alpha2-macroglobulin receptor/low-density-lipoprotein-receptor-related protein. *Eur. J. Biochem.* **248,** 270–281.

Kounnas, M. Z., Church, F. C., Argraves, W. S., and Strickland, D. K. (1996). Cellular internalization and degradation of antithrombin III-thrombin, heparin cofactor II-thrombin, and alpha 1-antitrypsin-trypsin complexes is mediated by the low density lipoprotein receptor-related protein. *J. Biol. Chem.* **271,** 6523–6529.

Kruger, O., Ladewig, J., Koster, K., and Ragg, H. (2002). Widespread occurrence of serpin genes with multiple reactive centre-containing exon cassettes in insects and nematodes. *Gene* **293,** 97–105.

Lazzaro, B. P., Sceurman, B. K., and Clark, A. G. (2004). Genetic basis of natural variation in *D. melanogaster* antibacterial immunity. *Science* **303,** 1873–1876.

Leclerc, V., Pelte, N., El Chamy, L., Martinelli, C., Ligoxygakis, P., Hoffmann, J. A., and Reichhart, J. M. (2006). Prophenoloxidase activation is not required for survival to microbial infections in *Drosophila*. *EMBO Rep.* **7,** 231–235.

Lee, J. E., and Edery, I. (2008). Circadian regulation in the ability of *Drosophila* to combat pathogenic infections. *Curr. Biol.* **18,** 195–199.

Lemaitre, B., Kromer-Metzger, E., Michaut, L., Nicolas, E., Meister, M., Georgel, P., Reichhart, J. M., and Hoffmann, J. A. (1995). A recessive mutation, immune deficiency (imd), defines two distinct control pathways in the *Drosophila* host defense. *Proc. Natl. Acad. Sci. USA* **92,** 9465–9469.

Lemaitre, B., Nicolas, E., Michaut, L., Reichhart, J. M., and Hoffmann, J. A. (1996). The dorsoventral regulatory gene cassette spatzle/Toll/cactus controls the potent antifungal response in *Drosophila* adults. *Cell* **86,** 973–983.

Lemaitre, B., Reichhart, J. M., and Hoffmann, J. A. (1997). *Drosophila* host defense: Differential induction of antimicrobial peptide genes after infection by various classes of microorganisms. *Proc. Natl. Acad. Sci. USA* **94,** 14614–14619.

Leulier, F., Parquet, C., Pili-Floury, S., Ryu, J. H., Caroff, M., Lee, W. J., Mengin-Lecreulx, D., and Lemaitre, B. (2003). The *Drosophila* immune system detects bacteria through specific peptidoglycan recognition. *Nat. Immunol.* **4,** 478–484.

Levashina, E. A., Langley, E., Green, C., Gubb, D., Ashburner, M., Hoffmann, J. A., and Reichhart, J. M. (1999). Constitutive activation of toll-mediated antifungal defense in serpin-deficient *Drosophila*. *Science* **285,** 1917–1919.

Ligoxygakis, P., Pelte, N., Ji, C., Leclerc, V., Duvic, B., Belvin, M., Jiang, H., Hoffmann, J. A., and Reichhart, J. M. (2002). A serpin mutant links Toll activation to melanization in the host defence of *Drosophila*. *EMBO J.* **21,** 6330–6337.

Ligoxygakis, P., Roth, S., and Reichhart, J. M. (2003). A serpin regulates dorsal–ventral axis formation in the *Drosophila* embryo. *Curr. Biol.* **13,** 2097–2102.

Lomas, D. A., and Carrell, R. W. (2002). Serpinopathies and the conformational dementias. *Nat. Rev. Genet.* **3,** 759–768.

Lomas, D. A., and Mahadeva, R. (2002). Alpha1-antitrypsin polymerization and the serpinopathies: Pathobiology and prospects for therapy. *J. Clin. Invest.* **110,** 1585–1590.

Lomas, D. A., Evans, D. L., Finch, J. T., and Carrell, R. W. (1992). The mechanism of Z alpha 1-antitrypsin accumulation in the liver. *Nature* **357,** 605–607.

Lomas, D. A., Belorgey, D., Mallya, M., Miranda, E., Kinghorn, K. J., Sharp, L. K., Phillips, R. L., Page, R., Robertson, A. S., and Crowther, D. C. (2005). Molecular mousetraps and the serpinopathies. *Biochem. Soc. Trans.* **33,** 321–330.

Lung, O., Tram, U., Finnerty, C. M., Eipper-Mains, M. A., Kalb, J. M., and Wolfner, M. F. (2002). The *Drosophila melanogaster* seminal fluid protein Acp62F is a protease inhibitor that is toxic upon ectopic expression. *Genetics* **160,** 211–224.

Michel, T., Reichhart, J. M., Hoffmann, J. A., and Royet, J. (2001). *Drosophila* Toll is activated by Gram-positive bacteria through a circulating peptidoglycan recognition protein. *Nature* **414,** 756–759.

Misra, S., Hecht, P., Maeda, R., and Anderson, K. V. (1998). Positive and negative regulation of Easter, a member of the serine protease family that controls dorsa–lventral patterning in the *Drosophila* embryo. *Development* **125,** 1261–1267.

Mueller, J. L., Ravi Ram, K., McGraw, L. A., Bloch Qazi, M. C., Siggia, E. D., Clark, A. G., Aquadro, C. F., and Wolfner, M. F. (2005). Cross-species comparison of *Drosophila* male accessory gland protein genes. *Genetics* **171,** 131–143.

Mueller, J. L., Page, J. L., and Wolfner, M. F. (2007). An ectopic expression screen reveals the protective and toxic effects of *Drosophila* seminal fluid proteins. *Genetics* **175,** 777–783.

Nehme, N. T., Liegeois, S., Kele, B., Giammarinaro, P., Pradel, E., Hoffmann, J. A., Ewbank, J. J., and Ferrandon, D. (2007). A model of bacterial intestinal infections in *Drosophila melanogaster*. *PLoS Pathog.* **3**, e173.

Pelte, N., Robertson, A. S., Zou, Z., Belorgey, D., Dafforn, T. R., Jiang, H., Lomas, D., Reichhart, J. M., and Gubb, D. (2006). Immune challenge induces N-terminal cleavage of the *Drosophila* serpin Necrotic. *Insect Biochem. Mol. Biol.* **36**, 37–46.

Providence, K. M., and Higgins, P. J. (2004). PAI-1 expression is required for epithelial cell migration in two distinct phases of in vitro wound repair. *J. Cell. Physiol.* **200**, 297–308.

Ram, K. R., and Wolfner, M. F. (2007). Sustained post-mating response in *Drosophila melanogaster* requires multiple seminal fluid proteins. *PLoS Genet.* **3**, e238.

Reichhart, J. M. (2005). Tip of another iceberg: *Drosophila* serpins. *Trends Cell Biol.* **15**, 659–665.

Richer, M. J., Keays, C. A., Waterhouse, J., Minhas, J., Hashimoto, C., and Jean, F. (2004). The Spn4 gene of *Drosophila* encodes a potent furin-directed secretory pathway serpin. *Proc. Natl. Acad. Sci. USA* **101**, 10560–10565.

Robertson, A. S., Belorgey, D., Lilley, K. S., Lomas, D. A., Gubb, D., and Dafforn, T. R. (2003). Characterization of the necrotic protein that regulates the Toll-mediated immune response in *Drosophila*. *J. Biol. Chem.* **278**, 6175–6180.

Sasaki, T. (1991). Patchwork-structure serpins from silkworm (*Bombyx mori*) larval hemolymph. *Eur. J. Biochem.* **202**, 255–261.

Scherfer, C., Tang, H., Kambris, Z., Lhocine, N., Hashimoto, C., and Lemaitre, B. (2008). *Drosophila* Serpin-28D regulates hemolymph phenoloxidase activity and adult pigmentation. *Dev. Biol.* **323**, 189–196.

Schmidt, R. L., Trejo, T. R., Plummer, T. B., Platt, J. L., and Tang, A. H. (2008). Infection-induced proteolysis of PGRP-LC controls the IMD activation and melanization cascades in *Drosophila*. *FASEB J.* **22**, 918–929.

Silverman, G. A., Bird, P. I., Carrell, R. W., Church, F. C., Coughlin, P. B., Gettins, P. G., Irving, J. A., Lomas, D. A., Luke, C. J., Moyer, R. W., Pemberton, P. A., Remold-O'Donnell, E., Salvesen, G. S., Travis, J., and Whisstock, J. C. (2001). The serpins are an expanding superfamily of structurally similar but functionally diverse proteins. Evolution, mechanism of inhibition, novel functions, and a revised nomenclature. *J. Biol. Chem.* **276**, 33293–33296.

Soukup, S. F., Culi, J., and Gubb, D. (2009). Uptake of the necrotic serpin in *Drosophila melanogaster* via the lipophorin receptor-1. *PLoS Genet.* **5**, e1000532.

Stenbak, C. R., Ryu, J. H., Leulier, F., Pili-Floury, S., Parquet, C., Herve, M., Chaput, C., Boneca, I. G., Lee, W. J., Lemaitre, B., and Mengin-Lecreulx, D. (2004). Peptidoglycan molecular requirements allowing detection by the *Drosophila* immune deficiency pathway. *J. Immunol.* **173**, 7339–7348.

Swanson, W. J., Clark, A. G., Waldrip-Dail, H. M., Wolfner, M. F., and Aquadro, C. F. (2001). Evolutionary EST analysis identifies rapidly evolving male reproductive proteins in *Drosophila*. *Proc. Natl. Acad. Sci. USA* **98**, 7375–7379.

Tang, H., Kambris, Z., Lemaitre, B., and Hashimoto, C. (2008). A serpin that regulates immune melanization in the respiratory system of *Drosophila*. *Dev. Cell* **15**, 617–626.

Teixeira, L., Ferreira, A., and Ashburner, M. (2008). The bacterial symbiont Wolbachia induces resistance to RNA viral infections in *Drosophila melanogaster*. *PLoS Biol.* **6**, e2.

Tzou, P., Ohresser, S., Ferrandon, D., Capovilla, M., Reichhart, J. M., Lemaitre, B., Hoffmann, J. A., and Imler, J. L. (2000). Tissue-specific inducible expression of antimicrobial peptide genes in *Drosophila* surface epithelia. *Immunity* **13**, 737–748.

Vodovar, N., Vinals, M., Liehl, P., Basset, A., Degrouard, J., Spellman, P., Boccard, F., and Lemaitre, B. (2005). *Drosophila* host defense after oral infection by an entomopathogenic Pseudomonas species. *Proc. Natl. Acad. Sci. USA* **102**, 11414–11419.

Zhang, Q., Buckle, A. M., Law, R. H., Pearce, M. C., Cabrita, L. D., Lloyd, G. J., Irving, J. A., Smith, A. I., Ruzyla, K., Rossjohn, J., Bottomley, S. P., and Whisstock, J. C. (2007). The N terminus of the serpin, tengpin, functions to trap the metastable native state. *EMBO Rep.* **8,** 658–663.

Zhou, W., Rousset, F., and O'Neil, S. (1998). Phylogeny and PCR-based classification of Wolbachia strains using wsp gene sequences. *Proc. Biol. Sci.* **265,** 509–515.

Zou, Z., Lopez, D. L., Kanost, M. R., Evans, J. D., and Jiang, H. (2006). Comparative analysis of serine protease-related genes in the honey bee genome: Possible involvement in embryonic development and innate immunity. *Insect Mol. Biol.* **15,** 603–614.

CHAPTER TWELVE

MODELING SERPIN CONFORMATIONAL DISEASES IN *DROSOPHILA MELANOGASTER*

Thomas R. Jahn,[*,†] Elke Malzer,[*,‡,§] John Roote,[*] Anastasia Vishnivetskaya,[*] Sara Imarisio,[*] Maria Giannakou,[*] Karin Panser,[*] Stefan Marciniak,[‡,§] and Damian C. Crowther[*,§]

Contents

1. Introduction	228
2. Why *Drosophila* Models of Serpinopathies?	229
3. Screening for Polymerogenic Mutations in Physiologically Important Fly Serpins	229
4. First Steps in the Generation of a Human Serpinopathy Model in *Drosophila*	230
5. Longevity Assays in Flies Expressing Human Serpins	233
6. Behavioral Assays in Flies Expressing Human Serpins	234
7. Microscopic Phenotyping of Flies Expressing Human Serpins	235
8. Using Genetic Screens to Identify Genes That Have a Role in Generating, or Suppressing Serpin-Induced Phenotypes	236
9. Conclusions	237
10. Method 1: Generating Transgenic *Drosophila*	237
10.1. Plasmid preparation	239
10.2. Microinjection method	239
10.3. Crossing scheme to identify transformants	241
11. Method 2: The *Drosophila* UAS/GAL4 Expression System	242
12. Method 3: Chemical Mutagenesis and X-ray Mutagenesis	242
12.1. EMS mutagenesis	243
12.2. X-ray mutagenesis	245
13. Method 4: Screening for Mutations Caused by Chemical and X-ray Mutagenesis	245
14. Method 6: Examination of the Eye Imaginal Disc	246
15. Method 7: Pseudopupil Assay	248
16. Method 8: Protein Extraction from Flies	250
17. Method 9: Genetic Backcrossing	251

[*] Department of Genetics, University of Cambridge, Cambridge, United Kingdom
[†] Department of Chemistry, University of Cambridge, Cambridge, United Kingdom
[‡] Department of Medicine, University of Cambridge, Cambridge, United Kingdom
[§] Cambridge Institute for Medical Research, University of Cambridge, Cambridge, United Kingdom

18. Method 10: Longevity Assays	252
19. Method 11: Locomotor Assays	253
20. Method 12: Immunostaining of Fly Brains	253
21. Method 13: P-element and RNAi Screening	254
22. Method 15: Deletion Kit Screening	255
References	255

Abstract

Transgenic *Drosophila melanogaster* have been used to model both the physiological and pathological behavior of serpins. The ability to generate flies expressing serpins and to rapidly assess associated phenotypes contributes to the power of this paradigm. While providing a whole-organism model of serpinopathies the powerful toolkit of genetic interventions allows precise molecular dissection of important biological pathways. In this chapter, we summarize the contribution that flies have made to the serpin field and then describe some of the experimental methods that are employed in these studies. In particular, we will describe the generation of transgenic flies, the assessment of phenotypes, and the principles of how to perform a genetic screen.

1. INTRODUCTION

The metastable state of the native conformation of most serpins is essential for the physiological function of these proteins but also makes them prone to pathological aggregation. The same conformational changes that accompany the inhibition of proteases also permit their self-association, aggregation, and tissue deposition. The resulting damage to the synthetic tissues may result in complications such as liver cirrhosis in α_1-antitrypsin (A1AT) deficiency and dementia in familial encephalopathy with neuroserpin inclusion bodies (FENIB). The consequent lack of circulating protease inhibitory activity may have systemic consequences such as emphysema in A1AT deficiency or more local effects such as the epilepsy that characterizes FENIB.

The *in vitro* understanding of the molecular mechanisms that underpin the physiological and pathological protein–protein interactions of the serpins have been studied in great detail and are described elsewhere in this volume. A major challenge of the coming decade is to advance our understanding of the biological consequences of serpin aggregation. There is enormous interest in the protective mechanisms that the cell employs to prevent protein misfolding in early life and also the changes that occur when the secretory machinery decompensates, resulting in protein aggregation and subsequently in cell death. The pathways of cell stress and death have not yet been elucidated and it is the progress in these areas that will help in the search for therapies to prevent a range of devastating diseases.

2. Why *Drosophila* Models of Serpinopathies?

The choice of model organism for the study of human disease is governed by two competing requirements. First, there should be good biological orthology between the model organism and the human, that is, the genetic make-up and the cellular and tissue functioning should be as similar as possible. It is on this criterion that the mouse modeling community has become predominant in many fields. However, when considering the second requirement, that experiments should be powerful without being slow and expensive, the fly has undoubted advantages. In this light working with invertebrate systems becomes much more attractive, and the fruit fly *Drosophila* particularly so, because of its close genetic orthology to humans (up to 70%; Rubin *et al.*, 2000).

Fly work can contribute to our understanding of the serpinopathies, principally because murine experiments are long and expensive and mouse serpins show layers of functional redundancy, for example, the mouse has five A1AT genes. In contrast, the fruit fly has a number of useful features. First, flies have endogenous protease-serpin networks that can be studied in their own right, allowing generic disease mechanisms to be developed. Second, flies can be made transgenic (method 1) for human genes and then express (method 2) serpins of interest using the GAL4–UAS system (Brand and Perrimon, 1993). Both the wild-type serpins and their corresponding disease-related isoforms may be expressed in the fly and the consequent phenotypic differences assessed. Expression of human serpins in the fly eye and/or brain has resulted in remarkable phenotypes including developmental deficits, degenerative behavioral changes, and biochemical abnormalities (Kinghorn *et al.*, 2006; Miranda *et al.*, 2008).

Third, the power of the genetic tool kits that have become available to the fly community (Cook *et al.*, 2010) has allowed the investigation of the genes that can modify model serpinopathies. Specifically, genetic screens have been performed, looking for genes that, when they are differentially expressed, can increase or decrease the accumulation of polymerogenic serpin variants in the whole fly or else modify the phenotypic consequences of serpin accumulation.

3. Screening for Polymerogenic Mutations in Physiologically Important Fly Serpins

The fly provides ample opportunities to study abnormal serpin aggregation thanks to the presence of at least 29 recognizable members of the superfamily in the *Drosophila* genome (Garrett *et al.*, 2009). The best characterized serpin functions in the fly include the regulation of the proteolytic cascades that underpin innate immunity (reviewed in Reichhart, 2005). The essential role played by serpins in fly homeostasis provides an opportunity for us to dissect how polymerogenic mutations, and consequent loss of protease inhibition, can cause disease. A good example of this approach is the study by Green et al. (2003) in which chemical and X-ray mutagenesis (method 3) were used to generate pathological mutations in the endogenous immune-regulating serpin, necrotic (nec; Green et al., 2000).

This screen (generic protocol in method 4) was particularly powerful because inactivation of the serpin, by whatever means, including polymerogenesis, resulted in unrestrained immune activation, and a foreshortened lifespan, accompanied by characteristic melanised necrotic degeneration of the cuticle. Among the mutations that resulted in the necrotic phenotype were a number that caused polymerisation of the serpin nec (Green et al., 2003). Indeed two mutations were described that exactly replicated pathogenetic mutations in the orthologous human serpin A1AT. This finding of conserved polymerogenic mutations that resulted in homologous biochemical abnormalities in serpin function was particularly encouraging and underpinned the value of *Drosophila* as a model to study human genetic disorders. In the same study, the investigators also created transgenic flies expressing *nec* variants that corresponded to yet other human disease-linked mutations and found that they too polymerized and failed to rescue the characteristic necrotic phenotype when expressed in the *nec*-null background. In contrast, transgenic expression of wild-type *nec* completely rescued the *nec*-null flies.

4. First Steps in the Generation of a Human Serpinopathy Model in *Drosophila*

The overexpression of novel aggregation-prone *Drosophila* serpin variants by Green and colleagues presaged the subsequent development of flies expressing disease-related, polymerogenic mutant forms of human serpins. To date, the fly has been used to characterize the *in vivo* behavior of one human serpin, namely neuroserpin. In this study, Kinghorn et al. (2006) assessed a range of phenotypes including both gross and microscopic abnormalities, behavioral deficits, and biochemical changes (Kinghorn et al., 2006). These phenotypes were caused either by the unregulated protease inhibition that results from transgenic serpin overexpression or, more interestingly, by the accumulation of serpin polymers. This study exemplifies how flies can be used as tools to understand the biological consequences of serpin action.

An initial insight into the general biological effects of a serpin can be achieved by expressing the protein in all the fly tissues. In the case of wild-type neuroserpin, ubiquitous expression (see Table 12.1) resulted in complete developmental lethality. This was determined by constructing a cross in which offspring inherit the GAL4 driver, the *UAS-neuroserpin* or both or neither. A count of all the offspring allows any missing, that is lethal, genotypes to be identified. Using the same approach, it was found that the inactivated serpin with Pro-Pro substitutions at P1-P1′ was no longer toxic (Kinghorn *et al.*, 2006). Detailed studies of the developmental stage at which lethality is observed are possible by microscopic analysis of the embryos, larvae, or pupae as appropriate.

When ubiquitous expression of a serpin is lethal then restricting the protein to a nonessential organ will often allow flies to develop and potentially exhibit interesting phenotypes. In this context, one of the most convenient tissues to study, particularly for models of neurological disease, is the retina of the compound eye. Using GMR-GAL4 driver lines, it is possible to get high-level expression of a serpin that is limited almost entirely to the developing eye imaginal disc and thereafter to the adult eye.

The expression of genes driven by GMR-Gal4 shows an interesting pattern of expression during the development of the eye (Ellis *et al.*, 1993). The adult eye is derived from the imaginal disc and the last step in the differentiation of the retinal cells occurs as a wave that sweeps from the posterior to the anterior. This wave is known as the morphogenetic furrow and as it advances across the disc it leaves mature photoreceptor cells; these cells are thereafter permissive for GMR-Gal4 expression and hence transgene transcription (Freeman, 1996). Dissection of the discs (Legent and Treisman, 2008) with appropriate immunolabeling (method 6) should allow the investigator to exploit this precise temporal and spatial organization to generate insights into the kinetics of protein expression and aggregation.

Serpin expression in the imaginal disc may be toxic and if the development of the eye is disturbed then the outcome will often be a

Table 12.1 Some *GAL4* driver lines and their pattern of expression

Name of driver	Tissue	Reference
Act5c	Ubiquitous	Ishikawa *et al.* (1999)
Daughterless	Ubiquitous	Wodarz *et al.* (1995)
GMR	Retinal neuronal tissue	Freeman (1996)
Sevenless	Subset of photoreceptors in the retina during eye development	Basler *et al.* (1989)
Elav	Neurones	Yao and White (1994)

"rough eye" phenotype in the adult fly (Tan et al., 2008; Wittmann et al., 2001). Such corneal defects in the eye are readily apparent at eclosion (when flies emerge from the pupal case) and represent a disruption of the regular hexagonal array of the 800 individual photosensitive units (ommatidia). More severe phenotypes include loss and fusion of ommatidia, distortion, and loss of the corneal bristles and even necrotic spots (Fig. 12.1; Kinghorn et al., 2006). In some model systems, the most severely affected eyes appear shrunken and depigmented (Leulier et al., 2006). The loss of pigment usually reflects degeneration of the underlying retinal tissue. Retinal damage may be apparent at eclosion or it may progress during adult life. In this context, the longitudinal assessment of retinal integrity offers an attractive model system for progressive degenerative disorders such as the dementia FENIB. To this end, the pseudopupil assay allows the investigator to measure ommatidial integrity throughout adult life (method 7) and is performed on the eyes of flies immediately after decapitation (Stark and Thomas, 2004). Each ommatidium contains photoreceptive cells (rhabdomeres), seven of which can be visualized in the healthy retina. The tabulation of the number of intact rhabdomeres per ommatidium with time allows the investigator to chart the temporal dependence of progressive retinal degeneration.

Ubiquitous or retinal expression is also usually sufficient for the initial biochemical characterization of serpins. Protein extraction is easily performed with extraction buffers of specified properties. Typically, PBS with SDS is sufficient to extract soluble proteins, whereas harsher conditions such as high guanidinium or urea concentrations may be required if a measurement of total (soluble and polymerized) serpin is required (method 8). The characterization of the serpins may be performed by immunoblotting of native and denaturing PAGE gels and transverse urea gradient gels (explained elsewhere in this volume), while size exclusion chromatography of the brain protein extracts using ELISA can provide a quantitative measure of serpin partitioning between monomer and polymer fractions (Fig. 12.2).

Figure 12.1 Rough eye phenotypes caused by neuroserpin expression (A). Expression of neuroserpin in the eye using GMR-GAL4 results in a severe, necrotic rough eye at 29 °C; at lower temperatures (25–18 °C) less severe phenotypes are observed (B–D; Kinghorn et al., 2006). Control eyes show a regular array of ommatidia (E). Scale bar = 100 μm.

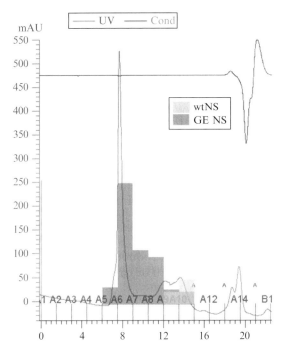

Figure 12.2 Size exclusion chromatography demonstrates partitioning of mutant neuroserpin into high molecular mass species in fly brain protein extracts. S200 size exclusion chromatography using ÄKTA equipment (GE Healthcare, UK) permits the fractionation of brain proteins from flies expressing wild-type human neuroserpin (wtNS, green bars) and the G392E variant (GE-NS, purple bars). ELISA quantitation of neuroserpin content shows that the polymerogenic serpins elute earlier, indicating a higher molecular mass. (See Color Insert.)

5. LONGEVITY ASSAYS IN FLIES EXPRESSING HUMAN SERPINS

A fly's longevity depends in part on the environment. At 29 °C, the median survival of wild-type flies is approximately 40 days, whereas at 18 °C this figure is in excess of 100 days. For this reason, when comparing the longevity of flies expressing different transgenes, it is important to ensure that all lines experience identical environmental conditions, in particular, they should be cultured on the same batch of food and be kept in the same incubator. The longevity of the flies is also sensitive to the genetic background (Partridge and Gems, 2007) and so crossing schemes should ensure that the only significant genetic differences between the lines studied are the different transgenes being compared. In some circumstances, this is easily achieved; however, in other circumstances, extensive genetic backcrossing (method 9) may be necessary to ensure that comparative data is reliable.

Longevity data is attractive because the live/dead assessment is trivial, the statistical analysis is robust and median survival is widely accepted as a measure of general health. However, the disadvantages of longevity assays are that they are long and relatively expensive because of the regular changes in fly food. Also many observations must be performed on the flies and the consequent data needs to be efficiently tracked and then processed according to the particular statistical tests that are required.

Online software is now available (http://flytracker.gen.cam.ac.uk) that allows the investigator to track survival data for barcode-labeled tubes of flies (method 10). When all the flies die the data can be downloaded and processed using software such as GraphPad Prism (GraphPad Software, Inc., USA) or SPSS (IBM Inc., USA). A popular statistical approach is to use Kaplan–Meier plots to show the fractional survival with time. Log-rank statistical analysis then provides a robust way of determining whether two survival profiles are significantly different. However, the log-rank tests whether the survival profiles are different rather than whether one group lives longer than another. To avoid the potential paradox of having two populations of flies with significantly different longevities but with the same median survival, a more conservative statistical approach is often used. In this case, each tube of 10 flies is considered as an individual subpopulation, so allowing the investigator to make 5–10 estimates of the median survival of the whole population. Mann–Whitney analysis is then used ($n = 5$–10) to determine whether the median survivals of the two fly populations are significantly different. In the study by Kinghorn et al. (2006), the longevity of flies expressing the Pro-Pro P1-P1′ neuroserpin was not reduced when assessed in these ways. Surprisingly, the polymerogenic mutants of neuroserpin that could be clearly seen accumulating as polymers in the brain of the flies did not reduce the longevity of the flies. This is in contrast to other models of protein aggregation diseases such as the modeling of Aβ peptide toxicity in Alzheimer's disease (reviewed in Crowther et al., 2006) where protein aggregation and deposition in the brain strongly affect longevity phenotypes (Luheshi et al., 2007; Rival et al., 2009). This points to profound mechanistic differences in the way that serpins cause cell death as compared to other aggregating polypeptides and proteins.

6. Behavioral Assays in Flies Expressing Human Serpins

There are a plethora of possible behavior assays that can be used in *Drosophila* model systems. These include Pavlovian conditioning (Iijima et al., 2004, 2008), behavioral habituation (Asztalos et al., 1993), diurnal actimetry (Pokrzywa et al., 2007), flying assays, larval crawling assays, and locomotor (climbing) assays (Martin, 2003). Of these, the climbing assay is most commonly used in the assessment of fly models of human disease (see supplementary video in Moloney et al., 2009). Normal flies show positive phototaxis (movement toward light) and negative geotaxis (movement upward), however, given a constant light source the flies will normally climb rapidly to the top of a tube. For flies that may be blind, the assays can be performed in a dark room, however, in practice this precaution is rarely necessary. Typically, the experiments are performed in relatively narrow tubes (<2 cm diameter) as this inhibits flying and allows the measurement of walking/climbing behavior. Such locomotor behavior may be affected by the brain-specific expression of aggregation-prone neuroserpin variants using *elav-GAL4*. In these experiments, the severity of the age-related movement abnormalities were shown to be directly correlated with the level of serpin accumulation. Classically, locomotor behavior has been quantified manually with the behavior represented by a performance index (PI; Rival et al., 2004), essentially a measure of the proportion of flies that behave normally (method 11). However, the need for greater throughput and the desire for more quantitative measures of behavioral disturbance have inspired several methods for the automatic accumulation and analysis of locomotor data (method 11).

7. Microscopic Phenotyping of Flies Expressing Human Serpins

Drosophila embryos, larvae, and adults can be studied by both light and electron microscopy. Light microscopy, often using fluorophore-tagged transgenic proteins can be undertaken in live larvae because their body walls are transparent and so fixing and mounting of tissues is not necessary. In the adult, wax embedding of fixed tissues or alternatively cryosectioning of unfixed material is also commonly undertaken. Wax sections of adult brains have been studied in neuroserpin-expressing flies (Wan et al., 2011) and when stained with a polymer-specific antibody (described elsewhere in this issue) can demonstrate progressive protein accumulation (Fig. 12.3B, method 12). The intact adult brain can also be fixed and stained with specific antisera and imaged using confocal microscopy.

Electron microscopy is particularly useful for the characterization of rough eye phenotypes. Figure 12.1 shows how scanning electron micrographs provide superb images of rough eye phenotypes.

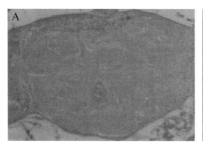 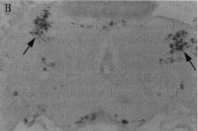

Figure 12.3 Wax sections of control (A) and neuroserpin-expressing (B) flies show specific protein accumulation in neuronal subpopulations (arrows). Image courtesy of Ian Macleod and David Lomas, University of Cambridge.

8. Using Genetic Screens to Identify Genes That Have a Role in Generating, or Suppressing Serpin-Induced Phenotypes

Genetic modifier screens can be used to find mutations that can alter the severity of a particular phenotype of interest (reviewed in Greenspan, 2004; St Johnston, 2002). For example, a screen to look for mutations that increase (enhance) or decrease (suppress) the severity of the necrotic phenotype could be performed using one of several approaches. In each case, a library of mutagenized flies is generated and then the variant chromosomes are introduced into a fly that expresses the phenotype of interest. Whether the mutagenized chromosomes enhance or suppress the phenotype of interest is then assessed in the offspring. There are a number of ways to induce mutations, X irradiation and chemical mutagenesis have already been introduced (method 3); however, mobile-genetic elements, genomic deletions, and RNAi constructs can also provide genetic diversity that can be used in a screening protocol.

Of the many mobile-genetic elements that have been described, this review will only discuss the use of P-elements (method 13). All engineered mobile elements remain stably inserted until an enzyme is expressed that allows their mobilization; in the case of P-elements, this is called P-transposase. In the presence of transgene-encoded transposase, the P-element is removed from its original site and pseudorandomly inserted elsewhere in the fly genome. When a P-element lands near or within a gene the result is either (i) the knockdown of the gene itself (all the mobile elements can do this) or (ii) upregulation of neighboring genes (usually only EP-elements (Mata *et al.*, 2000; Rorth, 1996) and GS elements (Aigaki *et al.*, 2001; Toba *et al.*, 1999) can do this). Although significant labor is required to generate P-element insertions, the great advantage over equivalent chemical or

X-ray approaches is that the genomic site of the mutation can be quickly determined by DNA sequencing. However, the availability of large collections of well-characterized fly lines carrying single transposons (http://flystocks.bio.indiana.edu/Browse/in/GDPtop.htm, https://drosophila.med.harvard.edu/, and http://kyotofly.kit.jp/cgi-bin/stocks/index.cgi) means that most investigators no longer need to generate their own libraries.

Genomic deletions are a source of genetic diversity that can be used to screen for genes that interact with a phenotype of interest. Deficiency kits are available that provide maximal coverage of the genome with the minimal number of stocks. The Bloomington Deficiency Kit (http://flystocks.bio.indiana.edu/Browse/df/dfextract.php?num=all&symbol=bloomdef) covers almost 98% of the genome with less than 500 molecularly defined deletion stocks. As modifiers are identified using the "low-resolution" kit then progressively smaller deletion lines can be screened to refine the search until only a few candidate modifier genes remain (method 14). At this stage, the investigator may use fly stocks with mutations for the particular genes or use RNAi expression to knockdown individual genes (method 13). An important strength of RNAi screens is that they can generate dominant phenotypes caused by gene product loss of function and so provide targets for pharmacological inhibitors.

9. CONCLUSIONS

Animal models of serpinopathies will be essential to understanding the complex pathogenesis of diseases as diverse as the cirrhosis of A1AT deficiency and the dementia FENIB. The fruit fly is an attractive model system that will continue to show its relevance to serpin biology and pathology.

10. METHOD 1: GENERATING TRANSGENIC DROSOPHILA

There are broadly two methods for generating transgenic *Drosophila*. The classical approach most commonly uses plasmids such as pUAST with transient P-element transposase expression to achieve pseudorandom insertion of a construct into the fly genome, yielding a GAL4-responsive transgene. The transposase is encoded by a helper plasmid, typically Δ2-3, that is coinjected with pUAST. Microinjection is performed on early fly embryos from a white-eyed (*white*$^-$) stock while they are still at the syncytial stage, into the location where the gonads will develop. This favors incorporation of the construct into the gonads, and hence the sperm or eggs, of the developing fly. The presence of the *mini-white*$^+$ gene in the pUAST plasmid means that transgenic offspring of the injected flies can be detected by looking for the occasional colored (red-to-orange) eyed flies, the otherwise white-eyed offspring. The intensity of the eye color is usually a good indicator of how strongly the transgene will be expressed when it is subsequently driven with a GAL4 "driver" line. The presence of pale orange eyes indicates that the construct has inserted in a relatively inactive area of the genome, while increasingly dark-eyed offspring are likely to have either single insertions of pUAST in actively transcribed genomic loci or else have multiple insertions of the construct. One enduring attraction of the "classical" pUAST transgenesis approach is precisely the variation in expression levels that one can achieve. When one considers A1AT deficiency and the pathology generated in the liver it is likely that we will need high levels of expression to model the pathology effectively. In this regard, a series of pUAST lines with high expression levels and/or multiple insertions may prove invaluable in generating a sufficiently strong phenotype. When fine-tuning a phenotype in flies expressing serpins there are a number of variables that can provide a dynamic range of expression intensity of approximately 100-fold (Table 12.2).

Targeted transgenesis in *Drosophila* has using the phiC31 system allows the single insertion of a construct into a known position in the fly genome

Table 12.2 Serpin expression can be modulated approximately 100-fold in transgenic flies

Factor	Mechanism	Fold contribution to expression levels (approximate)
Site of pUAST genomic insertion	The transcriptional activity of the surrounding genome affects transgene expression.	1–10×
pUAST copy number	Not uncommon to have two inserts on a chromosome. Can incorporate 1–4 transgenes into a fly without difficulty.	1–4×
Environmental temperature	The pUAST promoter has heat-shock elements that make GAL4 transcription stronger at higher temperatures (from 19 to 29 °C).	1–4×
Choice of GAL4 driver line	GAL4-driver constructs drive transgenic expression with varying intensity	1–4×

(Bischof et al., 2007). This approach has the principal advantage of producing any number of fly lines that have similar, if not identical, levels of expression of their respective transgenes. Consequently, the comparison between, for example, flies expressing wild-type and mutant serpins is simplified because genomic insertional effects are no longer relevant. Targeted transgenesis has become a commercial enterprise undertaken by a number of companies, notably by BestGene Inc., USA. The transgene of interest is cloned into pUAST-attB, a *mini-white$^+$* marked plasmid derived from pUAST that also carries an attB site. In the presence of phiC31 integrase, this attB sequence mediates the efficient and site-specific insertion of the plasmid into an attP acceptor site in the fly genome. *White$^-$* fly stocks carrying a single attP acceptor site and a transgene for the phiC31 integrase are the source for the microinjected embryos. Upon emerging, the injected flies are crossed to a *white$^-$* stock so that transgenic offspring can be identified by their pigmented eyes. It is also good practice to select, usually by the loss of a body color marker such as yellow, for flies that have lost the integrase transgene (Bischof et al., 2007).

10.1. Plasmid preparation

The pUAST or pUAST-attB plasmids that carry the transgene of interest and any helper plasmid (pπ25.7 Δ2-3 wc) are purified using Qiagen or Promega kits. Twenty micrograms of pUAST is mixed with 2 µg of helper plasmid, that transiently expresses transposase, greatly increasing the efficiency of transgenesis. The DNA solution is centrifuged at 18,000×g on a bench top microcentrifuge for 10 min to remove any particulate matter that might block the injection needle. The supernatant is retained and the DNA ethanol precipitated to remove unwanted salt, washed with 70% ethanol, and redissolved in Spradling buffer (5 mM KCl, 0.1 mM phosphate buffer, pH 7.8 Rubin and Spradling, 1982). The DNA solution is centrifuged again at 18,000×g for 10 min just prior to loading into the microinjection needle. The concentration of the construct DNA should exceed 1 µg/µl.

10.2. Microinjection method

Typical microinjection apparatus consists of an Eppendorf transjector 5246 and a micromanipulator 5171 combined with Femtotips II needles. The *Drosophila* embryos are harvested from population cages of white-eyed flies (w^{1118}) comprising 2000 males and females that are 2–7 days old. At least 1 day prior to collection, the flies are transferred from their usual cornmeal media into large collection cages on plates containing grape juice agar with some fresh yeast at 25 °C. During the 2 h prior to embryo collection, the plates are changed half-hourly to remove older embryos and to synchronize egg laying. During the collection period, the flies are allowed to lay for 30 min on freshly yeasted grape juice agar plates. This ensures that all embryos can be injected during the syncytial blastoderm stage, facilitating spread of the injected DNA into the future germ cells. Working at 18 °C (to slow down embryo development and increase survival), the embryos are removed from the agar plates using tap water and a paintbrush and collected into a fine meshed sieve. The embryos, in the sieve, are placed in 50% (w/v) bleach (sodium hypochlorite) for 2 min to remove the tough chorion membrane. The sieve is then rinsed well in tap water to remove the bleach and blotted on tissue to remove excess water to enable easy handling of the embryos.

Working quickly for no longer than 20 min, about 180 embryos are lined up on a cover slip (18 mm × 18 mm), immobilized on a thin strip of nontoxic double-sided adhesive tape (e.g., Scotch #665) with a gap of one embryo width between each. The embryos are placed on their sides with the posterior pole of each embryo (the end without the micropyle) orientated toward the mounted needle (Fig. 12.4). The coverslips are mounted on slides and the embryos are left to desiccate for about 5–15 min by placing the slides in a sealed chamber containing silica gel. Drying the embryos allows maximum uptake of DNA while reducing leakage and bursting of embryos following injection. The embryo is optimally desiccated when it loses its shiny appearance and becomes slightly wrinkled. After dehydration, the embryos are covered in a dense halocarbon oil (e.g., Voltalef 10S) to improve injection efficiency and aid healing of the wound.

The loaded needle is mounted on the micromanipulator and visualized under the microinjection microscope. The tip of the needle is broken by pushing it against a glass cover slip to give a fine, slightly angled end. Mild positive pressure is applied using a syringe and tubing attached to the needle such that when the needle is in the oil, a small bubble of DNA solution leaks out. The size of this aqueous bubble within the oil should be a quarter of the width of the embryo; the size of the bubble is adjusted by varying the pressure applied to the syringe. A slide of embryos is placed on the microscope stage and the embryos injected one-by-one by moving the microscope stage toward the tip of the needle until it inserts through the posterior pole of the embryo; the stage is further advanced until the tip of the needle rests 20% of the way into the embryo. Successful injection results in a small clear patch within the posterior portion of the embryo. Once all embryos have been injected, those that are too old (cellularized or older embryos) or unfertilized or those that are leaking excessively are discarded. The coverslip carrying the remaining embryos is then transferred to a humid grape juice agar plate with some fresh yeast and left at 18 °C until the embryos develop into larvae. Care should be taken to ensure the plates are left level to prevent the halocarbon oil from draining off the

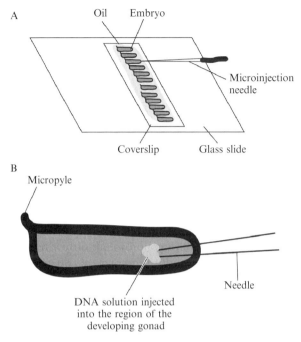

Figure 12.4 The generation of transgenic flies by DNA microinjection. Young (<55 min postlaying) dechorionated *Drosophila* embryos are immobilized using sticky tape on a coverslip under halocarbon oil (A). The embryos are orientated so the micropyle (the anterior end) is facing away from the microinjection needle (B). The coverslip is manipulated so that the mounted microinjection needle enters the embryo to a depth of one-third of its length from the posterior. The injected DNA solution appears as a pale patch.

embryos. The larvae are counted and transferred to a tube of normal fly food with yeast and cultured at 25 °C until adults emerge.

10.3. Crossing scheme to identify transformants

The flies that emerge (F_0 generation) will all have white eyes but in some of flies the DNA construct of interest will have been incorporated into the gonads and will be carried by either sperm or eggs. To harvest somatically transgenic flies, single F_0 males are crossed, in separate vials, to four virgin w^{1118} (white-eyed) females. Similarly, single F_0 virgin females are crossed to four w^{1118} males. In the subsequent F_1 offspring, flies with any shade of red, orange, or yellow eyes carry the transgene construct and are retained to allow mapping of the insert to a chromosome and to establish a new transgenic fly line.

11. METHOD 2: THE *DROSOPHILA* UAS/GAL4 EXPRESSION SYSTEM

The standard expression system used in *Drosophila* is the bipartite *GAL4–UAS* system first described by Brand and Perrimon (1993). This elegant and robust approach relies on the availability of so-called GAL4 driver lines. These flies are often freely available from stock centers (http://flystocks.bio.indiana.edu/ and http://kyotofly.kit.jp/cgi-bin/stocks/index.cgi) and typically carry a construct containing the core promoter sequences from a gene that exhibits tissue-specific expression. These promoter elements are coupled to the coding sequence for GAL4. The driver line exhibits tissue-, and/or temporally, restricted expression of the yeast transcriptional activator GAL4. In most cases, GAL4 has no biological activity in insect cells, although in some cases, such as high-level expression in the retina, using GMR-GAL4, may cause mild artifactual phenotypes, possibly due to aggregation of the GAL4 at high concentrations. The transgene of interest is placed downstream of a GAL4 upstream activating sequence that supports transcription of the construct only in the presence of GAL4. In this way, the pattern of transgenic protein expression can be tailored, according to the driver line used, to the requirements of the investigator.

A typical crossing scheme is described in Fig. 12.5. For most, if not all, genetic crosses the female flies should be virgins. This is because females store sperm, thus the paternity of any offspring becomes uncertain when a male is subsequently introduced to the female. The most widely used protocols rely on the fact that males remain infertile, and hence any coexistent females remain virgins, for 8 h at 25 °C and 16 h at 18 °C (see Table 12.3). If virginity is essential, the females are kept in groups of 10 flies per tube for 2 days at 25 °C to ensure that there are no larvae on the food before crossing to the males. It is often most convenient to source the virgins from the GAL4-driver stock that can be expanded to supply females for the several experimental and control crosses, rather than collecting virgins from many stocks.

12. METHOD 3: CHEMICAL MUTAGENESIS AND X-RAY MUTAGENESIS

Chemical mutagens such as ethylmethanesulfonate (EMS) and X-rays can both be used to induce pseudorandom mutations into the genome of the fly. EMS typically causes single base-pair changes while X-rays often result in deletions and gross chromosomal rearrangements (Greenspan, 2004).

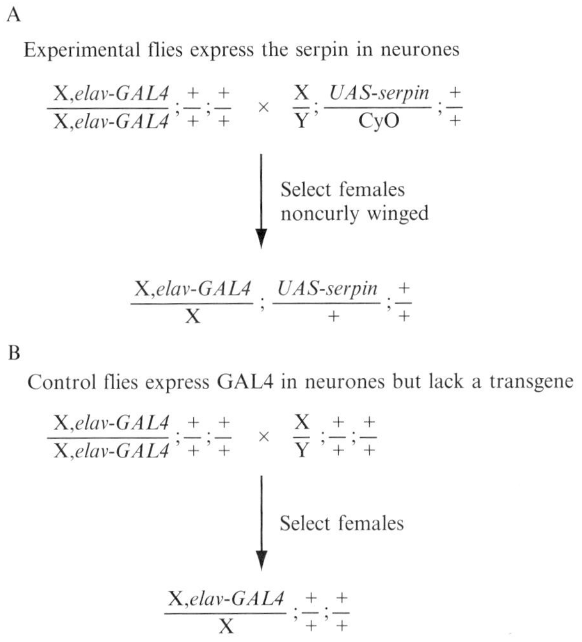

A

Experimental flies express the serpin in neurones

$$\frac{X, elav\text{-}GAL4}{X, elav\text{-}GAL4} ; \frac{+}{+} ; \frac{+}{+} \quad \times \quad \frac{X}{Y} ; \frac{UAS\text{-}serpin}{CyO} ; \frac{+}{+}$$

↓ Select females noncurly winged

$$\frac{X, elav\text{-}GAL4}{X} ; \frac{UAS\text{-}serpin}{+} ; \frac{+}{+}$$

B

Control flies express GAL4 in neurones but lack a transgene

$$\frac{X, elav\text{-}GAL4}{X, elav\text{-}GAL4} ; \frac{+}{+} ; \frac{+}{+} \quad \times \quad \frac{X}{Y} ; \frac{+}{+} ; \frac{+}{+}$$

↓ Select females

$$\frac{X, elav\text{-}GAL4}{X} ; \frac{+}{+} ; \frac{+}{+}$$

Symbols	Meaning
X	X chromosome
Y	Y chromosome
+	wild-type chromosome
elav-GAL4	Construct driving GAL4 in neuronal tissues
UAS-serpin	GAL4-dependent serpin transgene
X/Y;2;3	Sex and each autosome separated by semicolons

Figure 12.5 A *Drosophila* crossing scheme that allows the investigator to compare the consequences of ubiquitous neuronal expression of a serpin (A) to be compared to the appropriate control flies (B). In these examples only female offspring are compared.

12.1. EMS mutagenesis

One to two hundred male flies are aged on normal food for 2 days after they emerge and are then starved for 6–8 h in bottles containing only Kleenex tissue moistened with tap water. Pure EMS is insoluble in water, however, 250 μl can be dispersed into 100 ml of a 1% (w/v) sucrose solution by repeated pipetting with a P5000. Two sheets of Kleenex tissue paper are placed at the bottom of otherwise empty fly culture bottles and 10 ml of the EMS/sucrose solution is dispensed onto the paper. Fifty of the male flies are added to each bottle and left at room temperature in a fume hood for 24 h. After treatment, still EMS-contaminated, the flies are transferred using a funnel to a bottle of fresh fly food. This protocol will usually generate one

Table 12.3 Five things you need to know about *Drosophila* genetics to be able to create a serpin model

Biological observation	Practical consequences for the fly modeler
Female flies store sperm after mating.	Crosses should usually be set up using males and virgin females of known genotypes so that the parentage of the offspring can be determined.
Male flies remain infertile for 8 h at 25 °C and 16 h at 18 °C.	If a *Drosophila* culture bottle is emptied at 9 a.m. and left at 25 °C until 5 p.m. then all the females that emerge during that period will be virgins. Likewise, if a culture bottle is emptied at 5 p.m. and left at 18 °C until 9 a.m., then all the females will still be virgins.
There is no genetic recombination during spermatogenesis.	If two DNA constructs are linked on the same chromosome in the male, then they will remain linked in the offspring. Likewise, if two DNA constructs are present on different copies of a chromosome in a male then recombination will never bring them onto the same chromosome in the offspring.
A marker chromosome helps with fly genotyping because the presence or absence of the marker phenotype allows the investigator to infer the presence or absence of other genetic elements or transgenes.	Marker chromosome must be used with caution in females because they do not prevent recombination during oogenesis. Marker chromosomes are usually homozygous lethal.
A balancer chromosome prevents recombination and usually carries a dominant marker.	Balancer chromosomes are essential when transgenic and other chromosomes of interest must pass through the male and female germline. Balancer chromosomes are usually homozygous lethal or female sterile for X chromosome balancers. A balancer chromosome may also carry a selection marker such as *hs-hid* that is lethal if developing flies are heat shocked (Grether *et al.*, 1995).

lethal mutation in a gene of interest once in every 1000 flies tested. This is thought to be an optimal intensity as higher mutation frequencies are accompanied by unacceptable reductions in male fertility.

EMS is highly toxic and mutagenic for humans and all manipulations should be performed in a fume hood with the appropriate protection. Excess EMS solution and contaminated equipment can be made safe by overnight incubation in a 2-l solution of 10% (w/v) sodium thiosulphate in a fume hood.

12.2. X-ray mutagenesis

X irradiation of *Drosophila* is commonly performed using a Torrex 150 kV cabinet because the high intensity of the beam (>1.5 Gy min^{-1} through a 1.5-mm Al filter) means that experiments can be performed relatively quickly with less dehydration for the flies. Males are typically mutagenized with a total dose of 40 Gy, the target being the mature sperm.

Offspring of the EMS- or X-ray-mutagenized males are tested for the presence of relevant mutations using the screening protocol described in method 4.

13. METHOD 4: SCREENING FOR MUTATIONS CAUSED BY CHEMICAL AND X-RAY MUTAGENESIS

It is often convenient to induce mutations in males carrying a marked chromosome such as *cinnabar (cn)*, an eye color marker on chromosome II. This allows mutations to be recovered and tested by following *cn* in subsequent generations. The mutagenized males are crossed to virgin females carrying a balancer chromosome corresponding to the marked male chromosome being tested. If, for example, the second chromosome is being screened then virgin females with the *CyO* (Curly of Oster) balancer are crossed in equal numbers with the mutagenized males. Male offspring with both *cn* and *CyO* are crossed in single male crosses with three to four virgin females carrying a deletion or null allele for the gene of interest. This allows the investigator to find which mutagenized chromosomes cannot complement the null allele. It can be seen therefore that most screens can only be performed if homozygous disruption of a gene results in lethality or some other strong phenotype. For the necrotic screen, Green *et al.* (2003) relied on the fact that homozygous *nec*^{null} is rapidly lethal to flies. Similarly in the crossing scheme in Fig. 12.6 (see also Table 12.3 for *hs-hid*) where homozygous loss of function of the serpin in question is lethal then a mutation (★) that inactivates the serpin will result in only curly flies in the final generation when present in the serpin^{null} background. In practice,

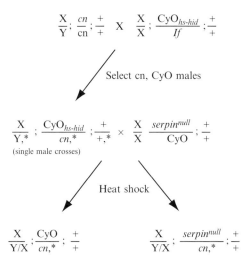

Figure 12.6 A *Drosophila* crossing scheme to screen for recessive serpin mutations that segregate with the mutagenized chromosome carrying the cinnabar (cn) marker. The first cross allows a mutagenized cn chromosome from a single male to be balanced over a heat-shock-lethal chromosome ($CyO_{hs\text{-}hid}$). If, following heat shock, the resulting offspring are all CyO, then it is likely that the mutation in the cn chromosome has caused a recessive lethal mutation at the serpin locus.

tubes containing *only* curly winged flies are very likely to have mutations that inactivate the serpin and the genomic DNA should be sequenced to confirm.

14. Method 6: Examination of the Eye Imaginal Disc

The larval tissue destined to become the adult antenna and eye forms a unit called the antennal-eye imaginal disc (AED). These structures are dissected from the rest of the third instar larva in PBS under a dissecting microscope (Legent and Treisman, 2008). First, the mouth hook is

Modeling Serpin Conformational Diseases in *Drosophila melanogaster* 247

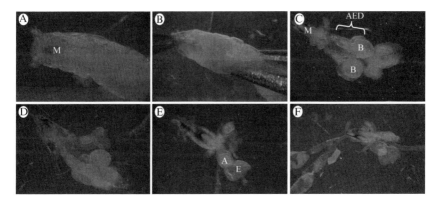

Figure 12.7 Dissection of the eye imaginal disc from a third instar larva. Step by step dissection of the third instar larva. The mouth hook (M) is used to separate the antennal-eye imaginal disc complex (AED) along with the brain (B) from the rest of the larva. Further dissection yields antennal and eye imaginal discs that are ready for immunostaining (A and E). Scale bar = 1 mm for all images.

identified as a black structure at the anterior end of the larva (Fig. 12.7A). The mouth hook is grasped with forceps and pulled while the larva is held halfway down with another pair of forceps (Fig. 12.7B). This will release a complex of tissues including the AED and brain (Fig. 12.7C and D, labeled "B"). After removing all surrounding tissues, including the brain the AED remains attached to the mouth hook (Fig. 12.7E and F). Again grasping the mouth hook with forceps the tissues are fixed for 20 min in a snap top tube containing 1 ml of 4% (w/v) formaldehyde in PBS.

To immunolabel the AED for a protein of interest, the tissue is first washed three times with 1 ml PBT (PBS, 0.3%, v/v, triton X-100) to remove the formaldehyde and permeabilize the cells. With each wash, the solution is gently mixed and the waste buffer removed with a pipette, taking care to avoid aspiration of the tissue. The primary antibody is incubated with the tissue in PBT either for 1 h at room temperature, or overnight at 4 °C, with gentle tipping of the tube. A murine anti-elav antibody (e.g., the monoclonal *elav-9F8A9, Developmental Studies Hybridoma Bank, University of Iowa, USA*) provides a useful positive control as it stains mature photoreceptor cells posterior to the morphogenetic furrow, allowing the zone of GMR expression to be visualized. A negative control should always be included, for example, one set of tissue should be incubated with PBT alone, without the primary antibody, but treated identically thereafter If significant background signal is observed in the negative control then a blocking agent such as bovine serum albumin (1%, w/v) may be added to the antibody solutions, however, in practice this is usually unnecessary. Following the primary antibody incubation, the tissue is washed three times in PBT, as before, followed by two prolonged washing steps for 10 min on a gentle

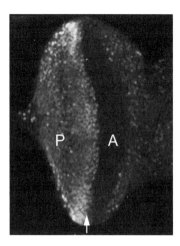

Figure 12.8 Immunolabeling of the eye imaginal disc. The eye imaginal disc costained for a positive control protein (elav, green) and a *GMR-GAL4* protein of interest (red). Red staining is seen at the morphogenetic furrow (arrow) and posterior (P) to this band of cells. Anterior (A) to the furrow expression has not yet occurred. (See Color Insert.)

rotary shaker, at room temperature in the dark. The tissue is now incubated in a similar way with the appropriate fluorescently labeled secondary antiserum and washed again in PBT as before but in the dark.

The tissue is then carefully transferred onto a glass slide, using either dissection forceps or a pipette and mounted under a coverslip with Vectashield solution (Vector laboratories Ltd, Peterborough, UK). The slide may now be stored in the dark in the fridge until required for fluorescence microscopic examination using either a standard or confocal microscope (Fig. 12.8).

15. Method 7: Pseudopupil Assay

The pseudopupil technique provides a quantitative assessment of retinal integrity and permits the rate of degeneration to be measured over a time-course. This approach is particularly useful for comparing the rate of degeneration in the presence or absence of putative pharmacological or genetic modifiers.

New born flies of the desired genotype should be collected 4 h apart, on at least four occasions, to exclude any systematic bias that may come from the normal, senile degeneration of the retina. At a specific time after collection, for example, 3 days, the flies are decapitated using a blade under the dissecting microscope and the heads are put on a drop of nail-polish on a slide. The slide is mounted on the stage of an optical microscope and the aperture of the diaphragm is reduced until only the interior of the fly head is brightly illuminated. Under these conditions the light that passes through intact rhabdomeres, and out from the cornea of the compound eye, can be visualized against a dark field using a 60–100× oil immersion objective lens. In a healthy retina, each ommatidium should have seven visible rhabdomeres (Fig. 12.9A). Only ommatidia that are on a plane perpendicular to the objective lens are viewed and for each ommatidium assessed the plane of focus is moved up and down until the maximal number of rhabdomeres is visible. The number of visible rhabdomeres decreases as retinal degeneration progresses (Fig. 12.9B). This maximal number is recorded for 15 ommatidia per eye for 5–10 flies, from each of at least three separate genetic crosses on four separate occasions. All measurements should be performed blind to the genotype of the flies and all microscopic observations should be performed within 20 min, so as to avoid artifactual postmortem degeneration.

For each experimental condition, the data should be analyzed to give (i) the average number of rhabdomeres per ommatidium and (ii) the frequency of rhabdomere count per ommatidium. It may be helpful to calculate these values for each session as well as the overall experiment to ensure that the results are robust and consistent throughout.

Most investigators will perform a pilot assay before embarking on a full-scale pseudopupil experiment. The goals are to determine, first, whether males or females should be used (they should never be mixed in assays) and second what conditions of temperature and age provide for the optimal severity of retinal degeneration. In most cases, the baseline conditions should provide for mild-to-moderate degeneration, thus allowing the pharmacological or genetics modifier the best chance to show an effect.

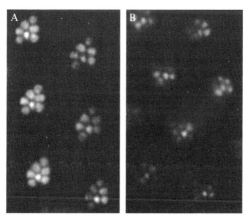

Figure 12.9 Seven rhabdomeres may be visualized in a healthy retina (A), however, with retinal degeneration the number declines (B).

16. Method 8: Protein Extraction from Flies

Proteins can be extracted from whole flies where ubiquitous expression is expected. However, in many cases, for example, when modeling FENIB, the expression may be limited to the head—either to the brain (*elav-GAL4* driven) or the retina (*GMR-GAL4* driven).

Where extracts are to be taken from the head alone then efficient decapitation can be achieved by freezing groups of 10–25 flies in liquid nitrogen in 1.5-ml snap top tubes. The tube is vigorously vortexed for 15 s before repeating, once more, the freezing/vortexing cycle. Alternatively, several tubes of flies can be frozen in liquid nitrogen and then placed in a nonsealed box that is half filled with dry ice. The box is then shaken vigorously for 2 min. After this treatment, the contents of each snap top tube, still frozen, are emptied onto a Petri dish or tissue paper that is chilled over dry ice. When handling more than 100 flies, a brass sieve, chilled in liquid nitrogen immediately before use, can be used to separate heads from bodies. The heads should just pass through the sieve and the bodies will then be retained. In either case, the frozen fly heads are collected with a small paint brush and transferred into a new snap top tube and chilled on dry ice until ready to proceed with the protein extraction protocol.

In some cases, the protein will be present in the hemolymph, the circulating fluid that is the *Drosophila* equivalent of blood. In this case, the fluid is released by puncturing each of 10–20 flies either using a blade to decapitate the flies or using a needle to pierce the abdominal wall. The flies are loaded into a 0.5-ml snap top tube that has itself been punctured at its tip using a needle. The small tube is loaded into a clean 1.5-ml snap top tube and the combination placed in a microcentrifuge and centrifuged at $18,000 \times g$ for 5 min. The smaller tube containing the fly bodies is discarded and the hemolymph, appearing as a clear fluid at the tip of the larger snap top tube, is collected for analysis.

When working with either whole flies or heads then the tissues must be mechanically disrupted with a subsequent, optional, sonication step. When few samples are being processed then 10–20 flies can be conveniently homogenized in a 1.5-ml snap top tube with a well-fitting single-use plastic pestle and an extraction buffer comprising 100 mM Tris, pH 7.4 with 2% (w/v) SDS. The tubes are sealed and can then be transferred to a chilled, bath sonicator and sonicated for 5 min. Following centrifugation at $18,000 \times g$ for 5 min the supernatant is removed and retained for analysis.

When larger numbers of samples require processing then a 96-well format is more convenient. In this case, up to 15 flies or 50 fly heads are placed in each well of a deep (1.2-ml well capacity, ABgene) microtiter plate. A 3-mm brass grinding ball is added to each well and the plate is sealed using PCR sealing film (ABgene). A rotatory shaker is used to vigorously agitate the plates at 20 cycles/s for 1 min. The plate is then pulsed to $3500 \times g$ in a centrifuge before 150 µl of extraction buffer (100 mM Tris, pH 7.4 with 0.5%, w/v, SDS) is added to each well. The wells are resealed and the plate is shaken at 20 cycles/s for a further 2 min. The pulse centrifugation and shaking are repeated once. Finally, the plate is centrifuged at $3500 \times g$ for 15 min. The supernatant in each well is retained for analysis of soluble material. Optionally, 150 µl or more extraction buffer supplemented with a chaotropic agent, such as 5 M urea or 6 M guanidinium HCl, can be added and the cycles of shaking and centrifugation repeated. In this way, differential extraction of soluble and insoluble, potentially polymerised, serpin is made possible.

Proteolytic degradation of the serpins is minimized by working at 4 °C and by supplementing the buffers with EDTA (10–100 mM) and/or broad-spectrum protease inhibitor cocktails (Roche Complete).

17. METHOD 9: GENETIC BACKCROSSING

When studying behavioral phenotypes and longevity, differences in the genetic background of different fly strains can generate spurious results. When a particular crossing scheme cannot control for the differences between strains then it becomes essential to bring the genes of interest into the same genetic background. This is a lengthy procedure requiring at least six generation of meiotic recombination down the female germline.

The first requirement is that the genetic feature of interest should be identifiable throughout the backcrossing scheme. For transgenic flies, this is usually easy because they carry the $white^+$ gene that permits selection by eye color. For mutant alleles or deficiencies, it may be necessary to use PCR to detect the desired genetic feature.

The first generation of back crossing typically pairs a male with many virgins from a $white^-$ reference stock such as w^{1118}. When the transgene of interest is marked with $white^+$ then virgin female offspring are chosen with pigmented eyes. These are crossed to $white^-$ reference stock males. The selection of red-eyed virgin females and mating with $white^-$ males is repeated a further five times.

Where the genetic feature of interest is not marked by $white^+$ then single-fly genomic PCR may be used to identify the correct females. This approach requires that a number of virgin females be placed in individual vials with several males from the reference stock. After the female has been fertilized and has laid eggs, she is genotyped by PCR. Where a female is shown to carry the correct DNA construct then vials containing her offspring are retained for further crossing.

Once backcrossing has been completed then male offspring are crossed to virgins carrying appropriate balancer chromosomes such that further recombination is suppressed.

18. METHOD 10: LONGEVITY ASSAYS

According to preference of the investigator, longevity assays are usually performed on virgin or mated females or males but never a mixture. To generate mated females, all flies from a cross are collected each day and placed together in a vial at 25 °C for 24 h. The females are then sorted into groups of 10 and placed in individual vials. A total of 50–100 flies are sufficient for most longevity assays. If mortality trajectories are to be calculated then up to 1000 flies may be required to achieve noise-free data. The number of viable flies in each tube is counted, and the flies transferred to fresh vial, on days 1, 3, and 5 of a 7-day cycle. More frequent counting may be appropriate if the flies are expected to have a particularly short lifespan (< 10 days) or around the median survival time when the rate of death is maximal. The investigator also records when flies escape or are otherwise erroneously lost to follow-up.

When all flies have died, the data are tabulated as the age of each fly when it was observed as dead, or loss to follow-up. The survival of each group of flies is visualized as a Kaplan–Meier plot (Fig. 12.10A) and the statistical significance of any differences is calculated in two main ways. First, the log-rank test provides the most powerful approach; however, this approach assumes that each of the 5–10 subpopulations of 10 flies behaves identically to the pooled population of 50–100 flies. A more conservative approach is to assume that each tube of 10 flies provides an estimate of the population's median survival. This provides an $n = 5\text{–}10$ series of median survival ages that can be analyzed by the Mann–Whitney test (Fig. 12.10B).

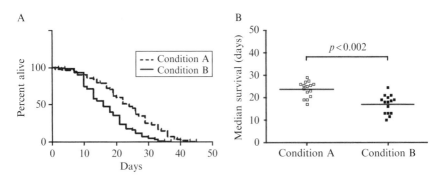

Figure 12.10 Survival plot. Survival data may be displayed as (A) Kaplan–Meier plots or (B) as scatter plots showing multiple estimates of the median survival.

19. METHOD 11: LOCOMOTOR ASSAYS

The classical manual locomotor assay tests the flies' climbing behavior in a tall cylinder such as a 25-ml pipette (Bainton et al., 2000; Feany and Bender, 2000; Ganetzky and Flanagan, 1978). In preparation for the assay, the tops are removed from clean pipette tubes to facilitate the loading of 15 flies; cotton wool is then used as a stopper. Experiments are performed either in a dark room or else in the open lab provided that the lighting is stable. At the beginning of the assay, the flies are tapped to the bottom of the tube and then they are given a set time (20–60 s) to climb. At the end of the time, the number above a particular level (say the 25-ml mark) and those below another level (say the 0-ml mark) are counted and the results tabulated. The experiment is repeated three times for each set of 15 flies and on three independent sets of flies.

A PI is calculated from the mean ± S.E.M. of the number of flies at the top (n_{top}) and at the bottom (n_{bottom}), expressed as the percentages of the total number of flies (n_{tot}). The PI is therefore defined as $\frac{1}{2}((n_{tot} + n_{top} - n_{bottom})/n_{tot})$. Statistical analysis is then performed on the PIs of different fly lines using the Student's t-test.

The best locomotor data are generated when the measurements are taken at the same time of day, in stable light conditions, using clean tubes. The flies should be equilibrated to the ambient temperature and more than one-hour postanesthesia.

Recently, more advanced technologies have been put forward to automatically assess the locomotor behavior in flies. The *Drosophila* Activity Monitor from Trikinetics Inc., USA uses infrared beam breaking to measure circadian locomotor activity in single flies that are confined to a capillary tube. Other methods use computer vision to track the movement of flies either in two or three dimensions (Grover et al., 2009). Here, flies are placed in a glass tube and the locomotor trajectories are constructed. This allows the calculation of quantitative parameters that describe the motion of the flies and permits subtle changes to be detected early.

20. METHOD 12: IMMUNOSTAINING OF FLY BRAINS

The expression and accumulation of proteins in the brain of *Drosophila* can be visualized using a number of well-established immunohistochemical techniques (Corrigall et al., 2007; Wu and Luo, 2006). The first step is the isolation of the adult brain by dissection of the fly head, followed by the fixation of the brain. Imaging can be generated from tissue sections or from the intact brain using confocal microscopy. Eight to 10 flies of the appropriate genotype are anesthetized and placed in a dish containing PBS-T buffer (PBS + 0.1 %, v/v, Triton X-100) under a dissecting microscope. Using two forceps and starting by removing one eye, the cuticle and underlying membranes are gently removed from the brain. Fixation is performed in 4% (w/v) paraformaldehyde with gentle agitation for 30 min at room temperature, followed by three 20 min washing steps with PBS-T buffer. Before staining, the brains are incubated in 0.5-ml blocking solution (5%, v/v, normal goat serum in PBS-T) for 30 min at room temperature. The brains are then removed from the blocking solution and placed in 0.5 ml of the primary antibody solution and incubated with gentle agitation for 24 h at 4 °C. The brains are then washed three times in PBS-T and then incubated with the secondary antibody with gentle agitation for 24 h at 4 °C. The precise incubation times and antibody dilutions must be optimized for each application. Following three washes with PBS-T, the brains are ready to be mounted on a glass slide and imaged using a confocal microscope.

21. METHOD 13: P-ELEMENT AND RNAi SCREENING

The simplest P-elements produce simple disruptive insertions that usually have no observable effect, presumably because the function of the DNA can be complemented by the nonmutagenized chromosome. However, the effect of a P-element insertion is likely to be a loss-of-function mutation. Because the consequent phenotypes are usually recessive the screening strategy therefore resembles that for chemical and X-ray mutagenesis (method 3).

Subsequent generations of engineered P-elements can have two effects; they can either disrupt the gene at the landing site or else cause activation of the gene by bringing into close proximity of the GAL4-UAS sequences that are found within such P-elements. These EP or GS elements need to be activated by a GAL4 driver line and often result in dominant phenotypes that can be assessed in the first generation of offspring flies. Similarly, the expression of RNAi using GAL4 driver lines allows tissue-specific knockdown of particular genes. The development of whole genome libraries of RNAi lines allows the systematic screening for the dominant effects of gene loss of function. The crossing schemes for the EP, GS, and RNAi lines are similar; an example is given in Fig. 12.11, where the phenotype of females with curly wings (and not carrying a GS element) is compared to females without curly wings (and so carrying the GS element). Any interaction between the element and the phenotype would provide *prima facie* evidence that a gene near the GS element is modifying the serpin-dependent phenotype.

$$\frac{X, elav\text{-}GAL4}{Y}; \frac{UAS\text{-}serpin}{UAS\text{-}serpin}; \frac{+}{+} \quad \times \quad \frac{X}{X}; \frac{GS\text{-}element}{CyO}; \frac{+}{+}$$

noncurly winged females ↙   ↘ curly winged females

$$\frac{X, elav\text{-}GAL4}{X}; \frac{UAS\text{-}serpin}{GS\text{-}element}; \frac{+}{+} \qquad \frac{X, elav\text{-}GAL4}{X}; \frac{UAS\text{-}serpin}{CyO}; \frac{+}{+}$$

Symbols	Meaning
CyO	Balancer chromosome marked with curly wings
elav-GAL4	Construct driving GAL4 in neuronal tissues
UAS-serpin	GAL4-dependent serpin transgene
GS-element	Chromosome with GS-element insertion

Figure 12.11 A *Drosophila* crossing scheme to allow the screening of P-element modifiers of serpin-related phenotypes. In this cross, female CyO offspring do not have the P-element insertion while non-CyO flies do. If a serpin-related phenotype is modified in non-CyO flies specifically, then this provides evidence for a genetic interaction between the gene at the site of insertion of the P-element and the serpin under investigation.

22. Method 15: Deletion Kit Screening

Deficiency kit screens can be used to find genes that, when present as only one copy rather than two, can modify a phenotype of interest. A widely used kit is supplied by Bloomington (http://flystocks.bio.indiana.edu/Browse/df/dfextract.php?num=all&symbol=bloomdef) as a library of 467 fly lines each with hemizygous deletions of the X chromosome and the autosomes (chromosomes II, III, and IV). When a large deletion is seen to interact with the phenotype of interest then further lines with smaller deletions may be ordered, allowing more precise identification of the responsible locus. If the smallest deletion available that contains the modifying locus still contains several genes then each of these genes needs to be investigated using RNAi and mutant alleles that may be available in the stock centers.

REFERENCES

Aigaki, T., Ohsako, T., Toba, G., Seong, K., and Matsuo, T. (2001). The gene search system: Its application to functional genomics in *Drosophila melanogaster*. *J. Neurogenet.* **15,** 169–178.

Asztalos, Z., von Wegerer, J., Wustmann, G., Dombradi, V., Gausz, J., Spatz, H. C., and Friedrich, P. (1993). Protein phosphatase 1-deficient mutant Drosophila is affected in habituation and associative learning. *J. Neurosci.* **13,** 924–930.

Bainton, R. J., Tsai, L. T., Singh, C. M., Moore, M. S., Neckameyer, W. S., and Heberlein, U. (2000). Dopamine modulates acute responses to cocaine, nicotine and ethanol in Drosophila. *Curr. Biol.* **10**, 187–194.

Basler, K., Siegrist, P., and Hafen, E. (1989). The spatial and temporal expression pattern of sevenless is exclusively controlled by gene-internal elements. *EMBO J.* **8**, 2381–2386.

Bischof, J., Maeda, R. K., Hediger, M., Karch, F., and Basler, K. (2007). An optimized transgenesis system for Drosophila using germ-line-specific phiC31 integrases. *Proc. Natl. Acad. Sci. USA* **104**, 3312–3317.

Brand, A. H., and Perrimon, N. (1993). Targeted gene expression as a means of altering cell fates and generating dominant phenotypes. *Development* **118**, 401–415.

Cook, K. R., Parks, A. L., Jacobus, L. M., Kaufman, T. C., and Matthews, K. A. (2010). New research resources at the Bloomington Drosophila Stock Center. *Fly (Austin)* **4**, 88–91.

Corrigall, D., Walther, R. F., Rodriguez, L., Fichelson, P., and Pichaud, F. (2007). Hedgehog signaling is a principal inducer of Myosin-II-driven cell ingression in Drosophila epithelia. *Dev. Cell* **13**, 730–742.

Crowther, D. C., Page, R., Chandraratna, D., and Lomas, D. A. (2006). A Drosophila model of Alzheimer's disease. *Methods Enzymol.* **412**, 234–255.

Ellis, M. C., O'Neill, E. M., and Rubin, G. M. (1993). Expression of Drosophila glass protein and evidence for negative regulation of its activity in non-neuronal cells by another DNA-binding protein. *Development* **119**, 855–865.

Feany, M. B., and Bender, W. W. (2000). A Drosophila model of Parkinson's disease. *Nature* **404**, 394–398.

Freeman, M. (1996). Reiterative use of the EGF receptor triggers differentiation of all cell types in the Drosophila eye. *Cell* **87**, 651–660.

Ganetzky, B., and Flanagan, J. R. (1978). On the relationship between senescence and age-related changes in two wild-type strains of *Drosophila melanogaster*. *Exp. Gerontol.* **13**, 189–196.

Garrett, M., Fullaondo, A., Troxler, L., Micklem, G., and Gubb, D. (2009). Identification and analysis of serpin-family genes by homology and synteny across the 12 sequenced Drosophilid genomes. *BMC Genomics* **10**, 489.

Green, C., Levashina, E., McKimmie, C., Dafforn, T., Reichhart, J.-M., and Gubb, D. (2000). The *necrotic* gene in Drosophila corresponds to one of a cluster of three Serpin transcripts mapping at 43A1.2. *Genetics* **156**, 1117–1127.

Green, C., Brown, G., Dafforn, T. R., Reichhart, J. M., Morley, T., Lomas, D. A., and Gubb, D. (2003). Drosophila necrotic mutations mirror disease-associated variants of human serpins. *Development* **130**, 1473–1478.

Greenspan, R.J. (2004). Fly Pushing: The Theory and Practice of *Drosophila* Genetics. 2nd edition. Cold Spring Harbor Laboratory Press: Cold Spring Harbor, NY.

Grether, M. E., Abrams, J. M., Agapite, J., White, K., and Steller, H. (1995). The head involution defective gene of *Drosophila melanogaster* functions in programmed cell death. *Genes Dev.* **9**, 1694–1708.

Grover, D., Yang, J., Ford, D., Tavare, S., and Tower, J. (2009). Simultaneous tracking of movement and gene expression in multiple *Drosophila melanogaster* flies using GFP and DsRED fluorescent reporter transgenes. *BMC Res. Notes* **2**, 58.

Iijima, K., Liu, H. P., Chiang, A. S., Hearn, S. A., Konsolaki, M., and Zhong, Y. (2004). Dissecting the pathological effects of human A{beta}40 and A{beta}42 in Drosophila: A potential model for Alzheimer's disease. *Proc. Natl. Acad. Sci. USA* **101**, 6623–6628.

Iijima, K., Chiang, H. C., Hearn, S. A., Hakker, I., Gatt, A., Shenton, C., Granger, L., Leung, A., Iijima-Ando, K., and Zhong, Y. (2008). Abeta42 mutants with different aggregation profiles induce distinct pathologies in Drosophila. *PLoS ONE* **3**, e1703.

Ishikawa, T., Matsumoto, A., Kato, T., Jr., Togashi, S., Ryo, H., Ikenaga, M., Todo, T., Ueda, R., and Tanimura, T. (1999). DCRY is a Drosophila photoreceptor protein implicated in light entrainment of circadian rhythm. *Genes Cells* **4**, 57–65.

Kinghorn, K. J., Crowther, D. C., Sharp, L. K., Nerelius, C., Davis, R. L., Chang, H. T., Green, C., Gubb, D. C., Johansson, J., and Lomas, D. A. (2006). Neuroserpin binds abeta and is a neuroprotective component of amyloid plaques in Alzheimer disease. *J. Biol. Chem.* **281**, 29268–29277.

Legent, K., and Treisman, J. E. (2008). Wingless signaling in Drosophila eye development. *Methods Mol. Biol.* **469**, 141–161.

Leulier, F., Ribeiro, P. S., Palmer, E., Tenev, T., Takahashi, K., Robertson, D., Zachariou, A., Pichaud, F., Ueda, R., and Meier, P. (2006). Systematic in vivo RNAi analysis of putative components of the Drosophila cell death machinery. *Cell Death Differ.* **13**, 1663–1674.

Luheshi, L. M., Tartaglia, G. G., Brorsson, A. C., Pawar, A. P., Watson, I. E., Chiti, F., Vendruscolo, M., Lomas, D. A., Dobson, C. M., and Crowther, D. C. (2007). Systematic in vivo analysis of the intrinsic determinants of amyloid beta pathogenicity. *PLoS Biol.* **5**, e290.

Martin, J. R. (2003). Locomotor activity: A complex behavioural trait to unravel. *Behav. Processes* **64**, 145–160.

Mata, J., Curado, S., Ephrussi, A., and Rorth, P. (2000). Tribbles coordinates mitosis and morphogenesis in Drosophila by regulating string/CDC25 proteolysis. *Cell* **101**, 511–522.

Miranda, E., MacLeod, I., Davies, M. J., Perez, J., Romisch, K., Crowther, D. C., and Lomas, D. A. (2008). The intracellular accumulation of polymeric neuroserpin explains the severity of the dementia FENIB. *Hum. Mol. Genet.* **17**, 1527–1539.

Moloney, A., Sattelle, D. B., Lomas, D. A., and Crowther, D. C. (2009). Alzheimer's disease: Insights from *Drosophila melanogaster* models. *Trends Biochem. Sci.* **35**(4), 228–235.

Partridge, L., and Gems, D. (2007). Benchmarks for ageing studies. *Nature* **450**, 165–167.

Pokrzywa, M., Dacklin, I., Hultmark, D., and Lundgren, E. (2007). Misfolded transthyretin causes behavioral changes in a Drosophila model for transthyretin-associated amyloidosis. *Eur. J. Neurosci.* **26**, 913–924.

Reichhart, J. M. (2005). Tip of another iceberg: Drosophila serpins. *Trends Cell Biol.* **15**, 659–665.

Rival, T., Soustelle, L., Strambi, C., Besson, M. T., Iche, M., and Birman, S. (2004). Decreasing glutamate buffering capacity triggers oxidative stress and neuropil degeneration in the Drosophila brain. *Curr. Biol.* **14**, 599–605.

Rival, T., Page, R. M., Chandraratna, D. S., Sendall, T. J., Ryder, E., Liu, B., Lewis, H., Rosahl, T., Hider, R., Camargo, L. M., Shearman, M. S., Crowther, D. C., et al. (2009). Fenton chemistry and oxidative stress mediate the toxicity of the beta-amyloid peptide in a Drosophila model of Alzheimer's disease. *Eur. J. Neurosci.* **29**, 1335–1347.

Rorth, P. (1996). A modular misexpression screen in Drosophila detecting tissue-specific phenotypes. *Proc. Natl. Acad. Sci. USA* **93**, 12418–12422.

Rubin, G. M., and Spradling, A. C. (1982). Genetic transformation of Drosophila with transposable element vectors. *Science* **218**, 348–353.

Rubin, G. M., Yandell, M. D., Wortman, J. R., Gabor Miklos, G. L., Nelson, C. R., Hariharan, I. K., Fortini, M. E., Li, P. W., Apweiler, R., Fleischmann, W., Cherry, J. M., Henikoff, S., et al. (2000). Comparative genomics of the eukaryotes. *Science* **287**, 2204–2215.

St Johnston, D. (2002). The art and design of genetic screens: *Drosophila melanogaster*. *Nat. Rev. Genet.* **3**, 176–188.

Stark, W. S., and Thomas, C. F. (2004). Microscopy of multiple visual receptor types in Drosophila. *Mol. Vis.* **10**, 943–955.

Tan, L., Schedl, P., Song, H. J., Garza, D., and Konsolaki, M. (2008). The Toll–> NFkappaB signaling pathway mediates the neuropathological effects of the human Alzheimer's Abeta42 polypeptide in Drosophila. *PLoS ONE* **3**, e3966.

Toba, G., Ohsako, T., Miyata, N., Ohtsuka, T., Seong, K. H., and Aigaki, T. (1999). The gene search system. A method for efficient detection and rapid molecular identification of genes in *Drosophila melanogaster*. *Genetics* **151,** 725–737.

Wan, L., Nie, G., Zhang, J., Luo, Y., Zhang, P., Zhang, Z., and Zhao, B. (2011). beta-Amyloid peptide increases levels of iron content and oxidative stress in human cell and *Caenorhabditis elegans* models of Alzheimer disease. *Free Radic. Biol. Med.* **50,** 122–129.

Wittmann, C. W., Wszolek, M. F., Shulman, J. M., Salvaterra, P. M., Lewis, J., Hutton, M., and Feany, M. B. (2001). Tauopathy in Drosophila: Neurodegeneration without neurofibrillary tangles. *Science* **293,** 711–714.

Wodarz, A., Hinz, U., Engelbert, M., and Knust, E. (1995). Expression of crumbs confers apical character on plasma membrane domains of ectodermal epithelia of Drosophila. *Cell* **82,** 67–76.

Wu, J. S., and Luo, L. (2006). A protocol for dissecting *Drosophila melanogaster* brains for live imaging or immunostaining. *Nat. Protoc.* **1,** 2110–2115.

Yao, K. M., and White, K. (1994). Neural specificity of elav expression: Defining a Drosophila promoter for directing expression to the nervous system. *J. Neurochem.* **63,** 41–51.

CHAPTER THIRTEEN

Using *Caenorhabditis elegans* to Study Serpinopathies

Olivia S. Long, Sager J. Gosai, Joon Hyeok Kwak, Dale E. King, David H. Perlmutter, Gary A. Silverman, *and* Stephen C. Pak

Contents

1. Introduction	260
2. Considerations for Transgenesis	261
2.1. Fluorescent Protein (FP) fusions	264
2.2. DNA origin	264
2.3. Promoter choice	264
2.4. *C. elegans* expression vectors	264
2.5. Transgenesis methods	265
2.6. Coinjection markers	266
3. Microinjection	268
3.1. Microinjection equipment and reagents	268
3.2. Preparation of plasmid DNA for injections	269
3.3. Microinjection method	270
3.4. Isolation and propagation of transgenic worms	271
3.5. Stable integration of transgenes	271
3.6. Characterizing new transgenic lines	273
4. High-Content Drug Screening	273
4.1. Equipment, consumables, and reagents	274
4.2. OP50 food preparation for drug screening	274
4.3. Compound library preparation	275
4.4. Animal preparation and sorting using the COPAS BIOSORT	275
4.5. Automatic image acquisition	277
4.6. Considerations for HCS	278
Acknowledgments	279
References	279

Department of Pediatrics, Cell Biology and Physiology, University of Pittsburgh School of Medicine, Children's Hospital of Pittsburgh of UPMC, Pittsburg, Pennsylvania, USA

Abstract

Protein misfolding, polymerization, and/or aggregation are hallmarks of serpinopathies and many other human genetic disorders including Alzheimer's, Huntington's, and Parkinson's disease. While higher organism models have helped shape our understanding of these diseases, simpler model systems, like *Caenorhabditis elegans*, offer great versatility for elucidating complex genetic mechanisms underlying these diseases. Moreover, recent advances in automated high-throughput methodologies have promoted *C. elegans* as a useful tool for drug discovery. In this chapter, we describe how one could model serpinopathies in *C. elegans* and how one could exploit this model to identify small molecule compounds that can be developed into effective therapeutic drugs.

1. INTRODUCTION

A serpinopathy is a conformational disease in which a genetic mutation predisposes a serpin protein to misfold and polymerize. The accumulation of abnormal proteins within cells lead to tissue injury and degenerative disease (Belorgey *et al.*, 2007; Gooptu *et al.*, 2009; Lomas and Carrell, 2002; Silverman *et al.*, 2007). Examples include α1-antitrypsin (AT) deficiency, hereditary angioedema, familial encephalopathy with neuroserpin inclusion bodies (FENIB), and thrombophilia. Mutations occur most commonly in or near the hinges of the reactive site loop and the shutter region that underlies the opening of the β-sheet A. One of the best-studied serpinopathies is AT deficiency. AT deficiency is most commonly caused by a mutation that results in a substitution of the amino acid, glutamic acid at position 392 to a lysine (known as the Z mutation). This change in charge, destabilizes the hinge region making the AT protein more prone to polymerization. These protein polymers are poorly secreted and accumulate within the endoplasmic reticulum (ER) of liver cells (Perlmutter, 2002). Since AT is the predominant serine peptidase inhibitor in extracellular fluids, decreased AT secretion results in a loss-of-function phenotype manifest by peptidase-inhibitor imbalance, connective tissue matrix destruction, and susceptibility to chronic obstructive lung disease. In contrast, the accumulation of aggregation-prone α1-antitrypsin Z (ATZ) in liver cells leads to a gain-of-toxic-function phenotype characterized by liver failure and carcinogenesis (Rudnick and Perlmutter, 2005). Indeed, AT deficiency is the most common genetic cause of pediatric liver disease and the most frequent genetic diagnostic indication for liver transplantation during childhood. Curiously, only ~10% of all ATZ homozygous individuals develop severe liver disease (Sveger, 1988). This observation suggests that other genetic and/or environmental factors play key roles in determining disease severity and outcome. Our current understanding of these "other" factors is rudimentary and effective therapeutic options are very limited.

Model systems have provided valuable insights into various human diseases. For example, transgenic worms and fruit flies have enhanced our understanding of Alzheimer's, Parkinson's, Huntington's, Duchenne muscular dystrophy, diabetes, cancer, and a plethora of other complex diseases. Thus, simple model systems may be useful for unraveling previously undiscovered information about serpinopathies and provide insights into the cellular mechanisms that protect against protein misfolding.

Caenorhabditis elegans is a well-established model organism with a proven track record. Their small size (~ 1.5 mm as an adult), short life-cycle (2–3 days), high fecundity (~ 300 progeny in ~ 3 days), easy genetic manipulation, and ease of culture, makes them an excellent, low-cost, alternative to other model organisms. Transgenic animals can be generated with relative ease and well-established reverse and forward genetic tools are readily available. Recently, techniques for large-scale liquid worm culture, high-throughput worm sorting, and automated high-content imaging have been described (Gosai *et al.*, 2010), paving the road for the use of *C. elegans* in drug discovery.

A comprehensive description of basic *C. elegans* methodology is beyond scope of this chapter. We refer interested readers to *C. elegans*: A Practical Approach (Stiernagle, 1999), for further information. This chapter will focus on methods relating to the study of serpinopathies in *C. elegans* with particular emphasis given to the generation of transgenic animals and high-content drug screening.

2. Considerations for Transgenesis

Transgenic animal approaches can provide a wealth of information regarding how a gene is regulated and what function it performs within the context of a whole organism. In a null mutant background, altered (mutated) copies of a gene can be easily reintroduced to study structure–function relationships. In some cases where a serpin null fails to yield an overt phenotype, overexpression of a gene may provide insights into protein function by perturbing the serpin–proteinase balance (Pak *et al.*, 2004). Transgenes fused to one or more fluorescent proteins (FPs) or affinity tags can be exploited for the study of protein–protein interactions and target protein identification (refer Chapter 14 for details on using *C. elegans* to identify *in vivo* serpin targets). Transgenic worms can also be used to study toxicology and as a tool for drug discovery.

A wide array of *C. elegans* strains, plasmids, promoters, reporter genes, and RNAi reagents are readily available thanks to the generosity of the *C. elegans* research community (Fire *et al.*, 1990). A list of useful *C. elegans* resources and their websites is shown in Table 13.1.

Table 13.1 List of useful *C. elegans* resources

Source	Mission	URL for searching information	Comments
Wormbase	Major repository for *C. elegans* information including genomic, genetic, anatomy, and literature	http://www.wormbase.org	Very helpful resources including: • Gene expression patterns • Genetic map information • Mutant phenotypes • Complete Genome sequence • Developmental lineage of the worm
Caenorhabditis Genetics Center (CGC)	Collects, maintains, and distributes stocks of *C. elegans*	http://www.cbs.umn.edu/CGC/	• Thousands of strains available • Constantly being updated • Strains available to research community for nominal fee
Caenorhabditis elegans Gene Knockout Consortium (OMRF)	Production of deletion alleles of specified gene targets	http://celeganskoconsortium.omrf.org/	• Investigators can submit request for generation to knockout mutants • Mutants deposited into CGC • Free with MTA
C. elegans National Bioresource Project of Japan (NBRP)	Collect nematode bioresources including: making and distributing deletion mutants	http://www.shigen.nig.ac.jp/c.elegans/index.jsp	• Simple website to search for deletions • Over 4000 deletions available
NCBI GenBank, Blast	Repository of sequences from many diverse organisms with extensive blast services	http://www.ncbi.nlm.nih.gov/Genbank/index.html http://www.ncbi.nlm.nih.gov/BLAST/	• Wealth of information • Free access • Centralized collective archive • Blast easily used to find similarities between sequences

Name	Description	URL	Notes
WormAtlas	Provides anatomical information	http://www.wormatlas.org	• Simple text search • Excellent handbook on worm anatomy
C. elegans Movies	Provides a visual introduction to a C. elegans	http://www.bio.unc.edu/faculty/goldstein/lab/movies.html	• Good introductory videos about worm development, phenotypical studies, and techniques
Addgene	Archives and distributes plasmids	http://www.addgene.org	• Large variety of plasmids available • Vector kit available contain 288 vectors designed for C. elegans research
BioScience Life Science (formerly GeneService)	Provides RNAi feeding library originally constructed by Julie Ahringer's group	http://www.lifesciences.sourcebioscience.com/	• Library covers 87% of C. elegans genes • Library can be purchased whole or in part
BioScience Life Science	C. elegans Fosmid Library	http://www.lifesciences.sourcebioscience.com/	• Offers complete clone set or individual clones • C. elegans FOS finder tool that allows you to search for your gene or region of interest
C. elegans www server	C. elegans resource center	http://elegans.swmed.edu/	• Offers links to important C. elegans resources such as literature searches, software, genome, etc.

2.1. Fluorescent Protein (FP) fusions

Before attempting to generate a model of a serpinopathy in *C. elegans*, a number of important factors must be carefully considered. For example, if the transgenic line will be used to study protein synthesis and disposition, an FP tag may be helpful for visualization of the transgene (Chalfie *et al.*, 1994). FP tags obviate the need for fixation and antibody staining, and allows one to monitor proteins in real time in live animals. In addition, FP tags are useful in screens looking for genetic mutations or small molecules that alter misfolded protein accumulation. Alternatively, if the transgenic animals will be used to study protein–protein interactions, it may be useful to add one or more affinity tags to aid detection and purification of protein complexes (Hobert and Loria, 2005).

2.2. DNA origin

The source of DNA that will be used to generate the transgenic animal is another important consideration. If a *C. elegans* homologue can be readily identified, the endogenous gene can be used to minimize complications arising from cross-species differences in the coding region. If a clear worm homologue cannot be readily identified, human DNA may be used, however, modifications may be required to ensure efficient expression. Due to variations in RNA splicing and the large size of most human genes it is often not possible to directly express human genomic DNA in worms. However, transgene expression in *C. elegans* is enhanced by the presence of intronic sequences (Okkema *et al.*, 1993). As such, it is often necessary to clone several synthetic introns into human cDNA sequences to ensure efficient transgene expression.

2.3. Promoter choice

The promoter used to drive expression of the transgene is another important consideration. Transgene expression should be directed in cells or tissues that most closely resemble the biological environment in which the human protein is normally expressed. A large number of *C. elegans* promoters with various spatial and temporal specificities have been characterized. A few examples are listed in Table 13.2.

2.4. *C. elegans* expression vectors

A wide variety of modular *C. elegans* expression vectors have been generated and made available to the *C. elegans* research community by Fire *et al.* (1990). A schematic of the canonical expression vector is shown in Fig. 13.1. All expression plasmids contain an *Escherichia coli* ampicillin

Table 13.2 Common promoters used for transgenesis

Promoters	Expression location
nhx-2, vha-6	Intestinal cells
myo-2	Pharyngeal muscles
myo-3	Body muscles
pes-10	Embryo
hsp-4, hsp-16	Heat/stress-inducible expression
pie-1	Germline
cdh-3	Anchor cells
ttx-4	Sensory neurons
unc-119	Pan-neuronal
unc-122	Coelomocytes
lin-26	Epithelial tissues, gonad, and uterus

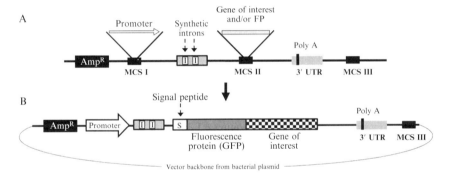

Figure 13.1 A schematic representation of a *C. elegans* expression plasmid. (A) The canonical vector used to build all expression vectors (Okkema et al., 1993). (B) An expression vector for expression of an aggregation-prone serpin protein in *C. elegans*. I, synthetic intron; UTR, untranslated region.

resistance gene, multiple cloning sites, introns, and a 3′UTR (Fig. 13.1A). Numerous vectors with built-in tissue-specific promoters, various fluorescent reporter proteins with nuclear localization, secretion, and specific organelle (e.g., mitochondria) targeting signals are also available. A typical expression vector used to generate transgenic animals expressing an aggregation-prone serpin (such as ATZ) fused to GFP is shown in Fig. 13.1B.

2.5. Transgenesis methods

The most widely used method for transgenesis in *C. elegans* is microinjection. DNA is injected into the gonads of a young adult hermaphrodite (Mello and Fire, 1995). DNA is then taken up by the germ cells and transgene expression is seen in developing worms within 24–48 h. Transgenes are typically injected as plasmids, however, PCR products can also be used obviating the need for cloning (Hobert, 2002). Injected DNA forms large extrachromosomal arrays that are passed on to subsequent generations in a non-Mendelian inheritance pattern (Stinchcomb et al., 1985). Transmission rates can vary anywhere from 5% to 95%, so multiple independent transgenic lines should be isolated.

Microparticle bombardment is an alternate method for *C. elegans* transgenesis. DNA is bound to small microbeads (usually gold particles) and ballistically bombarded into worms. Bombardment is commonly used to generate animals carrying low copy numbers of the transgene. This is a particularly important consideration if one is concerned that an overexpression of a transgene might be toxic to the animal. In some cases, stable integrated transgenic lines can be obtained (Praitis et al., 2001). However, bombardment is more expensive, labor intensive, and time-consuming than microinjection.

2.6. Coinjection markers

If the transgene of interest is not readily identifiable (i.e., does not have a fluorescent tag or is anticipated to be expressed at low levels), it may be necessary to use a coinjection marker to facilitate selection of positive transgenic lines following injection. Coinjection markers typically recombine with other DNA in the mix and are incorporated concomitantly into the extrachromosomal array. A coinjection marker can therefore be used as an indirect indicator of transgenesis. Coinjection markers often make use of a promoter that is specific to an anatomical structure remote from the region of interest and is tagged with an FP of a different color to reduce interference during multicolor imaging. If a fluorescent microscope is not available, one may utilize a coinjection marker that induces a clear dominant phenotypic change or rescues a mutant phenotype. For example, *rol-6* induces a unique "rolling" movement and transgenic animals can be easily identified under a standard dissecting microscope. Some commonly used coinjection markers are described in Table 13.3.

A flow chart of the transgenesis procedure is illustrated in Fig. 13.2. In brief, the serpin gene of interest is cloned into a *C. elegans* expression vector under the control of an appropriate promoter (Fig. 13.2A). The plasmid DNA (along with a suitable coinjection marker) is microinjected into the gonad of a young adult hermaphrodite (Fig. 13.2B). Positive transgenic lines are identified and propagated (Fig. 13.2C). Stable integrated lines are generated by exposure to ionizing radiation (Fig. 13.2D). Stable integrants are outcrossed to remove nonspecific mutations that may have been introduced by exposure to radiation.

Table 13.3 Coinjection markers for DNA transformations in *C. elegans*

Coinjection marker	Description	Advantages	Disadvantages
myo-2::FP	Drives FP expression in pharynx	• Dominant visible marker • Transgenic worms easily identified under a fluorescent dissecting scope • Useful for genetic screens and mating • Promoter expressed through all stages of development	• Requires fluorescent dissecting scope • High concentration of plasmid can cause expression of fluorescent protein in coelomocytes
myo-3::FP	Drives FP expression in body wall muscles	• Same as above	• Requires fluorescent dissecting scope • Could interfere with expression studies of other fluorescent fusions
rol-6	Dominant roller	• Transgenic worms will have a dominant *roller* phenotype • No need for a fluorescent dissecting scope	• Suppressed in some mutant backgrounds (uncoordinated and dumpy) • Roller phenotype not visible before L2 stage
unc-119 (rescue)	Used in conjunction with *unc-119(ed3)* worms	• Nontransgenic worms are uncoordinated and do not move • Only transgenic animals will move freely	• Requires the use of *unc-119* mutant strain
pha-1 (rescue)	Used in conjunction with *pha-1*(E2123)	• Nontransgenic worms embryonic lethal at 25 °C • Only transgenic worms will survive at 25 °C	• Requires the use of *pha-1* mutant strain
unc-22	Dominant twitcher	• Transgenic worms will display a dominant *twitching* phenotype	• Phenotype can be subtle and sometimes difficult to detect

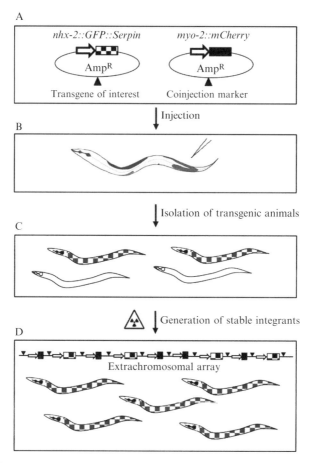

Figure 13.2 Transgenesis flow chart. (A) Plasmid DNA for injection are prepared. (B) DNA is injected into young adult hermaphrodites. (C) Transgenic progeny are isolated and enriched. (D) Stably integrated transgenic lines are established by exposure to irradiation.

3. Microinjection

3.1. Microinjection equipment and reagents

- *Injection table*: A heavy marble table or air table to minimize vibrations.
- *Inverted DIC microscope*: Zeiss Axiovert 100 or equivalent instrument, equipped with standard 5× and 40× objectives. The microscope should have a flat, free-sliding glide stage to allow for easy manipulation of slides.

- *Micromanipulator*: Narishige MMN-1 or equivalent. Microinjections require a manipulator that can hold and position a needle with fine mobility in the X (forward and back), Y (side to side), and Z (vertical) planes and allows for easy adjustments of needle angle and location.
- *Pressurized injection system*: FemtoJet (Eppendorf; Haupauge, NY catalog # 920010504) is an injection needle system in which the needle is placed in a holder with a tight seal collar, which is attached by plastic tubing to a regulated pressure source.
- Standard dissecting microscope.
- *Dehydrated agarose pads*: Provide a "sticky" surface for immobilizing worms during microinjection.
- *Microinjection needles*: Femtotips (Eppendorf, catalog # 930000035) are a high-quality injection needle.
- *Injection oil*: Halocarbon Oil 700 (Sigma; St. Louis, MO catalog # H8898).
- *Recovery buffer*: Phosphate buffered saline (PBS).

3.1.1. Agarose pads

1. A drop ($\sim$100 µl) of hot 2% agarose (dissolved in H_2O) is placed onto a 24 × 100 mm microscope glass coverslip (avoid creating air bubbles).
2. A second coverslip is immediately placed on top of the agarose.
3. When the agarose has hardened ($\sim$10 s), the top cover slip is gently removed.
4. Steps 1–3 is repeated until desired number of agarose pads are made.
5. Agarose pads are placed onto a tray, covered with aluminum foil and allowed to dry for a period of 1–3 days.
6. Agarose pads are stored in the original glass coverslip box for many months.

3.2. Preparation of plasmid DNA for injections

1. *E. coli* clones harboring the appropriate expression vectors are grown overnight.
2. Plasmid DNA is purified using the Qiagen (Valencia, CA) mini prep kit (or similar).
3. DNA injection solution: 10 µl final volume.
   a. Transgene plasmid final concentration $\sim$50–75 ng/µl (if gene is toxic, the plasmid concentration can be lowered).
   b. Coinjection marker final concentration $\sim$5–30 ng/µl.
   c. Total DNA concentration $\sim$80 ng/µl (Bluescript plasmid DNA is added as required to adjust the final DNA concentration).

3.3. Microinjection method

1. The injection mix is centrifuged for 10 min at 13,000 rpm to pellet particulate matter that may clog the needle.
2. 2 µl of DNA solution is carefully placed on the top of the injection needle and allowed to fill the tip of the needle by capillary action.
3. The micromanipulator is then depressurized to allow mounting of the loaded injection needle.
4. Once the needle is mounted, a worm is placed on an agarose pad and covered with a drop of halocarbon oil to prevent desiccation.
5. Using the 5× objective, the position of the worm is adjusted to ensure correct orientation for injection. The dorsal surface of the worm should be at an ∼45–75° angle to the needle (see Fig. 13.3(1)).
6. Upon focusing on the worm, coarse controls on the micromanipulator are used to position the needle near the gonad of the worm.
7. Using the 40× objective, final adjustments are made to ensure that the gonad and the tip of the injection needle are in the correct focal plane.
8. The needle is then gently inserted into the cytoplasmic core of the gonad (see Fig. 13.3(2)).
9. DNA is then injected by applying short bursts of pressure (1300 psi; see Fig. 13.3(3)).
10. If performed correctly, DNA should move freely through the gonad syncytium without disrupting any germ cell precursors (see Fig. 13.3(4)).

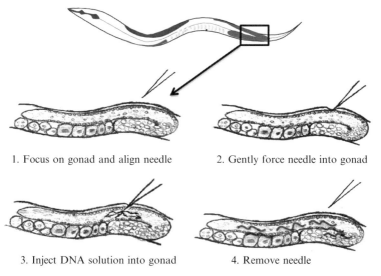

Figure 13.3 Schematic of microinjection into *C. elegans* gonad.

11. Immediately following injections, the agarose pad carrying the injected animal is transferred to the dissecting microscope.
12. The injected worm is recovered using a small drop (20 µl) of PBS buffer.
13. Once the worm is freely moving about, it is transferred to a fresh Nematode Growth Medium (NGM) agar plate. [NGM: 3.0 g NaCl, 2.5 g Bacto-peptone, 20.0 g Agar dissolved in 1.0 l of double-distilled H_2O; solution is autoclaved and then cooled to 55 °C to which the following solutions are added: 1.0 ml cholesterol (5 mg/ml in Ethanol), 1.0 ml 1 M $CaCl_2$, 1.0 ml 1 M $MgSO_4$, 25.0 ml 1 M KH_2PO_4, pH 6.00] (Hope, 1999).

Steps 6–13 should be performed as quickly as possible to prevent desiccation of the worm. To ensure sufficient transgenic animals are generated, at least 5–10 worms should be injected.

3.4. Isolation and propagation of transgenic worms

1. An injected worm is placed on a 60-mm NGM plate seeded with a lawn of *E. coli* OP50 strain. This is repeated for each injected worm. (Alternatively, two to three injected worms may be placed on one 60-mm seeded plate if one is injecting eight or more worms.)
2. The injected worm is allowed to recover and is examined daily for progeny expressing the desired transgene or coinjection marker.
3. Expressing progeny in the F1 generation are segregated to separate plates and propagated as distinct lines. It is recommended that as many expressers as possible in the F1 generation be selected and propagated as it is common for expression to be reduced or lost in the F2 generation of some strains.
4. Any subsequent progeny are transferred to a new seeded 60-mm plate and the progeny observed for continued transmission of the transgene or coinjection marker.
5. After approximately three generations, the expression and transmission of the transgene is relatively predictable and stable.

3.5. Stable integration of transgenes

Once relatively stable lines of transgenic animals have been isolated, they can be either maintained as extrachromosomal arrays or integrated into the genome depending on the intended use. Extrachromosomal arrays are generally simple to maintain but require constant diligence to ensure that animals expressing the desired transgene are propagated under the desired conditions. For example, if the transgene transmits a survival disadvantage compared with unaffected progeny, then care must be taken to not allow conditions of stress (starvation, temperature fluctuation, etc.) to affect the animals. This can be time-consuming and, for that reason, one might consider integrating the transgene.

Integration involves exposure of the transgenic animals containing extrachromosomal arrays to ionizing radiation that causes double-stranded DNA breaks (Mello et al., 1991). Upon repair of these breaks, the extrachromosomal array may be incorporated into a chromosome of the animal. The radiation may also cause unintended mutations in other genes, so outcrossing following isolation of stable integrated lines is required (Evans, 2006). Integration of extrachromosomal arrays requires a significant investment of time and resources but can result in a more stable and predictable transgenic line. When selecting lines to integrate, keep in mind that lines with a generally low transmission rate (less than 30% of progeny) will be significantly easier to integrate. In the end, the distinguishing feature of the integrated population is that it will transmit the transgene to 100% of progeny. This distinction will be easier to identify in lines with a low transmission rate to begin with.

During the integration procedure, lines are allowed to starve several times. This facilitates integration in a couple of ways. First, it allows for several generations to pass following irradiation and may help remove lethal spurious mutations introduced by the radiation. Second, it imparts a competitive advantage to transgenic animals with a high transmission rate. After starvation, recovered transgenic animals are more likely to have higher transmission rates.

Prior to beginning, prepare the following (per line to be integrated):

- 33 × 100 mm NGM plates seeded with OP50.
- 30 × 60 mm NGM plates seeded with OP50.
- 210 × 35 mm NGM plates seeded with OP50.

1. 10 transgenic adult worms are transferred to 2 × 100 mm NGM plates and allowed to grow for 2–3 days until there are maximum progeny at the late L4 stage. (For a more closely synchronized population, adults may be allowed to lay eggs for several hours, then removed and the eggs allowed to develop to late L4.)
2. On the day of irradiation, 200 late L4 stage transgenic worms are selected and placed on a plate.
3. Worms are exposed to 3500 rads (35 Gy) of radiation.
4. The irradiated worms are allowed to recover for several hours at room temperature (or overnight at 16 °C).
5. Two worms are transferred to one 100-mm plate. A total of 30 plates are inoculated.
6. These plates are incubated to starvation at room temperature (7–10 days).

7. The worms are washed off with 5 ml of PBS and transferred in 50-μl aliquots to a new 60-mm plate. This is repeated for each of the 30 plates to new 60-mm plates.
8. These plates are incubated to starvation at room temperature (4–5 days).
9. The worms are washed off with 1 ml of PBS and transferred to new 35-mm plates in 25-μl aliquots. Repeat for each of the 30 plates to new 35-mm plates.
10. The worms are allowed to develop for 2 days.
11. From each of the 30 plates, six transgenic worms are selected and placed one each on a new 35-mm plate (6 × 30 = 180 total plates).
12. The worms are allowed to grow for 3 days at room temperature.
13. The plates are examined for 100% transmission.
14. The integrated strains are outcrossed several times to remove spurious background mutations.

We recommend that several integrated lines be isolated and stored. Transgene integration sites are determined by SNP mapping. A description of mapping techniques is beyond the scope of this chapter. Readers interested in these techniques should refer to *C. elegans, a practical approach* (Hope, 1999) or www.wormbook.org for more information.

3.6. Characterizing new transgenic lines

Once a new transgenic line has been established, regardless of whether it is by extrachromosomal array or integrated, one can characterize the phenotype of that line. Basic studies include assays for postembryonic development (PED), brood size, and longevity. These techniques are described in detail in the following chapter. If one is using an FP, then thorough imaging should also be undertaken. One of the advantages of *C. elegans* is that it is a transparent multicellular organism and is very amenable to high-powered imaging. One can utilize coinjection with fluorescent markers to colocalize FPs to subcellular structures such as lysosomes or endoplasmic reticula.

4. High-Content Drug Screening

High-content screening (HCS) has been developed as a tool to expand mainstream drug development for cells in multiwell plates (Giuliano *et al.*, 2006; Gosai *et al.*, 2010; Haney *et al.*, 2006; Johnston and Johnston, 2002; Johnston *et al.*, 2007, 2008, 2009; Nickischer *et al.*, 2006; Trask *et al.*, 2006; Williams *et al.*, 2006). While unicellular models are extensively used for drug discovery, they lack phenotypic diversity and multiorgan complexity of whole organism models (Bleicher *et al.*, 2003; Hodgson, 2001). Thus, HCS campaigns utilizing a whole organism model can offer information regarding drug absorption, distribution, metabolism, excretion, or toxicity (ADMET) at the earliest stages of screening (Gleeson, 2008; Gleeson et al., 2009) and improve chances of identifying efficacious drugs (Gosai et al., 2010).

4.1. Equipment, consumables, and reagents

- VPrep (Velocity11; Menlo Park, CA): Automated liquid handling device for transferring liquids across microtiter plates.
- Flex Drop dispenser (PerkinElmer; Waltham, MA): Automated liquid dispenser for microtiter plates.
- COPAS BIOSORT (Union Biometrica; Holliston, MA): Large particle (animal) sorter and dispenser.
- Cellomics Arrayscan V^{TI} (Thermo Scientific; Pittsburgh, PA): Automated fluorescence imaging and analysis platform.
- Optical Bottom plates (Nalge Nunc International; Rochester, NY. 164730).
- Compound Library (Sigma-Aldrich; St. Louis, MO. LO1280-1KT).
- E. coli (OP50 strain).
- 10-cm seeded NGM plates.

4.2. OP50 food preparation for drug screening

1. A single OP50 colony is used to inoculate 3 ml of sterile LB.
2. Culture is incubated at 37 °C with vigorous shaking overnight.
3. The following morning, 1 ml of the culture is used to inoculate 1-l sterile LB broth prepared in a 2-l flask.
4. One liter culture is then incubated at 37 °C with vigorous shaking until $OD_{600} = 0.5$. This usually takes 2–4 h.
5. Bacteria is separated into 4×250 ml centrifuge tubes and pelleted by centrifugation at $4500 \times g$ for 5 min.
6. Pellet is washed twice with 200 ml PBS and a small aliquot is removed to measure OD_{600}.
7. Volume of PBS is adjusted accordingly so that a 1:100 dilution of the bacteria suspension has an $OD_{600} = 1.0$.
8. Unused bacteria suspension can be stored at -80 °C indefinitely by adding equal volume of 50% glycerol.
9. When ready to use, bacteria is thawed and pelleted by centrifugation for 10 min at $4500 \times g$.
10. Supernatant is removed and pellet is resuspended in half the original volume using PBS.

4.3. Compound library preparation

A diverse set of compound libraries are available from a number of academic and commercial sources (e.g., National Institutes of Health (NIH), ChemBridge, MayBridge, Prestwick, ChemDiv, and Sigma-Aldrich). When developing a screening method, it is important to assess the quality of the assay using a manageable compound library. The Library of Pharmacologically Active Compounds (LOPAC) is a modestly sized library considered to be a gold standard in assay validation. It consists of 1280 small molecule modulators that affect a diverse range of cellular processes and represents most major drug target classes. Compounds libraries are supplied in 96- or 384-well formats. Libraries are frequently prepared using DMSO as the primary solvent.

1. Compounds library is replicated from the master plate by transferring 2 μl of each compound to daughter plates using the VPrep liquid handling device. Space should be left of the plates to provide treatment free controls.
2. Plates are sealed with aluminum plate seals and stored at −80 °C.
3. When it is time to prepare assay plates, daughter plates are thawed at 37 °C and centrifuged at $50 \times g$ for 60 s.
4. Seals are removed and Flex Drop dispenser is used to add 98 μl of S-medium to each well.
5. Fifteen microliter of diluted compounds are then transferred to four sets of optical bottom 384-well assay plates (Nalge Nunc International; Rochester, NY).
6. Assay plates are then covered with aluminum plate seals and placed at −80 °C for long-term storage.
7. On the day of sorting, assay plates are thawed at 37 °C and centrifuged at $214 \times g$ for 60 s.
8. Fifteen microliters of fresh 4× assay medium (4.0 ml OP50, 25.4 ml sterile S-medium, 0.6 ml antibiotic antimycotic solution (100× stabilized BioReagent, sterile-filtered, with 10,000 units penicillin, 10 mg streptomycin, and 25 μg amphotericin B per milliliter. Sigma-Aldrich, A5955), 120 μl 100 mM (+)-5-fluoro-2′-deoxyuridine (FUDR) to prevent growth of progeny) is added to each well.

4.4. Animal preparation and sorting using the COPAS BIOSORT

To minimize variability in the assay, it is imperative to start a drug screen with a tightly synchronous population of worms expressing an uniform level of transgene. Although, animal preparation methods may vary slightly depending on the model and the specific assay being employed, we recommend using late L4–young adult stage worms. For a high-throughput drug screen, worm transfer via traditional hand-picking would not be a viable option. As such, we utilized a flow cytometer specifically designed for worms and other multicellular organisms called COPAS BIOSORT. This is a high-throughput, pressure driven system that has the capacity to perform multiparametric analysis and sorting. A screenshot of the COPAS BIOSORT interface is shown in Fig. 13.4. In this example, worms are initially gated to isolate a desired developmental stage using empirically determined time-of-flight (TOF) and coefficient of extinction (EXT) values. Animals are then sorted into assay plates based on GFP fluorescence intensity and TOF.

1. 6–12 transgenic L4 stage animals are picked by hand onto 12 10-cm NGM plates.
2. Plates are incubated at a permissive temperature for 1 week or until the F2 generation reaches adulthood.
3. Gravid adults are isolated using the COPAS BIOSORT to select highest GFP expression (~30% of the population).
4. 12,000 sorted animals are transferred to 10-cm seeded NGM plates, ~1000 animals per plate.
5. Animals are allowed to lay eggs for 5–8 h.

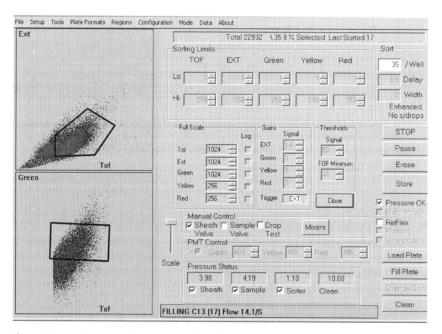

Figure 13.4 Automated worm transfer using COPAS BIOSORT. A screenshot of the COPAS BIOSORT interface is shown. Dots in the scatter plots represent an individual animal. Animals are gated based on their TOF and EXT values which are parameters for size and granularity, respectively (*polygon, upper* scatter plot). Animals are further gated based on their GFP expression (*polygon, lower* scatter plot). Only animals that satisfy both criteria are dispensed into wells for drug screening.

6. Adults are removed by gently washing plates with 2 ml of PBS. Eggs will remain stuck to the bacteria on the plates.
7. Eggs are hatched by 16-h incubation at a permissive temperature.
8. Hatchlings are washed off and transferred to ~100 seeded NGM plates. More or less, plates might be used depending on the total population size. Care must be taken to prevent starvation or overcrowding.
9. Animals are incubated for 36 h or until most of the animals on the plates are in the L4-young adult stages.
10. Worms are transferred to 50-ml conical tubes by PBS wash.
11. Animals are allowed to pellet by gravity for 5 min and supernatant is discarded.
12. S-medium (minus EDTA; 11.7 g NaCl; 2 g K_2HPO_4; 12 g KH_2PO_4; 10 mg cholesterol; 20 ml 1 M potassium citrate; 20 ml trace metal solution: [0.346 g $FeSO_4 \cdot 7H_2O$; 0.098 g $MnCl_2 \cdot 4H_2O$; 0.144 g $ZnSO_4 \cdot 7H_2O$; 0.012 g $CuSO_4 \cdot 5H_2O$; 500 ml H_2O, filter sterilize and shield from light]; 6 ml 1 M $CaCl_2$; 6 ml $MgSO_4$; 200 µl Triton X-100 in 2 l H_2O. filter sterilize) is used to adjust the worm suspension to a concentration of 400 animals per milliliter.
13. Prior to sorting, sheath and cleanout bottles are filled with filter-sterilized S-medium containing 0.01% Triton X-100.
14. Animals suspended in S-medium are added to the sample cup in 40-ml aliquots.
15. Flow rate of the BIOSORT is adjusted up to 25 worms per second to prevent the discarding of desired samples and coincidence check is enabled to enhance specificity.
16. 35 animals are sorted into the wells of the 384-well assay plate (containing drugs and OP50).
17. Sorted plates are sealed with ThinSeal (ISC Bioexpress; Kaysville, UT. T-2417-4) and incubated at room temperature for 24–48 h.

A starting population of ~45,000 animals is necessary to dispense 35 worms into every well of a 384-well plate given a 30% selection rate. This is to offset any worms that were discarded during the sorting process due to the coincidence check. Coincidence check detects when the BIOSORT cannot separate a desired and undesired animal flowing through the system and will discard both.

4.5. Automatic image acquisition

Images are acquired with the ArrayScan V^{TI} fitted with a 2.5× objective and a 0.63× coupler. ArrayScan V^{TI} is a high-content screening instrument designed for high-capacity automated fluorescence imaging. Quantitative data analysis is performed using the SpotDetector BioApplication optimized to identify FP accumulation in *C. elegans*. Transgenic animals expressing an

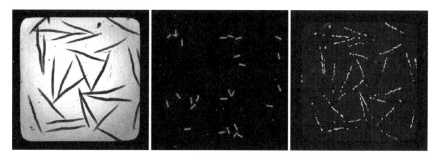

Figure 13.5 Image acquisition using the ArrayScanV^{TI}. Brightfield (*left*), mCherry (*center*), and GFP (*right*) images of transgenic worms expressing an aggregation-prone serpin–GFP fusion. The SpotDetector BioApplication is used to automatically quantify the number, area, and intensity of GFP-positive, serpin aggregates (green spots) within each worm. Images obtained (in part) from Gosai *et al.*, 2010. (See Color Insert.)

aggregation-prone serpin fused to GFP in the intestine and mCherry in the pharynx are generated for the drug screen (Fig. 13.5). A two-channel (mCherry and GFP) assay is used to acquire images of these transgenic worms. Images acquired in the GFP channel are converted to numerical data using complex algorithms to assess the extent of misfolded protein accumulation defined as "spots." Parameters that can be measured include total spot count, total spot area, and total spot intensity. Image acquisition and analysis of a 384-well plate is rapid and takes less than 45 min to complete. Data is subjected to sophisticated statistical analyses to identify hit compounds. In this example, a hit compound is defined as a compound that significantly reduces the total GFP spot area or intensity. Using this approach, we recently identified compounds that significantly reduced accumulation of misfolded ATZ in *C. elegans* (Gosai *et al.*, 2010).

1. 30 µl of 20 mM NaN$_3$ is added to each well to anesthetize worms.
2. Plate is resealed and mixed by inversion.
3. Plate is incubated for 7 min at room temperature to ensure all worms are anesthetized.
4. Seal is removed and plate is imaged using the ArrayScan V^{TI}.
5. Using Protocol Interactive, the exposure time of each channel is set to 25% saturation using the no treatment controls.
6. Sample wells are then selected and imaged using the Scan Plate function.
7. After the scan is finished the data is exported to excel for further analysis.

4.6. Considerations for HCS

It is important to understand the goals of the screen from the earliest stages of preparation. Transgenic animals that will maximize the assay quality must be isolated and chromosomal integration is essential for screening efficiency.

Further, it is also important to optimize incubation temperature, assay duration, and developmental stage as transgene expression and response varies heavily with these parameters. Optimization and assessment of a potential high-throughput screening (HTS) assay is accomplished by performing a series of experiments using a large number of positive and negative control samples ($n = 100$). Assay quality is quantified using Z'-factor analysis. The Z'-factor of an assay can be calculated using the equation: $Z' = 1-((3 \times (\sigma_p + \sigma_n))/(\mu_p - \mu_n))$ where σ is the standard deviation, μ is the mean and p and n are positive and negative controls, respectively. A Z'-factor between 1.0 and 0.5 indicates an excellent assay, between 0.5 and 0.0 indicates a good assay, and below 0.0 indicates a poor assay.

The screen should be performed in duplicate to increase selectivity. This will ensure that compounds with the most reproducible effects will be identified. Initial hits are subjected to secondary assays to confirm results. To validate the efficacy of a potential hit, dose response and time response experiments are performed. Compounds that provide promising dose- and time-dependent responses are then chosen for further validation in cell culture and mouse models.

ACKNOWLEDGMENTS

This work was supported by grants from NIH/NIDDK (DK079806 and DK084512) and Hartwell foundation.

REFERENCES

Belorgey, D., et al. (2007). Protein misfolding and the serpinopathies. *Prion* **1**(1), 15–20.
Bleicher, K. H., et al. (2003). Hit and lead generation: Beyond high-throughput screening. *Nat. Rev. Drug Discov.* **2**(5), 369–378.
Chalfie, M., et al. (1994). Green fluorescent protein as a marker for gene expression. *Science* **263**(5148), 802–805.
Evans, T. C. (2006). Transformation and microinjection. *In* "The *C. elegans* Research Community," (WormBook, ed.), WormBook, doi/10.1895/wormbook.1.108.1, http://www.wormbook.org.
Fire, A., Harrison, S. W., and Dixon, D. (1990). A modular set of lacZ fusion vectors for studying gene expression in *Caenorhabditis elegans*. *Gene* **93**(2), 189–198.
Giuliano, K. A., et al. (2006). Systems cell biology based on high-content screening. *Methods Enzymol.* **414**, 601–619.
Gleeson, M. P. (2008). Generation of a set of simple, interpretable ADMET rules of thumb. *J. Med. Chem.* **51**(4), 817–834.
Gleeson, P., et al. (2009). ADMET rules of thumb II: A comparison of the effects of common substituents on a range of ADMET parameters. *Bioorg. Med. Chem.* **17**(16), 5906–5919.
Gooptu, B., Ekeowa, U. I., and Lomas, D. A. (2009). Mechanisms of emphysema in alpha1-antitrypsin deficiency: Molecular and cellular insights. *Eur. Respir. J.* **34**(2), 475–488.

Gosai, S. J., Kwak, J. H., Luke, C. J., Long, O. S., King, D. E., Kovatch, K. J., Johnston, P. A., Shun, T. Y., Lazo, J. S., Perlmutter, D. H., Silverman, G. A., and Pak, S. C. (2010). Automated high-content live animal drug screening using *C. elegans* expressing the aggregation prone serpin alpha1- antitrypsin Z. *PLoS ONE* **5**(11), e15460.

Haney, S. A., *et al.* (2006). High-content screening moves to the front of the line. *Drug Discov. Today* **11**(19–20), 889–894.

Hobert, O. (2002). PCR fusion-based approach to create reporter gene constructs for expression analysis in transgenic *C. elegans*. *Biotechniques* **32**(4), 728–730.

Hobert, O., and Loria, P. M. (2005). Uses of GFP in *Caenorhabditis elegans*. *In* "Green Fluorescent Protein: Properties, Applications and Protocols," (Kain Chalfie, ed.). 2nd edn. John Wiley & Sons, Inc., Hoboken, New Jersey.

Hodgson, J. (2001). ADMET–turning chemicals into drugs. *Nat. Biotechnol.* **19**(8), 722–726.

Hope, I. (1999). *C. elegans*: A practical approach, Oxford University Press, New York.

Johnston, P. A., and Johnston, P. A. (2002). Cellular platforms for HTS: Three case studies. *Drug Discov. Today* **7**(6), 353–363.

Johnston, P. A., *et al.* (2007). HTS identifies novel and specific uncompetitive inhibitors of the two-component NS2B-NS3 proteinase of West Nile virus. *Assay Drug Dev. Technol.* **5**(6), 737–750.

Johnston, P. A., *et al.* (2008). Development of a 384-well colorimetric assay to quantify hydrogen peroxide generated by the redox cycling of compounds in the presence of reducing agents. *Assay Drug Dev. Technol.* **6**(4), 505–518.

Johnston, P. A., *et al.* (2009). Cdc25B dual-specificity phosphatase inhibitors identified in a high-throughput screen of the NIH compound library. *Assay Drug Dev. Technol.* **7**(3), 250–265.

Lomas, D. A., and Carrell, R. W. (2002). Serpinopathies and the conformational dementias. *Nat. Rev. Genet.* **3**(10), 759–768.

Mello, C., and Fire, A. (1995). DNA transformation. *Methods Cell Biol.* **48,** 451–482.

Mello, C. C., *et al.* (1991). Efficient gene transfer in *C. elegans*: Extrachromosomal maintenance and integration of transforming sequences. *EMBO J.* **10**(12), 3959–3970.

Nickischer, D., *et al.* (2006). Development and implementation of three mitogen-activated protein kinase (MAPK) signaling pathway imaging assays to provide MAPK module selectivity profiling for kinase inhibitors: MK2-EGFP translocation, c-Jun, and ERK activation. *Methods Enzymol.* **414,** 389–418.

Okkema, P. G., *et al.* (1993). Sequence requirements for myosin gene expression and regulation in *Caenorhabditis elegans*. *Genetics* **135**(2), 385–404.

Pak, S. C., *et al.* (2004). SRP-2 is a cross-class inhibitor that participates in postembryonic development of the nematode Caenorhabditis elegans: Initial characterization of the clade L serpins. *J. Biol. Chem.* **279**(15), 15448–15459.

Perlmutter, D. H. (2002). Liver injury in alpha1-antitrypsin deficiency: An aggregated protein induces mitochondrial injury. *J. Clin. Invest.* **110**(11), 1579–1583.

Praitis, V., *et al.* (2001). Creation of low-copy integrated transgenic lines in *Caenorhabditis elegans*. *Genetics* **157**(3), 1217–1226.

Rudnick, D. A., and Perlmutter, D. H. (2005). Alpha-1-antitrypsin deficiency: A new paradigm for hepatocellular carcinoma in genetic liver disease. *Hepatology* **42**(3), 514–521.

Silverman, G. A., Lomas, D. A., and Huckensack, N. J. (2007). Molecular and Cellular Aspects of the Serpionpathies and Disorders in Serpin Activity. World Scientific Publishing Co. Pte. Ltd., Singapore.

Stiernagle, T. (1999). Maintenance of *C. elegans*. *In* "*C. elegans*: A Practical Approach," (I. Hope, ed.), pp. 51–67. Oxford University Press, Oxford, UK.

Stinchcomb, D. T., *et al.* (1985). Extrachromosomal DNA transformation of *Caenorhabditis elegans*. *Mol. Cell. Biol.* **5**(12), 3484–3496.

Sveger, T. (1988). The natural history of liver disease in alpha 1-antitrypsin deficient children. *Acta Paediatr. Scand.* **77**(6), 847–851.

Trask, O. J., Jr., *et al.* (2006). Assay development and case history of a 32K-biased library high-content MK2-EGFP translocation screen to identify p38 mitogen-activated protein kinase inhibitors on the ArrayScan 3.1 imaging platform. *Methods Enzymol.* **414,** 419–439.

Williams, R. G., *et al.* (2006). Generation and characterization of a stable MK2-EGFP cell line and subsequent development of a high-content imaging assay on the Cellomics ArrayScan platform to screen for p38 mitogen-activated protein kinase inhibitors. *Methods Enzymol.* **414,** 364–389.

CHAPTER FOURTEEN

Using *C. elegans* to Identify the Protease Targets of Serpins *In Vivo*

Sangeeta R. Bhatia,* Mark T. Miedel,* Cavita K. Chotoo,[†] Nathan J. Graf,* Brian L. Hood,[‡] Thomas P. Conrads,[‡] Gary A. Silverman,*,[†] *and* Cliff J. Luke*

Contents

1. Introduction	284
1.1. General reagents used for nematode growth and maintenance	285
2. Methods for Identifying the Targets of Intracellular Serpins in *C. elegans*	286
2.1. Identifying serpin targets using genetics	287
2.2. Identifying serpin targets using biochemistry	291
Acknowledgments	297
References	297

Abstract

Most serpins inhibit serine and/or cysteine proteases, and their inhibitory activities are usually defined *in vitro*. However, the physiological protease targets of most serpins are unknown despite many years of research. This may be due to the rapid degradation of the inactive serpin:protease complexes and/or the conditions under which the serpin inhibits the protease. The model organism *Caenorhabditis elegans* is an ideal system for identifying protease targets due to powerful forward and reverse genetics, as well as the ease of creating transgenic animals. Using combinatorial approaches of genetics and biochemistry in *C. elegans*, the true *in vivo* protease targets of the endogenous serpins can be elucidated.

---

* Department of Pediatrics, Children's Hospital of Pittsburgh, University of Pittsburgh School of Medicine, Pittsburgh, Pennsylvania, USA
[†] Department of Cell Biology and Physiology, University of Pittsburgh School of Medicine, Pittsburgh, Pennsylvania, USA
[‡] Department of Pharmacology and Chemical Biology, University of Pittsburgh Cancer Institute, University of Pittsburgh School of Medicine, Pittsburgh, Pennsylvania, USA

1. INTRODUCTION

Apart from a few human extracellular serpins, the *in vivo* targets for most serpins have yet to be identified. This is particularly true for the intracellular serpins, whose natural *in vivo* serpin:protease complexes seem to be rapidly degraded, in part by the proteasome (Hirst *et al.*, 2003; Luke, C. J. and Silverman, G. A., unpublished data). Additionally, another hypothesis to explain the technical difficulties associated with serpin target identification is that the serpin and protease do not coexist within the same location within the cell until an external signal or stress is applied (e.g., Luke *et al.*, 2007). Thus, complexes can only be isolated under certain environmental and/or genetic conditions. Until recently, the genetic tools for phenotypic analysis of serpins in mammalian cell culture have been unavailable, and traditional pull-down techniques by coimmunoprecipitation have yielded little information. The use of model organisms to identify serpin function and protease targets has overcome some of the obstacles of human cell culture systems (Silverman *et al.*, 2010). One such model, *Caenorhabditis elegans*, was the first multicellular organism to have the entire genome sequenced (*C. elegans* Sequencing Consortium, 1998). From this data, the *C. elegans* genome was found to contain no extracellular but nine intracellular serpin sequences of which six were functional protease inhibitors (Luke *et al.*, 2006; Pak *et al.*, 2004; Whisstock *et al.*, 1999). Although these serpins are more evolutionarily conserved at the amino acid level to each other than to mammalian intracellular serpins (Luke *et al.*, 2006; Silverman *et al.*, 2001), there is *in vitro* functional homology to many mammalian intracellular serpins (Luke *et al.*, 2007; Pak *et al.*, 2004, 2006). The genetic controllability of *C. elegans*, as well as its short life cycle, ease of cultivation and transparent body, makes this organism an ideal tool to identify the physiological protease targets of the intracellular serpin subfamily.

C. elegans is a transparent, free-living, androdioecious nematode consisting mostly of a nervous system, muscle, intestine, and reproductive system. It is the simplest multicellular model organism routinely used in the laboratory. The worm has a short life cycle going from egg to adult in 3.5 days at 20 °C and is relatively inexpensive to maintain and grow. Despite the extreme morphological differences, ~35% of the genes in *C. elegans* have homology with human genes and most of the fundamental cellular biological processes are conserved. The most beneficial tool available to the *C. elegans* researcher is the genetic tractability of this model organism. Both forward and reverse genetic, as well as transgenic approaches, can be used to understand protein function *in vivo*. Although less often utilized by *C. elegans* researchers, the ability to rapidly and easily generate transgenic animals (see Chapter 13) also makes this organism ideal for biochemical approaches aimed at identifying protein:protein interactions within a multicellular organism.

1.1. General reagents used for nematode growth and maintenance

Unless otherwise stated, we use the following reagents for the growth and maintenance of the *C. elegans* strains (Brenner, 1974).

1.1.1. Preparation of nematode growth medium (NGM)

Standard NGM is prepared in the following manner:

1. NGM base (3.0 g/l NaCl, 20.0 g/l Bacto-Agar, 2.5 g/l Bacto-peptone) is sterilized using an autoclave on a liquid cycle.
2. After sterilization, the media is then cooled to 55 °C in a water bath.
3. The following sterile solutions are added in the order shown below. Be sure to mix the flask thoroughly between each addition. These solutions must be added after autoclaving to prevent precipitation.
   - 1.0 ml/l cholesterol (5 mg/ml in 95% Ethanol; Sigma–Aldrich)
   - 1.0 ml/l 1 M MgSO$_4$
   - 1.0 ml/l 1 M CaCl$_2$
   - 25.0 ml/l 1 M KPO$_4$ buffer pH 6
4. Media is then dispensed into non-vented petri dishes (Tritech Research).

1.1.2. Preparation of OP50 for nematode food stocks

The laboratory food source of *C. elegans* is the *Escherichia coli* strain OP50. To grow OP50, we use the following protocol:

1. From a frozen glycerol stock, OP50 is streaked onto an LB agar plate and allowed to grow at 37 °C overnight.
2. A single colony from this plate is used to inoculate 10 ml of LB broth and grown overnight at 37 °C with shaking at 200 rpm.
3. 1 L of LB broth is inoculated with the 10 ml overnight culture and grown at 37 °C in an orbital shaker at 200 rpm until an OD$_{600}$ ≈ 0.5 is reached (this usually takes 5–6 h).
4. The OP50 is then centrifuged at 6000×g for 5 min.
5. The supernatant is discarded and the resultant pellet is resuspended in 100 ml sterile M9 buffer (42.3 mM Na$_2$HPO$_4$, 22.0 mM KH$_2$PO$_4$, 85.6 mM NaCl, 1 mM MgSO$_4$·7H$_2$O).
6. The OP50 is then washed a further two times with 100 ml of sterile M9 buffer.
7. After the final wash, the pellet is resuspended in 100 ml of sterile M9 buffer plus 25% glycerol.
8. The OP50 is then dispensed into aliquots and frozen at −80 °C until needed.
9. When required, thaw the aliquot of frozen OP50 and pellet the bacteria by centrifugation at 6000×g for 15 min.
10. The supernatant is discarded and the pellet is washed three times with 6 ml of sterile M9 Buffer. The OP50 is now ready to use.

2. Methods for Identifying the Targets of Intracellular Serpins in *C. elegans*

Both biochemical and genetic approaches can be used to study the protein targets of serpins. Figure 14.1 depicts a flow chart of the scheme to identify protein targets. Both methodologies start with analyzing the genome. Using basic local alignment search tool (BLAST) analysis, *C. elegans* has nine intracellular serpin genes at the nucleotide level (Pak *et al.*, 2004; Whisstock *et al.*, 1999). Further analyses of cDNAs amplified from total RNA and phylogenetic analyses show that of these nine, only five probably serve to function as *bona fide* protease inhibitors; *srps-1*, *-2*, *-3*, *-6*, and *-7* (Luke *et al.*, 2006; Pak *et al.*, 2004; Whisstock *et al.*, 1999). Although these serpins are more closely related to each other than to any of the mammalian serpins and, using phylogenetic analysis, form their own clade (clade L; Silverman *et al.*, 2001), BLAST searches directed exclusively against the human genome reveal that they have primary homology to different human serpins (Table 14.1). However, based on their *in vitro* inhibitory profiles, their functional homologues may be very different. For example, SRP-2 most closely resembles human SERPINI1 (Neuroserpin) at the primary amino acid level. SERPINI1 is an extracellular serpin

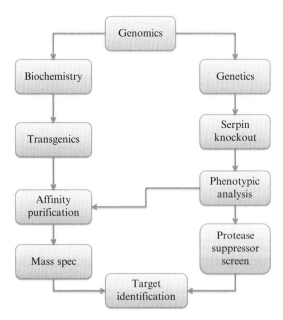

Figure 14.1 General schematic of the methodologies employed for identifying serpin targets in *C. elegans*.

Table 14.1 Functional *C. elegans* serpins

C. elegans serpin	*In vitro* protease targets[a]	Closest human homologue[b]	Knockout allele(s)[c]
SRP-1	Cathepsin K	SERPINA5	ok262
SRP-2	Granzyme B; lysosomal cysteine proteases	SERPINI1	ok350; tm744
SRP-3	Chymotrypsin-like	SERPINB1	ok1433; tm1401
SRP-6	Lysosomal cysteine proteases; calpains	SERPINI2	ok319; tm1994
SRP-7	Unknown	SERPINB4	ok1090

[a] Reviewed in (Silverman *et al.*, 2010).
[b] http://www.wormbase.org, release WS219, October 26, 2010.
[c] Alleles starting with ok are generated by the *C. elegans* Gene Knockout Consortium (http://celeganskoconsortium.omrf.org/); alleles starting with tm are generated by the National Bioresource Project for the Experimental Animal Nematode *C. elegans* (http://www.shigen.nig.ac.jp/c.elegans/index.jsp). Allelic information is deposited at http://www.wormbase.org, release WS219, October 26, 2010.

and inhibits tPA, uPA, and plasmin (Osterwalder *et al.*, 1998), whereas in *C. elegans*, SRP-2 is an intracellular serpin that inhibits granzyme B and lysosomal cysteine peptidases (Pak *et al.*, 2004). Therefore, SRP-2 is unlikely to be a functional homologue of SERPINI1, and only genetic and biochemical analyses will reveal if there is a human serpin with similar function to SRP-2.

2.1. Identifying serpin targets using genetics

The powerful genetic toolbox of *C. elegans* is one of the main attractions to using it as a model system. Even before a worm is handled, the powerful online database, Wormbase (www.wormbase.org), can provide a wealth of information for each specific serpin gene, such as genomic position, expression profiles, nucleotide and protein sequences, phenotypic information by RNAi and reagents available. Some of the reagents that can prove to be most useful for studying protein targets of serpins are knockout alleles. There are two centers which are dedicated to deleting all 19,735 protein coding genes: the *C. elegans* Gene Knockout Consortium (http://celeganskoconsortium.omrf.org/) and the National Bioresource Project for the Experimental Animal Nematode *C. elegans* (http://www.shigen.nig.ac.jp/c.elegans/index.jsp). For the five serpin genes, there is at least one knockout allele (Table 14.1). These knockout alleles have been deposited into a central resource: the Caenorhabditis Genetics Center (CGC; http://www.cbs.umn.edu/CGC/). The *C. elegans* strain containing the allele can be obtained through this central resource. However, most alleles are generated through random mutagenesis techniques, such as EMS (Brenner, 1974). These techniques will often cause mutation in other genes within the same genome and thus should be eliminated through backcrossing the mutant allele against the wild-type background and selecting for the serpin knockout allele by single worm PCR. These methodologies have been published previously by *C. elegans* genetic experts in Hess *et al.* (2006).

2.1.1. Phenotypic analysis

Identification of phenotypes associated with serpin knockouts in *C. elegans* can lead to insights into the targets of these genes. One approach to study protein function is to investigate the behavior of mutant animals in which one or more proteins are mutated to cause a visible macroscopic or microscopic phenotype. In such studies, it is necessary to characterize the general health of mutant animals relative to wild-type (N2) animals by carrying out longevity, brood size, and postembryonic development assays. These assays are carried out under standard conditions (Brenner, 1974), and the proper positive control is a population of the N2 Bristol strain that is tested under identical conditions. Each assay is repeated $n = 3$ times.

2.1.1.1. Longevity assays Under standard laboratory conditions and at 20 °C, N2 *C. elegans* survive an average of 2–3 weeks (Brenner, 1974). In many cases, protein mutations may not have adverse effects on longevity under standard laboratory conditions, but longevity may be affected by subjecting mutants to environmental stress factors. For example, in the case of *srp-6* null animals, the median survival is ∼1 min after immersion in water, whereas N2 animals survive for greater than 30 min under the same conditions (Luke *et al.*, 2007). Longevity has been described previously (Klass, 1977).

1. A population of 25 late L4 worms of a particular worm strain is placed on an NGM plate seeded with OP50.
2. The worms are transferred on a day-to-day basis to seeded NGM plates. Each day the percentage survival is recorded until the population is depleted. The data can be represented on a Kaplan–Meier survival curve, and longevities of various worm strains can be analyzed using a log rank test.

2.1.1.2. Development assays *C. elegans* adjusts its metabolism to survive a range of temperatures from 16 °C to 25 °C under standard laboratory conditions. At 20 °C, the embryo develops through its four larval stages, L1 through L4, to become an adult after 3.5 days. At 15 °C, the life cycle is increased to 5.5 days, and at 25 °C, it is decreased to 2.5 days. At 25 °C, *srp-2* null mutants develop at a comparable rate to N2 worms, while *srp-2*-overexpressing worms are developmentally delayed due to the subjects undergoing larval arrest and slow development (Pak *et al.*, 2004).

Postembryonic development assays have been described previously by Larsen et al. (1995).

1. A synchronized population of 25 young adult worms are placed onto a seeded NGM plate and allowed to lay eggs for 2 h at the required temperature, after which the adults are removed.
2. The developmental stage of the progeny is scored after 72 h at 20 °C or after 48 h at 25 °C. N2 worms develop to the adult stage and/or L4 larvae after this time period. Slowly developing worms will remain in the L1–L3 stages and will eventually develop to the L4 or adult stage after this time period, while larvally arrested worms will not develop into adults.

2.1.1.3. Brood size   C. elegans are classified as androdioecious nematodes, where males occur naturally one in 1000. Each N2 hermaphrodite lays approximately 300 eggs by self-fertilization and this facilitates mass culture of the organism. The brood size of a worm strain is determined by summing the total number of progeny produced by a worm during its fertility period (Singson et al., 1998).

1. Ten late L4 worms are individually placed onto seeded NGM plates on sequential days until the fertility period comes to an end. At 20 °C, the fertility period is 5 days, and at 25 °C, it is 4 days.
2. The number of eggs laid by a single subject is determined on each day or over the entire fertility period.

2.1.1.4. Tissue-specific phenotypes   The physiological processes of C. elegans can be macroscopically divided into seven major systems: epithelial, nervous, muscular, excretory, coelomocyte, alimentary, and reproductive. To study the functions of a protein in C. elegans, it is necessary to know the expression profile of the protein of interest. One method to do this is to tag the protein with a fluorescent marker protein and to create transgenic worm strains where the protein is expressed under its natural promoter. Often times, the system can be further reduced to study the function of the protein in a specific organ system. C. elegans offers the advantage of driving expression of genes under tissue-specific promoters. For example, SRP-6 has been shown to be expressed in intestinal cells, socket cells, vulva cells, and the pharyngeal-intestinal valve, but its function in intestinal cells can be studied by driving its expression under an NHX-2 promoter as described previously (Nehrke and Melvin, 2002).

2.1.2. The use of RNAi by feeding to study protease targets

Once a phenotype has been found, the protease targets of the serpin can be inferred by the suppressor of the phenotype using RNA-mediated interference (RNAi) of specific protease genes. RNAi was first described in

C. elegans (Fire *et al.*, 1998; Guo and Kemphues, 1995), a process in which introduction of double stranded RNA (dsRNA) results in the targeted silencing of a specific cognate gene. Advances in the underlying mechanism of RNAi (Grishok, 2005; Wang and Barr, 2005), and its experimental uses as well the availability of the complete genomic sequence of *C. elegans* (Consortium, 1998), have facilitated the functional analysis of both limited and large-scale gene sets involved in numerous biological pathways (Boutros and Ahringer, 2008; Kamath and Ahringer, 2003; Maine, 2008; Sugimoto, 2004). It is now common for groups to employ large-scale RNAi studies to systematically examine large sets of genes for specific knockdown phenotypes. Thus, RNAi is a reverse genetic approach where the examination of a particular gene is carried out by the knowledge of a specific sequence rather than by the presence of a mutant phenotype. In *C. elegans*, dsRNA can be introduced in four ways: injection of dsRNA into worms (Fire *et al.*, 1998), feeding animals with bacteria that produce a specific dsRNA (Kamath *et al.*, 2001), direct soaking of worms in solution containing the dsRNA (Tabara *et al.*, 1998), and the induced *in vivo* production of a specific dsRNA via an exogenous promoter (Tavernarakis *et al.*, 2000).

When using an RNAi-based approach, a key aspect of the study design is to consider the method of dsRNA delivery to the animals. Among dsRNA delivery methods, the feeding method is well suited to examine a potential large number of RNAis applicable to an entire library. The feasibility of inducing RNAi by feeding bacteria-expressing dsRNA specific for a gene of interest has been well characterized in *C. elegans* (Kamath *et al.*, 2001; Timmons and Fire, 1998). Once ingested by the worm, the dsRNA is processed and systemically spread, resulting in an RNAi-specific phenotype in both the original dsRNA recipients as well as subsequent progeny (Wang and Barr, 2005). Thus, advantages of RNAi by feeding include the ability to introduce dsRNA to a large number of animals at the same time, the ability to introduce dsRNA in a developmental stage-selective manner, and the ability to examine RNAi effects on successive generations.

The techniques used for RNAi in *C. elegans* have already been described in great detail (Kamath and Ahringer, 2003; Wang and Barr, 2005). To perform a large, systematic RNAi screen, a gene library is necessary to provide the template for the synthesis of a specific dsRNA. Gene libraries can be constructed based upon the use of genomic sequence or cDNA libraries. Specifically for the feeding method, the Ahringer RNAi library contains ~17,000 bacterial clones, covering ~87% of the *C. elegans* genome and has been employed in many large-scale and genome-wide RNAi screens (Fraser *et al.*, 2000; Kamath *et al.*, 2003; Timmons and Fire, 1998). The bacterial vector, L440, contains a genomic PCR fragment corresponding to the gene of interest flanked by T7 promoters in opposite orientations. The PCR fragment-containing plasmid is transformed into the

E. coli host HT115(DE3) which is both RNAse III deficient and contains an IPTG-inducible T7 polymerase.

The major application of using this methodology is to screen for and identify genes involved in a specific biological process or pathway. Compared to forward genetic screens, the major advantages of using an RNAi-based approach are the systematic nature by which genes in a particular pathway can be analyzed, and the speed by which the connection between gene identity and phenotype can be made. In addition to large-scale studies in which RNAi knockdown is performed throughout the entire genome, it is also possible to examine specific subsets of genes in a more targeted fashion.

Using this RNAi screening approach, we have begun to try to isolate specific targets of intracellular serpin activity. Specifically, we have employed the MEROPS (http://merops.sanger.ac.uk) peptidase database to identify ~400 peptidases found within the *C. elegans* genome to screen as potential serpin regulatory targets. In particular, we have begun to identify potential peptidase targets of SRP-6, whose antipeptidase activity has been demonstrated for both lysosomal and calpain peptidases and has been implicated in the regulation of necrotic cell death (Luke *et al.*, 2007). It is not well characterized which *C. elegans*-specific proteases may serve as regulatory substrates for SRP-6 in the regulation of necrosis. Therefore, the systematic examination of potential peptidase substrates via RNAi may serve as a useful tool in the identification of genetic SRP-6 targets.

2.2. Identifying serpin targets using biochemistry

While the powerful genetic techniques can provide a great deal of insight into the targets of the *C. elegans*, they do not directly identify a physical interaction. Therefore, a complementary approach is to use affinity purification techniques to coimmunoprecipitate potential targets from transgenic animals expressing a multiple epitope-tagged serpin. This technique, referred to as tandem affinity purification (TAP), was first developed in yeast (Rigaut *et al.*, 1999) and employs the use of two or more affinity tags, separated by a recombinant protease cleavage site fused to the protein of interest at the N- or C-terminus (Table 14.2). This protein can then be purified and concentrated from the complex worm lysate through two rounds of purification to yield the target protein and any potential interactors. This method is particularly useful for low-abundant proteins. Since serpins and the complexed protease are covalently bound (Egelund *et al.*, 1998), the serpin target should be relatively easy to purify. However, this is not usually the case. Two possible reasons have been proposed: (1) serpin: protease complexes contain both inactive protease and serpin and are rapidly turned over (Hirst *et al.*, 2003; Luke, C. J. and Silverman, G. A., unpublished data) and (2) the serpin and enzyme exist under normal conditions in

Table 14.2 Affinity tags used for tandem affinity purification in *C. elegans*

TAP tag name	Composition	Reference
	HA::8 × His::TEV::c-Myc	Polanowska *et al.* (2004)[a]
LAP	GFP::TEV::S·Tag	Cheeseman and Desai (2005)[b]
	Protein A::TEV::CBP	Gottschalk *et al.* (2005)[c]
SnAvi	SB1::SB2::2×TEV::GFP::Avi	Schaffer *et al.* (2010)[d]
TrAP	GFP::2×TEV::c-Myc::Pre::Flag	Bhatia, S. R. *et al.* (unpublished)

Abbreviations: HA, Human influenza hemagglutinin tag; His, Histidine tag; TEV, tobacco etch virus protease recognition site; GFP, green fluorescent protein; CBP, calmodulin-binding peptide; SB, SB1 monoclonal antibody epitope; Avi, Avidin; Pre, precission protease recognition site.
[a] Polanowska *et al.* (2004).
[b] Cheeseman and Desai (2005).
[c] Gottschalk *et al.* (2005).
[d] Schaffer *et al.* (2010).

two distinct cellular compartments or cell types and normally only interact under abnormal conditions, for example, SRP-6 and lysosomal cysteine proteases (Luke *et al.*, 2007). However, it is more likely that both hypotheses are true. Although, due to the ease of transgenic animal generation, *C. elegans* is an ideal organism for TAP methods, it has not been widely used to date.

2.2.1. Generation of TAP plasmids for use in *C. elegans*

Several considerations are required when designing a useful vector for TAP techniques: (1) Choice of promoter, (2) Choice of affinity tags, and (3) N- or C-terminal fusions. In general, the native gene promoter is used for TAP plasmids to avoid misexpression in tissues other than the native protein. This is achieved by amplifying ~2000 bp 5′ upstream to the start codon by PCR (discussed in Chapter 13). Although yields of the protein of interest may be lower than if a stronger promoter is used, this may reduce nonspecific interactions generated by overexpression of the gene of interest. The choice of TAP tag is also important. To date, five different TAP tag sequences are being used in *C. elegans* (Table 14.2). Of these, the LAP, SnAvi, and TrAP tags contain GFP, which can be used for both protein purification and protein localization. In general, most TAP tags are placed at the C-terminus of the protein of interest (Cheeseman and Desai, 2005). However, for serpin biochemistry, the C-terminus contains the reactive center loop, which is cleaved by the target protease, releasing a small 3–4 KDa C-terminal fragment. Thus, to purify serpin:protease complexes in context of the *C. elegans* intracellular serpins, the TAP tag is placed at the N-terminus. In our laboratory, we use the TrAP tag. For expression in *C. elegans*, the genomic coding sequence plus the 3′ untranslated region of the serpin is amplified by PCR from the genome and cloned in frame downstream of the TrAP tag.

2.2.2. Generation of transgenic lines for purification of serpin targets

Plasmid DNA microinjection is an essential tool for *C. elegans* research and is a straightforward technique for incorporating exogenous DNA into worms (Kimble *et al.*, 1982; Mello and Fire, 1995; Stinchcomb *et al.*, 1985). Transgenic animals are created by microinjection of the plasmid DNA directly into the distal arm of the gonad (Berkowitz *et al.*, 2008). The plasmid DNA encoding a given gene of interest is usually coinjected with another plasmid that serves as a transformation marker, providing either a physical (Kramer *et al.*, 1990) or visual (Gu *et al.*, 1998) phenotype by which to selectively identify transgenic animals. The plasmid DNA undergoes recombination to form extrachromosomal arrays that contain multiple copies of the coinjected DNAs in transgenic animals produced via microinjection (Mello *et al.*, 1991). However, extrachromosomal arrays are not passed on to all progeny. Therefore, to maximize the amount of serpin that can be purified, it is advisable to integrate the array into the genome of the animal. For purification of serpin:protease complexes, it is necessary to create transgenic lines expressing the TrAP tag::serpin fusion protein with the knockout allele, for example, TrAP::SRP-6 in the *srp-6(ok319)* knockouts. This will then maximize the yield of serpin:protease complexes obtained by eliminating the native serpin that will not be purified by the TAP procedure. The methods for creating transgenic lines and integration of the extrachromosomal array are discussed in detail in Chapter 13.

2.2.3. Purification of serpin::protease complexes from *C. elegans*

2.2.3.1. Large-scale C. elegans *growth*   Once an integrated transgenic line is generated, standard growth procedures will not produce enough worms to isolate sufficient protein for subsequent analysis. Therefore, in our laboratory, we use the following procedures for growing large quantities of worms.

1. The yolks from five large chicken eggs are separated from the egg whites and placed into a 500 ml sterile bottle and mixed with 200 ml of LB.
2. This mixture is incubated at 60 °C for 1 h and then cooled to room temperature rapidly on ice.
3. 20 ml of OP50 (grown up from an overnight culture) is added to media and then 5 ml is placed on top of 100 mm NGM plates.
4. The egg plates are allowed to settle overnight. The remaining liquid was poured off and the plates are dried. Egg plates can be stored at 4 °C for approximately 2–3 weeks until ready for use.

5. Onto 10 egg plates, 75 adult *C. elegans* are placed per plate. Worms are then grown at room temperature for 5–6 days.
6. Once the egg plates are full of gravid adult *C. elegans*, worms are washed off with M9 and pelleted via gravity on ice. Pellets are then frozen at −80 °C overnight in a 15-ml conical tube.

2.2.3.2. Preparing C. elegans *lysates* Once enough transgenic worms have been isolated, we use the following protocol for preparing worm lysates:

1. Worms are thawed on ice and resuspended in approximately 4 ml of lysis buffer (50 mM Tris–HCL, 150 mM NaCl, 1 mM MgCl$_2$, 1 mM CaCl$_2$, and Complete EDTA-free protease inhibitor cocktail (Roche)).
2. Resuspended worms are passed five times through a French press (Thermo Scientific) at 1000 PSI.
3. The resultant lysate is then centrifuged for 2 min at 700×g at 4 °C to pellet the worm cuticles as well as any worms that have not lysed. This supernatant is placed in new 1.7-ml microcentrifuge tube.
4. The lysate is then centrifuged at 16,000×g for 10 min to remove nuclei and mitochondria. The supernatant is then transferred to a fresh 1.7-ml microcentrifuge tube.
5. The total protein content of the lysate is determined by Bradford Protein Assay (BioRad).
6. Lysates can be immediately used for purification or can be frozen at −80 °C for several weeks before continuing.

2.2.4. Immunoprecipitation

To purify the protein, we employed a novel method in which the protein of interest was tagged with three purification epitopes: two epitopes to aid in purification and a third epitope to identify the protein (TrAP Tag). The first two epitopes were separated by a cleavage site, TEV; the second and third epitopes were separated by the Precission protease site (Table 14.2). All the antibodies used to purify the TrAP::serpin proteins were affinity-purified, desalted using Zeba Desalting columns (Promega), and coupled to CNBr-activated sepharose 4B (GE healthcare).

2.2.4.1. Coupling antibody to CnBr beads

1. Approximately 500 μl of CnBr sepharose powder is added to 10 ml of sterile PBS in a 15-ml conical tube. Five hundred microliters of dry beads will swell to approximately 1 ml of wet bead volume.
2. The sepharose beads are centrifuged at 500×g for 5 min at 4 °C, then washed three times with 10 ml of sterile PBS.

3. After the final wash, the beads are resuspended in 1 ml of PBS + 0.1% Tween-20 (PBST) to produce an ∼50% slurry.
4. Two hundred microliters of 50% bead slurry is transferred to new 1.7-ml microcentrifuge tube.
5. A total of 1 mg of affinity-purified desalted GFP or c-Myc antibody is then added to beads. The binding capacity of the CnBr beads is 6.5–13 mg/ml of bead volume (or per 2 ml of 50% slurry). However, if total saturation is achieved, this can cause clogging of the beads. The optimal amount of antibody bound should be around 80% of maximal binding.
6. The bead and antibody are nutated for 1 h at 4 °C (or until all the antibody is bound as determined by the lack of protein in the supernatant using the Bradford assay), centrifuged at $500 \times g$ in a tabletop microfuge, and the supernatant removed.
7. After the entire antibody has bound, 1 ml of PBST + 0.5 mg/ml BSA is added to beads and then nutated for another hour at 4 °C to block any used binding sites and reduce nonspecific interactions. The beads are again centrifuged at $500 \times g$ in a tabletop microfuge and the supernatant removed.
8. Any additional free sites are quenched by the addition of 1 ml of 1 M Tris (pH 7.4) + 0.1% BSA and nutated at 4 °C for 16 h.
9. Beads are then washed three times with 1 ml PBST.
10. Beads were stored as a 50% slurry of PBST + 0.5 mg/ml BSA + 0.01% sodium azide at 4 °C.

2.2.4.2. First and second round of purifications

1. The worm lysate (approximately 2 mg/ml) in lysis buffer (50 mM Tris–HCL, 150 mM NaCl, 1 mM MgCl$_2$, 1 mM CaCl$_2$, and EDTA-free protease inhibitor mixture (Roche)) was precleared by nutating for 2–4 h at 4 °C in the presence of Protein A beads (Immobilized rProtein A IPA 3000, Repligen).
2. The beads are then washed with lysis buffer three times to remove excess sodium azide. Approximately 40 μl of antibody-coupled CnBR beads were then added to the total worm lysate and nutated for 2–4 h at 4 °C.
3. The beads are then centrifuged at $500 \times g$ and washed a total of five times with 300 mM NaCl.
4. After these washes, the beads are then blocked with PBS and 5% BSA for 2 h at 4 °C to prevent the TrAP::serpin from nonspecifically sticking to the beads after TEV cleavage.
5. Beads were then resuspended in 30 μl of 300 mM NaCl and incubated with the TEV protease overnight at 4 °C.
6. The supernatant is removed to a new tube, and the beads were washed twice with small volume of 300 mM NaCl. With each wash, the wash was added to the original tube. An equal volume of water was added to

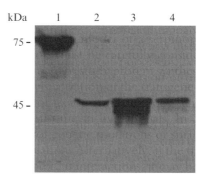

Figure 14.2 Western blotting using the FLAG monoclonal antibody of the rounds of purification in the general TAP procedure using the TrAP tag::serpin fusion. Lane 1 shows protein eluted after first round of purification with the GFP antibody. Lane 2 shows the amount of TrAP::serpin remaining on the beads and lane 3 is the amount of protein in the supernatant after cleavage with TEV. Lane 4 shows the amount of protein eluted after second round of purification with the c-Myc epitope antibody.

the Eppendorf tube bringing the final concentration of the protein in solution to 150 mM to facilitate binding of the beads for the second round of purification.

7. The protein solution is incubated with approximately 80 µl of the second antibody-coupled CnBr beads for 2–3 h at 4 °C.
8. The beads are then washed two times with PBST and resuspended in sample buffer, boiled for 5 min, and centrifuged.
9. Samples were loaded on 10% SDS-polyacrylamide gel (Bio-rad Criterion Tris–HCl) and run at 30 mA before transferring to nitrocellulose by standard Western blotting techniques.

An example of such a Western blot is shown in Fig. 14.2. The blot shows that TrAP::serpin is immunoprecipitated from the worm lysate using GFP polyclonal antibody-coupled beads, running at an apparent molecular mass of ~75 KDa (the calculated size of TrAP::serpin; Fig. 14.2, lane 1). Cleavage with the TEV protease results in most of the proteins being released into the supernatant (Fig. 14.2, lane 3). The supernatant from the TEV cleavage is then subjected to further purification using c-Myc epitope tag polyclonal antibody coupled to CnBr beads (Fig. 14.2, lane 4). The Western blot was developed using the FLAG monoclonal antibody, which is part of the TrAP tag. This data demonstrates the functionality of the TrAP tag which should lead to identification of the serpin targets by subsequent mass spectrometry.

ACKNOWLEDGMENTS

The authors would like to thank the entire *C. elegans* research community for their continued help and support for over a decade. This work was supported, in whole or in part, by National Institutes of Health Grants DK079806 and DK081422 (to G. A. S.). S. R. B. is supported through grant HD052892 Department of Human Health Services, Public Health Services. M. T. M. is supported by an individual NRSA F32DK086112 from the National Institute of Diabetes and Digestive and Kidney Diseases. C. K. C. is a graduate student, part of the Cell Biology and Molecular Physiology Graduate Program (CBMP) of the University of Pittsburgh and supported through the CBMP teaching fellowship.

REFERENCES

Berkowitz, L. A., Knight, A. L., Caldwell, G. A., and Caldwell, K. A. (2008). Generation of stable transgenic C. elegans using microinjection. *J. Vis. Exp.* **18,** pii: 833.

Boutros, M., and Ahringer, J. (2008). The art and design of genetic screens: RNA interference. *Nat. Rev. Genet.* **9,** 554–566.

Brenner, S. (1974). The genetics of Caenorhabditis elegans. *Genetics* **77,** 71–94.

Cheeseman, I. M., and Desai, A. (2005). A combined approach for the localization and tandem affinity purification of protein complexes from metazoans. *Sci. STKE* **2005,** pl1.

Egelund, R., Rodenburg, K. W., Andreasen, P. A., Rasmussen, M. S., Guldberg, R. E., and Petersen, T. E. (1998). An ester bond linking a fragment of a serine proteinase to its serpin inhibitor. *Biochemistry* **37,** 6375–6379.

Fire, A., Xu, S., Montgomery, M. K., Kostas, S. A., Driver, S. E., and Mello, C. C. (1998). Potent and specific genetic interference by double-stranded RNA in Caenorhabditis elegans. *Nature* **391,** 806–811.

Fraser, A. G., Kamath, R. S., Zipperlen, P., Martinez-Campos, M., Sohrmann, M., and Ahringer, J. (2000). Functional genomic analysis of C. elegans chromosome I by systematic RNA interference. *Nature* **408,** 325–330.

Gottschalk, A., Almedom, R. B., Schedletzky, T., Anderson, S. D., Yates, J. R., 3rd, and Schafer, W. R. (2005). Identification and characterization of novel nicotinic receptor-associated proteins in Caenorhabditis elegans. *EMBO J.* **24,** 2566–2578.

Grishok, A. (2005). RNAi mechanisms in Caenorhabditis elegans. *FEBS Lett.* **579,** 5932–5939.

Gu, T., Orita, S., and Han, M. (1998). Caenorhabditis elegans SUR-5, a novel but conserved protein, negatively regulates LET-60 Ras activity during vulval induction. *Mol. Cell. Biol.* **18,** 4556–4564.

Guo, S., and Kemphues, K. J. (1995). par-1, a gene required for establishing polarity in C. elegans embryos, encodes a putative Ser/Thr kinase that is asymmetrically distributed. *Cell* **81,** 611–620.

Hess, H., Reinke, V., and Koelle, M. (2006). Reverse genetics. In "WormBook," (J. Ahringer, ed.). WormBook, doi/10.1895/wormbook.1.7.1, http://www.wormbook.org.

Hirst, C. E., Buzza, M. S., Bird, C. H., Warren, H. S., Cameron, P. U., Zhang, M., Ashton-Rickardt, P. G., and Bird, P. I. (2003). The intracellular granzyme B inhibitor, proteinase inhibitor 9, is up-regulated during accessory cell maturation and effector cell degranulation, and its overexpression enhances CTL potency. *J. Immunol.* **170,** 805–815.

Kamath, R. S., and Ahringer, J. (2003). Genome-wide RNAi screening in Caenorhabditis elegans. *Methods* **30,** 313–321.

Kamath, R. S., Martinez-Campos, M., Zipperlen, P., Fraser, A. G., and Ahringer, J. (2001). Effectiveness of specific RNA-mediated interference through ingested double-stranded RNA in Caenorhabditis elegans. *Genome Biol.* **2**, RESEARCH0002.

Kamath, R. S., Fraser, A. G., Dong, Y., Poulin, G., Durbin, R., Gotta, M., Kanapin, A., Le Bot, N., Moreno, S., Sohrmann, M., Welchman, D. P., Zipperlen, P., et al. (2003). Systematic functional analysis of the Caenorhabditis elegans genome using RNAi. *Nature* **421**, 231–237.

Kimble, J., Hodgkin, J., Smith, T., and Smith, J. (1982). Suppression of an amber mutation by microinjection of suppressor tRNA in C. elegans. *Nature* **299**, 456–458.

Klass, M. R. (1977). Aging in the nematode Caenorhabditis elegans: Major biological and environmental factors influencing life span. *Mech. Ageing Dev.* **6**, 413–429.

Kramer, J. M., French, R. P., Park, E. C., and Johnson, J. J. (1990). The Caenorhabditis elegans rol-6 gene, which interacts with the sqt-1 collagen gene to determine organismal morphology, encodes a collagen. *Mol. Cell. Biol.* **10**, 2081–2089.

Larsen, P. L., Albert, P. S., and Riddle, D. L. (1995). Genes that regulate both development and longevity in Caenorhabditis elegans. *Genetics* **139**, 1567–1583.

Luke, C. J., Pak, S. C., Askew, D. J., Askew, Y. S., Smith, J. E., and Silverman, G. A. (2006). Selective conservation of the RSL-encoding, proteinase inhibitory-type, clade L serpins in Caenorhabditis species. *Front. Biosci.* **11**, 581–594.

Luke, C. J., Pak, S. C., Askew, Y. S., Naviglia, T. L., Askew, D. J., Nobar, S. M., Vetica, A. C., Long, O. S., Watkins, S. C., Stolz, D. B., Barstead, R. J., Moulder, G. L., et al. (2007). An intracellular serpin regulates necrosis by inhibiting the induction and sequelae of lysosomal injury. *Cell* **130**, 1108–1119.

Maine, E. M. (2008). Studying gene function in Caenorhabditis elegans using RNA-mediated interference. *Brief Funct. Genomic. Proteomic.* **7**, 184–194.

Mello, C., and Fire, A. (1995). DNA transformation. *Methods Cell Biol.* **48**, 451–482.

Mello, C. C., Kramer, J. M., Stinchcomb, D., and Ambros, V. (1991). Efficient gene transfer in C.elegans: Extrachromosomal maintenance and integration of transforming sequences. *EMBO J.* **10**, 3959–3970.

Nehrke, K., and Melvin, J. E. (2002). The NHX family of Na+—H+ exchangers in Caenorhabditis elegans. *J. Biol. Chem.* **277**, 29036–29044.

Osterwalder, T., Cinelli, P., Baici, A., Pennella, A., Krueger, S. R., Schrimpf, S. P., Meins, M., and Sonderegger, P. (1998). The axonally secreted serine proteinase inhibitor, neuroserpin, inhibits plasminogen activators and plasmin but not thrombin. *J. Biol. Chem.* **273**, 2312–2321.

Pak, S. C., Kumar, V., Tsu, C., Luke, C. J., Askew, Y. S., Askew, D. J., Mills, D. R., Bromme, D., and Silverman, G. A. (2004). SRP-2 is a cross-class inhibitor that participates in postembryonic development of the nematode Caenorhabditis elegans: Initial characterization of the clade L serpins. *J. Biol. Chem.* **279**, 15448–15459.

Pak, S. C., Tsu, C., Luke, C. J., Askew, Y. S., and Silverman, G. A. (2006). The Caenorhabditis elegans muscle specific serpin, SRP-3, neutralizes chymotrypsin-like serine peptidases. *Biochemistry* **45**, 4474–4480.

Polanowska, J., Martin, J. S., Fisher, R., Scopa, T., Rae, I., and Boulton, S. J. (2004). Tandem immunoaffinity purification of protein complexes from Caenorhabditis elegans. *Biotechniques* **36**, 778–780, 782.

Rigaut, G., Shevchenko, A., Rutz, B., Wilm, M., Mann, M., and Seraphin, B. (1999). A generic protein purification method for protein complex characterization and proteome exploration. *Nat. Biotechnol.* **17**, 1030–1032.

Schaffer, U., Schlosser, A., Muller, K. M., Schafer, A., Katava, N., Baumeister, R., and Schulze, E. (2010). SnAvi–a new tandem tag for high-affinity protein-complex purification. *Nucleic Acids Res.* **38**, e91.

Silverman, G. A., Bird, P. I., Carrell, R. W., Church, F. C., Coughlin, P. B., Gettins, P. G., Irving, J. A., Lomas, D. A., Luke, C. J., Moyer, R. W., Pemberton, P. A., Remold-O'Donnell, E., *et al.* (2001). The serpins are an expanding superfamily of structurally similar but functionally diverse proteins. Evolution, mechanism of inhibition, novel functions, and a revised nomenclature. *J. Biol. Chem.* **276,** 33293–33296.

Silverman, G. A., Whisstock, J. C., Bottomley, S. P., Huntington, J. A., Kaiserman, D., Luke, C. J., Pak, S. C., Reichhart, J. M., and Bird, P. I. (2010). Serpins flex their muscle: I. Putting the clamps on proteolysis in diverse biological systems. *J. Biol. Chem.* **285,** 24299–24305.

Singson, A., Mercer, K. B., and L'Hernault, S. W. (1998). The C. elegans spe-9 gene encodes a sperm transmembrane protein that contains EGF-like repeats and is required for fertilization. *Cell* **93,** 71–79.

Stinchcomb, D. T., Shaw, J. E., Carr, S. H., and Hirsh, D. (1985). Extrachromosomal DNA transformation of Caenorhabditis elegans. *Mol. Cell. Biol.* **5,** 3484–3496.

Sugimoto, A. (2004). High-throughput RNAi in Caenorhabditis elegans: Genome-wide screens and functional genomics. *Differentiation* **72,** 81–91.

Tabara, H., Grishok, A., and Mello, C. C. (1998). RNAi in C. elegans: Soaking in the genome sequence. *Science* **282,** 430–431.

Tavernarakis, N., Wang, S. L., Dorovkov, M., Ryazanov, A., and Driscoll, M. (2000). Heritable and inducible genetic interference by double-stranded RNA encoded by transgenes. *Nat. Genet.* **24,** 180–183.

The C. elegans Sequencing Consortium (1998). Genome sequence of the nematode C. elegans: A platform for investigating biology. *Science* **282,** 2012–2018.

Timmons, L., and Fire, A. (1998). Specific interference by ingested dsRNA. *Nature* **395,** 854.

Wang, J., and Barr, M. M. (2005). RNA interference in Caenorhabditis elegans. *Methods Enzymol.* **392,** 36–55.

Whisstock, J. C., Irving, J. A., Bottomley, S. P., Pike, R. N., and Lesk, A. M. (1999). Serpins in the Caenorhabditis elegans genome. *Proteins* **36,** 31–41.

CHAPTER FIFTEEN

VIRAL SERPIN THERAPEUTICS: FROM CONCEPT TO CLINIC

Hao Chen,* Donghang Zheng,* Jennifer Davids,*
Mee Yong Bartee,* Erbin Dai,* Liying Liu,* Lyubomir Petrov,[†]
Colin Macaulay,[‡] Robert Thoburn,* Eric Sobel,* Richard Moyer,[§]
Grant McFadden,[§] *and* Alexandra Lucas*,[§]

Contents

1. Introduction	302
1.1. Background	302
2. From Viral Pathogenesis to Identification of Immunomodulatory Potential	305
2.1. Serp-1 and viral pathogenesis—Identification of a secreted anti-inflammatory viral serpin	305
2.2. Serp-2—Discovery of viral cross-class anti-inflammatory and antiapoptotic serpins	306
2.3. Rationale for choosing viral serpins for analysis as potential immunotherapeutics	307
3. Testing Biological Potential	307
3.1. Mouse model of arterial angioplasty injury-induced atherosclerosis and aneurysm	307
3.2. Analysis of cellular mechanisms of anti-inflammatory activity	311
3.3. Detection analysis of altered gene expression in cells isolated after arterial injury with and without serpin treatment	318
4. Assessing Clinical Therapeutic Potential for a Viral Serpin in Clinical Trial: Trial of Serp-1 Treatment in Patients with Acute Coronary Syndrome and Stent Implant	319
4.1. Rationale	319
4.2. Stages for proceeding into clinic	320
4.3. Clinical trial findings	321
References	326

* Department of Medicine, Divisions of Cardiovascular Medicine and Rheumatology, University of Florida, Gainesville, Florida, USA
[†] Robarts' Research Institute, University of Western Ontario, London, Ontario, Canada
[‡] Viron Therapeutics, Inc., London, Ontario, Canada
[§] Department of Molecular Genetics and Microbiology, University of Florida, Gainesville, Florida, USA

Methods in Enzymology, Volume 499
ISSN 0076-6879, DOI: 10.1016/B978-0-12-386471-0.00015-8
© 2011 Elsevier Inc.
All rights reserved.

Abstract

Over the past 19 years, we have developed a novel myxoma virus-derived anti-inflammatory *ser*ine *p*rotease *in*hibitor, termed a *serpin*, as a new class of immunomodulatory therapeutic. This review will describe the initial identification of viral serpins with anti-inflammatory potential, beginning with preclinical analysis of viral pathogenesis and proceeding to cell and molecular target analyses, and successful clinical trial. The central aim of this review is to describe the development of two serpins, Serp-1 and Serp-2, as a new class of immune modulating drug, from inception to implementation.

We begin with an overview of the approaches used for successful mining of the virus for potential serpin immunomodulators in viruses. We then provide a methodological overview of one inflammatory animal model used to test for serpin anti-inflammatory activity followed by methods used to identify cells in the inflammatory response system targeted by these serpins and molecular responses to serpin treatment. Finally, we provide an overview of our findings from a recent, successful clinical trial of the secreted myxomaviral serpin, Serp-1, in patients with unstable inflammatory coronary arterial disease.

1. INTRODUCTION

1.1. Background

1.1.1. Serpins

>What is the time when all thought
>Becomes a flame
>All flame
>a flight, a lark, a game
>
>Serpentine, gibberose,
>paths to thought
>slowly met, or sudden sight
>So long sought
>
>Silent thought
>An inspiration
>Serpent wings
>Gem of creation
>
>Dragon dreams
>
>Anonymous, 2010

Serpins are an ancient class of protease regulators present from the time of the dinosaurs to the present day, found in prokaryotes, viruses, worms, birds, and mammals (Silverman *et al.*, 2010). Serpins regulate a large spectrum of protease functions with remarkable efficiency. Serpins modulate mammalian clot forming (thrombosis) and clot dissolving (thrombolysis) pathways, complement, immune responses, connective tissue breakdown, apoptosis, hormone transport, neuronal function, and blood pressure. Serpins are reported to represent from 2% to 10% of plasma proteins.

Serpins are suicide inhibitors, acting as bait for a target serine protease. During an attempt to cleave the P1–P1′ site in the reactive site loop (RSL), target proteases become covalently bonded to the RSL and are dragged across the face of the serpin to the opposite pole of the serpin, rendering both inactive (Arnold *et al.*, 2006; Kiefer *et al.*, 2009; Peitsch *et al.*, 1995; Whisstock *et al.*, 2010; Fig. 15.1).

When a serpin pathway is dysfunctional, there is often significant pathologic consequence, providing testimony to the importance of serpin regulation of proteases. Serpins such as plasminogen activator inhibitor-1 (PAI-1) and antithrombin III (AT-III) regulate thrombosis and thrombolysis and when over- or underactive can cause severe clotting or bleeding problems (Gramling and Church, 2010). Serpinopathies, or serpin-based pathologies, occur with mutations of selected serpins; mutation of alpha-1 antitrypsin causes emphysema, cirrhosis, and pancreatitis, whereas mutation of neuroserpin causes early dementia and increased tendency to seizures. Serpinopathies, as elegantly described by Lomas, Whistock, and others, occur when the normal function of a serpin is disrupted and interactions between serpins form aggregates, produced by insertion of the RSL into the beta sheet of an adjacent serpin (Ekeowa *et al.*, 2010; Gooptu and Lomas, 2008, 2009; Whisstock *et al.*, 2010). Conversely, in a normal state, serpins

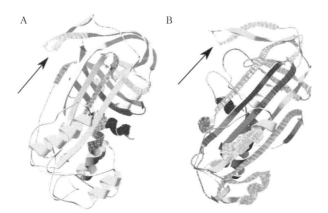

Figure 15.1 Hypothetical models of Serp-1 (left, A) and Serp-2 (right, B) based on known crystallized homologous protein structures, PAI-1 (PBD ID: 3CVM) for Serp-1 and CrmA (PBD ID: 1C8O) for Serp-2, respectively. The arrows point to the reactive site loops (RSL).

maintain a functional balance in multiple protease pathways. Treatments for various diseases have now been designed to enhance serpin functions. For example, treatment with the glycosaminoglycan (GAG) heparin activates the serpin AT-III and provides a highly effective therapy for treatment of clotting disorders.

Mammalian serpins have been developed for the clinic, primarily as replacements for those who lack or have low levels of the endogenous proteins. Many of these are simply isolated from blood products, but one is a recombinant molecule isolated from transgenic goats. These clinical serpins include Aralast (alpha-1 antitrypsin), Zemaira (alpha-1 antitrypsin), and Thrombate (AT-III) isolated from blood products as well as Atryn (recombinant antithrombin), made from milk of transgenic goats (Berry *et al.*, 2009; Bouchecareilh *et al.*, 2010; Kowal-Vern *et al.*, 2001; Mordwinkin and Louie, 2007). Thus tapping into the viral reservoir of potential serpin therapeutics is a natural extension of ongoing work with mammalian serpin replacement therapies.

1.1.2. Viral serpins

Viruses have commandeered the pathways regulated by mammalian serpins through engineering serpins of their own design. Current theory suggests that either these serpins are derived from mammals and modified for the benefit of the virus or conversely that mammalian serpins originate from viruses (Finlay and McFadden, 2006; Lucas *et al.*, 2009). Either way, viral serpins provide highly effective inhibitory reagents that block host attacks against viral proliferation, spread, and invasion in an infected host.

Serp-1 is a secreted 55 kDa myxomaviral serpin that binds proteases in the thrombolytic and thrombotic pathways. Serp-1 inhibits tissue- and urokinase-type plasminogen activators (tPA and uPA, respectively), plasmin in the thrombolytic cascade, and factor Xa (fXa) in the thrombotic cascade (Dai *et al.*, 2003; Lomas *et al.*, 1993; Nash *et al.*, 1998). When expressed and active in a myxomaviral infection, Serp-1 leads to a lethal infection in 90% of European rabbits, whereas mutation of Serp-1 leads to a benign infection that is cleared within weeks (Macen *et al.*, 1993; Upton *et al.*, 1990). Serp-2 is an intracellular myxomaviral serpin that targets granzyme B, a serine protease, and caspases 1 and 8, cysteine proteases (Messud-Petit *et al.*, 1998a,b; Nathaniel *et al.*, 2004; Turner *et al.*, 1999; Turner and Moyer, 2001). The cross-class serpins Serp-2 and CrmA, the latter from cowpox virus (Komiyama *et al.*, 1994), modify apoptosis and inflammatory responses and Serp-2, as for Serp-1, alters viral pathogenesis. In prior cellular and animal model work, both Serp-1 and Serp-2 have been identified as possessing anti-inflammatory activity:

- With this review, we define the process whereby these two potential viral serpin immunotherapeutics were identified, isolated, and tested for activity in animal models of vascular disease. We begin with the discovery process through which viral serpins with immunomodulatory activity are identified, followed by approaches to the initial testing of viral serpins for immune modulating actions in animal models of disease. We will then describe the animal surgical models and cell-based assays in detail.

2. From Viral Pathogenesis to Identification of Immunomodulatory Potential

2.1. Serp-1 and viral pathogenesis—Identification of a secreted anti-inflammatory viral serpin

- The discovery that viruses and other pathogens encode gene products that resemble elements of the host immune system dates back over 25 years and has generated the discipline of "anti-immunology," which emphasizes the coevolution of the vertebrate immune system with the counteractive strategies evolved by pathogens to subvert its functions (Finlay and McFadden, 2006). In the case of viruses, many of these anti-immune proteins are secreted from virus-infected cells and these have been categorized as either virokines (secreted molecules that resemble host cytokines, growth factors, or extracellular immune regulators) or viroceptors (cell surface or secreted viral molecules that resemble cellular receptors; McFadden, 1995, 2003; McFadden and Murphy, 2000). Poxviruses encode a broader variety of virokines and viroceptors than any other virus family and are the only known viruses to encode bioactive serpins. Of these, only one viral serpin is secreted from virus-infected cells, namely Serp-1 of myxoma virus, which is modified by both host and viral glycosyltransferases and expressed as a secreted 55-60 kDa glycoprotein (Nash et al., 2000).
- Myxoma virus is a rabbit-specific poxvirus that causes a severe disease known as myxomatosis in European rabbits but is completely nonpathogenic for all other known vertebrate hosts (Kerr and McFadden, 2002; McFadden, 2005). In studies to investigate the basis for the extreme lethality of myxoma virus in rabbits, a variety of viral virulence genes have been identified by targeted gene knockout analysis showing that the knockout virus exhibits attenuated pathogenesis in infected rabbits (Johnston and McFadden, 2004; Stanford et al., 2007). In the case of the myxoma virus Serp-1 gene, which is also called M008.1 (Cameron et al., 1999), targeted deletion of the gene caused severe attenuation of virus-induced disease symptoms in infected rabbits and was characterized by an exacerbated early inflammatory response to the infection (Macen et al., 1993; Upton et al., 1990). Specifically, infection with the Serp-1 knockout virus construct resulted in a more rapid inflammatory cell infiltration into primary lesions, a robust resolution of disease symptoms,

and pronounced attenuation of the disease in immunocompetent hosts (Macen *et al.*, 1993; Upton *et al.*, 1990). The observation of increased myeloid cell infiltration into Serp-1 knockout viral lesions led to the model that Serp-1 functions as an anti-inflammatory reagent which antagonizes the migration and activation of innate immune cells into sites of damaged or infected tissues (Lucas *et al.*, 1996). Biochemical studies indicated that the secreted Serp-1 protein was, as predicted by its serpin motif sequences, a *bona fide* inhibitor of host serine proteinases, including uPA and tPA, respectively, as well as plasmin and fXa (Lomas *et al.*, 1993; Nash *et al.*, 1998). Further, biologically active Serp-1 protein could be expressed independently of the other viral gene products and shown to still possess anti-inflammatory properties as a stand-alone reagent following intravenous infusion (Lucas *et al.*, 1996). Further, Serp-1 shares homology with well-known mammalian serpins, PAI-1 and neuroserpin which inhibit some shared proteases (tPA and uPA) and have demonstrated both anti- and proinflammatory functions in man and animal models (Gooptu and Lomas, 2008, 2009; Munuswamy-Ramanujam *et al.*, 2010). This has led to the realization that Serp-1, like certain other viral anti-immune proteins, could be adapted as novel anti-inflammatory drugs to treat diseases of hyperactive inflammation or exacerbated immune reactions (Lucas and McFadden, 2004).

2.2. Serp-2—Discovery of viral cross-class anti-inflammatory and antiapoptotic serpins

A second Myxoma virus-derived serpin, Serp-2, has very different biological targets, inhibiting the cysteine proteases, caspases 1 and 8, and the serine protease, granzyme B (Messud-Petit *et al.*, 1998a,b; Nathaniel *et al.*, 2004; Petit *et al.*, 1996; Turner *et al.*, 1999; Turner and Moyer, 2001). Serp-2 from myxoma virus and CrmA from cowpox virus (Komiyama *et al.*, 1994; MacNeill *et al.*, 2006; Turner and Moyer, 2001; Turner *et al.*, 1999) are related but distinct cross-class serpins that similarly inhibit granzyme B and caspases 1 and 8 *in vitro*. CrmA has greater inhibitory activity *in vitro*, whereas Serp-2 has more inhibitory activity in viral infections *in vivo* (Messud-Petit *et al.*, 1998a,b; Nathaniel *et al.*, 2004; Turner and Moyer, 2001; Turner *et al.*, 1999). Granzyme B and caspase 8 activate the apoptotic cascade, while caspase 1 is associated with the inflammasome (Messud-Petit *et al.*, 1998a, b). Granzyme B is considered an extracellular mediator of apoptosis, thus inhibition of granzyme B modulates apoptosis by blocking activation of downstream executioner caspases. Inhibiting caspase 1 blocks the activation of inflammatory responses, specifically preventing activation of interleukin 1 beta (IL-1β) and IL-18. The inflammasome is also able to initiate pyroptosis (inflammation driven macrophage cell death; Franchi *et al.*, 2009). Serp-2 shares homology with two mammalian serpins, murine serine protease inhibitor 6 (SPI-6) and human protease inhibitor 9 (PI-9), that target granzyme B and protect cells from cytotoxic T lymphocyte (CTL)-induced apoptosis (Medema *et al.*, 2001).

2.3. Rationale for choosing viral serpins for analysis as potential immunotherapeutics

The rationale for identifying viral proteins as potential immune modifying therapeutics is thus based upon the following factors:

1. Demonstrated role in viral pathogenesis as verified by infectious disease models and viral gene mutation
2. Related serpin functions in a broad range of species
3. Binding to and/or inhibition of proteases associated with innate immune responses
4. Correlation with mammalian serpins with similar protease targets and/or function
5. Proteins that are secreted or proteins with potential for extracellular function
6. Inhibitory activity for human inflammatory cells, *in vitro*
7. Proven efficacy in animal models of vascular inflammation.

The following sections will describe one of the several mouse models of vascular inflammatory disease used to study anti-inflammatory actions of viral serpins, studies used to examine the cell and molecular targets and the mechanisms of serpin activity, and a recent successful clinical trial of a viral serpin in man.

3. Testing Biological Potential

3.1. Mouse model of arterial angioplasty injury-induced atherosclerosis and aneurysm

The innate immune, or inflammatory, response occurs during interactions between cells in the lining of the arterial wall with cells in the circulating blood and vascular tissues. Specifically, interactions between the endothelial cells, T cells, macrophages, and in some cases, neutrophils (Dollery, *et al.*, 2003; Lucas *et al.*, 2006; Munuswamy-Ramanujam *et al.*, 2006; Serhan *et al.*, 2008a,b) drive the initial inflammatory reaction. These cells make up the inflammatory response and are often activated by circulating factors, for example, cytokines, chemoattractants or chemokines in the blood, serine proteases in the clotting and thrombolytic pathways, and serine and cysteine proteases in the apoptotic/inflammasome pathways.

Activation of an inflammatory response in any organ system (including the vasculature) requires transport of innate immune system cells through the circulating blood to the site in need, and from there, the cells must ingress into the arterial, or more precisely capillary, wall to reach the organ which has ongoing infection or trauma. Thus the arterial vasculature is central to all inflammatory responses. Prior research has attempted the use of vascular explant outgrowths *in vitro* in culture, but this cannot reproduce in entirety the vascular inflammatory responses. There is no replacement or alternative approach for the use of animals in these studies, wherein there is a physiologic and pathogenic interactive response between the vascular wall and circulating inflammatory cells in the blood stream.

A mechanical stretch injury model in the mouse aorta has been reported in previous publications from other researchers (Petrov *et al.*, 2005). We have further developed this mouse angioplasty injury model in mice that express high cholesterol levels in the blood, apolipoprotein E knockout (ApoE$^{-/-}$). ApoE$^{-/-}$ mice develop increased atherosclerotic arterial occlusions and also aneurysmal dilatations with long-term follow-up (Daugherty and Cassis, 2004). With angioplasty injury, we have produced accelerated atheroma as well as rapid aneurysmal dilatation. Using this mouse angioplasty model, we can obtain insight into the structure and cellular mechanisms leading to accelerated atherosclerosis, restenosis, and aneurysm growth.

Observing the characteristics of knockout mice provides a method to investigate how similar genes may cause or contribute to disease in humans. In our study, angioplasty injury, in a series of knockout (KO) mice with differing genetic deficiencies (KO mouse models), was assessed for effects on inflammation, plaque growth, and aneurysmal dilatation. Use of the KO mouse models allow us to precisely identify key molecules involved in plaque growth and aneurysmal dilation. This then enables understanding of the pathways through which these proteins alter atherosclerotic plaque growth after mechanical angioplasty injury and during aneurysm development, *in vivo*.

3.1.1. Mouse angioplasty injury model
3.1.1.1. Preoperative preparation

1. Mouse type: Male—ApoE$^{-/-}$, Apolipoprotein E deficient (B6.129P2-Apoe^{tm1Unc}/J); CCR2$^{-/-}$, CC chemokine receptor deficient (B6.129S4-Ccr2^{tm1Ifc}/J); PAI-1$^{-/-}$, plasminogen activator inhibitor-1 deficient (B6.129S2-Serpine1^{tm1Mlg}/J); Parp-1$^{-/-}$, poly(adenosine 5′-diphosphate-ribose) polymerase-1 (129S-Parp1^{tm1Zqw}/J), and wild-type (WT) control mice (C57BL/6J).
2. Mouse age: 12–14 weeks
3. Mouse diet: "Western diet" (WD): 16% fat, 1.25% cholesterol, and 0.5% Na-Cholate (Harlan Teklad, Madison, USA) starting 14 days before operation.

3.1.1.2. Operation

1. Anesthesia and analgesia for mice perioperative: ketamine/xylazine mixture: 1.1 ml Xylazine (100 mg/ml stock) is added to a 10-ml vial of Ketamine (100 mg/ml stock; Webster Veterinary Supply Inc., Chicago, USA) and this cocktail mixture has a concentration of 95 mg/ml Ketamine and 5 mg/ml Xylazine, and is then diluted 1 part premix plus 3 parts sterile saline. For a 25 g mouse, we give 0.1 ml premix intraperitoneal (i.p.) injection; 0.1 ml/25 g = 0.004 ml/g.
2. Isofluorane gas: Isofluorane gas anesthetic given by mask with initial anesthetic delivered at 2% with oxygen and then titrated to efficacy at 1–3%.
3. Analgesia: At the beginning of the surgical procedure, each animal receives buprenorphine at the dosage of 0.05–0.1 mg/kg subcutaneously (SC).
4. Skin preparation: Wash and shave the surgical area with a three-stage betadine soap/alcohol/betadine topical wash.
5. Make a long median abdominal laparotomy incision ending above the pubic symphysis and displace the abdominal contents to expose the right iliac artery.
6. Dissect free the right iliac artery from the right iliac vein and ligate proximally and distally using 8-0 suture line (Fig. 15.2).

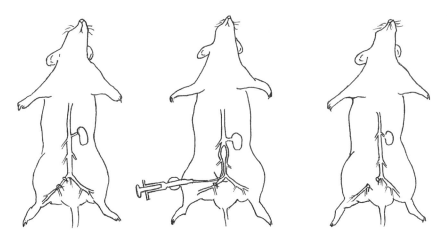

Figure 15.2 Illustration of aortic balloon angioplasty injury in mouse model of accelerated atherosclerosis and arterial aneurysmal dilatation. The right iliac artery is dissected and ligated proximally and distally (left panel). A microcatheter is inserted and advanced into the aorta and the balloon is inflated and then pulled back along the vessel (middle panel). Balloon passage is repeated twice and then the catheter is withdrawn and the iliac artery ligated (right panel).

7. Place a retraction suture on the front wall of the right iliac artery and below then make a tiny cut using microsurgical scissors.
8. Insert a microcatheter balloon (a 0.62-mm caliber polyurethane tube with a latex balloon; MED PLUS perfecseal, Inc., Oshkosh, WI) into the right common iliac artery (Figs. 15.2 and 15.3), loosen the proximal ligature, and advance the catheter retrograde into the aorta as far as the aortic arch (for thoracic vascular injury) or just below the renal branches (for abdominal vascular injury only).
9. Inflate the angioplasty balloon (2.1 mm diameter for thoracic, 1.6 mm for abdominal injury) with saline (9 µl for thoracic, 7 µl for abdominal injury), delivered via a microsyringe, and the balloon is then pulled back along the vessel. Repeat balloon passage twice and then withdraw the catheter. The iliac artery is ligated.
10. Close the inner muscle and connective tissue with sterile 3-0 absorbable Maxon suture and the dermal layers of the abdominal wall are closed with either interrupted or continuous sterile 4-0 nonabsorbable nylon suture.

3.1.1.3. Postoperative care

1. Temperature control: After surgery, animals are kept warm with a certified temperature controller, ATC 1000. Monitor animals every 5–20 min followed by every half hour intervals until stable.

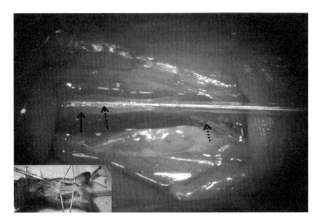

Figure 15.3 Picture of balloon angioplasty in mouse. Lower left inserted picture shows abdominal incision and surgical exposure of the aorta for angioplasty balloon insertion and injury. Mag 10×. ••••▶, represents the right iliac artery into which the balloon is inserted; ⟶▶, represents the abdominal aorta; – –▶, represents the balloon inserted into the right iliac artery and advanced into the abdominal aorta where it is inflated with saline and dragged back and forth in the aorta to stretch the aorta and to induce damage. (See Color Insert.)

2. Analgesia: Buprenophine at the dosage of 0.05–0.1 mg/kg is given every 12 h for a minimum of 48 h postoperatively or until veterinarian examination deems that analgesic administration is no longer necessary.
3. Nonabsorbable suture removal: Remove sutures at 7–10 days postsurgery.
4. Follow-up: Monitor animals daily for 4 weeks following the surgery. Any animals that show evidence of complications as a result of the procedure, such as nerve damage, hemorrhage (usually seen up to 5 days), infection, or difficulty with normal locomotion are euthanized. In some cases, if a mouse is lame temporarily on one side due to poor blood flow, the animal will be monitored for 24–48 h. At 4 weeks postoperatively, the animals are euthanized by pentobarbital sodium 150 mg/kg i.p. injection.

In follow-up analysis, the aorta and various other tissues such as heart, spleen, liver, bone marrow, and blood are harvested and isolated for histology and inflammatory/immune cell analysis. Tissue specimens, specifically aorta, are fixed in neutral buffered formalin, embedded cut, and stained using hematoxylin and eosin (H & E) or Masson's trichrome staining. Each aorta is cut into 2-3 equal length sections at measured intervals to ensure reproducible nonbiased sample analysis. Plaque area, aortic diameter, and cell invasion are measured by morphometric and immunohistochemical analysis of plaque area and cell number per microscopic high power field, as previously described.

3.1.1.4. Angioplasty injury in ApoE$^{-/-}$, but not WT (C57Bl/6) or other KO models, accelerates atheromatous plaque growth and aneurysmal dilatation Compared to the other knockout mice, the plaque area and the internal elastic lamina (IEL) diameter were dramatically increased in the ApoE$^{-/-}$ mice 4 weeks postballoon angioplasty (Figs. 15.4 and 15.5). No increase in plaque area or aneurysmal dilatation was observed in CCR2$^{-/-}$ ($N = 5$), PAI-1$^{-/-}$ ($N = 12$), Parp1$^{-/-}$ ($N = 6$), and C57BL/6 ($N = 19$) mice at 4 weeks follow-up after angioplasty injury (Fig. 15.5). All told, this indicates that PAI-1, CCR2, and Parp1 do not alter the development of atherosclerotic plaque or aneurysm formation in this balloon angioplasty model. ApoE deficiency plays an important role in the development of atherosclerotic plaque or aneurysm formation after balloon angioplasty in this mouse model.

3.2. Analysis of cellular mechanisms of anti-inflammatory activity

Once an effect is confirmed with serpin treatment in the animal models, one of the first questions that arises is "what particular stage of the inflammatory response is the serpin targeting?" Flow cytometry provides insight into cell

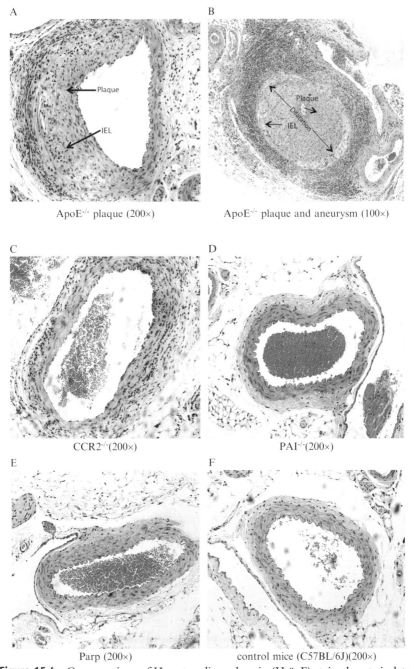

Figure 15.4 Cross-sections of Hematoxylin and eosin (H & E) stained aorta isolated from mice 4 weeks after balloon angioplasty injury. Large areas of intimal plaque growth (arrows, A. Mag 200×) and aneurysmal dilatation (B. Mag 100×) were detected in ApoE$^{-/-}$ mice, but not in CCR2$^{-/-}$ (C. Mag 200×), PAI-1$^{-/-}$ (D. Mag 200×), Parp1$^{-/-}$ (E. Mag 200×), nor WT C57Bl/6 (F. Mag 200×) mice. A lower magnification picture is used in panel B to illustrate the marked increase in internal elastic lamina (IEL) diameter as marked by a double sided arrow demonstrating aneurysmal dilatation.

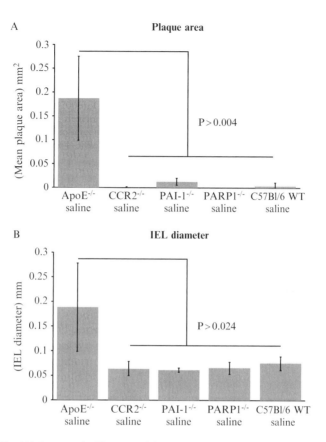

Figure 15.5 (A) Bar graphs illustrate differing plaque size with each mouse model. ApoE$^{-/-}$ mice with balloon injury have significantly larger plaque area ($P < 0.004$) than the other KO (knock out) mouse models or WT mice. (B) Bar graphs demonstrate increased mean internal elastic lamina (IEL) diameter in ApoE$^{-/-}$ mice when compared to WT and KO mouse models after angioplasty injury, evaluated at 4 weeks postangioplasty.

responses through an analysis of changes in representative populations, both in response to ongoing vascular disease and serpin modulation of the disease state. We assess cells isolated from the bone marrow, circulating blood, spleen, and lymph nodes for changes in monocytes, dendritic cells (DCs), T cells, B cells, and stem cells. In prior work, we have focused on pro- and anti-inflammatory monocytes (MC1 and MC2, respectively) and pro- and anti-inflammatory T lymphocytes (Th1, Th17, and Th2, Treg, respectively, in a very broad sense). In current work, we expand our analyses to incorporate a broader range of potential cell targets. With eight-color flow cytometry, we can analyze a set of cell types using unique antibodies that allow for a clear separation of cell populations. In the following sections, we will describe cell isolation and labeling for flow cytometry as well as subsequent analysis and differentiation of cell types (Shapiro, 1995; Wang et al., 2007).

3.2.1. Flow cytometric immunophenotyping of splenocytes

As a part of lymphatic system, the spleen hosts a large quantity of monocytes, T cells, B cells, and other leukocytes. Immunophenotyping of these cells sheds light on how the immune system responds to selected diseases as well as to serpin treatment. Prior work has demonstrated that the viral serpin, Serp-1, and the mammalian serpin, neuroserpin, have effects on a variety of cell types in inflammatory vascular transplant disease models (Hausen et al., 2001; Miller et al., 2000; Munuswamy-Ramanujam et al., 2010; Viswanathan et al., 2009). To extend our understanding of the disease pathogenesis and serpin-mediated regulation of innate and acquired immune responses, we have expanded our analyses of cell responses to incorporate more cell types. In our current studies in mouse models of vascular diseases, we examine monocyte, B cell, T cell subgroups, natural killer, dendritic, and progenitor cells using flow cytometry. The cell types and the markers we have chosen for the cells are summarized in Table 15.1. Using this approach, we are able to determine the timing and direction of change in selected cell types in the immune response to disease.

In our experiments, antibodies conjugated with different fluorochromes are accommodated in an antibody mixture to allow simultaneous labeling of multiple cell types in the sample. Due to the spectral overlap between different fluorochromes, it is necessary to adjust the amount of specific fluorescence emission by subtracting the "spillover" (overlap) component from the selected emission wavelength, a process called compensation. To perform these assays, a series of samples single-stained with each fluorochrome are utilized as controls (for more detailed discussion of compensation methodology, please refer to Shapiro, 1995). To rule out nonspecific staining, we also include isotype control staining using antibodies that have no specificity for the markers of interest in order to assess for nonspecific, irrelevant staining.

3.2.1.1. Preparation of single splenocyte suspension

1. Place mouse spleen on a 70-μm nylon mesh in a Petri dish with 5 ml of RPMI media with 10% FBS and 1% penicillin and streptomycin; mash the spleen with the rubber end of a plastic syringe plunger to release the cells.
2. Discard nylon mesh and the cell clumps; transfer the cell suspension in the Petri dish to a centrifuge tube.
3. Centrifuge cells at $400 \times g$ for 5 min and discard the supernatant.

Table 15.1 Immune system cell types and corresponding fluorochrome-labeled antibodies

Cell types	Marker
Cytotoxic T cell	Anti-CD3-PerCP-Cy5.5; Anti-CD8-APC-eFluor780
T helper cell	Anti-CD3-PerCP-Cy5.5; Anti-CD4-PE-Cy7
Th1 cell	Anti-IFNγ-FITC
Th2 cell	Anti-IL4-PE
Th17 cell	Anti-IL17a-AF647
Treg cell	Anti-FoxP3-eFluor450
B cell	Anti-CD19-Cy7
Hematopoietic stem cell	Anti-CD34-PerCP-Cy5.5
NK cell	Anti-NK1.1-eFluor450
Monocyte	Anti-CD11c-APC-eFluor780
Dendritic cell (Mature)	Anti-CD83-PE
Dendritic cell (Immature)	Anti-CD206-FITC
Memory T cell	Anti-CCR6-APC

4. To lyse erythrocytes, 5 ml of 0.84% ammonium chloride is added to resuspend the cells and then incubated at room temperature for 10 min.
5. Centrifuge at $400 \times g$ for 5 min, discard the supernatant.
6. Wash the cell pellet twice by resuspending cells in phosphate-buffered saline (PBS) and centrifuging at $400 \times g$ for 5 min; discard the supernatant.
7. Suspend cells in PBS at a concentration of 5×10^6 cells/ml.

3.2.1.2. Labeling of cells with fluorescent-conjugated antibodies

1. Transfer cells to 96-well plate total volume of 100 µl/well.
2. Centrifuge at $400 \times g$ for 5 min, discard the supernatant.
3. Antibodies to surface or intracellular antigens are diluted to 2 ng/µl in antibody dilution buffer (PBS + 2% fetal bovine serum (FBS)). Add 25 µl of surface antibody mix to each well; incubate for 30 min in the dark at room temperature.
4. Add 150 µl of PBS to each well, centrifuge at $400 \times g$ for 5 min; discard the supernatant.
5. If no intracellular antigen staining is planned, resuspend labeled cells in each well with 150 µl of PBS and transfer them to individual plastic tube, proceed to flow cytometry analysis, or add equal volume of 4% paraformaldehyde and store for later analysis.

6. If staining of intracellular antigens is desired, resuspend cell pellet from step 4 in 500 μl of fixation/permeabilization buffer (eBioscience, San Diego, CA) and transfer the cell suspension to a polystyrene tube, incubate in dark for 45 min at 4 °C.
7. Add 500 μl of 1× permeabilization buffer (eBioscience), centrifuge at $400 \times g$ for 5 min at 4 °C, and discard the supernatant. Repeat one more time.
8. Add 25 μl of intracellular antibody mix, shake gently, and incubate in dark for 30 min at 4 °C.
9. Add 500 μl of PBS to each tube, vortex gently, centrifuge at $400 \times g$ for 5 min at 4 °C, and discard the supernatant.
10. Resuspend labeled cells in 150 μl of PBS, proceed to flow cytometry analysis, or add same volume of 4% paraformaldehyde and store for later analysis.

3.2.1.3. Flow cytometry analysis We performed flow cytometry using a CyAn ADP Analyzer (Dako, Ft Collins, CO). The data were analyzed with Gatelogic (eBioscience). The percentage of a specific cell population is illustrated with a two-dimensional histogram. Figure 15.6 is a sample histogram that provides the percentage of T helper cell population (CD3+, CD4+) in a splenocyte sample.

3.2.2. Detection of serpin inhibitory activity for human cells, *in vitro*: Analysis of adhesive activity of human leukocytes

While the mouse models provide an initial analysis of serpin immune modulating activity in a disease state and also an assessment of cells targeted in the mouse model, as has been noted frequently, mice and men differ significantly (Kennedy *et al.*, 2007; Munuswamy-Ramanujam *et al.*, 2010). We thus analyze serpin effects on human cells, *in vitro*, using cells isolated from patients with aneurysms and other vascular diseases, as well as assessing effects of serpins on individual human endothelial cell, monocyte, and T cell lines. As a preliminary screen for effects on human cell activation, we use cell adhesion assays with and without activators and serpin treatment. In our prior studies, both peripheral blood mononuclear cells (PBMCs) and cultured cell lines were used for this purpose. (Viswanathan *et al.*, 2006, 2009; Zalai *et al.*, 2001).

3.2.2.1. Isolation of PBMCs

1. Isolate PBMCs by density gradient centrifugation on Ficoll-Paque PLUS (GE Healthcare, Piscataway, NJ).
2. 10 ml of heparinized blood is centrifuged at $400 \times g$ for 8 min.
3. Remove the serum layer on the top and resuspend the cells with an equal volume (∼5 ml) of PBS +2% FBS.

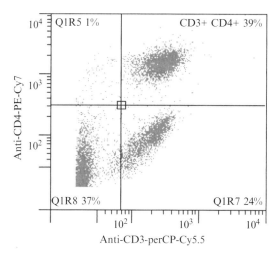

Figure 15.6 Fluorescence flow cytometric assay of mouse spleen cell isolates with dot plot displaying CD3-PerCP-Cy5.5 on the x-axis and CD4-PE-Cy7 on the y-axis.

4. Aliquot Ficoll-Paque PLUS into a 50-ml tube. The volume of Ficoll-Paque PLUS is equal to the cell suspension.
5. Carefully overlay the Ficoll-Paque PLUS with cell suspension, minimizing the interruption of cell to Ficoll interface.
6. Centrifuge at $400 \times g$ for 30 min at room temperature and wait for the rotor to stop without applying the brake.
7. PBMCs form a band between the upper buffer layer and the lower Ficoll layer. Carefully retrieve the PBMCs and transfer to a new centrifuge tube with a pipette, avoiding contamination by the buffer layer or Ficoll layer.
8. Wash the PBMCs twice with 10 ml of PBS and centrifuge at $400 \times g$ for 5 min. Resuspend the cell with proper media for further use.

3.2.2.2. In vitro *adhesion assay after serpin treatment*

1. Coat black 96-well plate with 100 µl of 50 ng/µl of fibronectin (33010–018, Invitrogen, Carlsbad, CA) per well.
2. After incubation at 37 °C for 1 h, discard the content of each well.
3. Wash each well twice with cold PBS.
4. Resuspend PBMCs at a concentration of 5×0^6 cells/ml in RPMI media without serum.
5. Label PBMCs with calcein acetoxymethyl (C1430, Invitrogen, Carlsbad, CA); add 5 µl of 1 mg/ml calcein acetoxymethyl per milliliter of cells, incubate for 1 h at 37 °C in the dark.
6. Wash cells with RPMI media and centrifuge at $400 \times g$ for 5 min; resuspend cells in RPMI media at 5×10^6 cells/ml.

7. To activate PBMCs, add Phorbol 12-myristate 13-acetate (PMA) to a final concentration of 10 μg/ml. At the same time, serpin can be added to treat the PBMCs at desired dose.
8. Transfer 100 μl of PBMCs to each fibronectin-coated well for cell assay repeated in triplicate. Set up standard curve by serial dilution of the calcein-labeled cells.
9. Incubate at 37 °C for 1 h.
10. Remove unattached cells by gentle washing 4× with RPMI media.
11. Add 100 μl of PBS to each well and measure the fluorescence at 527 nm emission during 485nm excitation.

3.3. Detection analysis of altered gene expression in cells isolated after arterial injury with and without serpin treatment

3.3.1. PCR arrays

In order to identify a cohort of genes involved in Serp-2's purported antiapoptotic effects and differentiate Serp-2 effects from the effects of antithrombotic serpin Serp-1 and chemokine modulating protein M-T7, three human cell lines were studied from the American Type Culture Collection (ATCC); Human umbilical endothelial vein cells (HUVECs, ATCC CRL-1730, passages 2–5), THP-1 monocytes (ATCC TIB-202), and Jurkat T cells (E6.1 clone, ATCC TIB-152). Prior studies have demonstrated a difference in apoptosis between Serp-2 and CrmA effectiveness in apoptosis induced by camptothecin (CPT), a topoisomerase I inhibitor, but not for any other apoptosis inducing agent.

3.3.1.1. RNA isolation

1. Treat cells with saline, 10 μM CPT in DMSO (Sigma), or CPT and 500 ng Serp-1, Serp-2, or M-T7 per million cells for either 30 min or 4.5 h. CPT and protein treatments are concurrent and continue until cell lysis.
2. Lyse cells and purify RNA using the Qiagen RNeasy kit (Valencia, CA).
3. Measure RNA concentration using a NanoDrop spectrofluorometer (Nanodrop, Wilmington, DE), and synthesize 500 ng of RNA into cDNA by SABioscience's First Strand kit (Frederick, MD).
4. Mix cDNA with SABioscience's premixed SYBR green/ROX fluor and PCR mix, and apply to the premanufactured plate containing apoptosis gene-specific primers in the bottom of each well.
5. Run plates on an ABI 7300 Real-time PCR Machine (Applied Biosystems, Foster City, CA) according to the recommended protocol—Stage 1: 10 min at 95 °C; Stage 2: 15 s at 95 °C, 1 min at 60 °C, 40 cycles; Stage 3: 15 s at 95 °C, 1 min at 60 °C, 15 s at 95 °C. Collect data points at Stage 2, Step 2 for each cycle.

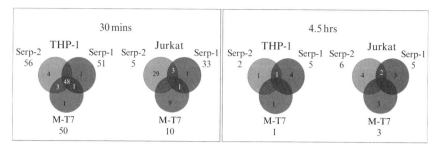

Figure 15.7 Dramatic differences between the number of genes regulated by all treatments in THP-1 monocytes at 30 min and Jurkat T cells, then again with the total number of genes regulated at 4.5 h. A set of apparently shared target genes is detectable in THP-1 cells at 30 min after treatment with each of the viral proteins.

6. Analyze gene expression fold changes using the ABI software. Changes deemed significant if they score $P = 0.005$ or lower on ANOVA analysis with Fisher's protected least significant difference (PSLD) *post hoc* testing provided by Statview statistics program (SAS Institute, Cary, NC).

For later studies, primers were manufactured to replicate the premanufactured plates, thus increasing the cost effectiveness of the technique. To increase the speed of the PCR array plate assembly, we recommend making a "master" primer plate, which contains ~ 200 µl of each primer pair in the correct well location, then use a multichannel pipette to transfer 5 µl of each primer pair into a fresh plate that will be used for sample analysis.

Early studies reveal that serpin and M-T7 treatments had the greatest effect in THP-1 monocytes at 30 min treatment (Fig. 15.7). Jurkat and HUVEC cell lines at both time points and THP-1 at 4.5 h demonstrated regulation of only a few concurrent genes (HUVEC not shown). This data underscores the significance of viral proteins in early regulation of monocyte responses to insult, particularly in regulation of host cell apoptosis.

4. Assessing Clinical Therapeutic Potential for a Viral Serpin in Clinical Trial: Trial of Serp-1 Treatment in Patients with Acute Coronary Syndrome and Stent Implant

4.1. Rationale

In order to assess the efficacy of a new therapeutic, in this case the viral serpin as a new class of immunotherapeutic, the ultimate analysis is a clinical trial. While clinical trial assessment of this new class of virus-derived serpins has been reported recently and is beyond the scope of this review, we would like to briefly present an overview of the steps used to complete the clinical trial of Serp-1 in acute coronary syndromes (ACS). The usual process from initial inception and testing of a new therapeutic and clinical trial (early stages) requires 15–20 years, on average. We provide here a brief outline of the steps required for clinical trial of a new class of immunotherapeutic protein and then provide some limited data on the findings in the Serp-1 Phase 2A clinical trial in patients with ACS and stent implant. This trial demonstrates safety, efficacy in reducing markers of myocardial damage, and trends toward modulation of inflammation.

4.2. Stages for proceeding into clinic

Taking a new class of drug into clinical testing in humans requires multiple preparatory steps in order to safely test a new protein immunotherapeutic. An outline of the requisite steps is provided below. While this outline provides an overview toward an approach to clinical testing, this is neither a detailed nor a comprehensive review of the approaches to be used to take this serpin protein into clinic.

1. Good lab practice (GLP) testin—Test potential immunomodulatory protein for toxicity prior to testing in man. Toxicity testing of the purified protein therapeutic is performed under highly controlled and reproducible conditions assessing for adverse effects, in a facility not connected to the initial research lab making the discovery.
2. Expression system for good manufacturing protocol (GMP) production—a production system for expression and purification of the final protein to be used for clinical testing is developed and established. In the case of the serpins tested, we use a cell line approved for protein expression by the Food and Drug Administration (FDA). The serpin gene is inserted into a CHO cell line and expressed from this cell line as a secreted protein. Alternative expression systems can include either *Escherichia coli* or Baculoviral expression vectors. Proteins are tested for purity, contaminants, and activity. A serpin activity assay is established to measure serpin inhibitory activity both during the isolation and purification steps and in blood samples for testing during clinical testing.
3. GMP product development—expression and purification of protein is then performed at a site experienced in expressing and purifying clinically active and safe drug.
4. Final GMP product is again tested for safety and toxicity, as well as half-life, in animal safety trials. Toxicity testing involves both broad spectrum analysis of potential adverse effects on test animal health, organ damage (liver, lung, spleen, etc.), and also selective testing for toxicity related to the disease and/or organ system designated for clinical testing. In some cases, this can require primate testing at an experienced lab.

5. FDA approval for Phase 1 safety trial—FDA approval is required for initial testing in man. This requires review by the FDA of all efficacy and also toxicity data as well as the protein therapeutic and its manufacture. A Phase 1 safety trial is designed in consultation with a clinical trials group and the FDA. We performed our initial safety trial using a group that specializes in Phase 1 trials, GFI (Go For It) in Indiana. The initial trial is monitored by the FDA, and progress to a true clinical trial is assessed and potentially approved by the FDA if the Phase 1 testing demonstrated safety.
6. FDA approval of a Phase 2A trial in patients with the targeted disease—Safety, with and without efficacy, of the serpin therapeutic is assessed in patients with clinical disease. In the case of Serp-1, a randomized, dose escalating, double blind trial was designed and performed and managed through a clinical regulatory organization as well as physicians designated as the site leaders for each hospital involved. A data safety monitoring board (DSMB) is established to monitor for any potential adverse effects of the drug tested.
7. Efficacy testing—This Serp-1 Phase 2A trial design allowed for both measurement of the circulating serpin half-life ($t_{1/2}$), monitoring for any changes in serum markers of organ dysfunction (liver and renal and cardiac as well as hematological markers), and also collection of serum markers for effects of serpin treatment on markers of inflammation.

4.3. Clinical trial findings

Serp-1 was tested as a potential anti-inflammatory therapeutic for reduction of inflammation in patients with unstable coronary syndromes, so-called acute coronary syndromes, which include unstable angina and non-ST elevation MI. The details of this trial have been recently published. Forty-eight patients with coronary lesions appropriate for stent implant were enrolled, after obtaining informed consent, and randomized in a 3:1 pattern for either Serp-1 or saline control treatments; Serp-1 at two doses, with dose escalation. Coronary stenosis (narrowing) appropriate for stent implant as well as efficacy of stent implant was assessed by contrast angiography and intravascular ultrasound (IVUS). With this Phase 2A, clinical trial Serp-1 treatment was given for three consecutive days starting immediately after stent implant and then as an intravenous bolus every 24 h.

Of interest, this trial demonstrated excellent safety for serpin treatment when used in this clinical population with unstable coronary plaque. Major adverse cardiovascular events (MACE) were low and no significant adverse effects were demonstrated (MACE—control 2 of 12 patients with placebo, 5 of 19 in low-dose 5 μg/kg×3 days Serp-1 treatment group, and 0 of 17 patients with high-dose 15 μg/kg×3 days Serp-1 ($P = 0.058$)). In analysis

Table 15.2 Inflammatory and Thrombotic markers in patients separated into those treated with statins and those without stating treatment prior to stent implant

Treatment/ biomarker	0 h All	0 h No statin	0 h Statin	8 h All	8 h No statin	8 h Statin	16 h All	16 h No statin	16 h Statin	24 h All	24 h No statin
Pl/MCP-1	28.18 + 2.46	25.41 + 4.18	30.16 + 3.04	36.29 + 3.53	29.48 + 5.21	41.97 + 3.68	37.54 + 6.08	29.63 + 9.09	42.06 + 8.00	33.24 + 3.12	26.94 + 7.14
S-1 5 µg/ MCP-1	28.03 + 4.40	27.85 + 6.48	28.17 + 6.29	30.28 + 3.38	33.63 + 6.31	27.85 + 3.71★	29.23 + 2.73	30.53 + 4.46	28.28 + 3.58★	28.45 + 2.53	33.20 + 4.90
S-1 15 µg/ MCP-1	28.82 + 4.29	26.75 + 6.57	26.87 + 5.89	31.05 + 3.75	35.35 + 7.28	28.71 + 4.31★★	34.92 + 4.40	31.85 + 6.13	36.77 + 6.18	26.68 + 3.08	24.94 + 4.21
Pl/MPO	85.96 + 23.99	71.46 + 29.26	96.32 + 38.86	19.51 + 3.77	15.22 + 1.86	23.08 + 6.65	28.76 + 9.43	19.07 + 3.75	34.30 + 14.66	24.61 + 2.95	34.66 + 1.65
S-1 5 µg/ MPO	68.63 + 15.1	61.68 + 26.91	74.20 + 17.79	25.16 + 4.21	29.05 + 8.08	22.33 + 4.46	31.73 + 5.19	31.27 + 9.36	32.07 + 6.26	47.74 + 13.12	67.04 + 27.4
S-1 15 µg/ MPO	90.29 + 21.25	90.78 + 32.02	90.00 + 29.37	40.34 + 22.69	88.25 + 62.79★	14.21 + 2.18	29.59 + 6.43	36.66 + 9.48	25.35 + 8.67	38.31 + 12.20	46.78 + 29.0
Pl/DD	485.3 + 267.4	179.6 + 48.8	703.7 + 452.3	664.4 + 320.0	263.4 + 86.1	989.6 + 566.9	857.5 + 239.5	735.3 + 415.6	927.4 + 314.1	939.2 + 288	611.9 + 256
S-1 5 µg/ DD	223.6 + 43.3	177.4 + 37.8	260.5 + 71.7★	373..5 + 74.9	456.2 + 139.4	313.3 + 81.8★	532.4 + 100.5	629.1 + 206.9	462.0 + 91.7★★	538.7 + 114.1	705.1 + 261
S-1 15 µg/ DD	202.7 + 51.2	202.4 + 53.5	202.9 + 77.7★★	433.7 + 91.8	577.4 + 186.6	355.4 + 97.9★	406.7 + 103★	450.9 + 189.8	380.1 + 126.3★★	601.0 + 184.3	957.7 + 446

Variable reductions in circulating markers of inflammation are detectable at follow up
★ P ≤ 0.05 comparison Serp-1 treatment to Placebo by ANOVA with Fisher's PLSD;
★★ P ≤ 0.10 comparison Serp-1 treatment to Placebo by ANOVA with Fisher's PLSD.
All values expressed as mean + SE, Pl – Placebo; S-1 – Serp-1; DD - D dimer, MCP-1 - monocyte chemoattractant protein - 1, MPO – myeloperoxidase.

Table 15.3 Inflammatory and Thrombotic markers in patients separated into those treated with Bare Metal Stent (BMS) and those with Drug Eluting Stent (DES) implants

Treatment/ biomarker	0 h All	0 h BMS	0 h DES	8 h All	8 h BMS	8 h DES	16 h All	16 h BMS	16 h DES	24 h All	24 h BMS	24 h DES
Pl/MCP-1	28.18 + 2.46	26.68 + 3.72	30.27 + 3.04	36.29 + 3.53	35.06 + 5.39	38.45 + 5.92	37.54 + 6.08	30.22 + 5.78	46.33 + 10.87	33.24 + 3.12	30.39 + 3.33	35.24 + 8.74
S-1 5 µg/ MCP-1	28.03 + 4.40	28.61 + 6.44	26.88 + 3.79	30.28 + 3.38	27.72 + 4.06	35.82 + 5.92	29.23 + 2.73	31.58 + 3.58	24.13 + 3.24★	28.45 + 2.53	30.69 + 3.22	23.60 + 3.52★
S-1 15 µg/ MCP-1	28.82 + 4.29	23.97 + 4.68	39.18 + 8.48	31.05 + 3.75	31.25 + 4.71	30.39 + 5.27	34.92 + 4.40	33.93 + 5.31	37.91 + 8.61	26.68 + 3.08	25.73 + 3.87	29.50 + 4.67
Pl/CRP	3.49 + 0.76	2.86 + 0.67N		4.36 + 1.59	3.85 + 0.76	2.81 + 0.57	5.68 + 1.54	3.92 + 0.95	3.43 + 1.36	4.51 + 1.44	3.65 + 0.82	2.82 + 0.50	5.33 + 2.42
S-1 5 µg/ CRP	3.15 + 0.80	3.45 + 0.96	2.53 + 1.52	3.11 + 0.69	2.86 + 0.73	3.64 + 1.59★		3.18 + 0.68	3.04 + 0.77	3.48 + 1.49	3.78 + 0.75	4.23 + 0.99	2.81 + 1.04
S-1 15 µg/ CRP	2.76 + 0.70	3.21 + 0.81	0.79 + 0.36	2.25 + 0.63	2.65 + 0.79	0.92 + 0.29	3.75 + 1.05	4.18 + 1.25	2.47 + 2.05	2.65 + 0.74	3.04 + 0.94	1.50 + 0.92★★
Pl/DD	485.3 + 267.4	241.4 + 37.7	826.8 + 647.3	664.4 + 320.0	447.5 + 142	1044 + 890.3	857.5 + 239.5	871.0 + 259	841.4 + 463.8	939.2 + 288.0	693.9 + 138	1446 + 944
S-1 5 µg/ DD	223.6 + 43.3	217.2 + 49.9	236.3 + 89.97	373..5 + 74.9	328.1 + 84.4	471.8 + 155.2	532.4 + 100.5	531.8 + 125	533.6 + 183.7	538.7 + 114.1★★	474.4 + 117	656.5 + 252.3
S-1 15 µg/ DD	202.7 + 51.2	203.3 + 611.1	200.3 + 88.33	433.7 + 91.8	489.0 + 74.9	254.2 + 62.08	406.7 + 103★	461.5 + 134.2	242.2 + 34.01	601.0 + 184.3	720.7 + 238	241.7 + 31.9★

Variable reductions in circulating markers of inflammation are detectable at follow up.
★P ≤ 0.05 comparison Serp-1 treatment to Placebo by ANOVA with Fisher's PLSD;
★★ P = 0.10 comparison Serp-1 treatment to Placebo by ANOVA with Fisher's PLSD.
All values expressed as mean + SE, Pl - Placebo; S-1 – Serp-1; DD - D dimmer, CRP - C reactive protein, MCP-1 - monocyte chemoattractant protein -1.

Table 1

	48 h			54 h			336 h			612 h		
Statin	All	No statin	Statin	All	No statin	Statin	All	No statin	Statin	All	No statin	Statin
35.94 + 3.08	32.06 + 5.57	31.21 + 9.65	32.91 + 7.15	34.56 + 3.31	39.83 + 6.28	31.55 + 3.65	39.49 + 5.64	46.54 + 15.38	35.46 + 2.88	40.86 + 7.35	52.19 + 18.53	34.38 + 4.91
25.00 + 2.22*	23.92 + 2.71	25.76 + 3.05	22.59 + 2.12	26.29 + 2.26	25.46 + 5.10	26.89 + 1.61	31.14 + 3.22	33.76 + 7.28	29.23 + 2.07	30.32 + 3.16	29.80 + 6.07*	30.74 + 3.32
27.72 + 4.37	24.28 + 3.50	21.68 + 4.25	25.58 + 4.90	25.69 + 3.80	20.72 + 3.92**	28.44 + 5.44	27.08 + 3.31*	28.22 + 3.55**	26.45 + 4.91	33.15 + 4.04	33.59 + 6.86	32.98 + 5.17
20.30 + 2.81	17.60 + 3.40	22.70 + 5.84	12.50 + 1.54	24.59 + 3.83	30.92 + 6.92	20.97 + 4.33	23.18 + 4.18	25.65 + 3.94	21.77 + 6.36	22.54 + 3.62	31.88 + 5.42	17.21 + 3.58
33.70 + 10.21	26.25 + 5.24	30.47 + 11.72	23.19 + 3.64	24.28 + 3.55	25.81 + 4.95	23.16 + 5.14	30.96 + 9.37	18.93 + 2.28	39.72 + 15.87	20.63 + 3.50	18.01 + 2.06	22.74 + 6.15
33.23 + 10.39	22.24 + 3.33	19.71 + 3.05	23.51 + 4.83	25.36 + 6.90	17.29 + 1.79	29.84 + 10.60	16.70 + 1.84	19.66 + 2.68	15.05 + 2.36	36.43 + 9.70	37.84 + 20.83	35.87 + 11.50
1080 + 397	785.4 + 158	825.4 + 286	745.4 + 181	1125 + 251	797.1 + 336	1312 + 340.6	965.0 + 351.8	315.1 + 73.9	1336 + 509.2	645.4 + 201.5	213.8 + 47.8	892.0 + 278.9
422.1 + 62.5*	698.0 + 136	554.5 + 173	802.3 + 199	524.1 + 69.6*	540.8 + 140	511.9 + 71.3*	405.4 + 61.8*	441.3 + 89.2	379.3 + 87.5*	281.3 + 53.8**	246.9 + 45.2	308.9 + 91.6
387.0 + 103*	583.1 + 109	749.5 + 233	500.0 + 116	438.5 + 72.8*	505.5 + 72.8*	401.2 + 85.1*	283.7 + 55.5*	410.5 + 115	213.3 + 49.0*	216.6 + 36.8*	288.0 + 96.6	184.9 + 31.2*

Table 2

	48 h			54 h			336 h			612 h		
	All	BMS	DES	All	BMS	DES	All	BMS	DES	All	BMS	DES
	32.06 + 5.57	33.34 + 8.55	29.92 + 6.49	34.56 + 3.31	36.75 + 4.39	31.945.32	39.49 + 5.64	39.23 + 8.93	39.93 + 3.49	40.86 + 7.35	39.19 + 10.87	43.77 + 8.79
	23.92 + 2.71	26.01 + 3.74	19.41 + 2.29	26.29 + 2.26	26.58 + 3.26	25.64 + 1.79	31.14 + 3.22	34.37 + 4.45	24.12 + 0.89	30.32 + 3.16	29.12 + 4.04	32.71 + 5.33
	24.28 + 3.50	23.58 + 4.23	27.07 + 5.43	25.69 + 3.80	25.68 + 4.45	25.71 + 2.81	27.08 + 3.31*	28.35 + 4.17	22.42 + 1.03	33.15 + 4.04**	33.33 + 5.08	32.51 + 4.71
	3.51 + 1.02	2.11 + 0.53	5.84 + 2.11	3.95 + 1.09	2.64 + 1.16	5.52 + 1.83	2.29 + 0.62	1.99 + 0.75	2.80 + 1.18	1.11 + 0.19	0.82 + 0.05	1.63 + 0.42
	4.52 + 0.79	4.45 + 0.94**	4.68 + 1.55*	4.41 + 0.83	4.39 + 0.95	4.45 + 1.77	2.07 + 0.53	2.18 + 0.73	1.84 + 0.69	0.97 + 0.15	1.04 + 0.02	0.83 + 0.23
	2.53 + 0.49	2.76 + 0.60	1.60 + 0.33**	2.90 + 0.62	3.09 + 0.71	1.80 + 0.01	1.78 + 0.43	1.88 + 0.75	1.39 + 0.53	1.08 + 0.21	1.17 + 0.25	0.77 + 0.37
	785.4 + 157.5	756.8 + 243	833.1 + 183	1125 + 251	713.7 + 228	1618 + 394.5	965.0 + 351.8	824.6 + 490	1211 + 512.9	645.4 + 201.5	348.7 + 124	1165 + 420.7
	698.0 + 136.0	725.0 + 179	639.4 + 207	524.1 + 69.6	557.4 + 91*	451.8 + 103.4*	405.4 + 61.8*	444.5 + 76.9	320.7 + 103.4**	281.3 + 53.8**	297.9 + 74.1	248.2 + 70.9**
	583.1 + 109.3	557.8 + 118	684.4 + 322.5	438.5 + 72.8*	451.9 + 83.6*	357.6 + 111*	283.7 + 55.5*	318.5 + 66.8*	156.3 + 39.2**	216.6 + 36.8**	241.4 + 44.4	134.1 + 34.1**

of markers, a dose- and time-dependent reduction in markers of myocardial damage, specifically in troponin I levels with Serp-1 at 8, 16, 24, and 54 h ($P < 0.05$) and in CK-MB levels at 8, 16, and 24 h postdose ($P < 0.05$), was detected. (Tardif et al., 2010).

On further analysis of inflammatory biomarkers, both trends as well as some significant changes in markers of inflammation at selected times postinfusion and PCI were also detected (Tables 15.2 and 15.3). The markers tested included PAI-1, C reactive protein (CRP), myeloperoxidase (MPO), monocyte chemoattractant protein (MCP-1), D dimmer (DD), and brain natriuretic peptide (BNP). These changes in circulating markers of inflammation were most pronounced when data was separated into patients treated with or without statins at the time of stent implant (Table 15.2) or into groups of patients receiving either bare metal stents (BMS) or drug eluting stents (DES) (Table 15.3). While these analyses did not achieve dose and time-dependent significance, as detected for markers of myocardial damage (Troponin I or CK-MB), particularly when both statins and stent types were incorporated into the analysis, this lack of significance is thought to be in part limited by the small number of patients in the trial ($N = 48$) for all treatment groups, making detection of significance more challenging.

In separating patient treatment groups into treatment with statins and serpin dose at the time of stent implant, there are apparent reductions in MCP-1 and DD beginning at 8 h follow-up in the statin-treated patients (comparison to a combined placebo group). MPO had minimal demonstrated change in expression in the non statin-treated patients, and this change is only detectable at 8 h. By later follow-up times (54 and 336 h), DD was significantly reduced in both the total groups and in patients treated with statins (Table 15.2).

Similar trends were detected for patients when separated into those receiving different stent types, BMS or DES. In these analyses, MCP-1 and DD again showed more significant changes, with CRP showing only one limited change at the 8 h follow-up time point in patients with DES. Of interest, again MCP-1 and more significantly DD were significantly reduced, DD showing greater reductions in the whole group of patients and in patients with BMS 54–336 h follow-up (Table 15.3). The changes in DD are illustrated as a time course in bar graph format in Fig. 15.8, separated by placebo, serpin treatment and dosage, and time to follow-up. DD has been correlated with inflammatory, unstable coronary syndromes (ACS), as well as with pulmonary emboli (Empana et al., 2008; Hamaad et al., 2009) and is a by-product of fibrin degradation (fibrinolysis), the pathways targeted in part by the serpin treatment.

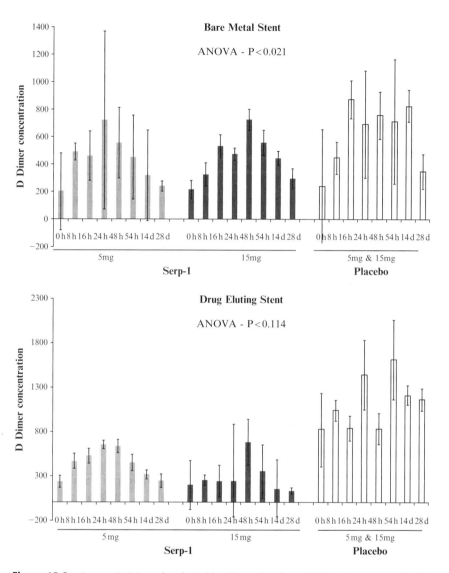

Figure 15.8 Serum D Dimer levels in blood samples from ACS patients treated with Serp-1 at 5 or 15 µg/kg or placebo after either Bare Metal Stent (BMS, A) or Drug Eluting Stent (DES, B) coronary implants after treatment with 5 or 15 mg of Serp-1 or Placebo.

REFERENCES

Arnold, K., Bordoli, L., Kopp, J., and Schwede, T. (2006). The SWISS-MODEL Workspace: A web-based environment for protein structure homology modelling. *Bioinformatics.* **22,** 195–201.

Berry, L. R., Thong, B., and Chan, A. K. (2009). Comparison of recombinant and plasma-derived antithrombin biodistribution in a rabbit model. *Thromb. Haemost.* **102,** 302–308.

Bouchecareilh, M., Conkright, J. J., and Balch, W. E. (2010). Proteostasis strategies for restoring {alpha}1-antitrypsin deficiency. *Proc. Am. Thorac. Soc.* **7,** 415–422.

Cameron★, C., Hota-Mitchell★, S., Chen, L., Barrett, J., Cao, J. X., Macaulay, C., Willer, D., Evans, D., and McFadden, G. (1999). The complete DNA sequence of myxoma virus. *Virology* **264,** 298–318. (★denotes co-authorship).

Dai, E., Guan, H., Liu, L., Little, S., McFadden, G., Vaziri, S., Cao, H., Ivanova, I. A., Bocksch, L., and Lucas, A. R. (2003). Serp-1, a viral anti-inflammatory serpin,regulates cellular serine proteinase and serpin responses to vascular injury. *J. Biol. Chem.* **278,** 18563–18572.

Daugherty, A., and Cassis, L. A. (2004). Mouse models of abdominal aortic aneurysms. *Arterioscler. Thromb. Vasc. Biol.* **24,** 429–434.

Dollery, C. M., Owen, C. A., Sukhova, G. K., Krettek, A., Shapiro, S. D., and Libby, P. (2003). Neutrophil elastase in human atherosclerotic plaques: Production by macrophages. *Circulation* **107,** 2829–2836.

Ekeowa, U. I., Freeke, J., Miranda, E., Gooptu, B., Bush, M. F., Pérez, J., Teckman, J., Robinson, C. V., and Lomas, D. A. (2010). Defining the mechanism of polymerization in the serpinopathies. *Proc. Natl. Acad. Sci. USA* **107,** 17146–17151.

Empana, J. P., Canoui-Poitrine, F., Luc, G., Juhan-Vague, I., Morange, P., Arveiler, D., Ferrieres, J., Amouyel, P., Bingham, A., Montaye, M., Ruidavets, J. B., Haas, B., et al. (2008). Contribution of novel biomarkers to incident stable angina and acute coronary syndrome: The PRIME study. *Eur. Heart. J.* **29,** 1966–1974.

Finlay, B., and McFadden, G. (2006). Anti-immunology: Evasion of the host immune system by bacterial and viral pathogens. *Cell* **124,** 767–782.

Franchi, L., Eigenbrod, T., Muñoz-Planillo, R., and Nuñez, G. (2009). The inflammasome: A caspase-1-activation platform that regulates immune responses and disease pathogenesis. *Nat. Immunol.* **10,** 241–247.

Gooptu, B., and Lomas, D. A. (2008). Polymers and inflammation: Disease mechanisms of the serpinopathies. *J. Exp. Med.* **205,** 1529–1534.

Gooptu, B., and Lomas, D. A. (2009). Conformational pathology of the serpins: Themes, variations, and therapeutic strategies. *Annu. Rev. Biochem.* **78,** 147–176.

Gramling, M. W., and Church, F. C. (2010). Plasminogen activator inhibitor-1 is an aggregate response factor with pleiotropic effects on cell signaling in vascular disease and the tumor microenvironment. *Thromb Res.* **125,** 377–381.

Hamaad, A., Sosin, M. D., Blann, A. D., Lip, G. Y., and MacFadyen, R. J. (2009). Markers of thrombosis and hemostasis in acute coronary syndromes: Relationship to increased heart rate and reduced heart-rate variability. *Clin. Cardiol.* **32,** 204–209.

Hausen, B., Boeke, K., Berry, G. J., and Morris, R. E. (2001). Viral serine proteinase inhibitor (SERP-1) effectively decreases the incidence of graft vasculopathy in heterotopic heart allografts. *Transplantation* **72,** 364–368.

Johnston, J. B., and McFadden, G. (2004). Technical knockout: Understanding poxvirus pathogenesis by selectively deleting viral immunomodulatory genes. *Cell. Microbiol.* **6,** 695–705.

Kennedy, S. A., van Diepen, A. C., van den Hurk, C. M., Coates, L. C., Lee, T. W., Ostrovsky, L. L., Miranda, E., Perez, J., Davies, M. J., Lomas, D. A., Dunbar, P. R., and Birch, N. P. (2007). Expression of the serine protease inhibitor neuroserpin in cells of the human myeloid lineage. *Thromb. Haemost.* **97,** 394–399.

Kerr, P., and McFadden, G. (2002). Immune responses to myxoma virus. *Viral Immunol.* **15**, 229–246.

Kiefer, F., Arnold, K., Künzli, M., Bordoli, L., and Schwede, T. (2009). The SWISS-MODEL Repository and associated resources. *Nucleic Acids Res.* **37**, D387–D392.

Komiyama, T., Ray, C. A., Pickup, D. J., Howard, A. D., Thornberry, N. A., and Peterson, E. P. (1994). Inhibition of interleukin-1-beta converting enzyme by the cowpox virus serpin CrmA. An example of cross-class inhibition. *J. Biol. Chem.* **269**, 19331–19337.

Kowal-Vern, A., Walenga, J. M., McGill, V., and Gamelli, R. L. (2001). The impact of antithrombin (H) concentrate infusions on pulmonary function in the acute phase of thermal injury. *Burns* **27**, 52–60.

Lomas, D. A., Evans, D. L., Upton, C., McFadden, G., and Carrell, R. W. (1993). Inhibition of plasmin, urokinase, tissue plasminogen activator and C1S by a myxoma virus serine proteinase inhibitor. *J. Biol. Chem.* **268**, 516–521.

Lucas, A., and McFadden, G. (2004). Secreted immunomodulatory proteins as novel biotherapeutics. *J. Immunology.* **173**, 4765–4774.

Lucas, A., Liu, L., Macen, J., Nash, P., Dai, E., Stewart, M., Graham, K., Etches, W., Boshkov, L., Nation, P., Humen, D., Hobman, M. Z., *et al.* (1996). A virus-encoded serine protease inhibitor, SERP-1, inhibits atherosclerotic plaque development following balloon angioplasty. *Circulation* **94**, 2890–2900.

Lucas, A., Korol, R., and Pepine, C. (2006). Inflammation in Atherosclerosis. *Circulation* **113**, 728–732.

Lucas, A., Liu, L., Dai, E., Bot, I., Viswanathan, K., Munuswamy-Ramanujam, G., Davids, J. A., Bartee, M. Y., Richardson, J., Christov, A., Wang, H., Macaulay, C., *et al.* (2009). The Serpin Saga; Development of a New Class of Virus Derived Anti-inflammatory Protein Immunotherapeutics. *In* "Pathogen-Derived Immunomodulatory Molecules," (Padraic Fallon, ed.). Landes Bioscience. Vol. 666, pp. 132–156. Austin, TX.

Macen, J., Upton, C., Nation, N., and McFadden, G. (1993). SERP-1, a serine proteinase inhibitor encoded by myxoma virus, is a secreted glycoprotein that interferes with inflammation. *Virology* **195**, 348–363.

MacNeill, A. L., Turner, P. C., and Moyer, R. W. (2006). Mutation of the Myxoma virus SERP2P1-site to prevent proteinase inhibition causes apoptosis in cultured RK-13 cells and attenuates disease in rabbits, but mutation to alter specificity causes apoptosis without reducing virulence. *Virology* **356**, 12–22.

McFadden, G. (ed.) (1995). Viroceptors, Virokines, and Related Mechanisms of Immune Modulation by DNA Viruses, R.G. Landes Co, Austin, TX.

McFadden, G. (2003). Viroceptors: Virus-encoded receptors for cytokines and hemokines. *In* "Cytokines and Chemokines in Infectious Diseases Handbook," (M. Kotb and T. Calandra, eds.), pp. 285–299. Humana Press.

McFadden, G. (2005). Poxvirus tropism. *Nat. Rev. Microbiology.* **3**, 201–213.

McFadden, G., and Murphy, P. M. (2000). Host-related immunomodulators encoded by poxviruses and herpesviruses. *Curr. Opin. Microbiol.* **3**, 371–378.

Medema, J. P., Schuurhuis, D. H., Rea, D., van Tongeren, J., de Jong, J., and Bres, S. A. (2001). Expression of the serpin serine protease inhibitor 6 protects dendritic cells from cytotoxic T lymphocyte-induced apoptosis: Differential modulation by T helper type 1 and type 2 cells. *J. Exp. Med.* **194**, 657–667.

Messud-Petit, F., Geifi, J., Delverdier, M., Amardeilh, M. F., Py, R., and Sutter, G. (1998a). Serp2, an inhibitor of the interleukin-1beta-converting enzyme, is critical in the pathobiology of myxomavirus. *J. Virol.* **72**, 7830–7839.

Messud-Petit, F., Gelfi, J., Delverdier, M., Amardeilh, M. F., Py, R., and Sutter, G. (1998b). Serp2, an inhibitor of the interleukin-1-converting enzyme, is critical in the pathobiology of myxoma virus. *J. Virol.* **72**, 7830–7839.

Miller, L. W., Dai, E., Nash, P., Liu, L., Icton, C., Klironomos, D., Fan, L., Nation, P. N., Zhong, R., McFadden, G., and Lucas, A. (2000). Inhibition of transplant vasculopathy in a rat aortic allograft model after infusion of anti-inflammatory viral serpin. *Circulation* **101,** 1598–1605.

Mordwinkin, N. M., and Louie, S. G. (2007). Aralast: An alpha 1-protease inhibitor for the treatment of alpha-antitrypsin deficiency. *Expert Opin. Pharmacother.* **8,** 2609–2614.

Munuswamy-Ramanujam, G., Khan, K. A., and Lucas, A. R. (2006). Viral anti-inflammatory reagents: The potential for treatment of arthritic and vasculitic disorders. *Endocr. Metab. Immune Disord. Drug Targets* **6,** 331–343.

Munuswamy-Ramanujam, G., Dai, E., Liu, L. Y., Shnabel, M., Sun, Y. M., Bartee, M., Lomas, D., and Lucas, A. (2010). Neuroserpin, a thrombolytic serine protease inhibitor (serpin), blocks transplant vasculopathy with associated modification of T-helper cell subsets. *Thromb. Haemost.* **103,** 545–555.

Nash, P., Whitty, A., Handwerker, J., Macen, J., and McFadden, G. (1998). Inhibitory specificity of the anti-inflammatory myxoma virus serpin, SERP-1. *J. Biol. Chem.* **273,** 20982–20991.

Nash, P., Barry, M., Seet, B. T., Veugelers, K., Hota, S., Heger, J., Hodgkinson, C., Graham, K., Jackson, R. J., and McFadden, G. (2000). Post-translational modification of the myxoma virus anti-inflammatory serpin, SERP-1 by a virally encoded sialyltransferase. *Biochem. J.* **347,** 375–382.

Nathaniel, R., MacNeill, A. L., Wang, Y. X., Turner, P. C., and Moyer, R. W. (2004). Cowpox virus CrmA, Myxoma virus SERP2 and baculovirus P35 are not functionally interchangeable caspase inhibitors in poxvirus infections. *J. Gen. Virol.* **85,** 1267–1278.

Peitsch, M. C. (1995). *Protein modeling by E-mail Bio/Technology.* **13,** 658–660.

Petit, F., Bertagnoli, S., Gelfi, J., Fassy, F., Boucraut-Baralon, C., and Milon, A. (1996). Characterization of a myxoma virus encoded serpin-like protein with activity against interleukin-1 β-converting enzyme. *J. Virol.* **70,** 5860–5866.

Petrov, L., Laurila, H., Hayry, P., and Vamvakopoulos, J. E. (2005). A mouse model of aortic angioplasty for genomic studies of neointimal hyperplasia. *J. Vasc. Res.* **42,** 292–300.

Serhan, C. N., Chiang, N., and Van Dyke, T. E. (2008a). Resolving inflammation: Dual anti- inflammatory and pro-resolution lipid mediators. *Nat. Rev. Immunol.* **8,** 349–361.

Serhan, C. N., Yacoubian, S., and Yang, R. (2008b). Anti-inflammatory and proresolving lipid mediators. *Annu. Rev. Pathol.* **3,** 279–312.

Shapiro, H. M. (1995). Practical Flow Cytometry, 3rd edn., p. 164. Wiley-Liss, New York.

Silverman, G. A., Whisstock, J. C., Bollomley, S. P., Huntington, J. A., Kaiserman, D., Luke, C. J., Pak, S. C., Reichart, J. M., and Bird, P. I. (2010). Serpins flex their muscle I. Putting the clamps on proteolysis in diverse biological systems. *J. Biol. Chem.* **285,** 24299–24305.

Stanford, M., Werden, S., and McFadden, G. (2007). Myxoma virus in the European rabbit: Interactions between the virus and its susceptible host. *Vet. Res.* **38,** 299–318.

Tardif, J. C., L'Allier, P., Grégoire, J., Ibrahim, R., McFadden, G., Kostuk, W., Knudtson, M., Labinaz, M., Waksman, R., Pepine, C. J., Macaulay, C., Guertin, M. C., *et al.* (2010). A phase 2, double-blind, placebo-controlled trial of a viral Serpin (Serine Protease Inhibitor), VT-111, in patients with acute coronary syndrome and stent implant. *Circ. Cardiovasc. Intervent.* **3,** 543–548.

Turner, P. C., and Moyer, R. W. (2001). Serpins enable poxviruses to evade immune defenses. *Am. Soc. Microbiol. News.* **67,** 201–209.

Turner, P. C., Sancho, M. C., Thoennes, S. R., Caputo, A., Bleackley, R. C., and Moyer, R. W. (1999). Myxoma virus Serp2 is a weak inhibitor of granzyme B and interleukin- 1-converting enzyme in vitro and unlike CrmA cannot block apoptosis in cowpox virus-infected cells. *J. Virol.* **73,** 6394–6404.

Upton, C., Macen, J. L., Wishart, D. S., and McFadden, G. (1990). Myxoma virus and malignant rabbit fibroma virus encode a serpin-like protein important for virus virulence. *Virology* **179,** 618–631.

Viswanathan, K., Liu, L., Vaziri, S., Richardson, J., Togonu-Bickersteth, B., Vatsya, P., Christov, A., and Lucas, A. R. (2006). Myxoma viral serpin, Serp-1, a unique interceptor of coagulation and innate immune pathways. *Thromb. Haemost.* **95,** 499–510.

Viswanathan, K., Richardson, J., Bickersteth, B., Dai, E., Liu, L., Vatsya, P., Sun, Y., Yu, J., Ramunujam, G., Baker, H., and Lucas, A. R. (2009). Myxoma viral serpin, Serp-1. Inhibits human monocyte activation through regulation of Actin binding protein Filamin. *B. J. Leukoc. Biol.* **85,** 418–426.

Wang, F., Roberts, S. M., Butfiloski, E. J., Morel, L., and Sobel, E. S. (2007). Acceleration of autoimmunity by organochlorine pesticides: A comparison of splenic B-cell effects of chlordecone and estradiol in (NZBxNZW)F1 mice. *Toxicol. Sci.* **99,** 141–152.

Whisstock, J. C., Silverman, G. A., Bird, P. I., Bottomley, S. P., Kaiserman, D., Luke, C. J., Pak, S. C., Jean-Marc Reichhart, J. M., and Huntington, J. A. (2010). Serpins flex their muscle II. Structural insights into target peptidase recognition, polymerization, and transport functions. *J. Biol. Chem.* **285,** 24307–24312.

Zalai, C. V., Kolodziejczyk, M. D., Pilarski, L., Christov, A., Nation, P. N., Lundstrom-Hobman, M., Tymchak, W., Dzavik, V., Humen, D. P., Kostuk, W. J., Jablonsky, G., Pflugfelder, P. W., *et al.* (2001). Increased circulating monocyte activation in patients with unstable coronary syndromes. *J. Am. Coll. Cardiol.* **38,** 1340–1347.

CHAPTER SIXTEEN

HUMAN SCCA SERPINS INHIBIT STAPHYLOCOCCAL CYSTEINE PROTEASES BY FORMING CLASSIC "SERPIN-LIKE" COVALENT COMPLEXES

Tomasz Kantyka* and Jan Potempa*,†

Contents

1. Introduction — 332
2. Purification of Staphopains — 333
3. Purification of GST–SCCA1 and GST–SCCA2 Fusion Proteins — 333
4. Characterization of Inhibition — 334
   - 4.1. General inhibition assay — 334
   - 4.2. Determination of the stoichiometry of inhibitory complex formation — 335
   - 4.3. Determination of the rate of stable complex formation (k_{ass}) — 337
5. Detection of Serpin–Enzyme Complex — 338
   - 5.1. Sodium dodecyl sulfate-polyacrylamide gel electrophoresis — 338
   - 5.2. Western blot analysis — 340
6. Determination of an Interaction Site — 341
7. Summary — 343
Acknowledgments — 344
References — 344

Abstract

Proteolytic enzymes secreted by *Staphylococcus aureus* are considered important virulence factors. Here, we present data showing that staphylococci-derived cysteine proteases (staphopains) are efficiently inhibited by squamous cell carcinoma antigen 1 (SCCA1), a serpin abundant on the epithelial surfaces. The high association rate constant (k_{ass}) for inhibitory complex formation (1.9×10^4 and $5.8 \times 10^4\ M^{-1}\ s^{-1}$ for staphopain A and staphopain B interaction with SCCA1, respectively) argues that SCCA1 can restrain staphopain

---

* Department of Microbiology, Faculty of Biochemistry, Biophysics, and Biotechnology, Jagiellonian University, Krakow, Poland
† Department of Oral Health and Rehabilitation, University of Louisville Dental School, Louisville, Kentucky, USA

Methods in Enzymology, Volume 499 © 2011 Elsevier Inc.
ISSN 0076-6879, DOI: 10.1016/B978-0-12-386471-0.00016-X All rights reserved.

activity *in vivo* at epithelial sites colonized by *S. aureus*. The mechanism of staphopain inhibition by SCCA1 is apparently the same as for serpin interaction with target serine proteases. The formation of a covalent complex results in cleavage of the SCCA1 reactive site peptide bond, and it is associated with the release of the C-terminal peptide of 37 amino acid residues from the serpin. Significantly, the SCCA1 reactive site closely resembles a motif in the reactive site loop of natural *S. aureus*-derived inhibitors of the staphopains (staphostatins). Taking into account that SCCA1 is predominantly expressed in epithelial tissues, including respiratory pathways, hair follicles and skin [Kato, H. (1996). Expression and function of squamous cell carcinoma antigen. *Anticancer Res.* **16**, 2149–2153.], all of which are regularly colonized by *S. aureus*, the physiological relevance of SCCA1–staphopain B interaction as a defense mechanism seems to be very well substantiated.

1. INTRODUCTION

Squamous cell carcinoma antigen 1 and 2 (SCCA1 and SCCA2), members of intracellular SERPINB clade (ov-serpins) of protease inhibitors (Silverman *et al.*, 2004), are found in numerous healthy tissues of epithelial origin (Cataltepe *et al.*, 2000; Kato *et al.*, 1996). They are also present in biological fluids, including respiratory system mucus and saliva of healthy individuals, probably as a result of passive secretion during desquamation (Cataltepe *et al.*, 2000). SCCAs expression is stimulated by proinflammatory cytokines. Therefore, it is anticipated that they regulate the immune response to infections and participate in restoration of homeostasis via control of proteolysis in inflamed epithelial tissues (Cataltepe *et al.*, 2000; Suminami *et al.*, 2001).

Despite the high degree of amino acid sequence identity (92%), the SCCA serpins differ in their inhibitory specificity. SCCA1 targets papain-like cysteine proteases, including papain and cathepsins S, K, and L, while SCCA2 interacts with cathepsin G, mast cell chymase, and house dust mite proteases (Der p1 and Der f1) which are allergens (Sakata *et al.*, 2004; Schick *et al.*, 1997, 1998a). The broad spectrum of inhibited proteases of two catalytic classes suggests that the SCCA serpins may play a role in the control of bacterially derived proteases, such as those produced by *Staphylococcus aureus*, a frequent pathogen of epithelial surfaces. *S. aureus* secretes two cysteine proteases of the papain-like fold (family C47 of clan CA of cysteine peptidases). These enzymes can directly or indirectly damage the epithelium and underlying connective tissue as well as exert deleterious effect on cells of host immune system (Potempa and Pike, 2009; Smagur *et al.*, 2009a,b). Therefore, it is clear that staphopain local inhibition by the SCCA serpins may have beneficial effects. Accordingly, we present here the interaction of *S. aureus* cysteine proteases, staphopains, with epithelial-origin SCCA1, which is to our knowledge the first ever described example of the efficient inhibition of pathogen-derived proteases by human serpin. With k_{ass} in the range of $10^4 \, M^{-1} \, s^{-1}$ the inhibition of S. aureus cysteine proteases, SCCA1 may be fast enough to effectively abrogate staphopain activity in vivo.

2. Purification of Staphopains

Staphopains were purified, as described previously (Drapeau et al., 1972; Massimi et al., 2002; Potempa et al., 1988). Briefly, the S. aureus V8 BC10 strain was grown overnight in the TSB (Difco, Lawrence, KS, USA) medium supplemented with β-glycerophosphate (0.5%, w/v). Culture was centrifugated, and supernatant was collected. Soluble proteins were precipitated with ammonium sulfate and collected by centrifugation. Pellets were solubilized in 50 mM NaAc buffer, pH 5.5. After overnight dialysis versus the same buffer, the sample was loaded onto a Q-Sepharose column (GE Healthcare, Little Chalfont, UK). The column was washed extensively and then developed with the NaCl gradient from 0 to 1 M in the NaAc buffer. Flow through containing the ScpA activity was collected and concentrated using a membrane ultrafiltration system (Amicon, Millipore, Billerica, MA, USA). The ScpA protease was finally purified to homogeneity by chromatography on Mono S column (GE Healthcare, Little Chalfont, UK) equilibrated with the 50 mM phosphate buffer, pH 6.5.

Proteolytically active fractions (containing V8 protease and SspB), eluted from the Q-Sepharose column with the NaCl gradient, were pulled together and subjected to chromatography on a Phenyl-Sepharose column (GE Healthcare, Little Chalfont, UK) equilibrated with 25 mM Tris, pH 7.5, followed by separation on a Mono Q column. The purity of staphylococcal enzymes was confirmed by SDS-PAGE. The purified enzyme preparations were active-site titrated using E-64 (Sigma-Aldrich, St. Louis, MO, USA; ScpA) and α$_2$-macroglobulin (Biocentrum, Krakow, Poland; SspB). All enzyme concentrations described in this work correspond to active-site titrated molar concentrations of each enzyme, unless otherwise specified.

3. Purification of GST–SCCA1 and GST–SCCA2 Fusion Proteins

The recombinant GST–SCCA1 and GST–SCCA2 fusion proteins were synthesized in a bacterial expression system and purified using glutathione–Sepharose 4B beads (Amersham Biosciences, Piscataway, NY, USA) using modifications of previously described methods (Schick et al., 1997, 2004). Briefly, 50 ng of the pGEX–SCCA1 (or pGEX–SCCA2) plasmid was transformed into competent *Escherichia coli* BL21 cells. After overnight growth at 37 °C on LB agar plates supplemented with 100 μg ml^{-1} ampicillin, transformants were washed off using 5 ml of LB broth. The culture was then used to inoculate 1 l of LB broth supplemented with 100 μg ml^{-1} ampicillin and allowed to grow at 37 °C until the culture reached an OD of 0.5 at 600 nm. Following a 5-min incubation period in ice bath, protein expression was induced by the addition of isopropyl-1-thio-β-D-galactopyranoside (IPTG; Sigma-Aldrich, St. Louis, MO, USA) to a final concentration of 0.5 mM and incubated at 25 °C for 4 h. Cultures were harvested by centrifugation at 5000×g for 10 min and lysed using 60 ml of lysis buffer (100 mM NaCl, 100 mM Tris–HCl, pH 8.0, 50 mM EDTA, 2% Triton X-100, 1.5 mg ml^{-1} lysozyme and protease inhibitor cocktail). The bacterial lysate was clarified by centrifugation at 12,000×g for 20 min. Forty-eight milliliters of supernatant was transferred to a fresh tube containing 2 ml of 50% glutathione–Sepharose 4B beads (GE Healthcare, Little Chalfont, UK) and incubated for 30 min at 4 °C to facilitate binding. The beads were collected by centrifugation at 500×g and washed twice in lysis buffer (minus lysozyme) and twice in PBST (10 mM phosphate buffer, 27 mM KCl, 137 mM NaCl, 0.1% Tween 20, pH 7.4). The GST–SCCA1 and GST–SCCA2 proteins were eluted from the beads using three 1 ml washes of glutathione elution buffer (10 mM reduced glutathione, 50 mM Tris–HCl, pH 8.0). GST-SCCA concentrations in the eluates were determined by Bradford analysis, and protein purity was checked by SDS-PAGE analysis.

4. Characterization of Inhibition

4.1. General inhibition assay

For years, serpins have been assigned a potential role in controlling the activity of proteases from microbial pathogens; however, no data have yet been published to support this contention. Indeed, to the contrary, numerous reports show that serpins are readily inactivated by microbial proteases (Potempa and Pike, 2009). Therefore, in initial experiments, a general protease substrate was used, to determine effect of SCCA1 and SCCA2 on the staphopain activity.

1. The staphopains A and B were preactivated by incubation for 15 min at 37 °C in 0.1 M Tris–HCl, 5 mM EDTA, pH 7.6, freshly supplemented with DTT to 2 mM concentration.
2. Serpin solution in 0.1 M Tris–HCl, 5 mM EDTA, pH 7.6 was added to each sample, yielding 50 nM and 300 nM concentrations of an enzyme and an inhibitor, respectively (E:I = 1:6), in the total volume of 200 μl.

Control samples were prepared similarly, with buffer added instead of the inhibitor.
3. Samples were incubated at 37 °C for 30 min, allowing for interaction between an inhibitor and a protease.
4. One hundred microliters of 15 mg ml^{-1} azocoll (Calbiochem, San Diego, CA, USA) suspension in 0.5 M sucrose, 0.05% Tween-20 was added to each sample.
5. Samples were incubated for 30 min at 37 °C with shaking. Undigested substrate was removed by centrifugation (14,000 rpm), and 200 µl of supernatants were transferred into a 96-well microplate.
6. Absorbance at 520 nm was measured using a SpectraMAX microplate reader (Molecular Devices, Sunnyvale, CA, USA), and a residual enzyme activity in the presence of an inhibitor was calculated as percentage of activity of a protease alone.

Using this inhibition assay, it was found that at a sixfold molar excess, SCCA2 did not affect ScpA activity and only weakly inhibited SspB, while SCCA1 effectively inhibited both staphopains. The significant difference in residual activity of the two enzymes in the presence of SCCA1 indicated that SCCA1 is more efficient inhibitor of SspB than ScpA.

4.2. Determination of the stoichiometry of inhibitory complex formation

To further investigate the interaction between staphopains and SCCA1, we determined the stoichiometry of inhibitory complex formation, that is, the number of SCCA1 molecules necessary for formation of a stable staphopain–SCCA1 inhibitory complex.

1. The staphopains A and B were preactivated by incubation for 15 min at 37 °C in 0.1 M Tris–HCl, 5 mM EDTA, pH 7.6, freshly supplemented with DTT to 2 mM concentration and transferred to a black 96-well microplate (Nunc, Roskilde, Denmark).
2. Increasing SCCA1 concentrations were prepared in 0.1 M Tris, 5 mM EDTA, 2 mM DTT, pH 7.6 and added to wells of a microtitration plate, resulting in the final concentrations: 40 nM staphopain and SCCA1 concentration ranging from 0 to 200 nM in a final volume of 100 µl (molar ratio enzyme to inhibitor in the range from 0 to 5).
3. A plate was incubated for 30 min at 37 °C, allowing interaction between an enzyme and a protease.
4. Substrate solutions were prepared, separately for each staphopain. Each solution contained 20 µM of a fluorescent substrate specific for each staphopain in the reaction buffer containing 10% DMF (v/v).

5. Subsequently, 100 μl of prepared substrate solution was added to each sample yielding 20 nM staphopain, 10 μM substrate, and 5% DMF (v/v) concentration in the final volume of 200 μl. Abz-Glu-Ala-Leu-Gly-Thr-Ser-Pro-Arg-Lys(Dnp)-Asp-OH and Abz-Glu-Gly-Ile-Gly-Thr-Ser-Arg-Pro-Lys(Dnp)-Asp-OH synthesized by Polypeptide Laboratories (Wolfenbuettel, Germany) were used to measure activity of SspB and ScpA, respectively.
6. Reaction was monitored for 30 min at 37 °C using a SpectraMAX Gemini XS (Molecular Devices, Sunnyvale, CA, USA) microplate fluorimeter with excitation, and emission wavelengths set at 320 and 420 nm, respectively.
7. The initial velocity of substrate hydrolysis (V_{max}) was used to calculate the residual enzyme activity (mean ± SD) as a percentage of the activity (V_{max}) of uninhibited staphopain and plotted versus the staphopain/serpin molar ratio.
8. The stoichiometry of inhibition was determined by the linear regression of the experimental data points from three independent experiments.

The analysis revealed a stoichiometric index (SI) of 4.8 ± 0.3 for ScpA compared to only 2.7 ± 0.1 for SspB (Fig. 16.1). This difference explains the more efficient inhibition of SspB by SCCA1, compared to ScpA.

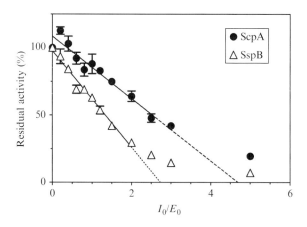

Figure 16.1 Stoichiometry of GST–SCCA1—staphopain A (closed circles) and staphopain B (open triangles) inhibition. The increasing amounts of GST–SCCA1 were preincubated for 30 min with constant concentration of staphopain, resulting in molar ratios of the inhibitor to enzyme in the range from 0 to 5. The residual enzyme activity was measured using fluorescent substrates specific for each staphopain. The data were plotted as the residual activity (V_i/V_0) versus the inhibitor:enzyme molar ratio. The stoichiometry of inhibitory complex formation (SI) was determined by linear regression to the initial points of inhibitory curves. The presented results are means of readings from triplicate experiments ± SD.

4.3. Determination of the rate of stable complex formation (k_{ass})

Apart from the stoichiometry, physiological efficacy of a serpin as an inhibitor of a targeted protease is dependent on the rate of stable complex formation. Therefore, to determine the potential biological relevance of staphopain inhibition by SCCA1, the kinetics of interaction of this serpin with staphopains was investigated.

1. Mixtures of constant (20 μM) substrate concentration (Abz-Glu-Ala-Leu-Gly-Thr-Ser-Pro-Arg-Lys(Dnp)-Asp-OH for ScpA and Abz-Glu-Gly-Ile-Gly-Thr-Ser-Arg-Pro-Lys(Dnp)-Asp-OH for SspB) with increasing concentration of SCCA1 in the range from 0 to 300 nM were prepared on 96-well black microplates (Nunc, Roskilde, Denmark) in a total volume of 100 μl.
2. Then, 100 μl of the staphopains preactivated in 0.1 M Tris, 5 mM EDTA, 2 mM DTT, pH 7.6, for 15 min at 37 °C, were added (0.5 nM final enzyme concentration), and the rate of substrate hydrolysis was recorded as the increase of fluorescence (λ_{ex} = 320 nm, λ_{em} = 420 nm) for 60 min using a SpectraMAX Gemini XS microplate fluorescence reader.
3. Data were analyzed by nonlinear fitting to the progress curve for the irreversible inhibition model, described by Eq. (16.1; Morrison and Walsh, 1988):

$$P = \frac{v_z}{k_{obs}}(1 - e^{-k_{obs}t}) \qquad (16.1)$$

where v_z denotes the initial velocity of reaction, k_{obs} is the apparent association constant, and t time. k_{obs} values determined for each inhibitor concentration were plotted against inhibitor concentration, and the apparent association constant k' was determined from the slope of a line fitted by the linear regression.

4. As an inhibitor competes with a substrate for a binding site and the reaction velocity is affected by the enzyme-inhibitor stoichiometry ratio, the final k_{ass} value was calculated from k' using Eq. (16.2; Morrison and Walsh, 1988):

$$k_{ass} = k'\left(1 + \frac{[S]}{K_m}\right)\text{SI} \qquad (16.2)$$

where SI stands for the stoichiometry ratio and K_m is the Michaelis constant for a given substrate.

5. The K_m for Abz-Glu-Ala-Leu-Gly-Thr-Ser-Pro-Arg-Lys(Dnp)-Asp-OH and for Abz-Glu-Gly-Ile-Gly-Thr-Ser-Arg-Pro-Lys(Dnp)-Asp-OH hydrolysis by ScpA and SspB was determined to be 79.4 and 53 μM, respectively.

Using progress curve analysis of the inhibitory reaction under pseudo-first-order conditions (Morrison and Walsh, 1988), the k_{ass} value was determined as $1.9 \pm 0.4 \times 10^4$ and $5.8 \pm 0.8 \times 10^4$ M^{-1} s^{-1} for ScpA and SspB inhibition by SCCA1, respectively (Fig. 16.2). According to the relationship described by Bieth (1984) (Eq. (16.3)), the time required for inhibition of 99.9% of enzyme by 1 μM GST–SCCA1 is 6 and 2 min for staphopains A and B, respectively, which is short enough to suggest that *in vivo* staphopain inhibition by SCCA1 has the physiological importance.

$$t_{1/2} = \ln \frac{2}{k_{ass}[I]} \qquad (16.3)$$

5. Detection of Serpin–Enzyme Complex

5.1. Sodium dodecyl sulfate-polyacrylamide gel electrophoresis

The reaction of a serpin with a target serine protease results in formation of a covalent complex in which a protease and an inhibitor are linked by an ester bond. Such complexes are resistant to separation during SDS-PAGE and stable during boiling in reducing conditions. With respect of complex stability, cysteine proteases differ significantly from serine protease since the thioester bond formed between the catalytic Cys and an inhibitor are labile and highly sensitive to reducing conditions. Therefore, to prevent breakdown of the complex, we have applied slightly modified SDS-PAGE procedure.

1. Staphopain B was preactivated for 15 min at 37 °C in the 0.1 M Tris–HCl, 5 mM EDTA, pH 7.6, freshly supplemented with 2 mM DTT.
2. The increasing concentrations of preactivated staphopain B (0–5 μM) were incubated in 0.1 M Tris–HCl, 5 mM EDTA, pH 7.6, together with 2 μM SCCA1 in separate Eppendorf tubes, resulting in SspB:SCCA1 molar ratio in the range from 0 to 2.5.
3. The samples were incubated for 30 min at 37 °C.
4. SDS-PAGE sample buffer (1:1 v/v) supplemented with 100 μM E-64, and devoid of a reducing agent was added to the samples.
5. Mixtures were incubated for additional 30 min at 37 °C.

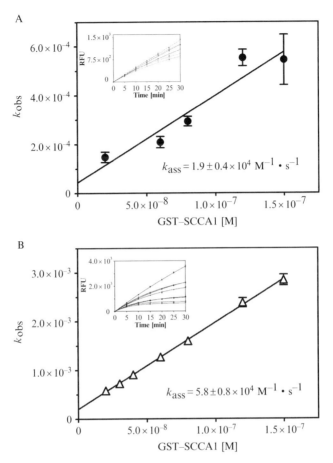

Figure 16.2 Determination of the secondary rate constant (k_{ass}) of staphopains inhibition by GST–SCCA1. The constant amount of staphopain A (A) and staphopain B (B) were added to samples, containing increasing concentration of GST–SCCA-1 and the fixed amount of substrate. Immediately, the samples were mixed, and time-dependent increase in fluorescence due to substrate hydrolysis was recorded (insets). The data were analyzed by the progress curve method, and obtained values of k_{obs} were plotted against inhibitor concentration in the sample (A and B). The secondary rate constant was determined by linear regression to the data points and corrected for substrate competition and the stoichiometry factor. The data represent results from triplicate experiments ± SD.

6. Samples were subsequently separated by 10%, T:C 29:1 SDS-PAGE in the Schagger–von Jagow tricine gel electrophoresis system.
7. After electrophoresis gels was stained with Coomassie brilliant blue G-250 (Serva).

The analysis revealed a characteristic banding pattern (Fig. 16.3). The intensity of the band corresponding to intact GST–SCCA1 (~70 kDa) gradually decreased proportionally to the SspB concentration and practically

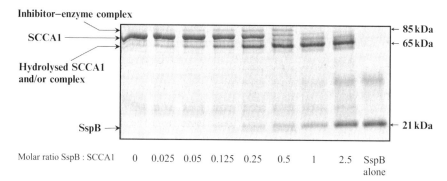

Figure 16.3 SDS-PAGE analysis of interaction between GST–SCCA1 and SspB. To identify the SDS-stable complex of GST–SCCA1–SspB, increasing concentrations of SspB were incubated together with constant amount of GST–SCCA1, resulting in molar ratios the inhibitor to the enzyme in the range from 1:0.025 to 1:2.5.

disappeared at 1:1 M ratio. This change correlated with proportional increase of intensity of 65 kDa band apparently corresponding to GST–SCCA1 cleaved within the RLS. Most notably, however, the band with molecular mass *circa* 85 kDa appeared in a manner proportionate to the SspB concentration. The intensity of this band was highest at optimal stoichiometric ratio of SspB–SSCA1 interaction (between 1:2 and 1:4), as determined in Section 4.2. At higher enzyme concentrations, the band disappeared. The molecular mass of the band (85 kDa) suggests that it represents the SDS-stable complex composed of SspB (21 kDa) and RLS-truncated GST–SCCA1 (65 kDa).

5.2. Western blot analysis

To confirm the presence of staphopain in the 85 kDa complex, we performed Western blot analysis using specific mouse monoclonal anti-SspB antibody.

1. For Western blot analysis, SDS-PAGE resolved proteins (as described above) were electrotransferred in 25 mM Tris, 192 mM glycine, 20% methanol in a semidry Western blot chamber (Bio-Rad) onto a PVDF membrane (Millipore, Billerica, MA, USA).
2. Nonspecific binding sites were blocked overnight in a solution containing 2% bovine serum albumin (Lab Empire, Rzeszow, Poland) and 10% goat serum, and the membrane was then incubated with the primary mouse monoclonal anti-SspB antibodies (clone 5E12.G9, UGA core facility for MAb development) for 3 h at room temperature.

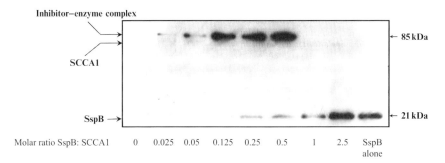

Figure 16.4 Western blot analysis of interaction between GST–SCCA1 and SspB using mouse monoclonal anti-SspB antibody.

3. Following a washing step with TBS–Tween, secondary goat anti-mouse IgG Fc fragment antibodies (Sigma-Aldrich, St. Louis, MO, USA) were applied for 1 h.
4. The membrane was developed using ECL + substrate (GE Healthcare, Little Chalfont, UK) and Kodak Biofilm plate (Eastman Kodak, Rochester, NY, USA).

As it is clearly shown in Fig. 16.4, the 85 kDa band strongly reacted with the mAb indicating the presence of SspB in the complex. The intensity of the reaction correlated with the Coomassie-stained band (Fig. 16.4) and was highest for 1:2 M mixture of SspB and SCCA1. This result unambiguously confirmed the presence of SspB in the higher molecular weight product resulting from the SspB–SCCA1 interaction, thus demonstrating formation of the SDS-stable serpin–enzyme complex.

6. Determination of an Interaction Site

Serpin–target protease complex formation involves hydrolysis of a specific peptide bond in the reactive site of an inhibitor within an exposed reactive site loop (RSL). The C-terminal peptide is released, while the large N-terminal domain remains covalently bound to a target enzyme. As serpin specificity is mainly dictated by a sequence in an exposed region of the RSL, it was essential to precisely determined the reactive site recognized by staphopains in SCCA1. In addition, the release of the SCCA1-derived C-terminal peptide during complex formation is yet another proof of the suicide substrate mechanism of interaction.

1. Preactivated staphopains (1.41 μM) were mixed with 7.04 μM GST–SCCA1 (1:5 M ratio) in 20 μl of 0.1 M Tris, 5 mM EDTA, 2 mM DTT, pH 7.6.

2. Control samples, containing same concentration of each staphopain and GST–SCCA1, were prepared separately in the same manner.
3. All mixtures were incubated for 30 min at 37 °C, and the reaction was stopped by addition of nonreducing SDS-PAGE sample buffer supplemented with 100 μM E-64.
4. Samples were resolved on a peptide gel using Schagger and von Jagow tricine gel electrophoretic system (Schagger and Von Jagow, 1987) and electrotransferred onto P polyvinylidene difluoride membranes in 10 mM CAPS, 10% methanol, pH 11 buffer using semidry blotting chamber.
5. Protein bands were visualized by Coomassie brilliant blue staining, excised, and analyzed by automated Edman degradation using a Procise 494HT amino acid sequencer (Applied Biosystems, Carlsbad, CA, USA).

As can be expected for the mechanism of protease inhibition by serpins, the complex formation between SspB and SCCA1 was accompanied with release of a 4.5-kDa peptide (Fig. 16.5) due to the RSL cleavage of the

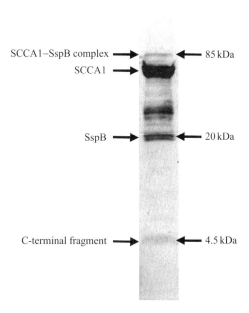

Figure 16.5 Determination of the cleavage site at the RSL during SCCA1 interaction with SspB. The staphopains were preincubated with SCCA1, reaction stopped by addition of nonreducing sample buffer, proteins separated by SDS-PAGE and then electrotransferred onto the PVDF membrane. Bands stained with CBB G250 corresponding to GST–SCCA1, the GST–SCCA1–SspB complex, and the released C-terminal peptide were excised and subjected to N-terminally sequence analysis.

```
SCCA1           -  T₃₄₇A  V  V  G  F  G  S  S  -  P  T  S  T₃₅₉-  -
Staphostatin A  -  -  -  -  E₉₃ A  L  G  T  S  -  P  R  M₁₀₁-  -  -
Staphostatin B  -  -  -  -  Q₉₅ G  I  G  T  S  R  P  I₁₀₃-  -  -  -
```

Figure 16.6 Alignment of the reactive site loop sequences of SCCA1 and staphostatins. The positions of the primary P1 and P1′ residues at cleavage sites by staphopains and cathepsin S are marked by open and filled block arrows, respectively (Rzychon et al., 2003; Schick et al., 1998b). The secondary, inactivating cleavage site, identified for papain in the SCCA1 RSL is an arrow (Masumoto et al., 2003).

inhibitor by the protease. Edman degradation analysis of the peptide yielded the same N-terminal sequence of Ser-Ser-Pro-Thr-Ser-Thr-Asn for SCCA1 interacting with either SspB or ScpA indicating that both enzymes cleaved the peptide bond between Gly354 and Ser355. This bond was identified as the P1–P1′ peptide bound of the inhibitor reactive site in the same manner as for inhibition of cathepsins (Schick et al., 1998b). Notably, the sequence (Phe-Gly-Ser-Ser) at the SCCA-1 RSL is strikingly similar to those in the reactive site of the staphostatins (Leu-Gly-Thr-Ser and Ile-Gly-Thr-Ser for staphostatins A and B, respectively; Fig. 16.6), which are natural inhibitors of the staphopain enzymes (Rzychon et al., 2003). This observation suggests the adaptation of the SCCA-1 serpin for inhibition of staphylococcal cysteine proteases.

7. Summary

Challenging reports that SCCA1 inhibits target proteases—cysteine cathepsins—via formation of the noncovalent enzyme–inhibitor complex (Masumoto et al., 2003; Sakata et al., 2004) we have clearly shown that staphopains inhibition occurs via the typical serpins suicide substrate mechanism involving formation of an 85 kDa covalent inhibitory complex which is stable in SDS-PAGE (Kantyka et al., 2011). Complex formation was accompanied by release of the C-terminal *circa* 4.5 kDa peptide generated by the cleavage of the RSL at the Gly354-Ser355 peptide bond (Kantyka et al., 2011); identified previously as the P1–P1′ residues for SCCA-1 interaction with human cathepsins (Schick et al., 1998b). The covalent mode of staphopain inhibition by the SCCA1 serpin is keeping with the observation that the cysteine protease inhibiting serpin MENT (Irving et al., 2002a) and an antitrypsin/SCCA-1 chimera (Irving et al., 2002b) also form classic "serpin-like" covalent complexes with cysteine proteases.

ACKNOWLEDGMENTS

This study was partially supported in part by the Jagiellonian University statutory funds (DS/9/WBBiB). J. P. and T. K. acknowledge support from Foundation for Polish Science (TEAM project DPS/424-329/10). The Faculty of Biochemistry, Biophysics, and Biotechnology of the Jagiellonian University is a beneficent of the structural funds from the European Union (Grant No. POIG.02.01.00-12-064/08—"Molecular biotechnology for health").

REFERENCES

Bieth, J. G. (1984). In vivo significance of kinetic constants of protein proteinase inhibitors. Biochem. Med. **32**, 387–397.

Cataltepe, S., Gornstein, E. R., Schick, C., Kamachi, Y., Chatson, K., Fries, J., Silverman, G. A., and Upton, M. P. (2000). Co-expression of the squamous cell carcinoma antigens 1 and 2 in normal adult human tissues and squamous cell carcinomas. J. Histochem. Cytochem. **48**, 113–122.

Drapeau, G. R., Boily, Y., and Houmard, J. (1972). Purification and properties of an extracellular protease of Staphylococcus aureus. J. Biol. Chem. **247**, 6720–6726.

Irving, J. A., Shushanov, S. S., Pike, R. N., Popova, E. Y., Brömme, D., Coetzer, T. H., Bottomley, S. P., Boulynko, I. A., Grigoryev, S. A., and Whisstock, J. C. (2002a). Inhibitory activity of a heterochromatin-associated serpin (MENT) against papain-like cysteine proteinases affects chromatin structure and blocks cell proliferation. J. Biol. Chem. **277**(15), 13192–13201.

Irving, J. A., Pike, R. N., Dai, W., Brömme, D., Worrall, D. M., Silverman, G. A., Coetzer, T. H., Dennison, C., Bottomley, S. P., and Whisstock, J. C. (2002b). Evidence that serpin architecture intrinsically supports papain-like cysteine protease inhibition: engineering alpha(1)-antitrypsin to inhibit cathepsin proteases. Biochemistry **41**(15), 4998–5004.

Kantyka, T., Plaza, K., Koziel, J., Florczyk, D., Stennicke, H. R., Thogersen, I. B., Enghild, J. J., Silverman, G. A., Pak, S. C., and Potempa, J. (2011). Inhibition of Staphylococcus aureus cysteine proteases by human serpin potentially limits staphylococcal virulence. Biol. Chem. **392**(5), 483–489.

Kato, H. (1996). Expression and function of squamous cell carcinoma antigen. Anticancer Res. **16**, 2149–2153.

Massimi, I., Park, E., Rice, K., Muller-Esterl, W., Sauder, D., and McGavin, M. J. (2002). Identification of a novel maturation mechanism and restricted substrate specificity for the SspB cysteine protease of Staphylococcus aureus. J. Biol. Chem. **277**, 41770–41777.

Masumoto, K., Sakata, Y., Arima, K., Nakao, I., and Izuhara, K. (2003). Inhibitory mechanism of a cross-class serpin, the squamous cell carcinoma antigen 1. J. Biol. Chem. **278**, 45296–45304.

Morrison, J. F., and Walsh, C. T. (1988). The behavior and significance of slow-binding enzyme inhibitors. Adv. Enzymol. Relat. Areas Mol. Biol. **61**, 201–301.

Potempa, J., and Pike, R. N. (2009). Corruption of innate immunity by bacterial proteases. J. Innate Immun. **1**, 70–87.

Potempa, J., Dubin, A., Korzus, G., and Travis, J. (1988). Degradation of elastin by a cysteine proteinase from Staphylococcus aureus. J. Biol. Chem. **263**, 2664–2667.

Rzychon, M., Sabat, A., Kosowska, K., Potempa, J., and Dubin, A. (2003). Staphostatins: An expanding new group of proteinase inhibitors with a unique specificity for the regulation of staphopains, *Staphylococcus* spp. cysteine proteinases. *Mol. Microbiol.* **49,** 1051–1066.

Sakata, Y., Arima, K., Takai, T., Sakurai, W., Masumoto, K., Yuyama, N., Suminami, Y., Kishi, F., Yamashita, T., Kato, T., Ogawa, H., and Fujimoto, K. (2004). The squamous cell carcinoma antigen 2 inhibits the cysteine proteinase activity of a major mite allergen, Der p 1. *J. Biol. Chem.* **279,** 5081–5087.

Schagger, H., and Von Jagow, G. (1987). Tricine-sodium dodecyl sulfate-polyacrylamide gel electrophoresis for the separation of proteins in the range from 1 to 100 kDa. *Anal. Biochem.* **166,** 368–379.

Schick, C., Kamachi, Y., Bartuski, A. J., Cataltepe, S., Schechter, N. M., Pemberton, P. A., and Silverman, G. A. (1997). Squamous cell carcinoma antigen 2 is a novel serpin that inhibits the chymotrypsin-like proteinases cathepsin G and mast cell chymase. *J. Biol. Chem.* **272,** 1849–1855.

Schick, C., Pemberton, P. A., Shi, G. P., Kamachi, Y., Cataltepe, S., Bartuski, A. J., Gornstein, E. R., Bromme, D., Chapman, H. A., and Silverman, G. A. (1998a). Cross-class inhibition of the cysteine proteinases cathepsins K, L, and S by the serpin squamous cell carcinoma antigen 1: A kinetic analysis. *Biochemistry* **37,** 5258–5266.

Schick, C., Brömme, D., Bartuski, A., Uemura, Y., Schechter, N., and Silverman, G. A. (1998b). The reactive site loop of the serpin SCCA1 is essential for cysteine proteinase inhibition. *Proc. Natl. Acad. Sci. USA* **95,** 13465–13470.

Silverman, G. A., Whisstock, J. C., Askew, D. J., Pak, S. C., Luke, C. J., Cataltepe, S., Irving, J. A., and Bird, P. I. (2004). Human clade B serpins (ov-serpins) belong to a cohort of evolutionarily dispersed intracellular proteinase inhibitor clades that protect cells from promiscuous proteolysis. *Cell. Mol. Life Sci.* **61,** 301–325.

Smagur, J., Guzik, K., Bzowska, M., Kuzak, M., Zarebski, M., Kantyka, T., Walski, M., Gajkowska, B., and Potempa, J. (2009a). Staphylococcal cysteine protease staphopain B (SspB) induces rapid engulfment of human neutrophils and monocytes by macrophages. *Biol. Chem.* **390**(4), 361–371.

Smagur, J., Guzik, K., Magiera, L., Bzowska, M., Gruca, M., Thøgersen, I. B., Enghild, J. J., and Potempa, J. (2009b). A new pathway of staphylococcal pathogenesis: apoptosis-like death induced by Staphopain B in human neutrophils and monocytes. *J. Innate Immun.* **1** (2), 98–108.

Suminami, Y., Nagashima, S., Murakami, A., Nawata, S., Gondo, T., Hirakawa, H., Numa, F., Silverman, G. A., and Kato, H. (2001). Suppression of a squamous cell carcinoma (SCC)-related serpin, SCC antigen, inhibits tumor growth with increased intratumor infiltration of natural killer cells. *Cancer Res.* **61,** 1776–1780.

CHAPTER SEVENTEEN

Plants and the Study of Serpin Biology

Thomas H. Roberts,[*] Joon-Woo Ahn,[†] Nardy Lampl,[‡] and Robert Fluhr[‡]

Contents

1. Introduction — 348
2. Detection of Serpins in Plant Extracts and Localization of Serpins in Plant Tissues and Cells — 349
   2.1. Detection using biotinylated proteases — 349
   2.2. Plant serpin antibody specificity — 350
   2.3. Subcellular localization of *Arabidopsis* serpins — 350
3. Purification of Serpins from Plant Tissues — 353
   3.1. Protection of serpins from cleavage by endogenous proteases during purification — 353
   3.2. Purification of serpins from mature seeds — 353
   3.3. Purification of serpins from vegetative organs — 359
   3.4. Monitoring of fractions during plant serpin purification — 359
   3.5. Separation and visualization of plant serpins using native-PAGE — 359
4. Production of Recombinant Plant Serpins — 360
5. Analysis of Plant Serpin–Protease Interactions — 360
   5.1. Reducing and nonreducing SDS-PAGE — 361
   5.2. Identification of target proteases for plant serpins — 361
Acknowledgments — 364
References — 364

Abstract

Serpins appear to be ubiquitous in the Plant Kingdom and have several unique properties when compared to the substantial number of other families of protease inhibitors in plants. Serpins in plants are likely to have functions distinct from those of animal serpins, partly because plants and animals

[*] Department of Chemistry and Biomolecular Sciences, Macquarie University, North Ryde, Australia
[†] Plant Systems Engineering Research Center, Korea Research Institute of Bioscience and Biotechnology (KRIBB), Yuseong-gu, Daejeon, South Korea
[‡] Department of Plant Sciences, Weizmann Institute of Science, Rehovot, Israel

Methods in Enzymology, Volume 499 © 2011 Elsevier Inc.
ISSN 0076-6879, DOI: 10.1016/B978-0-12-386471-0.00017-1 All rights reserved.

developed multicellularity independently and partly because most animal serpins are involved in animal-specific processes, such as blood coagulation and the activation of complement. To encourage and facilitate the discovery of plant serpin functions, here we provide a set of protocols for detection of serpins in plant extracts, localization of serpins in plant tissues and cells, purification of serpins from a range of organs from monocot and eudicot plants, production and purification of recombinant plant serpins, and analysis of plant–protease interactions including identification of *in vivo* target proteases.

1. INTRODUCTION

Plants and animals evolved from a common unicellular ancestor and developed multicellularity independently (Alberts *et al.*, 2008). We now appreciate that most proteins of the serpin family in animals function in animal-specific complex processes that are absent in both plants and unicellular eukaryotes (Law *et al.*, 2006; Silverman *et al.*, 2010) and that serpins are found in only a subset of unicellular eukaryotes (Roberts *et al.*, 2004). Thus, neither could the presence of serpins in plants be anticipated with confidence nor their biological functions inferred from those of animal serpins (Hejgaard and Roberts, 2007; Roberts and Hejgaard, 2008). Albumins (proteins soluble in water) of $M_r \sim 43$ kDa from the endosperm of barley (*Hordeum vulgare*) grain were given the collective name "protein Z" (Hejgaard, 1976) a decade before they were identified as homologous to human α_1-antitrypsin and related proteins (Hejgaard *et al.*, 1985)—the same year that the word "serpin" was coined (Carrell and Travis, 1985). Protein Z is still used by some scientists to describe cereal grain serpins (as well as serpins in beer proteomes in several recent studies) but this name can be confusing because (i) one of the pathological mutants of the archetypal serpin α_1-antitrypsin has been known as the "Z" type for decades (Fagerhol and Laurell, 1967); (ii) a mammalian vitamin K-binding protein (not a serpin) that serves as a cofactor for inhibition of coagulation factor Xa was also named "protein Z," or PZ (Prouse and Esnouf, 1977); and (iii) a PZ-dependent protease inhibitor (ZPI) was identified as a serpin (Han *et al.*, 1999).

Through the availability of large numbers of plant genome and transcript sequences, we can conclude that serpins are almost certainly present in all land plants (the Embryophyta). Indeed, serpins (and/or their genes) have been identified in species representing all major groups of land plants, including seed plants (the Spermatophyta), which include the flowering plants (Magnoliophyta) and other seed plants (e.g., conifers), bryophytes (e.g., the moss *Physcomitrella patens*), and in the relatives of land plants, the green algae (Chlorophyta; e.g., *Chlamydomonas reinhardtii*). The ubiquitous distribution of serpins in the Plant Kingdom is in stark contrast to their distribution in fungi, where serpins appear to be very rare, the only reported example being a serpin in the anaerobic fungus *Piromyces* sp. (Steenbakkers *et al.*, 2008).

Control of proteolysis in plants—both within cells (the symplast) and in the extracellular medium (the apoplast)—is critical for cellular function as well as for development and survival of the organism (van der Hoorn, 2008), as it is in animals. The fully sequenced genomes of several plant species have revealed hundreds of genes encoding proteolytic enzymes of a range of classes. Ongoing discovery of the properties and functions of plant serpins is important in furthering our understanding of the role of proteolysis and its control in plant biology (Roberts and Hejgaard, 2008). In this chapter, we describe how to purify serpins from seeds and vegetative tissues, separate and visualize plant serpins, detect serpins in complex plant extracts, analyze plant serpin–protease interactions and identify plant serpin target proteases.

2. Detection of Serpins in Plant Extracts and Localization of Serpins in Plant Tissues and Cells

2.1. Detection using biotinylated proteases

Serpins and other protease inhibitors in complex plant extracts can be detected using biotinylated proteases, provided the particular proteases used are able to form tight complexes (covalent or reversible) with the inhibitors in the plant extract. A suggested protocol is given here based on the detection of serpins (as well as a range of protease inhibitors of other families) from extracts of barley, wheat, rye, and oat grain using biotinylated trypsin, chymotrypsin, and subtilisin (Hejgaard and Hauge, 2002). Note that this method is suitable for seed extracts containing high concentrations of protease inhibitors but is unlikely to be effective for extracts of vegetative tissues, which normally contain much lower concentrations of these proteins.

1. Prepare the biotinylated protease using a standard biotinylation protocol (such as that offered through kits available from Pierce Biotechnology).
2. Separate the small- to medium-sized proteins in the plant extract at high resolution using SDS-PAGE with, for example, 10–20% Tris–Tricine gradient gels.
3. Immunoblot onto nitrocellulose membranes as described previously (Dahl *et al.*, 1996a).

4. Incubate the membrane for 1 h at 24 °C with biotinylated protease at a concentration of 50 mg ml^{-1}.
5. Treat the membrane with alkaline phosphatase-labeled streptavidin and stain under the same conditions as used for the immunoblotting.

2.2. Plant serpin antibody specificity

Western blotting and immunolocalization experiments to detect plant serpins are conducted using standard protocols, which are not repeated here, but sufficient numbers of such experiments have been performed using many different antibodies with extracts or tissues of barley (Roberts et al., 2003), wheat (Rosenkrands et al., 1994), apple (Hejgaard et al., 2005) and *Arabidopsis* (Lampl et al., 2010; Vercammen et al., 2006) to draw useful conclusions regarding plant serpin antibody specificity. The quality (potency and specificity) of the primary antibodies is key to success in these experiments. Consideration should be given to the aim of the experiment, which may be to obtain a general idea of the distribution of serpins in a tissue or (in contrast) to compare the specific localization of individual closely related serpins.

Some polyclonal antibodies raised against purified plant serpins, such as antibody R360 raised against recombinant barley serpin Zx (rBSZx), have been extremely potent and needed to be diluted as much as 40,000-fold in Western blotting experiments. Polyclonal antibodies raised in rabbits against purified plant serpins tend to cross-react with a subset of other serpins from both the same plant species and other species, even across the monocot–eudicot divide (Hejgaard et al., 2005). This did not appear to be the case with the polyclonal antibody raised in Guinea pigs against recombinant AtSerpin1 from *Arabidopsis*, as this (effective) antibody gave no signal on Western blots using an extract derived from an *atserpin1* (homozygous knockout) line (Lampl et al., 2010). Some monoclonal antibodies produced against plant serpins are highly specific, such as Mab11C7 raised against BSZ7b, which reacts against only this specific form of barley serpin (absent in some barley varieties; Evans and Hejgaard, 1999), while other monoclonal antibodies are less specific, such as Mab8E8, which was raised against BSZ7 but reacts strongly with rBSZx (Roberts et al., 2003).

2.3. Subcellular localization of *Arabidopsis* serpins

Subcellular localization of serpins in *Arabidopsis* has been performed using immunolabeling electron microscopy on ultrathin sections of root tips (Vercammen et al., 2006) and by confocal microscopy of *Arabidopsis* protoplasts expressing GFP–serpin fusion proteins (Ahn et al., 2009). Using the latter approach, the subcellular locations of AtSerpin1, AtSRP2, and

Plant Serpins

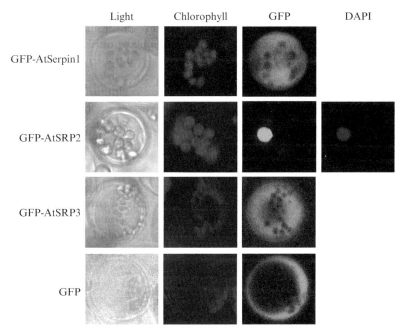

Figure 17.1 Subcellular localization of AtSerpin1, AtSRP2, and AtSRP3 in *Arabidopsis* protoplasts. GFP:AtSerpin1, GFP:AtSRP2, and GFP:AtSRP3 fusion constructs were each transformed into *Arabidopsis* protoplasts. Green fluorescence signals were detected by confocal laser scanning microscopy 24 h after transformation. Bright-field, chlorophyll, GFP, and DAPI images are shown. Expression of GFP alone was used as a control for cytosolic localization. DAPI served as a positive marker for nuclear localization. Protoplast preparation was from 2-week-old *Arabidopsis* seedlings. A reduced version of this figure was published previously (Ahn *et al.*, 2009).

AtSRP3 were found to be the cytosol, nucleus, and cytosol, respectively (Fig. 17.1), which reflected predictions made *in silico* (Ahn *et al.*, 2009). The protoplast preparation and transformation sections of the protocol used by Ahn *et al.* (2009) presented below are based largely on methods published previously (Abel and Theologis, 1994).

2.3.1. Generation of GFP–serpin fusion constructs

cDNA fragments of *AtSRP2* and *AtSRP3* corresponding to the full-length coding regions were amplified by PCR. Primers used were as follows: *AtSRP2*, forward primer 5′-ATCCCCGGGCAATGGATTCAAAAA-GAAAGAAC-3′ and reverse primer 5′-ACTGTCGACTTAGCCCG GTCCAACGCA-3′; *AtSRP3*, forward primer 5′-ATCCCCGGG-CAATGGATGTAAGAGAAGCT-3′ and reverse primer 5′-ACTGTC-GACTTAATAGTCATCTGAGTC-3′. Each PCR product was cloned into the 326-GFP vector (Lee *et al.*, 2001) using *Sma*I and *Sal*I. The *Arabidopsis* serpin cDNA was inserted into the 5'-end of *GFP* to avoid loss of a C-terminal signal peptide by the cleavage of the RCL near the C-terminus of the protein; however, the N-terminal fusion of the serpins with the GFP moiety means that that N-terminal targeting signals may be disrupted. Twenty micrograms of each plasmid DNA was prepared using Plasmid Midi Kit (Qiagen, USA) according to the manufacturer's instructions.

2.3.2. Preparation of *Arabidopsis* protoplasts

The following protocol is recommended for the preparation of *Arabidopsis* protoplasts:

1. Transfer seedlings grown on MS media for 10 days into 10 ml enzyme solution containing 2% cellulose (Yakult, Japan), 0.5% macerozyme (Yakult, Japan), 3 mM MES, 400 mM mannitol, and 1 mM $CaCl_2$. Incubate samples with shaking at 30 rpm at 24 °C for 10 h in the dark.
2. After incubation, remove cell debris using a 0.1-mm mesh and load the sample onto the top of an equal volume of 21% sucrose solution. Centrifuge at 500 rpm for 5 min. Collect the intact protoplasts found between the interface and the upper layer with a pipette.
3. Transfer the protoplast into 30 ml W5 solution (125 mM $CaCl_2$, 5 mM KCl, 150 mM NaCl, 5 mM glucose, 0.5 M mannitol, pH 5.8) and incubate for 1 h on ice.
4. Collect the protoplasts by centrifugation and transfer into 1-ml MaMg solution (15 mM $MgCl_2$, 400 mM mannitol, 5 mM MES-KOH, pH 5.6).
5. Perform cell counting using a haemocytometer with a light microscope. Use a minimum cell density of 5×10^6 cells ml^{-1} for transformation (below).

2.3.3. Transformation of the constructs and detection of GFP–serpin fusion proteins

The following protocol is recommended:

1. Mix 20 µg plasmid DNA of each fusion construct carefully with 250 µl protoplast suspension in MaMg solution (see above) containing 250 µl polyethylene glycol.
2. Incubate samples for 30 min at room temperature. Add 1 ml W5 solution to samples every 3 min for 21 min, collect the protoplasts by centrifugation, transfer to W5 solution and incubate at room temperature for 24 h in the dark.
3. To identify the cell nuclei, stain the protoplast samples with DAPI (4',6-diamidino-2-phenylindole) for 30 min.

4. Detect expression of fusion proteins using scanning confocal microscopy (for GFP: excitation 488 nm and emission 505–530 nm; for chlorophyll autofluorescence: excitation 488 nm and emission 650 nm).

2.3.4. Localization in extracellular medium based on Western blot analysis of exudates

To test whether AtSerpin1 from *Arabidopsis* could be found in the apoplast, Lampl *et al.* (2010) grew transgenic *Arabidopsis* plants overexpressing AtSerpin1-HA in aseptic liquid culture for 2 weeks. Total protein was extracted from leaves, roots, and the medium and analyzed on Western blots using anti-HA and anti-AtSerpin1 antibodies. The results of this experiment showed that AtSerpin1 can be detected in intra- and extracellular locations and enabled the researchers to conclude that a processed (lower molecular size) form of AtSerpin1 may be targeted to the secretory pathway (Lampl *et al.*, 2010), supporting conclusions from other experiments made earlier (Vercammen *et al.*, 2006).

3. Purification of Serpins from Plant Tissues

3.1. Protection of serpins from cleavage by endogenous proteases during purification

Some plant tissue extracts contain active proteases that, during purification, are capable of cleaving serpins, particularly in the exposed reactive center loop (RCL) of the serpin structure. This is not a major problem with serpin purification from mature cereal grains because protease activity is very low (although activity rises rapidly during germination). Nevertheless, it is recommended that protease inhibitor cocktails, such as those in Complete^{TM} Inhibitor tablets (Roche), be incorporated into extraction buffers during serpin purification.

3.2. Purification of serpins from mature seeds

Purification of particular serpins from mature seeds is facilitated by their high abundance. The following species are known to contain high concentrations of serpins in seeds: barley (*Hordeum* sp.; Hejgaard, 1982), wheat (*Triticum* sp.; Rosenkrands *et al.*, 1994), and rye (*Secale cereale*; Hejgaard, 2001), which are all members of the tribe Triticeae within the grass family (Poaceae); oat (*Avena sativa*; Hejgaard and Hauge, 2002), which is another grass but in the tribe Poeae; and apple (*Malus domestica*; Hejgaard *et al.*, 2005), a eudicot of the family Rosaceae. Immunohistochemistry has shown that most of the total serpin in mature barley grain is in the endosperm and subaleurone (Roberts *et al.*, 2003), while in apple seed, it is in the endosperm and cotyledons (Hejgaard *et al.*, 2005).

Interestingly, polypeptides representing serpins (more or less modified; all inactive as protease inhibitors) of barley origin can be readily purified from freeze-dried extracts of beer, where they are one of the dominant antigens present (Hejgaard and Kaersgaard, 1983). The contribution of serpin-derived polypeptides to beer quality continues to interest scientists working on the finer points of brewing (Curioni *et al.*, 1995; Evans *et al.*, 2003; Evans and Hejgaard, 1999; Evans and Sheehan, 2002; Garcia-Casado *et al.*, 2001). Serpin-derived polypeptides have also been examined in relation to allergic reactions to beer (Figueredo *et al.*, 1999; Garcia-Casado *et al.*, 2001; Herzinger *et al.*, 2004). While these studies are interesting, methods specific to the study of beer quality and beer anaphylaxis lie outside the scope of this chapter.

The model monocot species, rice (*Oryza sativa*) and maize (*Zea mays*), appear to contain much lower concentrations of grain serpins than other cereals of the Poaceae family (Jørn Hejgaard, personal communication). This difference is reflected in the representation of serpins among proteins identified in grain proteomics experiments performed with rice (Koller *et al.*, 2002) and maize (Mechin *et al.*, 2004) compared to those with barley (Finnie *et al.*, 2002; Ostergaard *et al.*, 2002) and wheat (Skylas *et al.*, 2000).

3.2.1. Thiol extraction

Serpins in the mature grains of barley, wheat (Fig. 17.2) and rye can be partially purified using "thiol extraction"; that is, extraction using a buffer containing a reducing agent such as dithiothreitol (DTT) following near-exhaustive "salt extraction" (extraction in a buffer with a substantial ionic strength but no reducing agent). Thiol extraction is based on the breaking up of disulfide-mediated protein aggregation, in the main involving serpins and β-amylases among the major salt-soluble proteins. Note that 40–60% of the serpin is lost in the salt extract as reflected in the results of SDS-PAGE analysis (see below). In addition, serpins without Cys residues available for intermolecular disulfide linkages, such as the "LR" serpin from barley, BSZx, will not be found in the thiol extract. Below is a suggested protocol for thiol extraction of barley, wheat, or rye grain:

1. Grind whole mature grains ($\sim$10 g) to a fine "flour" using a coffee grinder.
2. Transfer a sample of the flour (e.g., 2 g) to a centrifuge tube and stir on ice using a magnetic stirrer with extraction buffer (100 mM Tris–HCl, pH 8.0; 10 ml g^{-1} flour) for 30 min.
3. Centrifuge the suspension at 10,000$\times g$ for 20 min. Discard the supernatant (the first "salt extract") leaving a sample for analysis on SDS-PAGE.

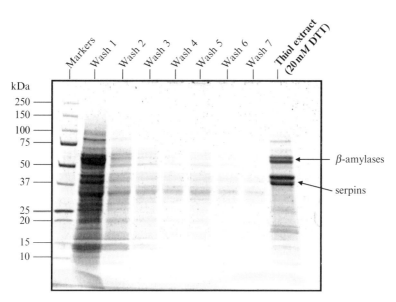

Figure 17.2 SDS-PAGE analysis of salt extracts and thiol extract of mature wheat grain. Mature grains of wheat (*Triticum aestivum* cv. Sunco) were ground to fine powder and proteins extracted as described in the text. The extract at each step was centrifuged and the supernatant run on reducing SDS-PAGE. The extraction/centrifugation step ("wash") was conducted seven times to remove almost all traces of salt-soluble proteins (normally only three washes are required for partial purification or native-PAGE analysis of the serpins). The thiol extract containing mostly serpins and β-amylases was obtained by addition of 20 mM DTT to the extraction buffer following exhaustive extraction with the extraction buffer alone. Experimental work and figure courtesy of Dr. Daniel J. Skylas.

Resuspend the pellet in extraction buffer, stir the suspension, and centrifuge as above.

4. After three salt extractions as above (more if desired, based on SDS-PAGE), resuspend the pellet in 20 ml extraction buffer containing 20 mM DTT, stir, and centrifuge the suspension. As an alternative to DTT, 0.2% (w/v) Na_2SO_3 (sodium sulfite) and 0.15% (w/v) $Na_2S_2O_5$ (sodium metabisulfite) can be added to the extraction buffer (Rosenkrands *et al.*, 1994). (Note that extraction with this mixture changes the charge of the serpins, which may affect a subsequent ion-exchange purification step.) The resulting supernatant is known as the "thiol extract," the proteins in which are mainly serpins and β-amylases.
5. Test the salt extracts and the thiol extract on reducing SDS-PAGE to confirm partial purification of the serpins (Fig. 17.2).

The thiol extraction protocol can be scaled down so that the extractions can be performed in Eppendorf tubes (e.g., for SDS-PAGE or 2D electrophoresis), in which case stirring with a magnetic flea is replaced by shaking.

3.2.2. Thiophilic adsorption chromatography

Further purification of serpins partially purified using thiol extraction (see above) can be achieved through thiophilic adsorption (T-gel) chromatography. T-gel is the β-mercaptoethanol (β-ME) derivative of divinylsulfone-activated agarose; it binds and releases proteins at high and low concentrations of lyotropic salt, respectively (Lihme and Heegaard, 1991). This method is recommended because a thiol (or sulfite) extract containing plant serpins can be applied directly to the T-gel without the need for desalting (indeed salt is added). The T-gel chromatography results in serpin fractions separated from β-amylases copurified during thiol extraction (Rosenkrands et al., 1994). A suggested protocol is as follows:

1. Prepare a T-gel column by coupling β-ME to a divinylsulfone-activated agarose matrix (Lihme and Heegaard, 1991).
2. Equilibrate the T-gel column (1.5×17 cm) with T-gel buffer (25 mM Tris–HCl, pH 8.0 containing 1 M Na_2SO_4 (sodium sulfate), 1 mM EDTA, and 20 mM β-ME or a combination of 0.2% (w/v) Na_2SO_3 and 0.15% (w/v) $Na_2S_2O_5$).
3. Bring the thiol (or sulfite) extract to 70% saturation with solid $(NH_4)_2SO_4$, redissolve the pellet in T-gel buffer, centrifuge, apply the supernatant to the column, and wash with T-gel buffer until the A_{280} returns to baseline levels. (An online calculator such as that provided at http://www.encorbio.com/protocols/AM-SO4.htm can be used to calculate the amount of $(NH_4)_2SO_4$ to add in g ml^{-1}.)
4. Elute adsorbed material using a 1–0 M gradient of Na_2SO_4 in T-gel buffer and collect, for example, 30 fractions.
5. Test the fractions with reducing SDS-PAGE to confirm separation of the serpins (∼40 kDa) from the β-amylases (∼60 kDa).

3.2.3. Anion-exchange chromatography

Further purification of serpins already partially purified can be achieved using anion-exchange chromatography. For the purification of serpins from wheat grain, for example, a three-step purification process involving thiol extraction, thiophilic adsorption chromatography, and anion-exchange chromatography using DEAE-Fractogel was used to yield essentially pure serpin fractions (Rosenkrands et al., 1994).

3.2.4. Purification from oat grain

Thiol extraction cannot be used as a partial purification of serpins from oat grain. A protocol that works well is as follows (Hejgaard and Hauge, 2002):

1. Grind oat flour (25 g) to a fine "flour" and extract on ice by stirring for 30 min with 500 ml 0.1 M Tris–HCl, pH 8.0, containing one Complete^{TM} inhibitor tablet (Roche).

2. Centrifuge the extract at 10,000×g for 20 min.
3. Add solid $(NH_4)_2SO_4$ to the supernatant to 35% saturation and centrifuge as above.
4. Precipitate the serpins in the supernatant with $(NH_4)_2SO_4$ at 65% saturation.
5. Collect both the pellet and the lipid-rich top-layer and suspend in 20 ml 25 mM Tris–HCl, pH 8.0, containing 1 M Na_2SO_4.
6. Centrifuge and filter the redissolved serpins and separate from other proteins by thiophilic adsorption chromatography (Fig. 17.3A).

The major serpin forms can then be separated by anion-exchange chromatography (e.g., using a MonoQ column, GE Healthcare) at pH 8.0 (Fig. 17.3B). Selected fractions can be pooled, treated with 20 mM DTT for 30 min at room temperature to dissolve any serpin–serpin and serpin–β-amylase aggregates (Hejgaard, 1976), and rechromatographed in 25 mM sodium phosphate, pH 8.0, on the same column using a 25–200 mM phosphate buffer gradient.

3.2.5. Purification from apple seed

A suggested purification protocol for serpins from apple seeds (Hejgaard et al., 2005) is presented here as an example of purification of eudicot seed serpins:

1. Decorticate seeds (3.0 g) from ripe apples and grind with a chilled mortar and pestle with 15 ml 0.1 M Tris–HCl, pH 8.0, containing 5 mM DTT and half a Complete™ inhibitor tablet.
2. Stir the slurry for 15 min on ice and centrifuge at 8000×g for 30 min at 4 °C and retain the supernatant.
3. Re-extract the pellet under the same conditions as above and precipitate the proteins in the combined supernatants adding solid $(NH_4)_2SO_4$ to 65% saturation at 4 °C.
4. Centrifuge as above and resuspend the pellet in 65% $(NH_4)_2SO_4$.
5. Dissolve the precipitate in 8 ml 0.1 M Tris–HCl, pH 8.0, containing 1 M Na_2SO_4.
6. Centrifuge and apply the supernatant to a T-gel (16×100 mm) as described above.
7. Dialyze the run-through from the T-gel column in 20 mM Tris–HCl, pH 9.0, and conduct anion-exchange chromatography using a MonoQ column (5×50 mm) equilibrated in 20 mM Tris–HCl, pH 9.0.
8. Apply a linear salt gradient from 0 to 0.2 M NaCl over 30 ml and then from 0.2 to 0.5 M over the next 15 ml.

This protocol should result in an effective separation of the serpins from each other and from contaminating proteins.

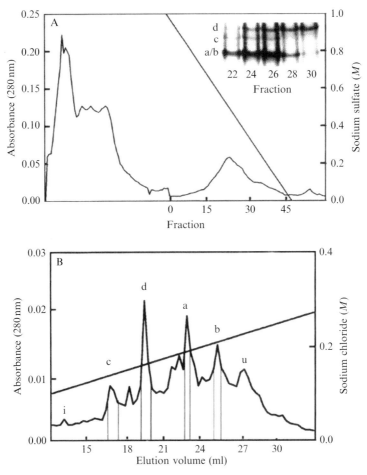

Figure 17.3 Purification of oat grain serpins. (A) Partial separation by thiophilic adsorption chromatography. The T-gel column (16 × 100 mm) was equilibrated as described in the text, and 20 ml protein concentrate from an $(NH_4)_2SO_4$ precipitation was applied. After washing with equilibration buffer, the adsorbed protein was eluted by a 1–0 M linear gradient of Na_2SO_4. Aliquots of selected 5-ml fractions (22–30) were subjected to native-PAGE followed by silver staining (inserted). Letters (a–d) mark bands representing the four oat serpin forms, which were subsequently separated by ion-exchange chromatography (B). The pooled fractions 22–26 from (A) were applied to a MonoQ column (5 × 50 mm) equilibrated as described in the text and eluted with a 0–0.3 M gradient of NaCl in the buffer. Letters a–d indicate the four serpin peaks, and vertical lines indicate the central peak fractions collected for anion-exchange rechromatography in a phosphate buffer system. Letters i and u indicate other protein peaks. Figure from Hejgaard and Hauge (2002) (Fig. 17.2).

3.3. Purification of serpins from vegetative organs

3.3.1. Purification from roots

In most plant species investigated, vegetative organs appear to contain much lower concentrations of serpins than are found in the mature seeds. There is evidence, however, that specific tissues or cell types within vegetative organs have relatively high serpin concentrations, such as the phloem sap of pumpkin (*Cucurbita maxima*; Yoo *et al.*, 2000) and the closely related species cucumber (la Cour Petersen *et al.*, 2005) as well as the root tips of barley (Roberts *et al.*, 2003). The following protocol was used for purification of serpins from whole roots of barley (Roberts *et al.*, 2003) but could be applied to root tips:

1. Harvest roots (20 g) from 7-day-old seedlings and homogenize at 4 °C with a mortar and pestle in 40-ml 25 mM Tris–HCl, pH 8.0, 1 mM EDTA (Buffer A).
2. Centrifuge the mixture at 4 °C to remove debris and then apply the supernatant to a DEAE Sephadex A-50 column (2.2×1.5 cm) previously equilibrated with Buffer A at room temperature.
3. Wash the column with Buffer A until the A_{280} returns to the baseline level and then apply a linear salt gradient from 0 to 1.0 M KCl in a total of 60 ml.
4. Identify fractions containing serpins, pool where appropriate, and remove salt by gel filtration through a PD-10 column (GE Healthcare).
5. Apply the sample to a MonoQ HR 5/5 anion-exchange column (GE Healthcare) equilibrated with Buffer A.
6. Wash the column with 5 ml Buffer A and elute the serpins with a gradient of 0–0.5 M KCl in a total volume of 40-ml Buffer A. Identify the serpin-containing fractions.

3.4. Monitoring of fractions during plant serpin purification

Methods used to monitor the presence and relative concentration of serpins in fractions during a purification protocol include SDS-PAGE (Rosenkrands *et al.*, 1994), native-PAGE (see below; Hejgaard, 2001; Hejgaard and Hauge, 2002; Ostergaard *et al.*, 2000; Roberts and Hejgaard, 2008), N-terminal sequencing (Roberts *et al.*, 2003), Western blotting (including dot-blotting directly onto membrane and the use of multiple antibodies of different specificity if appropriate; Roberts *et al.*, 2003) and inhibition assays conducted using microtiter plates (Ostergaard *et al.*, 2000).

3.5. Separation and visualization of plant serpins using native-PAGE

Most plant serpins have very similar molecular sizes (∼43 kDa); thus multiple serpins from a plant tissue extract cannot be visualized as distinct bands on SDS-PAGE. Fortunately, small differences in protein surface charge allow native-PAGE to provide clear separation and visualization of plant serpins, even those with a high degree of sequence identity (Ostergaard *et al.*, 2000). It is recommended that prior to native-PAGE (usually run with Tris–glycine gels containing 8% polyacrylamide), the sample containing the serpins be heated to 50 °C in native-PAGE Sample Buffer (125 mM Tris–HCl, pH 8.6, 50%, v/v glycerol, 0.025% bromophenol blue containing 20 mM DTT) for 15 min to break any intermolecular disulfide bonds. (Note that the native-PAGE Running Buffer (25 mM Tris, 200 mM glycine) need not contain any DTT or other reducing agent.)

Apart from its application in the monitoring of fractions during serpin purification, native-PAGE has facilitated the discrimination between cultivars of wheat and barley on the basis of the presence/absence of specific serpins in grain extracts (Jørn Hejgaard, personal communication).

4. Production of Recombinant Plant Serpins

Recombinant plant serpins encoded by cDNAs cloned from barley (Dahl *et al.*, 1996a), wheat (Dahl *et al.*, 1996a; Ostergaard *et al.*, 2000), pumpkin (Yoo *et al.*, 2000) and *Arabidopsis* (Lampl *et al.*, 2010) have been produced in *Escherichia coli* and purified for kinetic analysis. All these serpins were made as fusion proteins with either a His$_6$-tag at the N-terminus or an HA (hemagglutinin)-tag at the C-terminus. *E. coli* strains used successfully for recombinant plant serpin production include BL21(DE3; Lampl *et al.*, 2010) and BL21(DE3)pLysS (Dahl *et al.*, 1996a; Yoo *et al.*, 2000).

Yoo *et al.* showed that the phloem-sap-purified and N-terminally His$_6$-tagged recombinant forms of the pumpkin serpin CmPS-1 displayed very similar kinetic behavior; in this case, values for the association rate constant (k_a) and stoichiometry of inhibition (SI) for interaction *in vitro* with pancreatic elastase (Yoo *et al.*, 2000). Thus, kinetic analysis of recombinant plant serpins with this type of N-terminal fusion can be expected to provide realistic estimates of k_a and SI for the corresponding native proteins.

5. Analysis of Plant Serpin–Protease Interactions

Kinetic analyses of the interactions *in vitro* between plant serpins and proteases are conducted in the same manner as for serpins from other organisms. These methods have been documented elsewhere (Schechter and Plotnick, 2004) and are not included in this chapter. Plant serpin kinetic analyses (mainly of interactions with mammalian proteases) have been conducted for a range of species (Dahl *et al.*, 1996a,b; Hejgaard, 2001,

2005; Hejgaard et al., 2005; Hejgaard and Hauge, 2002; Ostergaard et al., 2000; Vercammen et al., 2006; Yoo et al., 2000), and summaries of the values for the kinetic constants have been made (Christeller and Laing, 2005; Roberts and Hejgaard, 2008).

5.1. Reducing and nonreducing SDS-PAGE

In a recent study, the major *Arabidopsis thaliana* serpin, AtSerpin1, was found to interact *in vivo* with an endogenous cysteine protease, RESPONSIVE TO DESICCATION-21 (RD-21; Lampl et al., 2010). In cases such as this, where a cysteine participates in the trapped acyl enzyme intermediate, the nature of the covalent linkage means the serpin–protease complex is very sensitive to reducing agents in the media. Thus, in all gel fractionation applications, when the Laemmli loading buffer contains 1% (v/v) β-ME as reducing agent, the complex will break down into cleaved serpin and free protease. In contrast, when samples are prepared without β-ME in the loading buffer, the serpin–protease complex is stable and can be isolated after fractionation in standard SDS-containing Laemmli polyacrylamide gels. Thus, the differential use of a reducing agent is a convenient way to follow serpin–protease interaction.

5.2. Identification of target proteases for plant serpins

5.2.1. Utilization of T-DNA knockout and HA-tagged overexpression mutants of *Arabidopsis*

The identification of endogenous target proteases for plant serpins is facilitated greatly by the use of transgenic expression of modified serpin, as demonstrated for the identification of the *in vivo* target protease for AtSerpin1 (Lampl et al., 2010). Transgenes can be expressed in plants using the viral 35S promoter (Kay et al., 1987). AtSerpin1-HA was constructed from an AtSerpin1 EST clone (accession number R65473) using forward primer 5′-TGCTCTAGAATGGACGTGCGTGAATCAATC-3′ and reverse primer 5′-AGGCCCGGGATGCAACGGATCAACAACTTG-3′. The clone was fused upstream to three repeats of the hemagglutinin (HA) epitope (YPYDVPDYA) in pPZP111 (Fass et al., 2002) under the control of the 35S promoter from the cauliflower mosaic virus by *Xba*I-*Sma*I and used to transform *Agrobacterium tumefaciens* strain EHA105 (Hood et al., 1993). Transgenic plants are isolated by the floral dip method (Clough and Bent, 1998) and homozygous lines are isolated.

5.2.2. Immunoprecipitation using anti-serpin antibodies

Immunoprecipitation and subsequent fractionation of the isolated complex facilitates the study of serpin complexes and identification of target proteases. For immunoprecipitation experiments involving wild-type plants or lines with serpin transgenes, the plants are grown in soil for 3 weeks in a 16-h light and 8-h dark regime at 21 °C. A tissue-culture source of plant material will suffice if only a small amount of material is necessary. In this case, seeds are germinated on sterile B5 Gamborg plates (Gamborg et al., 1976) and harvested after 1–2 weeks. The procedure detailed below was adapted for small-scale immunoprecipitation experiments using tissue culture extracts as a source.

1. For immunoprecipitation, plants are harvested rapidly from 14-day-old seedlings or 3-week-old leaf tissue and frozen in liquid nitrogen. A small-scale experiment might involve 0.3 mg of starting material. The material can be stored at -80 °C or used immediately. Plants are homogenized in liquid nitrogen with a mortar and pestle until a fine powder is obtained. Extraction buffer (20 mM Tris pH 8.0, 1 mM EDTA, and 50 mM NaCl (1 ml mg^{-1} of tissue)) with and without a cocktail of protease inhibitors (cat. # P9599, Sigma–Aldrich) is added and the frozen slurry further homogenized.

2. Immunoprecipitation was modified from a previously published method (Lee, 2007). In the following procedure, AtSerpin1 was immunoprecipitated using anti-AtSerpin1 antibody raised in Guinea pigs. The antibody was first reacted with Protein A Sepharose beads (cat. # 22811, Pierce Biotechnology). (Protein A shows strong binding affinity to Guinea pig antibodies and would need to be adapted if other antibodies were used.)

3. All procedures are performed with prechilled reagents either in a cold room or on ice. Prior to the immunoprecipitation purification, 10% of the extract is removed, mixed with one volume of 2×SDS sample buffer (100 mM Tris–HCl, pH 6.8, 4% SDS, 20% glycerol, 5% β-ME, 2 mM EDTA and 0.1 mg ml^{-1} bromophenol blue) and heated at 95 °C for 3 min. This sample serves as the control for "input."

4. For preloading of antibody, the beads (40 μl of a 50% slurry) are first equilibrated in 500 μl elution buffer (20 mM HEPES, pH 7.5, 100 mM NaCl, 0.05% Triton X-100, 1 mM DTT) and 33 μl plant protease inhibitor cocktail (P9599, Sigma–Aldrich) in 1 ml elution buffer for 15 min on a slowly rotating wheel (8 rpm) followed by centrifugation at 3000×g for 10 s. The beads are washed three times in 500 μl elution buffer and resuspended gently after each wash. After the final wash, the supernatant is removed and the beads are resuspended with two volumes of elution buffer.

5. The extract needs to be pretreated with washed beads to remove nonspecific adsorption. To this end, 10 μl of the equilibrated beads are added to the extract, incubated for 20 min at 4 °C on the rotating wheel, centrifuged at 12,000×g for 3 min, and the pretreated extract transferred to a new tube.

6. Coupling of antibody to beads involves adding the equilibrated beads (10 μl/reaction) to a microcentrifuge tube containing 300 μl of elution buffer. Antibody is added (3–5 μl whole antiserum or 0.5–1 μg affinity-purified antibody/reaction) to the tube and incubated at room temperature for 1 h on a rotating wheel. The sample is then centrifuged at $3000 \times g$ for 10 s and the supernatant removed. Washed beads treated only with elution buffer are also prepared as a control.
7. For isolation of protein, the pretreated protein extract is added to the tube containing the coupled antibody-protein A beads and incubated at 4 °C for 1–12 h in rotation. The sample is then centrifuged at $3000 \times g$ for 10 s and the supernatant removed. The sample is washed seven times at 4 °C with 1 ml elution buffer. In each wash, the reaction is incubated for 20 min with rotation, centrifuged at $3000 \times g$ for 10 s, and the supernatant removed. The beads are resuspended in 20 μl 2×sample buffer (described previously) and boiled at 95 °C for 3 min. Tubes are centrifuged at $12,000 \times g$ for 1 min and the sample loaded on SDS-PAGE.

Gel fractionation is carried out on 10% nonreducing SDS-PAGE (Laemmli, 1970). The loading buffer for nonreducing gels contains no reducing agent. The gels are blotted with a wet transfer onto polyvinylidene fluoride (PVDF, cat. # 1620177, Bio-Rad) membrane for 1.5 h at constant voltage (100 V). The membrane is developed with anti-AtSerpin1 antibodies (1:1000) and secondary anti-Guinea pig antibodies coupled with horseradish peroxidase (HRP, 1:3000).

5.2.3. Use of a mechanism-based probe that targets cysteine proteases

E-64 is an irreversible inhibitor of cysteine proteases that forms a thioether bond with the thiol of the cysteine in the active site. The biotinylated derivative of E-64, DCG-04, thus effectively covalently targets cysteine proteases of the papain family and facilitates their identification (Greenbaum et al., 2002). It was found that DCG-04 can compete effectively with AtSerpin1 for availability of target proteases (Lampl et al., 2010).

Plant cysteine proteases are labeled with DCG-04 in the following manner:

1. Plants (0.3 mg) are homogenized in liquid nitrogen with a mortar and pestle until a fine powder is obtained. Extraction buffer (1 ml for each milligram of tissue; 20 mM Tris, pH 8.0, 1 mM EDTA and 50 mM NaCl) is added and the frozen slurry further homogenized.
2. Protein extract (0.3 mg) is labeled in 0.125 ml total volume of a buffer containing 50 mM sodium acetate, pH 6.0, 10 mM DTT and 2 μM DCG-04. Control samples contain 0.2 mM E-64 (cat. # E3132, Sigma). Labeling is performed with gentle shaking for 3 h at room temperature.

3. Biotinylated proteins are separated on reducing 10% SDS gels. The gels are blotted with a wet transfer onto PVDF (cat. # 1620177, PVDF, Bio-Rad) for 1.5 h at constant voltage (100 V). The membrane is developed with streptavidin-HRP (1:3000, Jackson Laboratories).

ACKNOWLEDGMENTS

The authors thank Jørn Hejgaard (Technical University of Denmark, retired) for valuable discussions concerning the content of the chapter, Dr. Salla Marttila (Swedish University of Agricultural Sciences) for helpful suggestions on immunolocalization of plant serpins, and Daniel J. Skylas (BRI, Sydney) for contributing unpublished data (Fig. 17.2).

REFERENCES

Abel, S., and Theologis, A. (1994). Transient transformation of *Arabidopsis* leaf protoplasts: A versatile experimental system to study gene expression. *Plant J.* **5**, 421–427.

Ahn, J. W., Atwell, B. J., and Roberts, T. H. (2009). Serpin genes *AtSRP2* and *AtSRP3* are required for normal growth sensitivity to a DNA alkylating agent in *Arabidopsis*. *BMC Plant Biol.* **9**, 52.

Alberts, B., Johnson, A., Lewis, J., Raff, M., Roberts, K., and Walter, P. (2008). Molecular Biology of the Cell. Garland Science, New York.

Carrell, R. W., and Travis, J. (1985). Alpha1-antitrypsin and the serpins: variation and countervariation. *Trends Biochem. Sci.* **10**, 20–24.

Christeller, J., and Laing, W. (2005). Plant serine proteinase inhibitors. *Protein Pept. Lett.* **12**, 439–447.

Clough, S. J., and Bent, A. F. (1998). Floral dip: A simplified method for *Agrobacterium*-mediated transformation of *Arabidopsis thaliana*. *Plant J.* **16**, 735–743.

Curioni, A., Pressi, G., Furegon, L., and Peruffo, A. D. B. (1995). Major proteins of beer and their precursors in barley: Electrophoretic and immunological studies. *J. Agric. Food Chem.* **43**, 2620–2626.

Dahl, S. W., Rasmussen, S. K., and Hejgaard, J. (1996a). Heterologous expression of three plant serpins with distinct inhibitory specificities. *J. Biol. Chem.* **271**, 25083–25088.

Dahl, S. W., Rasmussen, S. K., Petersen, L. C., and Hejgaard, J. (1996b). Inhibition of coagulation factors by recombinant barley serpin BSZx. *FEBS Lett.* **394**, 165–168.

Evans, D. E., and Hejgaard, J. (1999). The impact of malt derived proteins on beer foam quality. Part 1. The effect of germination and kilning on the level of protein Z4, protein Z7 and LTP1. *J. Inst. Brew.* **105**, 159–169.

Evans, D. E., and Sheehan, M. C. (2002). Don't be fobbed off: The substance of beer foam—A review. *J. Am. Soc. Brew. Chem.* **60**, 47–57.

Evans, D. E., Robinson, L. H., Sheehan, M. C., Tolhurst, R. L., Hill, A., Skerritt, J. S., and Barr, A. R. (2003). Application of immunological methods to differentiate between foam-positive and haze-active proteins originating from malt. *J. Am. Soc. Brew. Chem.* **61**, 55–62.

Fagerhol, M. K., and Laurell, C. B. (1967). The polymorphism of "prealbumins" and alpha-1-antitrypsin in human sera. *Clin. Chim. Acta* **16**, 199–203.

Fass, E., Shahar, S., Zhao, J., Zemach, A., Avivi, Y., and Grafi, G. (2002). Phosphorylation of histone H3 at serine 10 cannot account directly for the detachment of human heterochromatin protein 1 gamma from mitotic chromosomes in plant cells. *J. Biol. Chem.* **277**, 30921–30927.

Figueredo, E., Quirce, S., del Amo, A., Cuesta, J., Arrieta, I., Lahoz, C., and Sastre, J. (1999). Beer-induced anaphylaxis: Identification of allergens. *Allergy* **54,** 630–634.

Finnie, C., Melchior, S., Roepstorff, P., and Svensson, B. (2002). Proteome analysis of grain filling and seed maturation in barley. *Plant Physiol.* **129,** 1308–1319.

Gamborg, O. L., Murashige, T., Thorpe, T. A., and Vasil, I. K. (1976). Plant tissue culture media. *In Vitro* **12,** 473–478.

Garcia-Casado, G., Crespo, J. F., Rodriguez, J., and Salcedo, G. (2001). Isolation and characterization of barley lipid transfer protein and protein Z as beer allergens. *J. Allergy Clin. Immunol.* **108,** 647–649.

Greenbaum, D., Baruch, A., Hayrapetian, L., Darula, Z., Burlingame, A., Medzihradszky, K. F., and Bogyo, M. (2002). Chemical approaches for functionally probing the proteome. *Mol. Cell. Proteomics* **1,** 60–68.

Han, X., Huang, Z. F., Fiehler, R., and Broze, G. J., Jr. (1999). The protein Z-dependent protease inhibitor is a serpin. *Biochemistry* **38,** 11073–11078.

Hejgaard, J. (1976). Free and protein-bound beta-amylases of barley grain: Characterization by two-dimensional immunoelectrophoresis. *Physiol. Plant* **38,** 293–299.

Hejgaard, J. (1982). Purification and properties of protein Z—A major albumin of barley endosperm. *Physiol. Plant* **54,** 174–182.

Hejgaard, J. (2001). Inhibitory serpins from rye grain with glutamine as P-1 and P-2 residues in the reactive center. *FEBS Lett.* **488,** 149–153.

Hejgaard, J. (2005). Inhibitory plant serpins with a sequence of three glutamine residues in the reactive center. *Biol. Chem.* **386,** 1319–1323.

Hejgaard, J., and Hauge, S. (2002). Serpins of oat (*Avena sativa*) grain with distinct reactive centres and inhibitory specificity. *Physiol. Plant* **116,** 155–163.

Hejgaard, J., and Kaersgaard, P. (1983). Purification and properties of the major antigenic beer protein of barley origin. *J. Inst. Brew.* **89,** 402–410.

Hejgaard, J., and Roberts, T. H. (2007). Plant serpins. *In* "Molecular and Cellular Aspects of the Serpinopathies and Disorders in Serpin Activity," (G. A. Silverman and D. A. Lomas, eds.), pp. 279–300. World Scientific, New Jersey.

Hejgaard, J., Rasmussen, S. K., Brandt, A., and Svendsen, I. (1985). Sequence homology between barley endosperm protein Z and protease inhibitors of the alpha-1-antitrypsin family. *FEBS Lett.* **180,** 89–94.

Hejgaard, J., Laing, W. A., Marttila, S., Gleave, A. P., and Roberts, T. H. (2005). Serpins in fruit and vegetative tissues of apple (*Malus domestica*): Expression of four serpins with distinct reactive centres and characterisation of a major inhibitory seed form, MdZ1b. *Funct. Plant Biol.* **32,** 517–527.

Herzinger, T., Kick, G., Ludolph-Hauser, D., and Przybilla, B. (2004). Anaphylaxis to wheat beer. *Ann. Allergy Asthma Immunol.* **92,** 673–675.

Hood, E. E., Gelvin, S. B., Melchers, L. S., and Hoekema, A. (1993). New *Agrobacterium* helper plasmids for gene transfer to plants. *Transgenic Res.* **2,** 208–218.

Kay, R., Chan, A., Daly, M., and McPherson, J. (1987). Duplication of CaMV 35S promoter sequences creates a strong enhancer for plant genes. *Science* **236,** 1299–1302.

Koller, A., Washburn, M. P., Lange, B. M., Andon, N. L., Deciu, C., Haynes, P. A., Hays, L., Schieltz, D., Ulaszek, R., Wei, J., Wolters, D., and Yates, J. R. (2002). Proteomic survey of metabolic pathways in rice. *Proc. Natl. Acad. Sci. USA* **99,** 11969–11974.

la Cour Petersen, M., Hejgaard, J., Thompson, G. A., and Schulz, A. (2005). Cucurbit phloem serpins are graft-transmissible and appear to be resistant to turnover in the sieve element-companion cell complex. *J. Exp. Bot.* **56,** 3111–3120.

Laemmli, U. K. (1970). Cleavage of structural proteins during the assembly of the head of bacteriophage T4. *Nature* **227,** 680–685.

Lampl, N., Budai-Hadrian, O., Davydov, O., Joss, T. V., Harrop, S. J., Curmi, P. M., Roberts, T. H., and Fluhr, R. (2010). *Arabidopsis* AtSerpin1: Crystal structure and *in vivo* interaction with its target protease RESPONSIVE TO DESICCATION-21 (RD21). *J. Biol. Chem.* **285,** 13550–13560.

Law, R. H. P., Zhang, Q. W., McGowan, S., Buckle, A. M., Silverman, G. A., Wong, W., Rosado, C. J., Langendorf, C. G., Pike, R. N., Bird, P. I., and Whisstock, J. C. (2006). An overview of the serpin superfamily. *Genome Biol.* **7**(5), 216.

Lee, C. (2007). Coimmunoprecipitation assay. *Methods Mol. Biol.* **362**, 401–406.

Lee, Y. J., Kim, D. H., Kim, Y. W., and Hwang, I. (2001). Identification of a signal that distinguishes between the chloroplast outer envelope membrane and the endomembrane system *in vivo*. *Plant Cell* **13**, 2175–2190.

Lihme, A., and Heegaard, P. M. (1991). Thiophilic adsorption chromatography: The separation of serum proteins. *Anal. Biochem.* **192**, 64–69.

Mechin, V., Balliau, T., Chateau-Joubert, S., Davanture, M., Langella, O., Negroni, L., Prioul, J. L., Thevenot, C., Zivy, M., and Damerval, C. (2004). A two-dimensional proteome map of maize endosperm. *Phytochemistry* **65**, 1609–1618.

Ostergaard, H., Rasmussen, S. K., Roberts, T. H., and Hejgaard, J. (2000). Inhibitory serpins from wheat grain with reactive centers resembling glutamine-rich repeats of prolamin storage proteins—Cloning and characterization of five major molecular forms. *J. Biol. Chem.* **275**, 33272–33279.

Ostergaard, O., Melchior, S., Roepstorff, P., and Svensson, B. (2002). Initial proteome analysis of mature barley seeds and malt. *Proteomics* **2**, 733–739.

Prouse, C. V., and Esnouf, M. P. (1977). The isolation of a new warfarin-sensitive protein from bovine plasma. *Biochem. Soc. Trans.* **5**, 255–256.

Roberts, T. H., and Hejgaard, J. (2008). Serpins in plants and green algae. *Funct. Integr. Genomics* **8**, 1–27.

Roberts, T. H., Marttila, S., Rasmussen, S. K., and Hejgaard, J. (2003). Differential gene expression for suicide-substrate serine proteinase inhibitors (serpins) in vegetative and grain tissues of barley. *J. Exp. Bot.* **54**, 2251–2263.

Roberts, T. H., Hejgaard, J., Saunders, N. F. W., Cavicchioli, R., and Curmi, P. M. G. (2004). Serpins in unicellular *Eukarya*, *Archaea*, and *Bacteria*: Sequence analysis and evolution. *J. Mol. Evol.* **59**, 437–447.

Rosenkrands, I., Hejgaard, J., Rasmussen, S. K., and Bjorn, S. E. (1994). Serpins from wheat grain. *FEBS Lett.* **343**, 75–80.

Schechter, N. M., and Plotnick, M. I. (2004). Measurement of the kinetic parameters mediating protease-serpin inhibition. *Methods* **32**, 159–168.

Silverman, G. A., Whisstock, J. C., Bottomley, S. P., Huntington, J. A., Kaiserman, D., Luke, C. J., Pak, S. C., Reichhart, J. M., and Bird, P. I. (2010). Serpins flex their muscle: I. Putting the clamps on proteolysis in diverse biological systems. *J. Biol. Chem.* **285**, 24299–24305.

Skylas, D. J., Mackintosh, J. A., Cordwell, S. J., Basseal, D. J., Walsh, B. J., Harry, J., Blumenthal, C., Copeland, L., Wrigley, C. W., and Rathmell, W. (2000). Proteome approach to the characterisation of protein composition in the developing and mature wheat-grain endosperm. *J. Cereal Sci.* **32**, 169–188.

Steenbakkers, P. J., Irving, J. A., Harhangi, H. R., Swinkels, W. J., Akhmanova, A., Dijkerman, R., Jetten, M. S., van der Drift, C., Whisstock, J. C., and Op den Camp, H. J. (2008). A serpin in the cellulosome of the anaerobic fungus *Piromyces* sp. strain E2. *Mycol. Res.* **112**, 999–1006.

van der Hoorn, R. A. (2008). Plant proteases: From phenotypes to molecular mechanisms. *Annu. Rev. Plant Biol.* **59**, 191–223.

Vercammen, D., Belenghi, B., van de Cotte, B., Beunens, T., Gavigan, J. A., De Rycke, R., Brackenier, A., Inze, D., Harris, J. L., and Van Breusegem, F. (2006). Serpin1 of *Arabidopsis thaliana* is a suicide inhibitor for Metacaspase 9. *J. Mol. Biol.* **364**, 625–636.

Yoo, B. C., Aoki, K., Xiang, Y., Campbell, L. R., Hull, R. J., Xoconostle-Cazares, B., Monzer, J., Lee, J. Y., Ullman, D. E., and Lucas, W. J. (2000). Characterization of *Cucurbita maxima* phloem serpin-1 (CmPS-1)—A developmentally regulated elastase inhibitor. *J. Biol. Chem.* **275**, 35122–35128.

Author Index

A

Aaronson, H., 92–93
Abdelnabi, R., 85–86
Abe, H., 177–178
Abitbol, M. M., 92–93
Abraham, S. N., 139, 158
Abramovitz, R., 114
Abrams, J. M., 244
Accurso, F. J., 139
Adachi, E., 137–139, 171–172
Adamini, N., 110–111
Adams, S. P., 26–27
Adler, A. J., 185–186
Agah, R., 91
Agapite, J., 244
Agostoni, A., 206
Ahern, S. M., 115
Ahmad, S. T., 212
Ahn, J. W., 350–351
Ahringer, J., 289–291
Aigaki, T., 236–237
Aitken, M. L., 139
Akassoglou, K., 81–82
Akhmanova, A., 348–349
Akiyama, Y., 118
Akool, el-S., 119
Alarco, A. M., 218
Al Balwi, M., 175
Alberdi, E. M., 190, 193, 194
Albert, P. S., 288–289
Alberts, B., 348
Alberts, M., 136–137
Albert, V., 92–93
Albornoz, F., 90, 91
Albright, S., 219
Aleksic, N., 81
Alessi, M. C., 79–81, 89
Alexander, C. M., 151
Alexander, S. J., 115
Almedom, R. B., 292
Almholt, K., 92–93
Alrasheed, S., 175
AlSwaid, A., 175
Altiok, O., 136–137
Alvarado, C. M., 84
Amaral, J., 199
Amardeilh, M. F., 304–305, 306–307
Ambros, V., 293

Ambrus, G., 64–65
Amouyel, P., 324
Ananthanarayanan, V. S., 169
Andersen, J. A., 81–82
Anderson, G., 84
Anderson, K. V., 81, 214
Anderson, R. G., 38
Anderson, S. D., 292
Andon, N. L., 354
Andreasen, P. A., 23, 27, 28, 78–79, 81–82, 92, 118, 213–214, 291–292
Andres, A. C., 151–152
Andrews, J. L., 108
Andrieu, J. P., 57
Anelli, T., 168
Anisowicz, A., 150, 154–155
An, J. K., 45, 48–49
Ansari, M. N., 88
Antalis, T. M., 112, 113, 115, 117, 119
Antal, J., 64–65
Aoki, K., 359, 360–361
Appella, E., 156–157
Apweiler, R., 229
Aquadro, C. F., 212–213
Arai, H., 177–178
Arandjelovic, S., 21
Argraves, K. M., 21–22, 28, 109–110
Argraves, W. S., 18, 20, 23, 24, 26, 28, 213–214
Arima, K., 108, 332–333, 343
Arkan, M. C., 113
Arkenbout, E. K., 91
Arlaud, G. J., 56–57
Arnman, V., 81
Arnold, W. V., 173–174
Arrieta, I., 354
Arthur, J. S., 113, 116–117
Arveiler, D., 324
Asada, S., 170–171
Ashburner, M., 211, 218–219
Ashcom, J. D., 18, 20, 24
Ashton-Rickardt, P. G., 284, 291–292
Askari, A. T., 90
Askew, D. J., 106, 136, 207–209, 284, 286–287, 288–289, 291–292, 332
Askew, Y. S., 136, 207–209, 284, 286–287, 288–289, 291–292
Aso, A., 170–171
Aso, Y., 114–116

367

Asztalos, Z., 234–235
Atkinson, J. B., 89
Atomi, Y., 176–177
Atwell, B. J., 350–351
Aufenvenne, K., 110–111
Auge, B., 212
Auron, P. E., 115
Avet-Rochex, A., 212
Avivi, Y., 361
Aymerich, M. S., 194

B

Bachinger, H. P., 171–174
Bachmann, F., 79–80, 81, 106–107, 108, 110
Bading, H., 114–116
Baek, M. J., 212
Baglin, T. P., 210
Baglioni, C., 113, 115, 119
Bagossi, P., 110
Baici, A., 286–287
Bainton, R. J., 253
Bajou, K., 92–93
Baker, H., 314
Baker, M. S., 107–108, 110, 112, 115–116
Baker, N. K., 84
Bakheet, T., 118–119
Balch, W. E., 304
Balczer, J., 64–65
Baldassarre, M., 173
Balliau, T., 354
Barcena, L., 209
Barker, J. H., 84
Barlage, S., 108, 115
Barnes, A. M., 175
Barnes, C., 111–112
Barnes, T., 115
Barnstable, C. J., 184
Barr, A. R., 354
Barrett-Bergshoeff, M. M., 109–110
Barrett, J., 305–306
Barr, M. M., 289–291
Barr, P. J., 106
Barry, M., 305–306
Barsky, S. H., 154–155
Barstead, R. J., 284, 288, 291–292
Bartee, M. Y., 304, 305–306, 314, 316
Bartuski, A. J., 332–334, 342–343
Baruch, A., 363
Basler, K., 231, 238–239
Basseal, D. J., 354
Basset, A., 217
Bass, R., 106
Bateman, J. F., 172–173, 174–175
Bates, E. J., 113
Batten, M. R., 170–171
Battey, F. D., 21–22, 26–27, 28, 109–110
Baudner, S., 79–80

Baumann, M., 139, 142–143, 144, 145
Baumann, U., 175
Baumeister, R, A., 292
Becerra, S. P., 184, 185, 187, 190–191, 194–195, 199
Becker, D., 175
Becker, L. M., 109
Becker, M., 81–82
Becker, P. B., 117–118
Beckett, C., 35, 36, 40, 41, 43, 48–49, 50
Beelen, V., 82, 83
Beer, H. D., 136–137
Beermann, F., 92–93
Beeton, M. L., 136–137, 139
Beglova, N., 23–24
Behl, R., 137–139
Behre, E. H., 23–24
Behringer, R., 153–154
Beinrohr, L., 56–57, 64–65, 66
Belaaouaj, A., 139
Belenghi, B., 350–351, 353, 360–361
Belin, D., 106–107, 110
Belle, A., 168
Bellido, M. L., 118
Belorgey, D., 206, 211, 260–261
Belvin, M., 212, 214, 216–218
Benarafa, C., 136–139, 140, 141, 142–143, 144, 145
Bendall, L. J., 145
Bender, W. W., 253
Bendtsen, L., 18
Bennett, B., 78–80, 81, 109–110
Bent, A. F., 361
Bentele, C., 213
Beretta, A., 206
Bergamaschini, L., 206
Berger, U., 81–82
Berget, S. M., 157
Bergonzelli, G. E., 81, 116–117
Berg, R. A., 170–171
Berguer, A. M., 84
Berkowitz, L. A., 293
Bern, H. A., 157
Bernspang, E., 35
Bernstein, E. F., 81
Berry, G. J., 314
Berry, L. R., 304
Berta, A., 110
Bertram, K. C., 108
Besson, M. T., 234–235
Best, A. K., 90
Betsuyaku, T., 145
Beunens, T., 350–351, 353, 360–361
Bhagat, G., 47–48
Bhatia, N., 91
Bicciato, S., 111–112
Bickersteth, B., 314
Bick, R. J., 2

Bienkowski, R. S., 175
Bieth, J. G., 338
Bijnens, A. P., 78–79
Binder, B. R., 81–82
Bingham, A., 324
Bingol-Karakoc, G., 136–137
Birch, N. P., 316
Bird, C. H., 284, 291–292
Bird, P. I., 57, 106, 108, 109, 137–139, 150, 206, 207–209, 284, 286–287, 287, 291–292, 302–304, 332, 348
Birman, S., 234–235
Bischof, J., 231, 238–239
Bissell, M. J., 151
Bitzer, J., 118
Biyikoglu, B., 110
Bjorn, S. E., 350, 353–355, 356
Bjornstad, C., 64–65
Bjorquist, P., 78–79
Blacher, S., 92–93
Blacklow, S. C., 21–22, 23–24
Black, S., 110
Blair, P. B., 157
Blann, A. D., 324
Blasi, F., 78–79
Bleackley, R. C., 304–305, 306–307
Bleicher, K. H., 273–274
Bloch Qazi, M. C., 212–213
Blomback, M., 81
Blomenkamp, K., 6, 48–49
Blom, N. R., 37–38
Blumenthal, C., 354
Boccard, F., 217
Bock, K. W., 113–114
Bock, S. C., 66
Bode, C., 81
Bodker, J. S., 81–82
Boeke, K., 314
Boggs, D. R., 142–143
Bogyo, M., 363
Boily, Y., 333
Boland, B., 40–41
Bollig, F., 118–119
Bollomley, S. P., 302–303
Bollrath, J., 113
Bolon, I., 111, 113
Bolsover, S., 139
Boman, K., 81
Boneca, I. G., 216–217
Bonfanti, L., 173
Bonifacino, J. S., 2
Bono, F., 91–92
Boot-Handford, R. P., 175
Booth, N. A., 78–80, 81, 108–110, 113
Bordin, G. M., 81
Borg, J. Y., 206
Borner, S., 213
Borregaard, N., 143

Borsboom, A. N., 81
Borstnar, S., 111–112
Boshkov, L., 305–306
Bos, I. G., 66
Bossenmeyer-Pourie, C., 142–143
Bot, I., 304
Botteri, F., 151, 154–155
Bottomley, S. P., 210, 284, 286–287, 287, 303–304, 343, 348
Bouchecareilh, M., 304
Bouchind'homme, B., 155
Bouck, N., 184
Boulton, S. J., 292
Boulynko, I. A., 343
Bouquet, C., 92–93
Boutros, M., 289–290
Bower, K., 168
Bowii, M. J., 81
Boxio, R., 142–143
Boyd, S. E., 57
Boyer, K., 212
Boyle, T. P., 89
Brackenier, A., 350–351, 353, 360–361
Bradley, P. P., 141
Bradshaw, C., 109
Brand, A. H., 215, 229, 242
Brandt, A., 348
Brantly, M., 12
Brashears-Macatee, S., 3, 12
Braun, A., 217
Brem, S., 92
Brennan, M. L., 90
Brennan, S. O., 35
Brenner, S., 285, 287–288
Breslow, J. L., 88–89
Bres, S. A., 305–306
Brew, S., 23–24
Brey, P. T., 212
Brickenden, A. M., 169
Brito-Robinson, T., 81–82, 92–93
Broadhead, M. L., 184
Broadhurst, K. A., 177–178
Brodsky, J. L., 2, 35, 174
Brodsky, S., 81–82
Brömme, D., 284, 286–287, 288–289, 332–333, 342–343
Bronson, R., 87
Brooks, P. C., 82, 92–93
Brorsson, A. C., 234
Brouwers, E., 85–86
Brown, A. J., 81
Brown, C. Y., 118–119
Brown, G., 229–230
Brown, J. L., 10–11
Brown, K. E., 177–178
Brownlie, J. C., 218–219
Brown, M. S., 38
Brown, N. J., 90, 91

Broze, G. J. Jr., 348
Bruce, D., 206
Bruckner, P., 172–173
Brueckner, S., 81
Brunette, E., 160
Bruning, M., 213
Brunner, A., 175
Brunner, N., 92–93
Brunt, E. M., 177–178
Bruyere, F., 92–93
Bucci, C., 37–38
Buccione, R., 173
Buchkovich, K., 155
Buchner, J., 168
Buckle, A. M., 210, 348
Budai-Hadrian, O., 350, 353, 360, 361, 363
Buduneli, N., 110
Buechler, C., 108, 115
Buessecker, F., 111–112
Buffat, C., 118
Bu, G., 21–22, 109–110
Bugge, T. H., 23, 26, 92–93
Bujard, H., 120–121
Bukau, B., 168
Bulleid, N. J., 170–171, 172–173
Bullock, D. W., 48–49
Bullock, J., 26–27
Bunn, C. L., 110, 115–116
Burgess, W. H., 18
Burlingame, A., 363
Burnand, K., 84
Burns, J. L., 139
Busato, F., 118
Bush, M. F., 303–304
Butfiloski, E. J., 314
Butt, T. M., 214, 216–217, 218
Buzza, M. S., 284, 291–292
Byers, P. H., 175
Bystrom, J., 115
Bzowska, M., 332–333

C

Cabral, C. M., 2, 4, 5–6, 7–8, 9, 10–11, 12–13, 14
Cabral, W. A., 175
Cabrita, L. D., 209, 210
Caccia, S., 64–65
Cadman, M., 136, 137
Caldelari, I., 214, 216–217
Caldwell, E. E., 64–65
Caldwell, G. A., 293
Caldwell, K. A., 293
Camani, C., 109–110
Camargo, L. M., 234
Camerer, E., 115
Cameron, C., 305–306
Cameron, P. U., 284, 291–292
Campana, W. M., 21

Campbell, L. R., 359, 360–361
Campbell, S. M., 151
Campos, M., 199
Canoui-Poitrine, F., 324
Cantera, R., 219
Cao, J. X., 305–306
Cao, W., 194–195
Cao, X. R., 112, 137–139
Capovilla, M., 216
Capuco, A. V., 155
Caputo, A., 304–305, 306–307
Carey, J., 150, 154–155
Carlson, J. A., 35, 48–49
Carmeliet, G., 92–93
Carmeliet, P., 87, 88, 89–90, 91–93
Caroff, M., 216–217
Carrell, R. W., 5–6, 108, 150, 206, 207, 210, 260–261, 284, 286–287, 304–306, 348
Carroll, A. S., 168
Carr, S. H., 293
Cartwright, G. E., 142–143
Casanova, G., 219
Cassis, L. A., 308
Castellino, F. J., 81–82, 90, 91, 92–93
Cataltepe, S., 106, 136–137, 332–334
Cates, G. A., 169
Cattoretti, G., 47–48
Cavicchioli, R., 348
Cayouette, M., 194–195
Cella, N., 106, 137–139
Cereghino, J. L., 65–66
Cesarman-Maus, G., 107–108
Cetinkalp, S., 110
Chader, G. J., 150
Chakrabarti, R., 81
Chalfie, M., 264
Chan, A. K., 304, 361
Chandraratna, D. S., 234
Chang, H. T., 229, 230, 231–232, 234
Chang, L. Y., 37–38
Chan, L., 23, 28, 213–214
Chapman, A., 84
Chapman, H. A., 332–333
Chappell, D. A., 28, 109–110
Chaput, C., 216–217
Charron, Y., 212
Chateau-Joubert, S., 354
Chatson, K., 332
Cheeseman, I. M., 292–293
Chelbi, S. T., 118
Chen, C. S., 110–111
Chen, C. Y., 84, 118–119
Cheng, X. F., 26–27
Chen, J., 81–82, 218
Chen, L., 305–306
Chen, W. T., 169
Cherry, J. M., 229
Chervenick, P. A., 142–143

Chesterman, C. N., 21
Chiang, A. S., 234–235
Chiang, H. C., 234–235
Chiang, N., 307
Chiba, T., 174
Chin, J. R., 136–137, 151
Chiti, F., 234
Chmielewska, J., 79–80, 81
Cho, A., 153–154
Chomczynski, P., 123–124
Choudhury, P., 2, 4, 5–6, 7–8, 9, 10–11, 12–13
Christeller, J., 360–361
Christensen, A., 23, 27, 28, 78–79, 81–82, 213–214
Christensen, R. D., 141
Christiansen, H. E., 175
Christov, A., 304, 316
Chu, J., 69–71
Church, F. C., 18, 23, 26, 26, 80, 108, 150, 206, 210, 213–214, 284, 286–287, 303–304
Cicardi, M., 64–65, 206
Cinelli, P., 286–287
Claes, C., 92–93
Clark, A. G., 212–213, 215
Clark, M. A., 115
Cleuren, A. C. A., 85
Clift, S. M., 48–49, 151
Clough, S. J., 361
Clouthier, D. E., 109–110
Coates, L. C., 316
Coetzer, T. H., 343
Cohen, R. L., 23
Cole, K., 213
Cole, M., 87
Coleman, S., 212–213
Cole, M. D., 78–79
Coles, M., 108
Collen, D., 79–80, 82, 83, 84, 87, 88, 89–90, 91–93
Collins, G. H., 109
Collins, K. L., 115
Colomb, M. G., 56–57
Colombo, M., 206
Combepine, C., 212
Compernolle, G., 78–79, 82
Condrey, L. R., 81
Conkright, J. J., 304
Connell, T., 156–157
Conner, D. A., 153–154
Contreras, A., 106
Cook, K. R., 229
Cooley, B. C., 84
Cooley, J., 136–137, 139
Copeland, L., 354
Cordwell, S. J., 354
Cornelissen, I., 91
Corrigall, D., 253–254

Costa, M., 116–117
Costelloe, E., 117
Coucke, P., 175
Coughlin, P. B., 108, 136, 145, 150, 206, 207–209, 284, 286–287
Cousin, E., 78–79, 81, 116–117
Coussens, L., 92–93, 154–155
Cox, C. R., 217
Cozen, A. E., 90
Crain, K., 88
Crandall, D. L., 87, 88, 90, 91
Crapo, J. D., 37–38
Cravens, J. L., 18, 20
Crawley, J., 91–92
Cregg, J. M., 65–66
Cremona, T. P., 142–143, 144, 145
Crespo, J. F., 354
Cressey, L. I., 113
Criqui, M. C., 215
Crisp, R. J., 27
Croll, A., 79–80
Crombie, P. W., 108–109, 113
Crook, T., 155
Croucher, D. R., 23, 109–110
Crowther, D. C., 206, 229, 230, 231–232, 234–235
Cruickshank, D. J., 110
Crum, B. E., 35
Cseh, S., 57
Csutak, A., 110
Cuervo, A. M., 174
Cuesta, J., 354
Cufer, T., 111–112
Cugno, M., 64–65, 206
Cui, J., 92–93
Culi, J., 213–214
Cumming, S. A., 66
Curado, S., 236–237
Curino, A., 92–93
Curioni, A., 354
Curmi, P. M. G., 108, 348, 350, 353, 360, 361, 363
Curtis, J. E., 21–22
Czekay, R.-P., 82

D

Dabelsteen, E., 81–82, 118
Dabelsteen, S., 81–82, 118
Dacklin, I., 234–235
Dafforn, T. R., 206, 207, 211, 212, 229–230
Daha, M. R., 57–58
Dahl, D., 78–79
Dahl, M. R., 57
Dahl, S. W., 349–350, 360–361
Dai, E., 304, 305–306, 314, 316
Dai, W., 343
Dai, Y. P., 21

Dalley, J. A., 172–173
Daly, M., 361
Damerval, C., 354
Daniel, C., 157
Danielli, A., 213
Dano, K., 78–79, 92–93
Darnell, G. A., 112
Darnell, J. E. Jr., 116
Darula, Z., 363
Daugherty, A., 308
Daughty, C., 137–139
Davanture, M., 354
Davids, J. A., 304
Davies, M. J., 229, 316
Davies, P. L., 136–137, 139
Davis, A. E. III., 57–58
Davis, R. L., 109, 229, 230, 231–232, 234
Davydov, O., 350, 353, 360, 361, 363
Dean, D. C., 112
Dear, A. E., 35, 115–117
De Bie, I., 118
Debrock, S., 85–86
de Bruin, E. C., 66
Debs, R., 160
Deciu, C., 354
Declerck, P. J., 78–82, 85–86, 89, 92–93
DeClerck, Y. A., 92–93
DeCoster, M. A., 194–195
de Cremoux, P., 155
de Faire, U., 81
De Gregorio, E., 212
de Groot, R., 81
Degrouard, J., 217
Deinum, J., 78–79, 82
de Jong, J., 305–306
del Amo, A., 354
Deligdisch, L., 110
Della, M., 206, 207
Dellas, C., 91
Delobel, J., 81
Delucinge-Vivier, C., 114–116
Delvenne, P., 92–93
Delverdier, M., 304–305, 306–307
DeMayo, F. J., 48–49
De Mol, M., 79–80, 88, 89–90
DeNardo, D., 154–155
Dendale, R. M., 155
Dennison, C., 343
Denny, P., 136, 137
Deome, K. B., 157
De Paepe, A., 175
Dephoure, N., 168
Derman, E., 116
de Ruig, C. P., 56–57
De Rycke, R., 350–351, 353, 360–361
Desai, A., 292–293
Descombes, P., 114–116
de Strihou, C. V., 86, 87

De Taeye, B
Detry, B., 92–93
Dettmar, P., 81–82
Devine, P. L., 115
Devy, L., 92–93
de Waard, V., 91
Dewar, K., 136, 207–209
Dewerchin, M., 79–80, 88, 91
Dewilde, M., 78–79, 82
Dhamija, S., 118–119
Diamond, L. E., 78–79
Dias, L., 150, 154–155
Diaz, J. A., 84
Di Bello, C., 111–112
Dichek, D. A., 90, 91
Dickerson, K., 18, 20
Dickinson, J. L., 113, 119
Diergaarde, P., 78–79
Dieval, J., 81
Dijkerman, R., 348–349
Dilley, R. B., 81
Dippold, R., 36, 40, 41, 43, 48–49, 50
Dishon, D. S., 3–4
Dittrich-Breiholz, O., 118–119
Ditzel, D. A., 114–116
Dixon, D., 261, 264–265
Dixon, J. D., 88
Dobi, E. T., 199
Dobó, J., 56–57, 64–65, 66
Dobson, C. M., 234
Dohmen, R. J., 108–109
Doi, T., 177–178
Dold, K. M., 115
Dolkas, J., 21
Doller, A., 119
Dollery, C. M., 307
Dolmer, K., 21
Domachowske, J. B., 115
Dombradi, V., 234–235
Dombrowski, K. E., 173–174
Donaldson, V. H., 66
Dong, H. Y., 115
Dong, Y., 290–291
Donnan, K., 117
Dorovkov, M., 289–290
Doskeland, S. O., 113
Dougherty, K. M., 107–108, 110, 112
Douglas, R., 78–79
Dournon, C., 142–143
Dragojlovic, N., 21
Drahn, B., 212–213
Drapeau, G. R., 333
Drinko, J., 90
Driscoll, M., 289–290
Driver, S. E., 289–291, 293
Drogemuller, C., 175
Drujan, D., 139

Dubin, A., 23, 28, 136–137, 213–214, 333, 342–343
DuBois, R. N., 155
Duffy, M. J., 78, 111–112
Duggan, C., 111–112
Duh, E. J., 187
Dunbar, P. R., 316
Dunn, W. A., 38
Dupont, D. M., 81–82
Durand, M. K., 81–82
Durbin, R., 290–291
Duvic, B., 212, 218
Dyson, N., 155
Dzavik, V., 316

E

Ebbesen, P., 108–109
Eberhardt, W., 119
Eberharter, A., 117–118
Eddy, R. L., 66, 115
Edery, I., 215
Egelund, R., 27, 291–292
Eigenbrod, T., 306–307
Eipper-Mains, M. A., 212–213
Eitzman, D. T., 83–84, 88, 89, 91–93
Ekengren, S. K., 219
Ekeowa, U. I., 260–261, 303–304
Ekstrom, J. O., 219
El Chamy, L., 214, 216–217, 218
Eldering, E., 56–57, 66
Ellgaard, L., 2, 12–13
Ellis, M. C., 231
Ellis, V., 106
Elokdah, H., 88, 90, 91
Elvin, P., 111–112
Empana, J. P., 324
Endo, S., 118
Endo, Y., 111–112
Engelbert, M., 231
Engel, J., 172–173
Enghild, J. J., 26, 78–79, 82, 136–137, 332–333
Ephrussi, A., 236–237
Eren, M., 88, 89
Ergun, S., 81–82
Erickson, J. R., 154–155
Erickson, L. A., 78–80, 89
Eskelinen, E.-L., 37–38, 47–48
Esmon, C. T., 210
Esnouf, M. P., 348
Etches, W., 305–306
Evans, D. E., 350, 354
Evans, D. L. I., 5–6, 206, 304–306
Evans, J. D., 210
Evans, T. C., 272
Ewan, P. W., 66
Ewbank, J. J., 217
Ewing, M., 35, 36, 40, 41, 43, 48–49, 50

Eyre, D. R., 175
Ezaki, J., 47–48, 174

F

Fagerhol, M. K., 348
Fahey, J., 118
Fahim, A. T., 89
Falck, J. R., 38
Fallon, R. J., 26–27
Falvo, J. V., 168
Fan, J., 108–109
Fan, L., 314
Farkkila, M., 110
Farrehi, P. M., 88, 91–92
Farris, D. M., 84, 85
Fass, E., 361
Faulkin, L. J. Jr., 157
Fay, W. P., 81, 88, 89, 91–92
Feany, M. B., 231–232, 253
Fehr, M., 175
Feiglin, D., 109
Felez, J., 80–81
Feng, J., 89
Fernandes, S. M., 57–58
Fernandez, A. M., 106
Ferrandon, D., 211, 214, 215, 216–217, 218
Ferrans, V. J., 12
Ferrante, A., 113
Ferreira, A., 218–219
Ferrell, G. A., 3–4, 7–8, 10–11, 12–13
Ferrer, P., 69–71
Ferrieres, J., 324
Ferro-Garcia, M. A., 12–13
Fertala, A., 173–174
Fewell, S. W., 2
Fichelson, P., 253–254
Fici, G. J., 89
Fidler, I. J., 159
Fiehler, R., 348
Field, L., 118
Fierer, J., 113
Figueredo, E., 354
Filleur, S., 184
Finch, J. T., 5–6, 206
Finegold, M. J., 2, 5–6, 48–49
Finlay, B., 304, 305–306
Finnerty, C. M., 212–213
Finnie, C., 354
Fire, A., 260–261, 264–266, 289–291, 293
Fischer, S., 219
Fisher, C., 23–24
Fisher, R., 292
Fish, R. J., 108–109, 112, 113, 114
Fitzgerald, D. J., 22
Fitzgerald, J., 175
Fitzpatrick, P. A., 106
Flanagan, J. R., 253

Fleischmann, W., 229
Fleming, M. L., 144
Fleming, T. J., 144
Florczyk, D., 343
Florke, N., 81
Fluhr, R., 350, 353, 360, 361, 363
Foekens, J. A., 111–112
Foellmer, B., 12–13
Fogo, A. B., 90, 91
Foidart, J. M., 92–93
Foieni, F., 64–65
Folsom, A. R., 81
Foltz, G., 118
Forlino, A., 175
Forsyth, S. L., 136
Fortini, M. E., 229
Fourquet, A., 155
Frakes, A., 118
Franchi, L., 306–307
Francis, C. W., 81
Frandsen, T. L., 92–93
Frank, M. M., 57–58
Franzen, C., 219
Fraser, A. G., 289–291
Frau, E., 92–93
Frederik, P. M., 159–161
Frederix, L., 107–108, 110
Freeke, J., 303–304
Freeman, M., 231–232
French, C. J., 90
French, R. P., 293
Friedrich, P., 234–235
Friedrich, S. O., 108, 115
Fries, J., 332
Fries, L. F., 57–58
Fritsch, E. F., 66
Frutkin, A. D., 91
Fryling, C. M., 22
Fuchs, H. E., 18
Fujii-Kuriyama, Y., 113–114
Fujii, S., 81
Fujimoto, K., 332–333, 343
Fujimura, T., 47–48
Fujisawa, K., 118
Fujitaka, K., 87
Fujita, N., 37–38, 174–175
Fujita, T., 57
Fujita, Y., 176–177
Fukumoto, H., 37–38
Fukumoto, M., 177–178
Fullaondo, A., 207–209, 210, 211, 212–213, 229–230
Furegon, L., 354
Furonaka, M., 87
Furth, P. A., 151, 154–155, 156–157
Furuya, N., 47–48
Fusella, A., 173
Fusenig, N. E., 92–93

G

Gabella, G., 139
Gabor Miklos, G. L., 229
Gaensler, K., 160
Gafforini, C., 120, 121–122
Gåfvels, M., 28, 109–110
Gaggar, A., 139
Gajdosik, V., 212
Gajkowska, B., 332–333
Gallagher, W. M., 78
Gál, P., 56–57, 64–65, 66
Gamborg, O. L., 361–363
Gamelli, R. L., 304
Ganan, S., 12–13
Ganetzky, B., 253
Ganrot, P. O., 18
Gao, C., 156–157
Gao, F. B., 137–139, 212
Gao, S., 81–82, 118
Gao, Z., 139
Garcia-Casado, G., 354
Gardan, L., 217
Gardner, J., 108–109, 112, 114
Garg, N., 81
Garneau, N. L., 118–119
Garrels, J. I., 169
Garrett, M., 207–209, 210, 211, 212–213, 253
Garza, D., 231–232
Gasco, M., 155
Gately, S., 92
Gatt, A., 234–235
Gaultier, A., 21
Gausz, J., 234–235
Gavigan, J. A., 350–351, 353, 360–361
Geifi, J., 304–305, 306–307
Gelehrter, T. D., 78–79
Gelfi, J., 304–305, 306–307
Gelvin, S. B., 361
Gems, D., 233–234
Genton, C. Y., 81, 106–107, 108, 110
Georg, B., 78–79
Georgel, P., 215, 216, 218
George, P. M., 35
Gerard, R. D., 91, 92–93
Gerke, L. C., 168
Gerlinger, P., 151–152
Gertsenstein, M., 153–154
Gething, M. J., 23, 109–110
Gettins, P. G., 108, 150, 206, 207, 284, 286–287
Gettins, P. G. W., 21
Geurts-Moespot, A., 110
Geuze, H. J., 37–38, 173
Geuze, J. H., 36
Ghaemmaghami, S., 168
Giammarinaro, P., 217
Gibson, H. L., 106
Gillard, A., 136, 137

Gilmore, M. S., 217
Gils, A., 78–79, 82, 85–86, 92–93
Ginsberg, M. H., 79–80
Ginsburg, D., 78–79, 81, 84, 88, 89, 91–93, 107–108, 110, 112
Ginsburg, J., 92–93
Giuliano, K. A., 273–274
Gjernes, E., 115
Gjertsen, B. T., 113
Glatstein, I., 110
Gleave, A. P., 350, 353–354, 357, 360–361
Gleaves, L. A., 89
Gleeson, M. P., 273–274
Gleeson, P., 273–274
Gliemann, J., 18, 23, 27, 108–109, 113
Gobert, V., 212, 214, 216–218
Goebel, J., 81, 88–89
Goetinck, P. F., 110–111
Goffinet, F., 118
Gohl, G., 113–114
Goktuna, S. I., 113
Goldbach, C., 36, 40, 41, 43, 48–49, 50
Goldstein, J. L., 38
Golic, K. G., 218, 219
Goligorsky, M. S., 81–82
Gondo, T., 332
Gong, J. H., 212
Gonias, S. L., 21
Gonzalez, F. J., 113–114
Gonzalez, R., 184
Goodall, G. J., 118–119
Goode, J., 113
Gooptu, B., 260–261, 303–304, 305–306
Gordon, D., 91–92
Gordon, P. B., 42
Gordon, S., 144
Gornstein, E. R., 332–333
Gorter, A., 57–58
Gosai, S. J., 36, 261, 273–274, 277–278
Gossen, M., 120–121
Goss, N. H., 108–109, 115
Gotkin, M. G., 175
Gotta, M., 290–291
Gottar, M., 214, 216–218
Gottschalk, A., 292
Gotz, J., 153–154
Gould, A. R., 108
Gowrishankar, G., 118–119
Graeff, H., 81–82
Graessmann, A., 154–155, 156–157
Graessmann, M., 154–155, 156–157
Grafi, G., 361
Gráf, L., 64–65
Graham, K. S., 5–6, 12, 305–306
Gramling, M. W., 80, 303–304
Granger, L., 234–235
Greenbaum, D., 363
Greenberg, M. E., 116, 123–124

Green, C., 211, 212, 229–230, 231–232, 234, 245–246
Greene, C. M., 139
Greenlee, W. F., 115
Greenspan, R.J., 236, 242
Grégoire, J., 321–324
Greten, F. R., 113, 116–117
Grether, M. E., 244
Greyson, M. A., 90
Griese, M., 139
Griffin, J. P., 90, 91
Griffiths, G., 37–38
Grignet-Debrus, C., 92–93
Grigoryev, S. A., 343
Grimsley, P. G., 21
Grishok, A., 289–290
Grob, J. P., 110
Groner, B., 151–152
Gropp, M., 81–82
Grossman, W. J., 137–139
Gross, S., 81
Gross, T. J., 111
Grover, D., 253
Gruca, M., 332–333
Grundstrom, T., 78–79
Gubb, D. C., 206, 207–209, 210, 211, 212–214, 229–230, 231–232, 234, 245–246
Gudinchet, A., 108, 110
Guertin, M. C., 321–324
Guhl, E., 154–155, 156–157
Guldberg, R. E., 291–292
Guo, S., 289–290
Gut, I., 118
Gutierrez, L. S., 81–82, 92–93
Guzik, K., 332–333
Guzman, K., 115

H

Haas, B., 324
Haase, B., 175
Habayeb, M. S., 219
Hack, C. E., 56–57, 66
Haenlin, M., 212
Hafen, E., 231
Hagelberg, S., 81
Hager, H., 108–109
Hajjar, K. A., 107–108
Hakker, I., 234–235
Hale, P., 35, 36, 40, 41, 43, 48–49, 50
Hallmans, G., 81
Halsall, D. J., 66
Hamaad, A., 324
Hamazaki, J., 174
Hammer, C. H., 57–58
Hammer, R. E., 109–110
Hammond, E., 10
Hamsten, A., 81

Handwerker, J., 304–306
Haney, S. A., 273–274
Hansen, M., 81–82
Hansen, S., 81–82
Han, S. J., 212
Han, S. W., 115
Han, X., 348
Hao, Y. Y., 69–71
Harada, T., 177–178
Hara, T., 174
Harbeck, N., 81–82
Harbour, J. W., 112
Harhangi, H. R., 348–349
Hariharan, I. K., 229
Harlow, E., 155
Harmat, V., 56–57, 66
Harpel, P. C., 18
Harrich, D., 112
Harris, B., 21–22
Harris, E. N., 84
Harris, J. L., 350–351, 353, 360–361
Harrison, S. W., 261, 264–265
Harrop, S. J., 108, 350, 353, 360, 361, 363
Harry, J., 354
Hartl, D., 139
Haruta, Y., 87
Haruyama, N., 153–154
Hasenfuss, G., 91
Hashimoto, C., 212–213, 214
Hashimoto, S., 115
Hasina, R., 111–112
Haslbeck, M., 168
Hassan, Z., 110
Hatano, M., 174–175
Hattori, M., 78–79
Hattori, N., 87
Hauert, J., 108, 110
Hauge, S., 349–350, 353–354, 356–357, 358, 359, 360–361
Hausen, B., 314
Haverkate, F., 81
Hawksworth, G. M., 109–110
Hawley, A. E., 84, 85
Hawley, R. S., 218, 219
Hawley, S. B., 23, 27
Hayakawa, Y., 119
Hayashi, M., 90
Hayashi-Nishino, M., 37–38
Hayes, M., 91–92
Haylock, D. N., 145
Haynes, P. A., 354
Hayrapetian, L., 363
Hayry, P., 308
Hays, L., 354
Hazelzet, J. A., 81
Hazen, S. L., 90
Hearn, S. A., 234–235
Heberlein, K. R., 90

Heberlein, U., 253
Hebert, D. N., 12–13
Hecht, P., 214
Hecke, A., 81, 88–89
Hedges, L. M., 218–219
Hediger, M., 231, 238–239
Heegaard, C. W., 23, 28, 213–214
Heegaard, P. M., 356
Heermeier, K., 154–155, 156–157
Heger, J., 305–306
Heimburger, N., 110
Heim, D. A., 81
Heintz, N., 47–48
Hejgaard, J., 348, 349–350, 353–355, 356–357, 358, 359, 360–361
He, K., 151, 154–155
Hekman, C. M., 80
Helenius, A., 2, 10, 12–13
Heller, D., 110
Hendershot, L. M., 170–171
Hendrix, M. J., 150, 154–155, 159
Hendy, J., 145
Henikoff, S., 229
Henke, P. K., 85
Henkin, J., 28
Hennighausen, L. G., 151–152, 154–155, 156–157
Henry, M., 81
Hepp, H., 81–82
Herbert, J. M., 91–92
Herscovics, A., 12–13
Herve, M., 216–217
Herzinger, T., 354
Herz, J., 22, 23, 56–57, 109–110
Hess, H., 287–288
Hession, C., 108, 115, 116
Hetru, C., 211
Hibino, T., 110–111
Hibshoosh, H., 47–48
Hider, R., 234
Hidvegi, T., 35, 36, 40, 41, 43, 48–49, 50
Hiemstra, P. S., 57–58
Higgins, P. J., 206
Higuchi, T., 118
Hill, A., 354
Hilpert, J., 23
Hinz, U., 231
Hippenmeyer, P., 5–6
Hirai, M., 90
Hirakawa, H., 332
Hirata, H., 176–177
Hirayama, N., 86, 87
Hirayoshi, K., 169, 170, 173–174, 176–178
Hirsh, D., 293
Hirsh, J., 81
Hirst, C. E., 284, 291–292
Hmadcha, A., 118
Hobert, O., 264, 265–266

Hobman, M. Z., 305–306
Hodgkin, J., 293
Hodgkinson, C., 305–306
Hodgson, J., 273–274
Hoekema, A., 361
Hoffmann, A., 113, 116–117
Hoffmann, J. A., 211, 212, 214, 215, 216–218
Hofler, H., 81–82
Hofmann, G. E., 110
Hofmann, K. J., 78–79
Hogan, B. L., 169
Hojima, Y., 173–174
Hojrup, P., 108–109
Hokland, P., 113
Holohan, P. D., 109
Holst-Hansen, C., 92–93
Holtmann, H., 118–119
Holzgreve, W., 118
Homma, D. L., 170–171
Hood, E. E., 361
Hope, I., 270–271, 273
Hop, W. C., 81
Hori, M., 155
Horvath, A. J., 136, 145
Horwich, A., 168
Hoseki, J., 174
Hosli, I., 118
Hosokawa, M., 137–139, 171–172
Hosokawa, N. H., 12–13, 43, 137–139, 169, 170, 171–172, 173–174, 176–177
Hota-Mitchell, S., 305–306
Hota, S., 305–306
Houard, X., 92–93
Houenou, L. J., 194–195
Hou, M., 108–109
Houmard, J., 333
Houwerzijl, E. J., 37–38
Howard, A. D., 304–305, 306–307
Howard, E. W., 151
Howley, P. M., 155
Howson, R. W., 168
Hoyer-Hansen, G., 92–93
Hoylaerts, M. F., 79–80, 85, 88, 91–92
Hsu, L. C., 113
Huang, W., 21
Huang, Y., 213
Huang, Z. F., 348
Huber, R., 66
Huckensack, N. J., 260–261
Hughes, R. C., 169
Huhtasaari, F., 81
Huh, W. K., 168
Huibregtse, J. M., 155
Hu, J., 154–155, 156–157
Hulett, K., 111–112
Hull, R. J., 359, 360–361
Hultmark, D., 219, 234–235
Humen, D. P., 305–306, 316

Huntington, J. A., 206, 207, 210, 284, 287, 302–304, 348
Hur, S. H., 115
Hutton, M., 231–232
Hwang, I., 351–352
Hwu, Y., 212
Hyland, J. C., 175

I

Ibrahim, R., 321–324
Iche, M., 234–235
Icton, C., 314
Iehara, N., 177–178
Igarashi, K., 113–114
Iijima-Ando, K., 234–235
Iijima, K., 234–235
Iino, S., 90
Ikeda, M., 177–178
Ikenaga, M., 231
Ikuta, T., 113–114
Imber, M. J., 18
Imler, J. L., 211
Inan, M., 69–71
Inden, Y., 90
Ingebrigtsen, M., 64–65
Ingham, K. C., 23–24
Inghardt, T., 78–79
Inze, D., 350–351, 353, 360–361
Iredale, J. P., 139
Irigoyen, J. P., 78–79
Irving, J. A., 106, 108, 150, 206, 207–209, 210–211, 284, 286–287, 332, 343, 348–349
Isakson, B. E., 90
Ishida, H., 86, 87
Ishida, Y., 171–172, 173–175
Ishiguro, K., 137–139
Ishii, M., 86–87
Ishikawa, N., 87
Ishikawa, T., 119, 231
Itohara, S., 137–139, 171–172
Ito, M., 90, 137–139
Ittner, L. M., 153–154
Iwamatsu, A., 169, 176–177
Iwanaga, S., 212
Iwase, M., 90
Iwata, J., 47–48, 174
Izuhara, K., 108, 343
Izuhara, Y., 86, 87

J

Jablonski, M. M., 194–195
Jablonsky, G., 316
Jackson, R. J., 305–306
Jacobs, H. L., 139
Jacobs, N., 92–93
Jacobus, L. M., 229
Jacquemin, M., 84

Jäger, S., 37–38
Jammes, H., 118
Janes, J., 21
Jang, W. G., 115
Janicke, F., 81–82
Jankova, L., 108
Janssens, S., 91
Jansson, J. H., 81
Jardine, D., 108
Jean, F., 213
Jean-Marc Reichhart, J. M., 303–304
Jenkinson, L., 170–171
Jensenius, J. C., 57
Jensen, J. K., 27, 81–82
Jensen, P. H., 108–109, 113
Jensen, P. J., 110–111
Jensen, T. G., 108–109
Jespersen, J., 81
Jetten, M. S., 348–349
Jiang, H., 210, 211, 212, 213, 218
Jiang, R., 212
Ji, C., 212, 218
Jin, S., 47–48
Johansson, J., 229, 230, 231–232, 234
Johansson, M., 154–155
Johnsen, M., 92–93
Johnson, A., 348
Johnson, J. J., 293
Johnson, K. N., 218–219
Johnson, L. V., 150
Johnson, M. A., 150
Johnsson, H., 81
Johnstone, R. W., 112
Johnston, J. B., 305–306
Johnston, P. A., 36, 261, 273–274, 277–278
Joslin, G., 26–27
Joss, T. V., 350, 353, 360, 361, 363
Jost, M., 92–93
Juhan-Vague, I., 79–81, 89, 324
Juneja, H., 81
Jung, A. C., 215

K

Kabeya, Y., 37
Kadel, J., 213
Kadomatsu, K., 137–139
Kaersgaard, P., 354
Kafatos, F. C., 213
Kaikita, K., 90
Kaiserman, D., 57, 109, 136, 137, 207–209, 284, 287, 302–304, 348
Kalb, J. M., 212–213
Kaltenbrun, E. R., 35
Kamachi, Y., 332–334
Kamada, Y., 38
Kamath, R. S., 289–291
Kambris, Z., 212–213
Kamimoto, T., 35–36, 43, 46–47, 50
Kanaji, S., 108
Kanapin, A., 290–291
Kandarpa, K., 81
Kanno, J., 113–114
Kanost, M. R., 210, 213
Kantyka, T., 332–333, 343
Kappler, M., 139
Karch, F., 231, 238–239
Kardesler, L., 110
Karin, M., 113, 116–117
Karuntu, Y. A., 66
Kasamatsu, A., 112
Kasza, A., 23, 28, 213–214
Katava, N., 292
Kato, H., 332
Kato, J., 177–178
Kato, T. Jr., 231, 332–333, 343
Kato, Y., 136, 207–209
Katsikis, J., 111
Kaufman, T. C., 229
Kaul, P. R., 64–65
Kawabata, S., 212
Kawajiri, K., 113–114
Kawano, S., 155
Kawano, T., 108
Kawano, Y., 177–178
Kawasaki, T., 79–80, 88, 91
Kay, R., 361
Keays, C. A., 213
Keegan, A. D., 92–93
Keeton, M. R., 81, 110–111
Kele, B., 217
Keller, M., 136–137
Kelly, E., 139
Kemp, C., 36, 40, 41, 43, 48–49, 50
Kemphues, K. J., 289–290
Kennedy, S. A., 111–112, 316
Kerr, F. K., 57
Kerr, P., 305–306
Kessler, J., 137–139
Khabar, K. S., 118–119
Khan, K. A., 307
Khush, R. S., 217
Kick, G., 354
Kidd, V. J., 3, 12
Kieckens, L., 87
Kiefer, M. C., 106, 154–155
Kienast, J., 81
Kikkawa, M., 47–48
Kikuchi, S., 84
Kimble, J., 293
Kim, C. H., 212
Kim, D. H., 351–352
Kim, D. R., 212, 214
Kim, E. H., 212
Kim, H. S., 115
Kim, J., 5–6

Kim, K. S., 139
Kimmig, R., 81–82
Kim, P. S., 21
Kim, Y. W., 351–352
King, D. E., 36, 261, 273–274, 277–278
King, G. C., 108
Kinghorn, K. J., 206, 229, 230, 231–232, 234
King, L. E., 89
Kinnby, B., 110
Kinter, J., 109
Kircher, P., 175
Kirisako, T., 37, 38
Kishi, F., 332–333, 343
Kitagawa, K., 170–171
Kitamura, A., 171–172, 173–175
Kita, T., 177–178
Kittrell, F. S., 156–157, 158
Kjelgaard, S., 81–82
Kjoller, L., 23, 27
Klass, M. R., 288
Klijn, J. G., 111–112
Klionsky, D. J., 174
Klironomos, D., 314
Knaggs, S., 136, 137
Knauer, D. J., 23, 27
Knauer, M. F., 23, 27
Knibbs, R., 84
Knight, A. L., 293
Knockaert, I., 78–79
Knoop, A., 81–82
Knudtson, M., 321–324
Knust, E., 231
Kobayashi, A., 176–177
Kobune, M., 177–178
Koelle, M., 287–288
Kohno, N., 87
Koide, T., 170–171
Koike, M., 174
Kojima, M., 64–65
Kojima, S., 82
Kojima, T., 90, 137–139
Koller, A., 354
Kolodziejczyk, M. D., 316
Komatsu, M., 47–48, 174
Kominami, E., 37–38, 47–48, 174
Komiyama, T., 304–305, 306–307
Kondo, K., 84
Kondo, T., 90
Konecny, G., 81–82
Konsolaki, M., 231–232, 234–235
Konstan, M. W., 139
Konstantinides, S. V., 81, 88–89, 91
Kopp, F., 89
Korkko, J., 175
Kornelisse, R. F., 81
Korol, R., 307
Korzus, E., 150
Korzus, G., 333

Koschnick, S., 88
Koskiniemi, M., 110
Kosowska, K., 342–343
Kostas, S. A., 289–291, 293
Koster, K., 213
Kostner, G. M., 108, 115
Kostuk, W. J., 316, 321–324
Kotecha, S., 136–137, 139
Kounnas, M. Z., 18, 22, 23, 26, 28, 213–214
Kouzu, Y., 112
Kovatch, K. J., 36, 261, 273–274, 277–278
Kowal-Vern, A., 304
Kozarsky, K. F., 28, 109–110
Kozawa, O., 88
Koziczak, M., 78–79
Koziel, J., 343
Kracht, M., 118–119
Krainick, U., 111–112
Kramer, J. M., 293
Kramer, M. D., 111–112
Krasnewich, D., 109
Kraus, K., 212–213
Krauss, J. C., 92–93
Krauter, K., 116
Kremen, M., 90
Krettek, A., 307
Kridel, S. J., 23, 27
Kristensen, T., 18
Krogdahl, A., 81–82, 118
Kromer-Metzger, E., 215, 216, 218
Kronmal, R. A., 139
Krueger, S. R., 21–22, 286–287
Kruger, O., 213
Kruithof, E. K., 78–80, 81, 106–107, 108, 109–110, 112, 113, 115–117
Kruse, K. B., 10–11, 35
Kubota, H., 171–172, 173–175, 176–178
Kubota, R., 86–87
Kudo, H., 169, 176–177
Kuhn, W., 81–82
Kulkarni, A. B., 153–154
Kuma, A. H., 42, 174–175
Kumar, A., 40–41
Kumar, S., 113
Kumar, V., 284, 286–287, 288–289
Kumatori, A., 177–178
Kuncl, R. W., 185
Kurkinen, M., 169
Kurokawa, K., 86, 87, 212
Kurz, K. D., 86
Kusugami, K., 137–139
Kutukculer, N., 110
Kuzak, M., 332–333
Kuze, K., 177–178
Kuznetsova, N., 173
Kwaan, H. C., 81–82, 92
Kwak, J. H., 36, 261, 273–274, 277–278
Kwon, H. M., 212

L

Labinaz, M., 321–324
Lacbawan, F., 109
Lackner, K. J., 108, 115
la Cour Petersen, M., 359
Lacroix, M., 57
Ladewig, J., 213
Laemmli, U. K., 58–61, 363
Laenkholm, A. V., 81–82
Lagnado, C. A., 118–119
Lahoz, C., 354
Laich, A., 64–65
Laing, W. A., 350, 353–354, 357, 360–361
Laissue, P., 118
L'Allier, P., 321–324
Lamande, S. R., 172–173, 174–175
Lamark, T., 64–65
Lambers, H., 78–79
Lambert, V., 92–93
Lambley, E., 114
Lampl, N., 350, 353, 360, 361, 363
Lange, B. M., 354
Langella, O., 354
Langendorf, C. G., 348
Langley, E., 211
Lanotte, M., 113
Lapaire, O., 118
Larsen, J. V., 27
Larsen, P. L., 288–289
Lasic, D. D., 159–161
Latha, K., 106
Laucirica, R., 155
Lau, D., 114–116
Laug, W. E., 92–93, 108–109, 112
Laurell, C. B., 18, 348
Laurila, H., 308
Lawrence, D. A., 23, 26–27, 78–79, 82, 87, 92–93
Law, R. H. P., 210, 348
Lawrie, L. C., 108–109, 113
Lazarus, G., 110–111
Lazo, J. S., 36, 261, 273–274, 277–278, 278
Lazzaro, B. P., 215
Le, A., 3–4, 5–6, 7–8, 10–11, 12–13
Le Beau, M. M., 115
Lebo, R. V., 78–79, 115
Le Bot, N., 290–291
Leclerc, V., 212, 214, 216–217, 218
Lecomte, J., 92–93
LeCuyer, T. E., 142–143, 144, 145
Lee, A., 81–82
Lee, B. L., 212
Leeb, T., 175
Lee, C., 361–363
Lee, D., 21–22
Leedman, P., 120
Lee, H., 118
Lee, I. K., 115
Lee, J. A., 12, 109–110, 212
Lee, J. E., 215
Lee, J. Y., 359, 360–361
Lee, K. N., 5–6
Lee, M. H., 81
Lee, S., 40–41, 176–177
Lees, J. F., 172–173
Lee, T. W., 316
Lee, W. J., 212, 216–217
Lee, Y. J., 351–352
Legent, K., 231, 246–247
Lehmkoster, T., 113–114
Lehoczky, J., 136, 207–209
Lehrbach, P. R., 115
Leijendekker, R., 36
Leik, C. E., 87
Leikina, E., 173
Leikin, S., 173
Leitinger, N., 90
Leivo, I., 110
Lejeune, A., 92–93
Lemaitre, B., 212–213, 215, 216–217, 218
LeMeur, M., 151–152
Leppla, S., 22
Le, Q.-Q., 3–4
Lesk, A. M., 150, 206, 207–209, 210–211, 284, 286–287
Letchworth, G. J., 67–68
Le, T. T., 114
Leulier, F., 216–217, 231–232
Leung, A., 234–235
Levashina, E. A., 211, 212, 229–230
Levesque, J. P., 145
Levine, A. J., 47–48
Levine, B., 47–48, 174
Lewis, B., 155
Lewis, H., 234
Lewis, J., 231–232, 348
Ley, T. J., 137–139
L'Hernault, S. W., 289
Lhocine, N., 212
Li, A., 150
Liang, R., 156–157, 158, 159
Liao, Y., 48–49
Libby, P., 307
Liegeois, S., 217
Liehl, P., 217
Liggitt, D., 160
Ligoxygakis, P., 212, 218
Lihme, A., 356
Li, J., 57–58
Lijnen, H. R., 78, 79–80, 82, 83, 85–86, 88, 89–90, 91, 107–108, 110
Lijnen, R. H., 91, 92–93
Liljestrom, P., 108, 109
Lilley, K. S., 211

Li, M., 151, 154–155, 156–157
Lindblad, D., 6
Lindblad-Toh, K., 175
Lindner, V., 91
Lingen, M. W., 111–112
Linhardt, R. J., 64–65
Link, D. C., 145
Linschoten, M., 78–79
Liou, W., 36
Li, P., 108–109
Lip, G. Y., 324
Li, P. W., 229
Li, S. H., 82
List, K., 23, 26
Li, S. W., 173–174
Liu, B., 234
Liu, D., 57–58
Liu, Γ., 145
Liu, H. P., 234–235
Liu, L. Y., 304, 305–306, 314, 316
Liu, X. W., 84
Liu, Y., 2, 4, 5–6, 7–8, 9, 10–11, 12–13, 14
Liva, S., 155
Li, X., 21
Li, Y., 22
Ljungberg, B., 81
Ljungner, H., 81
Lloyd, G. J., 210
Lobov, S., 109
Lockett, S., 21–22
Loethen, S., 45
Lomas, D. A., 5–6, 66, 108, 150, 206, 211, 229–230, 231–232, 234–235, 260–261, 284, 286–287, 303–305, 305–306, 314, 316
London, N. J., 81
Long, O. S., 36, 261, 273–274, 277–278, 284, 288, 291–292
Looft-Wilson, R. C., 90
Look, M. P., 111–112
Loos, M., 23, 56–57
Lopez, D. L., 210
Lorand, L., 108–109
Loria, P. M., 264
Lörincz, Z., 56–57, 66
Loskutoff, D. J., 78–80, 81, 82, 88–89, 90, 110–111
Louie, S. G., 304
Loukeris, T. G., 213
Lubbers, Y. T., 56–57
Lucas, A. R., 304, 305–306, 307, 314, 316
Lucas, W. J., 359, 360–361
Lucchesi, B. R., 84
Luc, G., 324
Ludolph-Hauser, D., 354
Lu, F., 57–58
Lu, H., 139
Luheshi, L. M., 234
Luini, A., 173

Luke, C. J., 36, 106, 108, 150, 206, 261, 273–274, 277–278, 284, 286–287, 288–289, 291–292, 302–304, 332, 348
Lu, L., 114–116
Lummer, M., 213
Lundgren, E., 234–235
Lund, J. E., 89
Lund, L. R., 78–79, 92–93
Lundstrom–Hobman, M., 316
Lung, O., 212–213
Luo, L., 253–254
Luo, Y., 235
Lupu, F., 81, 91–92
Luttun, A., 91–92
Lu, Z. H., 137–139
Lyons-Giordano, B., 110–111

M

Maass, N., 106, 151–152, 154–155
Mabbutt, B. C., 108
Macaulay, C., 304, 305–306, 321–324
MacCalman, C. D., 28, 109–110
MacDonald, K. P., 114
Macen, J. L., 304–306
MacFadyen, R. J., 324
MacGregor, I. R., 79–80
Mackintosh, J. A., 354
MacLeod, I., 229
MacNeill, A. L., 304–305, 306–307
Madani, R., 212
Madison, E. L., 109–110
Madsen, J. B., 78–79, 82
Madsen, P., 113
Maeda, R. K., 214, 231, 238–239
Maekawa, H., 26–27
Maesawa, C., 118
Magdelenat, H., 155
Magiera, L., 332–333
Magit, D., 106, 151, 154–155, 156–157
Magnusson, S., 66
Mahadeva, R., 206
Maillard, C. M., 92–93
Main, B. W., 86
Maine, E. M., 289–290
Major, L., 108–109, 112, 114
Makareeva, E., 173
Makarova, A., 23, 26
Ma, L., 90
Malek, T. R., 144
Mallampalli, R. K., 28
Mallya, M., 206
Maniatis, T., 66
Mann, M., 291–292
Maquoi, E., 89
Marcil, A., 218
Margaritondo, G., 212
Mariasegaram, M., 120, 121–122

Marini, J. C., 175
Markey, K., 114
Marotti, K. R., 89
Marrinan, J., 66
Marsh, J. C., 142–143
Martella, O., 173
Martial, J. A., 92–93
Martin, E., 155
Martinelli, C., 218
Martinez, A., 197, 198
Martinez-Campos, M., 289–291
Martinez-Menarguez, J. A., 173
Martin, J. R., 234–235
Martin, J. S., 292
Marttila, S., 350, 353–354, 357, 359, 360–361
Marutani, T., 171–172
Masos, T., 114
Masset, A., 92–93
Masset, A. M., 92–93
Massimi, I., 333
Masson, V., 92–93
Mast, A. E., 26
Masuda, H., 177–178
Masuda, T., 118
Masumoto, K., 332–333, 343
Mata, J., 236–237
Mathahs, M. M., 177–178
Mathiasen, L., 81–82
Matias, A. C., 108–109
Matskevich, A. A., 214, 216–218
Matsson, L., 110
Matsuda, Y., 110–111
Matsui, M., 174–175
Matsumoto, A., 231
Matsumoto, N., 47–48
Matsunaga, T., 177–178
Matsuno, H., 84, 88
Matsuoka, Y., 171–172
Matsuo, O., 82, 88
Matsuo, T., 236–237
Matsushima, K., 115
Matsushita, M., 57
Matsushita, T., 86–87, 137–139, 171–172
Mattaliano, R. J., 106–107
Matthews, A. Y., 57
Matthews, K. A., 229
Maurer, F., 111, 115–116, 120, 121–122
Maurice, N., 36, 40, 41, 43, 48–49, 50
Maxwell, N. C., 136–137, 139
Mayer, E. J., 78–79
Mayer-Hamblett, N., 139
McCarthy, R., 139
McCormick, F., 159
McCoy, R. D., 89
McCracken, A. A., 10–11, 35
McCrae, K. R., 28, 109–110
McDermott, E., 111–112
McDonagh, J., 81

McDonald, A. P., 85
McElvaney, N. G., 139
McFadden, G., 304–306, 314, 321–324
McGavin, M. J., 333
McGill, V., 304
McGowan, P. M., 78
McGowan, S., 348
McGraw, L. A., 212–213
McKeone, R., 136, 137
McKimmie, C., 212, 229–230
McMahon, B., 81–82
McMahon, G. A., 82, 92–93
McPherson, J., 361
Meade, T. W., 81
Meagher, M. M., 69–71
Medcalf, R. L., 78, 107–108, 110, 111, 115–117, 119, 120, 121–122
Medema, J. P., 305–306
Medina, D., 156–157, 158
Medzihradszky, K. F., 363
Meher, A., 90
Meier, P., 231–232
Meier, T. R., 85
Meins, M., 286–287
Meissenheimer, L. M., 78–79, 82
Meister, A., 108
Meister, M., 215, 216, 218
Melchers, L. S., 361
Melchior, S., 354
Melen, L., 92–93
Melen-Lamalle, L., 92–93
Melkko, T., 64–65
Mellgren, G., 113
Mello, C. C., 265–266, 272, 289–291, 293
Melvin, J. E., 289
Mengin-Lecreulx, D., 216–217
Mercer, K. B., 289
Mertts, M. V., 173
Messina, C. G., 139
Messud-Petit, F., 304–305, 306–307
Mestdagt, M., 92–93
Mesters, R. M., 81
Meyer, C., 190
Michalopoulos, G. K., 18, 36, 40, 41, 43, 48–49, 50
Michaut, L., 215, 216, 218
Michel, T., 217–218
Micklem, G., 207–209, 210, 211, 212–213, 229–230
Miething, C., 113
Mietz, J. A., 155
Migliorini, M. M., 18, 21–22, 23–24
Mignot, T. M., 118
Mihaly, J., 81–82
Mikhailenko, I., 21–22, 23, 26
Mikus, P., 108, 109
Milgrom, S., 175
Millan, J. L., 78–79

Author Index

Miller, A. D., 206, 207
Miller, G. J., 81
Miller, J. W., 212, 214
Miller, L. W., 314
Mills, D. R., 284, 286–287, 288–289
Mimura, J., 113–114
Mimuro, J., 78–80
Mineki, R., 47–48
Minhas, J., 213
Miranda, E., 206, 229, 303–304, 316
Mirnics, K., 35, 48–49
Mironov, A. A., Jr., 173
Mirza, G., 136, 137
Miskin, R., 114
Misra, S., 214
Mitola, D. J., 92–93
Miyanishi, K., 177–178
Miyata, N., 236–237
Miyata, T., 86, 87
Mizushima, N. Y., 35–36, 37, 38, 40–41, 42, 43, 46–48, 50, 174–175
Moayeri, M., 21–22
Modderman, P. W., 66
Moestrup, S. K., 18
Moffatt, B. E., 64–65
Molinari, M., 2, 12–13
Mollnes, T. E., 64–65
Molmenti, E., 5–6
Moloney, A., 234–235
Mondon, F., 118
Montaye, M., 324
Montero, L., 78–79
Montgomery, M. K., 289–291, 293
Montminy, M., 113, 116–117
Monzer, J., 359, 360–361
Moons, L., 91–92
Moore, K., 5–6
Moore, M. S., 253
Moore, N. R., 78–79, 110
Morange, P., 81, 89, 324
Mordwinkin, N. M., 304
Morehead, A., 90
Morel, L., 314
Moremen, K. W., 2, 5–6, 9–10, 12–13, 14
Moreno, S., 290–291
Morimoto, K., 108
Morisato, D., 212, 214
Morita, Y., 86–87
Moriwaki, H., 90
Morley, T., 229–230
Morrison, J. F., 337, 338
Morris, R. E., 314
Morton, P. A., 79–80
Mosesson, M. W., 108–109, 113
Mouchabeck, Z. M., 154–155
Moulder, G. L., 284, 288, 291–292
Moul, J. W., 156–157
Mounier, E., 81, 88–89

Moyer, R. W., 108, 150, 206, 284, 286–287, 304–305, 306–307
Muddiman, J., 117
Mueller, B. M., 112
Mueller, J. L., 212–213
Muenke, M., 109
Muensch, H., 3, 12
Muglia, L. J., 48–49
Muhammad, S., 26–27
Mukherjee, A., 36, 40, 41, 43, 48–49, 50
Mukherjee, P. K., 196–197
Muller-Berghaus, G., 79–80
Muller-Esterl, W., 333
Muller, K. M., 81, 88–89, 292
Mulligan, R. C., 87, 88
Mulnix, A. B., 213
Multanen, J., 110
Munaut, C., 92 93
Munoz-Canoves, P., 78–79
Muñoz-Planillo, R., 306–307
Munson, P. J., 24
Munuswamy-Ramanujam, G., 304, 305–306, 307, 314, 316
Munzel, P. A., 113–114
Murai, H., 87
Murakami, A., 332
Muramatsu, T., 137–139
Murano, A., 112
Murase, K., 177–178
Murashige, T., 361–363
Murata, S., 47–48, 174
Murhammer, D. W., 64–65
Murohara, T., 86–87, 90
Murphy, P. M., 305–306
Musil, D., 78–79
Myers, D. D., 84, 85
Myers, R. R., 21

N

Nabel, E. G., 84, 88, 91–92
Naessens, D., 78–79, 82
Nagaich, A. K., 156–157
Nagai, N., 79–80, 89–90, 137–139, 171–172
Nagai, S., 115
Nagamine, Y., 78–79, 120, 121–122
Nagashima, S., 84, 332
Nagata, K., 12–13, 137–139, 168, 169, 170–175, 176–178
Nagy, A., 153–154
Naitoh, M., 177–178
Nakai, A., 169, 170, 173–174, 176–177
Nakao, I., 343
Nakashima, M., 84
Nakashima, Y., 88–89
Nakaya, H., 174–175
Nakayama, Y., 137–139
Nalbone, G., 81

Nambi, P., 87
Nandan, D., 169
Nangaku, M., 86, 87
Narayanaswamy, M., 81
Narita, M., 109–110
Nash, P., 304–306, 314
Nathaniel, R., 304–305, 306–307
Nation, N., 304–306
Nation, P. N., 305–306, 314, 316
Natsume, T., 170
Naviglia, T. L., 284, 288, 291–292
Nawata, S., 332
Neckameyer, W. S., 253
Negroni, L., 354
Nehme, N. T., 217
Nehrke, K., 289
Nekarda, H., 81–82
Nelson, C. R., 229
Nelson, T., 118
Nerelius, C., 229, 230, 231–232, 234
Neuenhahn, M., 113
Neveu, M., 150, 154–155
Nguyen, G., 81, 109–110
Nichols, A., 111, 113
Nickischer, D., 273–274
Nicolas, E., 215, 216, 218
Nicolosa, G., 79–80
Nicoloso, G., 108, 110
Nie, G., 235
Nielsen, B. S., 81–82, 92–93, 118
Nielsen, E. W., 64–65, 66
Nielsen, L. S., 78–79
Niitsu, Y., 177–178
Niiya, K., 119
Nilsson, A., 81
Nilsson, I. M., 81
Nilsson, S. K., 145
Nilsson, T. K., 56–57, 81
Nishikawa, Y., 170–171
Nishito, Y., 174
Nitsche, M., 91
Niwa, M., 88
Nixon, R. A., 40–41
Nobar, S. M., 284, 288, 291–292
Noda, T., 37–38, 174–175
Noel, A., 92–93
Nomura, H., 112
Nordt, T. K., 81
Noria, F. A., 81–82, 91, 92–93
Norman, J., 81–82
Notari, L., 194–195, 196
Novelli, M., 139
Novoradovskaya, N., 12
Nukiwa, T., 115
Numa, F., 332
Numaguchi, Y., 86–87
Nuñez, G., 306–307
Nusse, O., 142–143

Nykjær, A., 23
Ny, T., 78–79, 108, 109

O

Obermoeller-McCormick, L. M., 22
Oborn, C. J., 157
Ogasawara, S., 118
Ogawa, H., 332–333, 343
Ogawara, K., 112
Ogbourne, S. M., 112, 117
Oguro, A., 176–177
Ohashi, S., 177–178
O'Higgins, N., 111–112
Ohishi, N., 177–178
Ohlsson, K., 18
Ohresser, S., 216
Ohsako, T., 236–237
Ohsumi, Y., 37, 38, 47–48, 174–175
Ohtani, K., 170–171
Ohta, S., 108
Ohtsuka, T., 236–237
Oji, M. E., 110–111
Oji, V., 110–111
Oka, K., 23, 28, 213–214
Okamoto, H., 177–178
Okawa, H., 113–114
Okkema, P. G., 264, 265
Okumura, K., 86–87
Olofsson, A., 110
Omor, H., 37–38
O'Neill, E. M., 231
O'Neill, S. L., 218–219
O'Neil, S., 218–219
Ono, K., 112
Op den Camp, H. J., 150, 348–349
Opolon, P. H., 92–93
Oppenheimer, C., 78–79
Ortego, J., 185
Orth, K., 22, 23, 109–110
Osaka, H., 22
Osaki, T., 212
O'Shea, E. K., 168
Oshima, M., 113–114
Osman, D., 212
Ostergaard, H., 359–361
Ostergaard, O., 354
Ostermann, H., 81
Osterwalder, T., 286–287
Ostrovsky, L. L., 316
Otaka, A., 170–171
Otsu, K., 113, 116–117
Otsuka, G., 91
Otter, M., 109–110
Overgaard, J., 81–82
Owen, C. A., 307
Owensby, D. A., 21, 79–80
Ozaki, C. K., 88

P

Pache, L., 81–82
Page, J. L., 212–213
Page, R. M., 206, 234
Painter, C. A., 89, 90
Pak, S. C., 36, 106, 261, 273–274, 277–278, 284, 286–287, 291–292, 302–304, 332, 348
Palmer, E., 231–232
Pang, I. H., 194–195
Pannekoek, H., 78–79, 91
Paques, E. P., 79–80
Park, E. C., 81, 293, 333
Parker, A. C., 81, 88, 91
Park, J. M., 113, 116–117
Park, J. W., 212
Park, K. G., 115
Park, K. S., 115
Parks, A. L., 229
Park, Y. B., 115
Parodi, A. J., 12–13
Parquet, C., 216–217
Partridge, L., 233–234
Pavlakis, G. N., 159–161
Pavloff, N., 154–155
Pawar, A. P., 234
Paxian, S., 113
Pearce, M. C., 210
Pearson, J. M., 107–108, 110, 112
Pedersen, K. E., 27, 78–79, 81–82
Pegram, M., 81–82
Peiretti, F., 81
Pelte, N., 211, 212, 218
Pemberton, P. A., 106, 108, 150, 154–155, 206, 284, 286–287, 332–334
Peng, H., 92–93
Peng, L., 91
Pennella, A., 286–287
Penn, M. S., 90
Pepine, C., 307
Pepine, C. J., 321–324
Pepin, M. G., 175
Pepinsky, B., 108–109
Peppel, K., 115, 119
Pérez, J., 229, 303–304, 316
Perez-Mediavilla, L. A., 187
Perlmutter, D. H., 5–6, 26–27, 34–36, 38, 40, 41, 42, 43, 45, 46–47, 46, 48–49, 50, 260–261, 273–274, 277–278
Perricaudet, M. J., 92–93
Perrimon, N., 215, 229, 242
Perry, D. J., 206
Peruffo, A. D. B., 354
Petersen, G., 212–213
Petersen, H. H., 23, 27, 28, 213–214
Petersen, L. C., 360–361
Petersen, S. V., 185
Petersen, T. E., 108–109, 291–292

Peters, H. A., 111–112
Peterson, E. P., 304–305, 306–307
Peters, W. H., 110
Petitclerc, E., 82, 92–93
Petitjean, M. M., 92–93
Petrov, L., 308
Petruzzelli, G. J., 111–112
Pfeilschifter, J., 119
Pflugfelder, P. W., 316
Pham, C. T., 139, 145
Phillips, R. L., 206
Phylactides, M., 139
Pichaud, F., 231–232, 253–254
Pickup, D. J., 304–305, 306–307
Pierangeli, S. S., 84
Pihan, A., 81–82
Pike, R. N., 57, 150, 206, 207–209, 210–211, 284, 286–287, 332–333, 334–335, 343, 348
Pilarski, L., 316
Pilatte, Y., 57–58
Pili-Floury, S., 206, 216–217
Pind, S., 7–8
Pittius, C. W., 151
Pitulainen, E., 35
Pizzo, S. V., 18, 26
Plantz, B. A., 69–71
Platt, F. M., 40–41
Platt, J. L., 214, 216
Plaza, K., 343
Ploegh, H. L., 36
Ploplis, V. A., 81–82, 90, 91, 92–93
Plotnick, M. I., 360–361
Plummer, T. B., 214, 216
Plump, A. S., 88–89
Pokrzywa, M., 234–235
Polanowska, J., 292
Polesello, C., 212
Pol, H-Wd., 37–38
Polites, H. G., 89
Poller, W., 23
Polverini, P. J., 198–199
Poole, A. R., 91–92
Popova, E. Y., 343
Porsch-Oezcueruemez, M., 108, 115
Posthuma, G., 37–38
Potempa, J., 136–137, 150, 332–333, 334–335, 342–343
Potma, E. O., 139
Poulin, G., 290–291
Pradel, E., 217
Prager, G. W., 81
Praitis, V., 266
Praus, M., 92–93
Preissner, K. T., 79–80
Prendergast, G. C., 78–79, 87
Priebat, D. A., 141
Priebe, G. P., 137–139, 140, 141
Pringle, S., 115

Prins, M. H., 81
Prioul, J. L., 354
Prockop, D. J., 170–171, 172–174, 175
Prouse, C. V., 348
Providence, K. M., 206
Prydz, H., 115
Przybilla, B., 354
Puls, M., 91
Pyke, S. D., 81
Pyott, S. M., 175
Py, R., 304–305, 306–307
Pytel, B. A., 115, 119

Q

Qin, G., 57–58
Qin, L., 106
Quan, G. M., 185
Qu, D., 5–6
Quinn, K. A., 21
Quirce, S., 354
Qu, X., 47–48

R

Radpour, R., 118
Radtke, K. P., 110
Radziejewska, E., 66
Rae, I., 292
Rafferty, U. M., 78–79
Raff, M., 348
Rafidi, K., 150, 154–155
Ragg, H., 169, 213
Raghunath, M., 110–111
Ragoussis, J., 136, 137
Raines, E. W., 88–89
Raja, K., 92–93
Rakic, J. M., 92–93
Ramirez-Castillejo, C., 184
Ram, K. R., 212–213
Ramon, R., 69–71
Ramos, P. C., 108–109
Ramsey, B. W., 139
Ramunujam, G., 314
Ranby, M., 79–80
Ranson, M., 23, 109–110
Raposo, G., 36
Rasmussen, L. K., 108–109
Rasmussen, M. S., 291–292
Rasmussen, S. K., 348, 349–350, 353–355, 356, 359, 360–361
Rathmell, W., 354
Ratner, J. A., 137–139
Ravi Ram, K., 212–213
Ray, C. A., 304–305, 306–307
Ray, M., 116
Razzaque, M. S., 177–178
Rea, D., 305–306
Read, R. J., 207

Ream, B., 87, 88
Rebourcet, R., 118
Redich, N., 118–119
Reeves, E. P., 139
Reichart, J. M., 302–303
Reich, E., 108
Reichhart, J.-M., 206, 207–209, 210, 211, 212, 214, 215, 216–218, 229–230, 284, 287, 348
Reilly, C. F., 78–79
Reinheckel, T., 92–93
Reinke, V., 287–288
Reith, A., 110
Remold-O'Donnell, E., 108, 136–139, 140, 141, 142–143, 144, 145, 150, 206, 284, 286–287
Rennke, S., 21–22
Resch, K., 118–119
Reuning, U., 81–82
Revell, P. A., 137–139
Ribeiro, P. S., 231–232
Riccio, A., 78–79
Rice, K., 333
Richardson, J., 304, 314, 316
Richardson, M., 108–109
Richer, M. J., 213
Richmond, A., 154–155
Richter, K., 168
Riddle, D. L., 288–289
Rigaut, G., 291–292
Rijken, D. C., 78, 109–110
Rincon, M., 91–92
Riordan, J. R., 7–8
Ripley, C. R., 175
Ripley, R. T., 90
Risberg, B., 81
Risseeuw-Appel, I. M., 81
Ritchie, H., 108–109, 113
Ritter, M., 108, 115
Rival, T., 234–235
Robbie, L. A., 81
Roberg, J. J., 64–65
Roberts, D. D., 159–161
Roberts, K., 348
Robertson, A. S., 206, 211
Robertson, D., 231–232
Roberts, S. M., 314
Roberts, T. H., 348, 349, 350–351, 353–354, 357, 359, 360–361, 363
Robinson, C. V., 303–304
Robinson, L. H., 354
Roch, F., 212
Rodbard, D., 24
Rodenburg, K. W., 27, 213, 291–292
Rodriguez, J., 354
Rodriguez, L., 253–254
Roepstorff, P., 354
Roes, E. M., 110
Roes, J., 139
Rogers, B. B., 48–49

Romer, M. U., 92–93
Romisch, K., 229
Rorth, P., 236–237
Rosado, C. J., 348
Rosahl, T., 234
Rosas, M., 144
Rose, C., 81–82
Rosenberg, H. F., 115
Rosen, J. M., 47–48, 106, 151
Rosenkrands, I., 350, 353–355, 356
Rosenwasser, L. J., 115
Rossi, V., 57
Rossjohn, J., 210
Ross, R., 88–89
Rothenbuhler, R., 108, 115, 116
Roth, S., 212
Rothstein, G., 141
Rousset, F., 218–219
Royet, J., 217–218
Rubben, H., 81–82
Rubin, G. M., 229, 231, 239
Ruddock, V., 81
Rudnick, D. A., 48–49, 260–261
Ruegg, A., 136–137
Ruegg, M., 117
Ruidavets, J. B., 324
Ruiz, J. F., 21–22
Rutschmann, S., 215, 217–218
Rutz, B., 291–292
Ruzyla, K., 210
Ryazanov, A., 289–290
Ryder, E., 234
Ryken, T. C., 118
Rymo, L., 81
Ryo, H., 231
Ryu, G. Y., 118
Ryu, J. H., 216–217
Ryu, K. H., 212
Rzychon, M., 342–343

S

Sabat, A., 342–343
Sacchi, N., 123–124
Sachchithananthan, M., 110
Sadler, J. E., 106–107, 115
Saelinger, C. B., 22
Saftig, P., 37–38
Saga, S., 169, 176–177
Sage, H., 169
Sagel, S. D., 139
Sager, R., 106, 150, 151–152, 154–155, 156–157
Saito, H., 90
Saito, K., 112, 118
Saito, S., 156–157
Sakai, J., 115
Sakamoto, Y., 112
Sakata, Y., 332–333, 343

Sakuragawa, N., 119
Sakurai, T., 176–177
Sakurai, W., 332–333, 343
Salcedo, G., 354
Salomaa, V., 81
Salvaterra, P. M., 231–232
Salvesen, G. S., 26, 206
Sambrook, J. F., 23, 66, 109–110
San Antonio, J. D., 175
Sanchez-Sanchez, F., 187
Sancho, M. C., 304–305, 306–307
Sanderowitz, J., 92
Sandford, R. M., 81
Sand, O., 18
Sandoval-Cooper, M. J., 91
Sandusky, G. E., 86
Saniabadi, A. R., 84
Sankaran, L., 151
Sanwal, B. D., 169
Sanz-Parra, A., 209
Sartorio, R., 78–79
Sasaki, T., 111–112, 213
Sastre-Garau, X., 155
Sastre, J., 354
Satoh, M., 169, 170, 173–174
Sato, N., 118
Sato, T., 177–178
Sato, Y., 177–178
Sattelle, D. B., 234–235
Sauder, D., 333
Saunders, D. N., 23, 108, 109–110
Saunders, N. F. W., 348
Savelkoul, H. F., 81
Sawa, H., 81
Sawaoka, H., 155
Sawdey, M. S., 78–80, 81
Sayers, R. D., 81
Scarff, K. L., 136, 137–139
Sceurman, B. K., 215
Schafer, A., 292
Schafer, K., 81, 88–89, 91
Schafer, W. R., 292
Schaffer, U., 292
Schagger, H., 341–342
Schatz, F., 110
Schechter, N. M., 136–137, 332–334, 342–343, 343, 360–361
Schedletzky, T., 292
Schedl, P., 231–232
Scherfer, C., 212
Scherrer, A., 110
Schevzov, G., 115
Schick, C., 332–334, 342–343, 343
Schieltz, D., 354
Schleef, R. R., 79–80, 81
Schleuning, W. D., 106–107, 108, 115, 116
Schlosser, A., 292
Schmalfeldt, B., 81–82

Schmeling, G., 84
Schmid, K. W., 81–82
Schmidt, B. Z., 35, 48–49
Schmidt, R. L., 214, 216
Schmidt, W. N., 177–178
Schmitt, M., 81–82
Schmitt, P. M., 92–93
Schmitz, G., 108, 115
Schmucker, P., 81
Schneider, D. J., 90, 91–92
Schneiderman, J., 81
Schneuwly, S., 219
Schoenhard, J. A., 90
Scholmerich, J., 219
Schonenberger, C. A., 151–152
Schonig, K., 120–121
Schoonjans, L., 87, 88
Schrenk, D., 113–114
Schrimpf, S. P., 286–287
Schroder, W. A., 108–109, 112, 114
Schroeder, J., 219
Schroeter, M. R., 91
Schuler, E., 108–109
Schulman, A., 81–82, 92–93
Schulman, S., 81
Schultz, L. D., 78–79
Schulz, A., 359
Schulze, E., 292
Schuurhuis, D. H., 305–306
Schwab, J. P., 84
Schwaeble, W. J., 57, 64–65
Schwartz, A. L., 22, 79–80, 109–110
Schwarze, U., 175
Scopa, T., 292
Scroyen, I., 85–86, 107–108, 110
Sebestyén, E., 57
Seeliger, F., 175
Seet, B. T., 305–306
Seftor, E. A., 150, 154–155, 159
Seftor, R. E., 159
Segal, A. W., 139
Seglen, P. O., 42
Seiffert, D., 79–80
Seigel, G. M., 196
Sekine, H., 113–114
Self, S. J., 109–110
Sendall, T. J., 234
Senior, R. M., 145
Senoo, T., 87
Seong, K. H., 236–237
Seraphin, B., 291–292
Serhan, C. N., 307
Serrahn, J. N., 64–65
Seth, P., 156–157
Shahar, S., 361
Shah, M., 118–119
Shaker, J. C., 7–8, 10–11, 12–13
Shami, S., 155

Shapiro, A. D., 81
Shapiro, H. M., 311–314
Shapiro, S. D., 136–137, 139, 307
Sharma, P. R., 90
Sharon, R., 114
Sharp, L. K., 206, 229, 230, 231–232, 232, 234
Shaw, A., 108, 115, 116
Shaw, J. E., 293
Shearman, M. S., 234
Sheehan, M. C., 354
Sheng, S., 150, 151–152, 154–155
Shen, J., 137–139
Shen, T., 89, 92–93
Shenton, C., 234–235
Shen, Y., 116–117
Shevchenko, A., 291–292
Shifman, M. A., 18
Shi, G. P., 332–333
Shih, J. L., 81
Shi, H. Y., 137–139, 151, 154–155, 156–157, 158
Shiiba, M., 112
Shimada, H., 112
Shinbo, M., 119
Shinozuka, K., 112
Shiraishi, H., 108
Shirane, H., 177–178
Shi, Y., 155, 156–157, 158, 159
Shnabel, M., 305–306, 314, 316
Shoji, S., 35–36, 43, 46–47, 50
Shows, T. B., 66, 115
Shrimpton, A. E., 109
Shulman, J. M., 231–232
Shun, T. Y., 36, 261, 273–274, 277–278, 278
Shushanov, S. S., 343
Shyu, A. B., 118–119
Sibenaller, Z., 118
Siegrist, P., 231
Siemens, H. J., 81
Sieron, A. L., 173–174
Siewert, J. R., 81–82
Sifers, R. N., 2, 3–4, 5–6, 7–8, 9–11, 12–13, 14, 48–49
Siggia, E. D., 212–213
Silver, D. M., 110
Silverman, G. A., 36, 106, 108, 136, 150, 206, 207–209, 260–261, 273–274, 277–278, 278, 284, 286–287, 287, 288–289, 302–304, 332–334, 342–343, 348
Simmons, P. J., 145
Simonovic, M., 150
Simon, R. H., 89
Simonsen, A. C., 81–82
Simpson, A. J., 78–80
Sim, R. B., 56–57, 64–65
Singh, C. M., 253
Singh, I., 84
Singson, A., 289
Sinha, J., 69–71

Sippel, A. E., 151
Siren, V., 110
Sitia, R., 168
Sitrin, R. G., 111
Sjoland, H., 91–92
Sjolin, L., 78–79
Skeldal, S., 27, 81–82
Skerritt, J. S., 354
Skobe, M., 92–93
Skottrup, P., 78–79, 82
Skriver, K., 66
Skylas, D. J., 354
Slamon, D., 81–82
Slot, J. W., 36, 37–38
Smagur, J., 332–333
Smale, S. T., 116
Smith, A. I., 84, 210
Smith, E., 92–93
Smith, J. E., 154–155, 284, 286–287, 293
Smith, L. H., 88
Smith, T., 293
Smolenaars, M. M., 213
Smyth, S. S., 82
Snyder, S. E., 115
Sobel, B. E., 81, 90, 91–92
Sobel, E. S., 314
Socher, S. H., 78–79
Soff, G. A., 92
Sohrmann, M., 290–291
Sonderegger, P., 109, 286–287
Song, H. J., 231–232
Sontag, M. K., 139
Sorensen, J. A., 81–82, 118
Sosin, M. D., 324
Sottrup-Jensen, L., 18
Soukup, S. F., 213–214
Sounni, N., 92–93
Sousa, M. C., 12–13
Soussi, T., 155
Soustelle, L., 234–235
Sou, Y. S., 174
Spathe, K., 81–82
Spatz, H. C., 234–235
Spellman, P., 217
Spiller, O. B., 136–137, 139
Spradling, A. C., 239
Srivastava, S., 156–157
Stacey, M., 144
Stacey, S. N., 78–79
Stanford, M., 305–306
Stark, W. S., 231–232
Stasinopoulos, S. J., 115–116, 120, 121–122
Stassen, J. M., 88
Steegers, E. A., 110
Steenbakkers, P. J., 150, 348–349
Stefansson, S., 26–27, 28, 82, 92–93
Steiber, Z., 110
Steinckwich, N., 142–143

Steiner, J. L., 7–8, 10–11, 12–13
Stein, P. E., 206, 207
Steller, H., 244
Stemme, S., 81
Stempien-Otero, A., 90
Stenbak, C. R., 216–217
Stennicke, H. R., 343
Stephens, R. W., 115
Sternlicht, M. D., 154–155
Stewart, M., 305–306
Stickeler, E., 157
Stieber, P., 81–82
Stiernagle, T., 261
Stillfried, G. E., 109–110
Stinchcomb, D. T., 265–266, 293
Stirling, Y., 81
St Johnston, D., 236
Stolley, J. M., 142–143, 144, 145
Stolorov, J., 212–213
Stolz, D. B., 284, 288, 291–292
Stoolman, L., 84
Storkebaum, E., 91–92
Storm, D., 23, 56–57
Strambi, C., 234–235
Strandberg, L., 78–79
Stratikos, E., 108, 187
Straub, A. C., 90
Strauss, J. F. III., 28, 109–110
Strech-Jurk, U., 151
Street, S., 114
Strelkov, S., 78–79, 82
St-Remy, J. M., 84
Strey, H. H., 159–161
Stribling, R., 160
Strickland, D. K., 18, 20, 21–22, 23–24, 26–27, 28, 109–110, 213–214
Strieter, R., 84
Stringer, B. W., 112
Stringham, J. R., 88
Stuurman, M. E., 57–58
Su, E. J., 87
Sugimoto, A., 289–290
Suhrbier, A., 108–109, 112, 114
Sukhova, G. K., 307
Suminami, Y., 332–333, 343
Sunamoto, M., 177–178
Sun, J., 137–139
Sunkar, M., 81
Sun, Y. M., 305–306, 314, 316
Suter, B., 218
Sutter, G., 304–305, 306–307
Sutter, T. R., 115
Suur, M. H., 81
Su, Y., 154–155
Suzuki, K., 38
Suzuki, M., 137–139, 176–177
Suzuki, S., 177–178
Suzuki, T., 115

Suzuki, Y., 89–90
Sveger, T., 260–261
Svendsen, I., 348
Svensson, B., 354
Svetic, B., 111–112
Svoboda, K., 81–82
Swanson, W. J., 212–213
Sweeney, N. T., 212
Sweeney, S. T., 212
Sweep, C. G., 110
Swinkels, W. J., 348–349
Swulius, M. T., 2, 5–6, 12–13
Symoens, S., 175
Sympson, C. J., 151
Szarvas, T., 81–82
Szema, L., 84
Szilágyi, K., 64–65

T

Taatjes, D. J., 91–92
Taatjes, H., 91–92
Tabara, H., 289–290
Tacke, M., 92–93
Tada, N., 176–177
Tagesen, J., 81–82, 118
Taguchi, T., 177–178
Takada, K., 177–178
Takagi, A., 137–139
Taka, H., 47–48
Takahara, Y., 170–171
Takahashi, K., 47–48, 231–232
Takahashi, S., 86, 87
Takahashi, T., 110–111, 177–178
Takai, T., 332–333, 343
Takamatsu, Y., 145
Takayama, T. K., 136–137, 177–178
Takechi, H., 169, 176–177
Takeda-Ezaki, M., 47–48
Takeda, N., 137–139
Takeoka, H., 177–178
Takeshita, K., 90
Takeuchi, M., 177–178
Takimoto, R., 177–178
Takita, H., 194–195
Takizawa, S. Y., 86, 87
Talhouk, R. S., 151
Tanaka, K., 47–48, 174
Tanaka, T., 177–178
Tang, A. H., 214, 216
Tang, H., 212–213
Tanida, I., 37–38, 47–48, 174
Taniguchi, T., 119
Tanimoto, T., 87
Tanimura, T., 231
Tan, L., 108–109, 231–232
Tanzawa, H., 112
Tapner, M., 21

Tardif, J. C., 321–324
Tartaglia, G. G., 234
Tasab, M., 170–171
Tavernarakis, N., 289–290
Taylor, A., 169
Taylor, P. R., 144
Tazawa, S., 119
Teckman, J. H., 5–6, 38, 42, 45, 46, 48–49, 303–304
Teixeira, L., 218–219
Templeton, N. S., 156–157, 158, 159–161
Tenev, T., 231–232
Tengborn, L., 81
Termine, D. J., 5–6, 9–10, 14
te Velthuis, H., 56–57
Thakur, B. K., 118–119
Theologis, A., 350–351
Thevenot, C., 354
Thibert, J. N., 85
Thielens, N. M., 57
Thiel, S., 57
Thinnes, T., 88
Thoennes, S. R., 304–305, 306–307
Thogersen, A. M., 81
Thogersen, H. C., 66
Thøgersen, I. B., 332–333, 343
Thomas, A. R., 57
Thomas, B., 144
Thomas, C. F., 231–232
Thomas, C. M., 110
Thomas, D. A., 137–139, 218
Thomas, J. D., 90
Thomas, R., 114
Thompson, G. A., 359
Thompson, S. G., 81
Thomson, C. A., 169
Thong, B., 304
Thor, A., 150, 154–155
Thornberry, N. A., 304–305, 306–307
Thorpe, T. A., 361–363
Tierney, M. J., 120
Tiller, S. E., 18, 20
Timmons, L., 290–291
Timpl, R., 172–173
Tipping, P., 91–92
Tipton, A. R., 154–155
Tkalcevic, J., 139
Toba, G., 236–237
Todo, T., 231
Togashi, S., 231
Togonu-Bickersteth, B., 316
Tokuhisa, T., 174–175
Tolhurst, R. L., 354
Tollefsen, D. M., 26–27
Tolstoshev, P., 115
Tombran-Tink, J., 150, 184
Tomisawa, Y., 118
Topol, E. J., 90

Topper, Y. J., 151
Tost, J., 118
Totsch, M., 81–82
Tower, J., 253
Toyoda, N., 115
Tozser, J., 110
Tram, U., 212–213
Tran-Thang, C., 108, 110
Trask, O. J. Jr., 273–274
Traupe, H., 110–111
Travers, K. J., 2
Travis, J., 136–137, 150, 206, 333, 348
Treisman, J. E., 231, 246–247
Trejo, T. R., 214, 216
Tremblay, L. O., 12–13
Trinder, P., 23, 56–57
Trombetta, S. E., 12–13
Troxel, A., 47–48
Troxler, L., 206, 207–209, 210, 211, 212–213, 229–230
Tsai, L. T., 253
Tsao, Y. P., 194–195
Tsatsaris, V., 118
Tsu, C., 284, 286–287, 288–289
Tsuji, H., 177–178
Tsujii, M., 155
Tsuji, S., 155
Turner, P. C., 304–305, 306–307
Turner, R. F., 136, 207–209
Tymchak, W., 316
Tyson, J., 91–92
Tzeng, Y. J., 154–155, 156–157
Tzou, P., 216

U

Uchiyama, Y., 47–48, 174
Ueda, R., 231–232
Uematsu, T., 84, 88
Uemura, Y., 108, 342–343
Ueno, T., 37–38, 47–48, 174
Ueshima, S., 82, 88
Ulaszek, R., 354
Ullman, D. E., 359, 360–361
Ullrich, H., 108, 115
Ulm, K., 81–82
Umans, K., 91
Umebayashi, K., 35–36, 43, 46–47, 50
Umemura, K., 84
Unemori, E. N., 151
Unger, T., 155
Ung, K. S., 137–139
Untch, M., 81–82
Upton, C., 304–306
Upton, M. P., 332
Urano, T., 108, 109
Urden, G., 81
Ushioda, R., 174
Uzawa, K., 112

V

Vaheri, A., 110
Vaiman, D., 118
Valero, F., 69–71
Vamvakopoulos, J. E., 308
Van Breusegem, F., 350–351, 353, 360–361
Van de Cotte, B., 350–351, 353, 360–361
Van de Craen, B., 78–79, 82, 85–86
Van den Hurk, C. M., 316
van den Oord, J. J., 88
Van der Drift, C., 348–349
Van der Hoorn, R. A., 349
van der Voort, E., 81
van der Want, J. J. L., 37–38
Van Diepen, A. C., 316
Van Dyke, T. E., 307
Van Es, L. A., 57–58
van Hoef, B., 79–80, 82, 83, 85, 89–90, 91
Van Houtte, E., 80
van Mourik, J. A., 78–79
van Nuffelen, A., 87
van Putten, W. L., 111–112
Van Rooijen, N., 113
van Santen, H. M., 36
van Tongeren, J., 305–306
van Vlijmen, B. J. M., 85
Vanzieleghem, B., 84
van Zonneveld, A. J., 78–79
Vasil, I. K., 361–363
Vassalli, J. D., 106–107, 111, 113, 212
Vatsya, P., 314, 316
Vaughan, D. E., 78–79, 80, 88, 89, 90, 91
Vaziri, S., 316
Veerman, H., 78–79
Vellenga, E., 37–38
Vembar, S. S., 174
Vendruscolo, M., 234
Verbeke, K., 78–79
Vercammen, D., 350–351, 353, 360–361
Vermylen, J., 79–80, 88, 91
Verrusio, E., 92
Verstreken, M., 79–80
Verweij, C. L., 78–79
Vetica, A. C., 284, 288, 291–292
Veugelers, K., 305–306
Vibhakar, R., 118
Viebahn, R., 113–114
Vinals, M., 217
Vincent Salomon, A., 155
Vintermyr, O. K., 113
Vintersten, K., 153–154
Virtanen, O. J., 110
Viswanathan, K., 304, 314, 316
Vodovar, N., 217
Volz, K., 150

Vom Dorp, F., 81–82
Von Jagow, G., 341–342
von Wegerer, J., 234–235
Voros, G., 89
Vosburgh, E., 81
Vreys, I., 91
Vrhovec, I., 111–112

W

Wada, I., 12–13
Wada, K., 118
Wadsworth, M., 91–92
Wagner, T., 81
Waguri, S., 47–48, 174
Wakamiya, N., 170–171
Wakefield, T. W., 84, 85
Waksman, R., 321–324
Waldrip-Dail, H. M., 212–213
Walenga, J. M., 304
Walker, T., 110–111
Walling, L., 116
Walsh, B. J., 354
Walsh, C. T., 337, 338
Walsh, J. D., 21–22
Walski, M., 332–333
Walter, P., 348
Walther, R. F., 253–254
Waltzer, L., 212
Wang, C., 214, 216–217, 218
Wang, F., 314
Wang, H., 89, 304
Wang, J., 81–82, 289–291
Wang, S. L., 289–290
Wang, X., 108–109
Wang, Y. X., 213, 304–305, 306–307
Wan, L., 235
Wardell, M. R., 206
Warley, A., 139
Warren, H. S., 284, 291–292
Washburn, M. P., 354
Washington, K., 88
Watanabe, Y., 111–112
Waterhouse, J., 213
Watkins, S. C., 36, 40, 41, 43, 48–49, 50, 284, 288, 291–292
Watson, I. E., 234
Webb, A. C., 115
Weeks, L., 91
Wegiel, J., 40–41
Weidle, U., 92–93
Wei, J., 354
Weiler, J. M., 64–65
Weinberger, C., 116
Weinehall, L., 81
Weinrauch, Y., 139
Weisberg, A. D., 90, 91
Weis, M. A., 175

Weiss, E., 91
Weiss, I., 92
Weiss, J., 139
Weiss, L. A., 212, 214
Weissman, A. M., 2
Weissman, J. S., 2, 168
Welchman, D. P., 290–291
Welinder, K. G., 78–79
Welti, H., 108, 110
Wenz, K. H., 110
Werb, Z., 151
Werden, S., 305–306
Werner, S., 136–137
Westrick, R. J., 83–84, 88, 91–93, 107–108, 110, 112, 113, 116–117
Whinna, H. C., 83
Whisstock, J. C., 57, 106, 109, 150, 206, 207–209, 210–211, 284, 286–287, 302–304, 332, 343, 348–349
White, K., 231, 244
Whiteway, M., 218
Whitman, I., 5–6
Whittaker, J. S., 108
Whitty, A., 304–306
Whyte, P., 155
Wight, T. N., 91
Wiiger, M., 115
Wijeyewickrema, L. C., 57
Wilczynska, M., 109
Willer, D., 305–306
Williams, B. R., 118–119
Williams, D. B., 7–8
Williams, R. G., 273–274
Williams, S. E., 18, 24
Willnow, T. E., 22, 23
Wilm, M., 291–292
Wilusz, C. J., 118–119
Wilusz, J., 118–119
Wiman, B., 56–57, 66, 79–81
Wind, T., 81–82
Wing, L. R., 109–110
Winkler, I. G., 145
Winkles, J. A., 22
Winn, M. E., 83–84
Wintrobe, M. M., 142–143
Winzen, R., 118–119
Wishart, D. S., 304–306
Wittmann, C. W., 231–232
Wittschier, M., 81–82
Wodarz, A., 231
Woessner, J., 50
Wohlwend, A., 106–107, 111, 113
Wolff, M. W., 64–65
Wolf, J. D., 37–38
Wolfner, M. F., 212–213
Wolters, D., 354
Wong, A., 113, 116–117
Wong, D. T., 106

Wong, M. K., 82, 92–93
Wong, W., 348
Woodrow, G., 108–109
Wood, S. A., 42
Woo, S. L. C., 2, 3, 5–6, 12, 48–49
Worrall, D. M., 343
Worrall, M. M., 209
Wortman, J. R., 229
Wright, K. C., 81
Wrigley, C. W., 354
Wrobleski, S. K., 84, 85
Wszolek, M. F., 231–232
Wu, D., 69–71
Wuillemin, W. A., 56–57
Wu, J. S., 253–254
Wu, K. K., 81
Wun, T. C., 79–80, 106–107, 108
Wu, S., 67–68
Wustmann, G., 234–235
Wu, Y. Q., 2, 5–6, 12–13, 14, 21–22, 185, 187
Wynn, T. A., 115

X

Xiang, Y., 359, 360–361
Xoconostle-Cazares, B., 359, 360–361
Xue, Y., 78–79
Xu, N., 118–119
Xu, S., 289–291, 293
Xu, Y., 113
Xu, Z., 90, 91–92

Y

Yacoubian, S., 307
Yagi, K., 177–178
Yamada, K. M., 169, 176–177
Yamamoto, A., 37–38, 171–172, 173–175
Yamamoto, H., 177–178
Yamamoto, K., 86–87, 137–139
Yamamoto, M., 118
Yamamoto, Y., 177–178
Yamamura, I., 176–177
Yamashita, T., 115, 332–333, 343
Yamazaki, C. M., 170–171
Yamazaki, N., 115
Yandell, M. D., 229
Yang, A. Y., 78–79, 107–108, 110, 112
Yang, C., 47–48
Yang, R., 307
Yao, K. M., 231
Yasuda, K., 176–177
Yasumatsu, C., 136–137
Yasumatsu, R., 136–137
Yates, J. R., III., 292, 354
Ye, R. D., 106–107, 115
Yokota, S., 169, 170, 173–174
Yokoyama, A., 87
Yoo, B. C., 359, 360–361

Yoon, J. G., 118
Yoon, K. H., 115
Yorihuzi, T., 170–171
Yoshida, H., 174
Yoshimori, T., 35–36, 37, 43, 46–47, 50, 172, 174–175
Yoshimori, Y., 37–38
Yoshimuri, T., 40–41
Yoshino, H., 111–112
You, Z., 12–13
Yue, Z., 47–48
Yu, H., 110, 111, 120, 137–139
Yu, J., 47–48, 314
Yu, M. H., 5–6
Yu, P., 21–22
Yu, W. H., 40–41
Yuyama, N., 332–333, 343
Yu, Y. B., 112
Yu, Z. X., 12

Z

Zachariou, A., 231–232
Zalai, C. V., 316
Zaman, A. K., 90
Zander, K., 118
Zanet, J., 212
Zanichelli, A., 64–65
Zarebski, M., 332–333
Závodszky, P., 56–57, 64–65, 66
Zdanovsky, A., 22
Zeheb, R., 78–79
Zemach, A., 361
Zeng, W., 137
Zhang, F., 64–65
Zhang, J., 212, 235
Zhang, M., 106, 137–139, 151–152, 154–155, 156–157, 158, 159, 284, 291–292
Zhang, P., 235
Zhang, Q. W., 210, 348
Zhang, S. J., 114–116
Zhang, S. L., 69–71
Zhang, W., 156–157, 158
Zhang, Y. Q., 108–109
Zhang, Z., 235
Zhao, B., 235
Zhao, J., 361
Zheng, X., 88, 89
Zhong, R., 314
Zhong, X. Y., 118
Zhong, Y., 234–235
Zhou, G. P., 21
Zhou, H. M., 111, 113
Zhou, W., 218–219
Zhou, X., 90
Zhuang, Y. P., 69–71
Zhu, Y. S., 88, 91–92, 108–109
Ziff, E. B., 116, 123–124

Zingale, L. C., 206
Zipperlen, P., 289–291
Zivy, M., 354
Zou, M., 114–116
Zou, Z., 150, 154–155, 156–157, 210, 211
Zundel, S., 57
Zusterzeel, P. L., 110
Zychlinsky, A., 139

Subject Index

A

Activity-regulated inhibitor of death (AID), 114–115
Acute coronary syndrome (ACS), viral serpin clinical trials
  biomarkers, inflammatory, 324
  inflammatory and thrombotic markers, 322
  major adverse cardiovascular events (MACE), 321–324
  patient treatment groups, 324
  Serp-1, 321
  serum D dimer levels, 325
  stent implant, 324
  proceeding stages
    efficacy testing, 320–321
    FDA approval, 320–321
    good manufacturing protocol (GMP), 320–321
    requisite steps, 320–321
AED. See Antennal-eye imaginal disc
AID. See Activity-regulated inhibitor of death
Antennal-eye imaginal disc (AED), 246–248
Antiangiogenic assays
  cell migration, 198–199
  chick embryo aortic arch, 197
  CNV, 199–201
  corneal pocket, 201–202
  directed *in vivo* angiogenesis assay (DIVAA), 198
Anti-inflammatory activity cellular mechanisms
  flow cytometry analysis, 316
  fluorescent-conjugated antibodies, 315–316
  hematoxylin and eosin, 312
  knock out mouse models, 313
  serpin inhibitory activity detection
    human cell activation, 316
    in vitro adhesion assay, 317–318
    PBMCs, 316–317
  single splenocyte suspension, 314–315
Antithrombin III (AT-III), 303–304
Arabidopsis
  subcellular localization
    AtSerpin1, AtSRP2 and AtSRP3, 351
    detection, GFP-serpin fusion proteins, 351–352
    GFP-serpin fusion constructs, 351–352
    immunolabeling electron microscopy, 350–351
  protoplasts preparation, 352
  Western blot analysis, 353
  and T-DNA knockout, 361
AREs. See AU-rich elements
Arterial angioplasty injury, mouse model
  cholesterol levels, 308
  inflammatory response, 307
  innate immune system cells, 308
  knockout mice characteristics, 308
  operation, 309–310
  postoperative care, 310–311
  preoperative preparation, 308
Arterial injury, serpin treatment
  RNA isolation, 318–319
  Serp-1 and Serp-2 effects, 318
Association rate constant
  C1-inhibitor, 57
  measurement, 63
AU-rich elements (AREs)
  mRNA stability
    destabilizing elements, 119
    structure, 118–119
  systematic mutagenesis screening approach, 122
Autolysosome, 38, 48
Autophagy role, alpha-1-antitrypsin (AT) deficiency
  analysis, cell line models
    DAMP, 38
    immune antigens, 37–38
    LC3 ultrastructural localization, 37
    plastic-embedded EM, 36
  assessment, long-lived protein degradation
    pulse-chase metabolic labeling, 41–42
    putative inhibitors, 42
  CBZ, 36
  COPD, 34–35
  deficient cell lines
    Atg5$^{-/-}$, 42, 43
    cellular model, 42
  genetic/environmental modifiers, 35
  hepatic injury detection
    BrdU labeling, 51
    OHP content, 50–51
    PiZ mouse model, 48–49
    Z mouse model, 49
  hepatic *in vivo* detection
    deficient mouse models, 47–48

Autophagy role, alpha-1-antitrypsin (AT) deficiency (cont.)
  electron microscopy, 45, 46
  fluorescence microscopy, 46–47
  LC3 conversion assay, 38–41
  mutation and liver disease, 34
  pulse–chase labeling
    immunoprecipitation, 44–45
    protocol, 43–44
    total protein content, TCA precipitation, 44
  vesicles and basal cellular proteostasis, 35–36

B

BCA assay, 39
Biochemical fractionation, PEDF protein
  ammonium sulfate, 187
  cation-exchange chromatography, 187
Biochemistry, serpin targets identification
  immunoprecipitation
    coupling antibody, CnBr beads, 294–295
    first and second round, purifications, 295–296
    western blotting, 296
  serpin-protease complexes purification
    C. elegans lysates preparation, 294
    large-scale *C. elegans* growth, 293–294
  TAP
    affinity tags, 291–292
    plasmids generation, 292–293
  transgenic lines, purification, 293

C

Caenorhabditis elegans, serpinopathy
  AT deficiency, 260–261
  description, 260–261
  HCS
    animal preparation and sorting, 275–277
    cells, multiwell plates, 273–274
    chromosomal integration, 278–279
    equipment, consumables and reagents, 274
    image acquisition, 277–278
    library preparation, 275
    OP50 food preparation, 274
    selectivity, 279
  microinjection
    equipment and reagents, 268–269
    isolation and propagation, transgenic worms, 271
    method, 270–271
    plasmid DNA preparation, 269
    stable integration, transgenes, 271–273
    transgenic lines characterization, 273

model systems, 261
transgenesis
  coinjection markers, 266–267
  DNA origin, 264
  expression vectors, 264–265
  fluorescent protein (FP) fusions, 264
  methods, 265–266
  promoter choice, 264
  resources and websites, 261, 262
  serpin-proteinase balance, 261
transgenic animals, 261
Carbamazepine (CBZ), 36, 43
Cardiovascular disease, genetically modified mice
  atherosclerosis-(re)stenosis
    bone marrow transplantation, 91–92
    description, 90
    hypercholesterolemia models, 91–92
    intima formation, 91
    TGF-β1 overexpression, 91
    vascular injury site, 91–92
  fibrosis, 90
  thrombosis
    cerebral hemorrhage and stroke, 89–90
    coronary arterial thrombosis, 89
    FeCl$_3$-induced carotid artery injury, 88–89
    L-NAME induced, model, 88
    MCA ligation, 89–90
    PAI-1 gene, 88–89
    preproendothelin-1 and metallothionein I promoter, 89
    rose bengal use, 88
CBZ. *See* Carbamazepine
Chemical and x-ray mutagenesis
  Drosophila, 245
  EMS, 243–245
  mutations screening, 246–248
  screening protocol, 245
  single base-pair changes, 242
  ubiquitous neuronal expression, 243
Chick embryo fibroblasts (CEFs), 169
Choroidal neovascularization (CNV)
  flatmount technique, 200
  laser-induced, 199
  subconjunctival injections, 201
Chronic obstructive pulmonary disease (COPD), 34–35
C1-inhibitor expression, *Pichia pastoris*
  fermentor-based
    batch medium, 69–71
    DO, 69–71
    proteases and carbon sources, 69–71
  flasks
    Erlenmeyer, 68
    SDS-PAGE analysis, 68
  methanol, 68
Clade B serpins, 150

Subject Index

CNV. *See* Choroidal neovascularization
COPD. *See* Chronic obstructive pulmonary disease

D

DAMP electron microscopy, 38
Deletion kit screening, 255
Dendritic cells (DCs), 311–314
Directed *in vivo* angiogenesis assay (DIVAA), 198
Dissolved oxygen (DO), 69–71
DIVAADirected *in vivo* angiogenesis assay
DNA footprinting, 116–117
DO. *See* Dissolved oxygen
Drosophila melanogaster
  assays
    locomotor, 253
    longevity, 252
    pseudopupil, 248–249
  deletion kit screening, 255
  EMS mutagenesis, 243–245
  eye imaginal disc, 246–248
  flies, human serpins
    behavioral assays, 234–235
    longevity assays, 233–234
    microscopic phenotyping, 235
  genetic backcrossing, 251–252
  human serpinopathy model
    biological effects, 231
    mature photoreceptor cells, 231
    neuroserpin expression, 232
    polymerogenic mutant forms, 230
    retinal expression, 232
    "rough eye" phenotype, 231
    size exclusion chromatography, 233
    ubiquitous expression, 231
    unregulated protease inhibition, 230
  immunostaining, fly brains, 253–254
  P-elements and RNAi screening, 253–254
  polymerogenic mutations screening, fly serpins
    fly homeostasis, 229–230
    inactivation, 230
  protein extraction, flies, 250–251
  screening, mutations, 242
  serpinopathies
    biological orthology, 229
    functional redundancy, 229
    genetic tool kits, 229
  suppressing serpin-induced phenotypes
    genetic modifier screens, 236
    genomic deletions, 237
    mobile-genetic elements, 236–237
    P-transposase, 236–237
  transgenic generation
    crossing scheme, transformants identification, 241
    microinjection method, 239–241
    P-element transposase expression, 237–238
    phiC31 system, 238–239
    plasmid preparation, 239
    pUAST approach, 237–238
    serpin expression, 237–238
    UAS/GAL4 expression system, 242
    x-ray mutagenesis, 245
Drosophila serpins, immunity and morphogenesis
  endemic infections control
    insect immunity, 218
    microsporidia, 219
    nora virus, 219
    Wolbachia, 218–219
  functions
    macroglobulins and tight-binding protease inhibitors, 209
    mammalian serpins, 207–209
    RCL sequences, 209
    terminal peptides and size conservation, 210–211
  genetic analysis
    hemolymph, 213–214
    innate immune response, 211, 212
    seminal fluid proteins, 212–213
    transcripts and multiple protease targets, 213
  hemolymph collection
    glass capillaries, 217–218
    prophenol oxydase activity, 218
  IMD pathways activation
    factors consideration, 215
    inducers, 214–215
    read-outs, 215
  immune challenge methods
    natural infection, 217
    pathogens, direct injection, 216–217
    septic injury, needles, 216
  noninhibitory proteins, 206
  "suicide-cleavage" mechanism, 207
  survival assays, 218

E

EIM model. *See* Electrolytic inferior vena cava model
Electrolytic inferior vena cava (EIM) model, 84
ELISA. *See* Enzyme-linked immunosorbent assay
Endemic infections control
  insect immunity, 218
  microsporidia, 219
  nora virus, 219
  Wolbachia
    description, 218–219
    Wsp primers, 218–219
Endoplasmic reticulum (ER)
  collagen-binding protein
    chick embryos, 169
    mammalian stress proteins, 169
    pH buffer, 169

Endoplasmic reticulum (ER) (cont.)
  procollagen aggregation/bundle formation, 169
  procollagen maturation
    Golgi apparatus, 173–174
    polypeptides, 172–173
    triple helix formation, 173
  resident stress, 168
Enzyme-linked immunosorbent assay (ELISA)
  ligand binding measurement, 24–25
  PEDF quantification, heterologous samples, 190
  quantitation, 233
Epigenetics
  described, 117–118
  histone deacetylase inhibitors, 118
ER-associated degradation (ERAD), 174
ER-Golgi intermediate compartment region (ERGIC), 169
Escherichia coli, 21, 64–65
Eye imaginal disc
  adult antenna and eye forms, 246–247
  AED, 246–247
  glass slide, 248
  PBT, 247–248

F

Familial encephalopathy with neuroserpin inclusion bodies (FENIB), 228
Flatmount technique, 200
Fly brains immunostaining
  confocal microscopy, 253–254
  immunohistochemical techniques, 253–254
  primary antibody solution, 253–254

G

Gene expression and regulation, PAI-2
  assessment, mRNA decay, 120–121
  AU-rich instability element role, 3'-UTR mRNA decay, 122
  sequence analysis, 121–122
  cellular, 116
  cloning, 115
  epigenetics
    described, 117–118
    histone deacetylase inhibitors, 118
  induction factors, 115–116
  mRNA stability, 118–119
  posttranscriptional
    induction and suppression, mRNA, 119
    regulatory domains, 120, 121
    use, transcription inhibitors, 119–120
  transcriptional
    nuclear "run-on" transcription assays, 116
    promoters and DNA footprinting, 116–117
    repressor element, 117
Genetically modified mice, PAI-1

cancer
  angiogenesis, 93
  growth and metastasis, B16 melanoma tumors, 92–93
  inhibitor, 92
  malignant keratinocytes, 92–93
cardiovascular disease
  atherosclerosis-(re)stenosis, 90, 91–92
  fibrosis, 90
  thrombosis, 88–90
Genetic backcrossing
  behavioral phenotypes and longevity, 251
  single-fly genomic PCR, 251
  virgin female offspring, 251
Glycoprotein endoplasmic reticulum-associated degradation (GERAD), 12–13
Good lab practice (GLP), 320–321

H

HCS. *See* High-content screening
Heat shock protein (HSP) 47, molecular chaperone
  collagen-binding protein, ER
    chick embryos, 169
    mammalian stress proteins, 169
    pH buffer, 169
    procollagen aggregation/bundle formation, 169
  definition, 168
  ER-resident stress, 168
  expression and clinical importance
    collagen-producing cells, 176–177
    promoter and regulatory element, 177
    therapy, collagen-related diseases, 177–178
    tissue-specific, 176–177
    vitamin A-coupled liposomes, 177–178
  interaction and recognition, collagen
    in vitro analysis, 170
    procollagen, 170
    synthetic peptide approach, 170–171
  knockout mice
    abnormal maturation, 172
    N-propeptides, 171–172
    procollagen accumulation, 172
    retardation and cardiac hypertrophy, 171
  molecular chaperones, 168
  null cells
    autophagy–lysosome pathway, 174
    Mov13 cells, 175
    point mutations, 175
    procollagen accumulation, 174
    proteasome inhibitors, 174–175
    unfolded protein response (UPR), 174
  procollagen maturation, ER
    Golgi apparatus, 173–174
    polypeptides, 172–173
    triple helix formation, 173

Subject Index

Hemolymph
  collection, 217–218
  melanization cascade, 218
  target proteases identification and serpin/
    protease complex degradation, 213–214
Heparan sulfate proteoglycan (HSPG), 191
High-content screening (HCS). *See also*
  Caenorhabditis elegans, serpinopathy
  automatic image acquisition
    ArrayScan V^{TI}, 277–278
    spots, 277–278
  COPAS BIOSORT
    animal sorting, 275–277
    automated worm transfer, 275–277
    coincidence checks, 277
    flow cytometer, 275–277
  high-throughput screening (HTS) assay,
    278–279
HSPG. *See* Heparan sulfate proteoglycan
Human plasma C1-inhibitor
  data interpretation, 64
  functional activity, purified C1-inhibitor
    association rate constant measurement, 63
    SDS-PAGE analysis, 62
    stoichiometry, protease inhibition, 62–63
  method, 64
  purification
    absorbance *vs.* fraction number, 58–61, 60
    Jacalin–agarose affinity chromatography, 60
    mono Q FPLC anion-exchange
      chromatography, 58–61
    PEG treatment, 58–61
    protease inhibitors, 58–61
    three-step procedure, 57–58
    Western blot analysis, 58–61
Hydroxyproline (OHP)
  hepatic fibrosis assessment
    acid hydrolysis, 50
    measurement, 50–51
  PiZ mouse model, 48–49

I

IEM. *See* Immuno electron microscopy
Immune deficiency (IMD) pathways activation
  factors consideration, 215
  inducers, 214–215
  read-outs, 215
Immunochemical assays
  antibodies, 188
  ELISA, 190
  immunoblot reaction, Ab rPEDF, 188
Immuno electron microscopy (IEM)
  gelatin and labeling, 37–38
  postembedding immunogold labeling, 37–38
  ultrastructural localization, LC3, 37–38
Internal elastic lamina (IEL), 312–313
Interphotoreceptor matrix (IPM)
  components, 185–186
  PEDF, 187
Intravascular ultrasound (IVUS), 321
IPM. *See* Interphotoreceptor matrix

J

Jacalin–agarose affinity chromatography, 60

K

Kaplan–Meier plot, 234, 252

L

LC3 conversion assay
  carboxyl terminal modification, 38–39
  gel electrophoresis, 39–40
  immunoblotting, 40
  interpretation, blots, 40–41
  lysosomal inhibitors, 40–41
  protein estimation, BCA assay, 39
  sample preparation, 39
  signal detection, 40
LDL receptor-related protein 1 (LRP1)
  hepatic receptor, 28
  ligand binding
    determinants, 23–24
    ELISA, 24–25
    quantitative measurements, 24
    serpin–enzyme complexes, 23
    surface plasmon resonance, 25–26
  α_2M–protease complexes, 18
  purification
    full length isolation, tissue extracts,
      18–20
    soluble forms, isolation, 21
  receptor fragments expression
    human glioblastoma U87 and COS-1 cells,
      21–22
    LDLa repeats, *Escherichia coli*, 21
    "minireceptors", 22
  serpin–enzyme complexes binding
    concentration, 26
    determinants, 26–27
    ^{125}I-labeled clearance, 27–28
    specificity, 26
  VLDL and LRP2/gp330, 28
Leukocyte elastase inhibitor (LEI), 136–137
Ligand binding, LRP1
  assays measurement
    ELISA, 24–25
    quantitative, ligand interaction, 24
    surface plasmon resonance, 25–26
  determinants
    Lys270 and Lys256, 23–24
    protease-inhibitor complexes and
      clusters, 23
    serpin–protease complexes report, 23

Locomotor assays
  climbing behavior, 253
  infrared beam breaking, 253
  stable light conditions, 253
  statistical analysis, 253
Longevity assays
  Kaplan–Meier plot, 252
  mated females, 252
LRP1. *See* LDL receptor-related protein 1
LRP1 purification, isolation
  full length
    affinity chromatography, 18, 20
    human placenta, 18
    α_2M and methylamine-reacted
      α_2M–sepharose, 18
    soluble forms, 21

M

Major adverse cardiovascular events (MACE), 321–324
Maspin investigation, breast cancer
  bioactive macromolecules breakdown, 150
  cell culture systems, 150–151
  clade B serpins, 150
  description, 150
  metastasis, 161–162
  serpins, 150
  syngeneic tumor model
    development, 157–159
    gene therapy, mouse, 159–161
  transgenic mouse models
    bitransgenic mice, 154–156
    limitations, 156–157
    WAP, 151–154
MCA. *See* Middle cerebral artery
Microinjection
  C. elegans gonad, 270
  equipment and reagents, 268–269
  isolation and propagation, transgenic worms, 271
  method, 270–271
    collection period, 239–240
    halocarbon oil, 241
    loaded needle, 240–241
    nontoxic double-sided adhesive tape, 240
    white-eyed flies, 239–240
  plasmid DNA preparation, 269
  postembryonic development (PED), 273
  stable integration, transgenes
    extrachromosomal arrays, 271–272
    lines starvation, 272
    materials, 272–273
    SNP mapping, 273
Microsporidia, 219
Middle cerebral artery (MCA), 89–90
Minimal dextrose (MD) plates, 67–68
Minimal methanol (MM) plates, 67–68
Monocyte neutrophil elastase inhibitor (MNEI), 136–137
Mono Q FPLC anion-exchange chromatography, 58–61
Mouse models
  autophagy-deficient, 47–48
  GFP-LC3, 46–47
  PiZ
    hepatocellular proliferation, liver, 48–49
    OHP content, 48–49
  Z
    GFP-LC3 bred, 50
    "target" strain, 49
Myeloperoxidase (MPO) activity assay
  azurophil granules, 142
  cell-free bronchoalveolar lavage granulocytes, 141
  sample preparation, 142
  standard and buffers, 141–142

N

Neurite-outgrowth analyses, 195
Neurotrophic assays
  neurite-outgrowth analyses, 195
  photoreceptor, 194–195
  protection, oxidative damage, 196–197
  retina cell survival, 196
Neutrophil homeostasis, bone marrow
  elastase activity measurement
    reagents and solutions, 145
    sample collection and staining, 145
    sample preparation and measurements, 145, 146
  granulocyte/macrophage progenitors, 143
  immunophenotyping
    cell surface markers, 143–144
    sample collection and staining, 137, 145
  serpinB1 role, 142–143
N-linked oligosaccharide contribution, serpin secretion
  chemical inhibitors, 13
  intracellular trafficking, biochemical assessment, 12
  modified, asparagine-linked oligosaccharides, 10–11
  N-glycan-mediated events
    Asn-X-Ser/Thr consensus sequences, 12–13
    GERAD signal, 12–13
    inhibitors, 12–13
    posttranslational removal, 13
  number and role, N-glycan, 10–11
  polar appendages, 10
Nora virus, 219

Subject Index

P

PAI-1. *See* Plasminogen activator inhibitor-1
PAI-2. *See* Plasminogen activator inhibitor type-2
PAI-1 gene
   deletion, 88–89
   genetic disruption, 88
   plasma PAI-1 concentrations, 81
PEDF. *See* Pigment epithelium-derived factor
P-elements
   GAL4-UAS sequences, 254
   nonmutagenized chromosome, 254
   serpin-related phenotype, 252
   subsequent generations, 254
Physically interacting proteins, serpin secretion
   alternative methods, 7–8
   caveats experiment, 8
   coimmunoprecipitation and identification
      bound protein complexes, 7
      interactions, 7
      mock-transfected cells, 7
Pichia pastoris
   advantages, 65–66
   C1-inhibitor expression
      fermentor based, 69–71
      flasks, 68, 70
      methanol, 68
   clone generation and isolation
      hexahistidine tagged serpin domain, C1-inhibitor, 66
      homologous recombination, 67–68
      MD and MM plates, 67–68
      sorbitol and YNB solution, 67–68
   vs. Saccharomyces cerevisiae, 65–66
Pigment epithelium-derived factor (PEDF)
   antiangiogenic assays, 197–202
   binding activities
      CPC precipitation, 190–191
      glycosaminoglycan-affinity column chromatography, 190
      heparan sulfate proteoglycan, 191
      receptor proteins, 193
      solid-phase assays, 192
      solution assays, 191–192
      surface plasmon resonance assays, 192–193
   description, 184
   immunochemical assays, 188–190
   importance, 184–185
   neurotrophic assays, 194–197
   protein purification
      aqueous humor, 186
      biochemical fractionation, 187–188
      description, 185
      interphotoreceptor matrix, 185–186
      IPM, 185–186
      plasma, 186
      recombinant, 187
      vitreous humor, 186
   receptor proteins
      ligand-affinity column chromatography, 194
      ligand blot, 194
      radiolabeled ^{125}I-PEDF binding assays, 193
Plant serpin–protease interactions
   kinetic analyses, 360–361
   SDS-PAGE, 361
   target proteases identification
      anti-serpin antibodies, 361–363
      Arabidopsis and T-DNA knockout, 361
      cysteine proteases, 363–364
      gel fractionation, 363
Plasminogen activator inhibitor-1 (PAI-1)
   cancer
      prognostic outcomes, clinical data, 81–82
      vitronectin, 82
   cardiovascular disease
      deficiency, 81
      deficient fibrinolytic response, 80–81
      level, increased and elevated, 81
   cell migration and tissue remodeling, 78
   description, 78
   genetically modified mice
      cancer, 92–93
      cardiovascular disease, 88–92
      in vivo role, 87
   knockout mouse, 93–94
   mouse models
      vs. human, 82, 83
      as putative therapeutic target, 82
   N-glycosylation sites and cDNAs encoding, 78–79
   nonreactive form, 80
   plasma and platelets, blood pools, 79–80
   plasminogen/plasmin system and extracellular matrix, 78, 79
   proteinase
      binding, 27
      complexes, 26–27
   structure, 80
   thrombosis and thrombolysis, 303–304
   wild-type mouse models
      cancer, 87
      cardiovascular disease, 83–87
      thrombosis, 83
      vessel size and blood flow, 83
Plasminogen activator inhibitor type-2 (PAI-2)
   activating system, 107–108
   apoptosis and innate immune response
      AhR knockout mice, 113–114
      Bacillus anthracis, 113
      HeLa cells, 113
      Th1 immunity, 114
   association, Rb, 112
   biological activities, 122–123
   clearance receptors, 109–110
   described, 106

Plasminogen activator inhibitor type-2 (PAI-2) (*cont.*)
  expression pattern, 110
  extracellular protease targets, 106–107
  gene expression and regulation
    assessment, mRNA decay, 120–121
    cellular, 116
    cloning, 115
    epigenetics, 117–118
    induction factors, 115–116
    mRNA stability, 118–119
    posttranscriptional, 119–120
    role, AU-rich instability elements, 121–122
    transcriptional, 116–117
  neuroprotective role
    AID genes, 114–115
    human/mouse monocytes, 114
  rapid run-on transcription assay protocol, 123–126
  role
    metastatic cancer, 111–112
    monocyte biology, 111
    skin, 110–111
  structure
    CD-loop, 108–109
    Ov-serpins, 108
    polymerization, 109
Plastic-embedded electron microscopy, 36
PN–1. *See* Protease nexin 1
Protease nexin 1 (PN–1), 27
Protease targets identification, serpins *in vivo*
  biochemical approaches, 284
  Caenorhabditis elegans, 284
  intracellular serpins, *C. elegans*
    biochemistry, 291–296
    BLAST analysis, 286–287
    genetics, 287–291
    methodologies, 286
    SRP-2, 286–287
  nematode growth and maintenance
    NGM preparation, 285
    OP50 preparation, nematode food stocks, 285
Protein extraction, flies
  freezing/vortexing cycle, 250
  hemolymph, 250
  PCR sealing film, 250–251
  proteolytic degradation, 251
  96-well format, 250–251
Protein interactions and role verification, serpin secretion
  caveats experiment, 10
  RNAi technology, 9–10
  sedimentation velocity centrifugation use, 9
  time course, detectable protein interactions
    antisera use, 8–9
    identical dishes, 8–9
Pseudopupil assay

degeneration rate, 248
pilot assay, 249
rhabdomeres, 249
senile degeneration, 248–249

R

Radioactive free assay, 191
Radioactivity binding assay, 191
Radiolabeled ^{125}I-PEDF binding assays
  attached cells, 193
  cells, suspension, 193
  determination, 193
Rapid run-on transcription assay protocol
  hybridization procedure, 124–126
  immobilized and hybridization, DNA, 124–126
  nuclei isolation, 123–124
  primary transcripts, 123
  RNA isolation, 123–124
  unincorporated ^{32}P-UTP removal, 124–126
Rb. *See* Retinoblastoma protein
Reactive center loop (RCL)
  segments, 213
  sequences, 209
  serpin structure, 207
  trypsin, 214
Reactive site loop (RSL), 341–342
Receptor fragments expression, LRP1
  human glioblastoma U87 and COS-1 cells, 21–22
  LDLa repeats, *Escherichia coli*, 21
  "minireceptors", 22
Recombinant C1-inhibitor production, yeast
  Escherichia coli, 64–65
  mutations, 64–65
  Pichia pastoris
    advantages, 65–66
    C1-inhibitor expression, flasks, 68
    cloning, 66
    fermentor-based C1-inhibitor, 69–71
    isolation, 67–68
    vs. Saccharomyces cerevisiae, 65–66
  purification
    elution profile, deglycosylated C1-inhibitor, 71–73
    endoglycosidase H. pool and SDS-PAGE, 71–73
    hyperglycosylation, 71–73
    imidazole, 71–73
    supernatant buffer, 71–73
Retinoblastoma protein (Rb), 112
RNA-mediated interference (RNAi)
  protease targets
    C. elegans, 289–290
    dsRNA delivery, 290
    SRP-6, 291
  screening, 254

Subject Index

serpin processing, 9–10
"Rough eye" phenotype, 231

S

SDS-PAGE. *See* Sodium dodecyl sulfate-polyacrylamide gel electrophoresis
Serpin biology, plants
  animal-specific complex process, 348
  genome and transcript sequences, 348–349
  native-PAGE, separation and visualization
    molecular sizes, 359–360
    wheat and barley, 360
  properties and functions, 349
  protease interactions
    in vitro, 360–361
    SDS-PAGE, 361
    target proteases identification, 361–364
  proteolysis control, 349
  purification, vegetative organs
    barley root tips, 359
    pumpkin, 359
  recombinant production, 360
  tissues and cells, detection
    antibody specificity, 350
    Arabidopsis subcellular localization, 350–353
    biotinylated proteases, 349–350
  tissues, purification
    anion-exchange chromatography, 356
    apple seed, 357
    endogenous proteases, 353
    Eppendorf tubes, 355
    mature seeds, 353–358
    oat grain, 356–358
    SDS-PAGE analysis, 355
    thiol extraction, 354–355
    thiophilic adsorption chromatography, 356
  Western blotting, 359
SerpinB1 knockout mouse
  characterization, 137
  description, 136
  embryonic stem cell targeting, 137–139
  lung infection and inflammation
    experimental design, 140
    intranasal inoculation, 140
    neutrophil proteases and inhibitors, 139–140
    tissue harvest, 141
  MNEI and LEI, 136–137
  neutrophil homeostasis
    bone marrow, 142–143
    cell surface markers, 143–144
    reagents and solutions, 145
    sample collection and staining, 145
    sample preparation and measurements, 145, 146
  plasmid construct design
    Cre-loxP technology, 137–139
  targeting vector, 138
  tissues and cell-free bronchoalveolar lavage
    assay buffers, 141–142
    MPO assay, 142
    MPO standard, 141
    sample preparation, 142
Serpin–enzyme complexes
  banding pattern, 339–340
  binding, LRP1
    concentration, increased, 26
    ^{125}I-labeled clearance, 27–28
    mutagenesis, 27
    PAI-1 complexes, 26–27
    PN–1, 27
    specificity, 26
  SDS-PAGE, 338–339
  SspB–SSCA1 interaction, 339–340
  Western blot analysis, 340–341
Serpin–enzyme receptors. *See* LDL receptor-related protein 1
Serpins and complement system
  action mechanisms, 56
  activation, 56
  human plasma C1-inhibitor
    data interpretation, 64
    functional activity, purified C1-inhibitor, 62–63
    method, 64
    purification, 58–61
    three-step procedure, 57–58
  lectin pathway activation, 57
  pathway initiation and heparin role, 56–57
  recombinant C1-inhibitor production, yeast
    Escherichia coli, 64–65
    mutations, 64–65
    Pichia pastoris, 66–71
    purification, 71–73
    Saccharomyces cerevisiae vs. *Pichia pastoris*, 65–66
Serpin secretion, misfolding and surveillance
  carbohydrate modifications, 2
  conformational maturation, 2
  interactions and role verification
    caveats experiment, 10
    RNAi technology, 9–10
    sedimentation velocity centrifugation, 9
    time course, detectable protein interactions, 8–9
  mammalian cell lines transfection
    cDNA, 3
    DNA restriction enzyme digestions, 3
    expression level, 3–4
    preblocking and signal detection, 3–4
    untransfected cells, 3–4
  N-linked oligosaccharide contribution
    chemical inhibitors, 13
    intracellular trafficking, biochemical assessment, 12

Serpin secretion, misfolding and surveillance (cont.)
  modified, asparagine-linked oligosaccharides, 10–11
  N-glycan-mediated events, 12–13
  number and role, N-glycan, 10–11
  polar appendages, 10
  physically interacting proteins
    alternative methods, 7–8
    caveats experiment, 8
    coimmunoprecipitation and identification, 7
  protein synthesis
    caveats experiment, 6
    immunoprecipitation and quantification, 5
    intracellular degradation system, 5–6
    metabolic pulse–chase radiolabeling, 4–5
  synthesis, 2
Serpins target identification
  biochemistry
    affinity tags, TAP, 291–292, 292
    immunoprecipitation, 294–296
    serpin-protease complexes, 293–294
    TAP plasmids generation, 292–293
    transgenic lines, purification, 293
  genetics
    brood size, 289
    development assays, 288–289
    knockout alleles, 287–288
    longevity assays, 288
    protein function, 288
    RNAi use, 289–291
    tissue-specific phenotypes, 289
Sodium dodecyl sulfate-polyacrylamide gel electrophoresis (SDS-PAGE)
  analysis
    fractions, 58–61
    protease inhibition stoichiometry, 62
  characteristic banding pattern, 339–340
  gels, immunoblotting, 58–61
  GST–SCCA1 and SspB, 340
  purified C1-inhibitor, 61
  serpin reaction, 338–339
Squamous cell carcinoma antigen 1 (SCCA1)
  cysteine proteases, 332–333
  GST purification, 333–334
  SERPINB clade, 332
Staphylococcal cysteine proteases
  GST–SCCA1 and GST–SCCA2 fusion proteins
    bacterial expression system, 333–334
    Escherichia coli BL21 cells, 333–334
  inhibition characterization
    complex formation, stoichiometry, 335–336
    linear regression, 336
    microbial proteases, 334–335
    SCCA2 and SCCA1, 335
    stable complex formation, 337–338

  interaction site determination
    Edman degradation analysis, 342–343
    reactive site loop sequences, 343
    RSL, 341–342
    SCCA1 interaction, 342
    serpins, protease inhibition, 342–343
  serpin–enzyme complex
    characteristic banding pattern, 339–340
    SDS-PAGE, 338–339
    SspB–SSCA1 interaction, 339–340
    Western blot analysis, 340–341
  staphopains purification
    phenyl-sepharose column, 333
    Q-Sepharose column, 333
Stent implant, viral serpin
  clinical trials
    biomarkers, inflammatory, 324
    inflammatory and thrombotic markers, 322
    major adverse cardiovascular events (MACE), 321–324
    patient treatment groups, 324
    Serp-1, 321
    serum D dimer levels, 325
    stent implant, 324
  proceeding stages
    efficacy testing, 320–321
    FDA approval, 320–321
    good manufacturing protocol (GMP), 320–321
    requisite steps, 320–321
"Suicide-cleavage" mechanism, 207
Surface plasmon resonance
  assays, 192–193
  measurements, 25–26
  recombinant mouse Hsp47, 170
Syngeneic tumor model, maspin
  development
    BALB/c implantation, 158, 159
    rationale experimental approach, 157
    TM40D maspin transfectants establishment, 157–158
  gene therapy
    delivery and tumor analysis, 160–161
    efficiency determination, liposome, 160
    liposome preparation, 159–160
    MMTV-PyV cells, FVB mice, 159
    rationale experimental approach, 159

T

TEM. *See* Transmission electron microscopy
Tetracycline (TET)-regulated expression systems, 120–121
Transgenesis
  C. elegans vectors
    aggregation-prone serpin protein, 264–265
    canonical expression, 264–265
    microparticle bombardment, 266

coinjection markers
  DNA transformations, *C. elegans*, 266, 267, 270
  fluorescent microscope, 266
  recombination, DNA, 266
  young adult hermaphrodite, 266, 268
fluorescent protein (FP) fusions, 264
methods, 265–266
microinjection, 265–266
microparticle bombardment, 266
promoter choice, 264
resources and websites, 261
RNA splicing, 264
serpin–proteinase balance, 261
Transgenic mouse models, maspin
  bitransgenic
    cross WAP-TAg male mice, 155
    mammary tissue biopsies and histology, 156
    metastasis, 154–155
    p53 gene mutations, 155
    rationale experimental approach, 154–155
    tumor progression, 155
  limitation, 156–157
  WAP-maspin, 151–154
Transmission electron microscopy (TEM)
  autophagosomes, 36
  DAMP EM, 38
  IEM, 37–38
  plastic-embedded EM, 36

U

Unfolded protein response (UPR), 174

V

Viral serpin therapeutics
  anti-inflammatory activity, 304–305
  biological potential tests
    altered gene expression, cells, 318–319
    angioplasty injury model, 307–311
    anti-inflammatory activity cellular mechanisms, 311–318
    atherosclerosis and aneurysm, 307–311
  bleeding problems, 303–304
  clinical therapeutic potential assessment
    proceeding stages, 320–321
    rationale, 319–320
    trial findings, 321–325
  endogenous proteins, 304
  hypothetical models, 303
  immunomodulatory potential identification
    anti-inflammatory and antiapoptotic, 306–307

potential immunotherapeutics, 307
secreted anti-inflammatory, 305–306
inhibitory reagents, 304
plasminogen activator inhibitor-1 (PAI-1), 303–304
suicide inhibitors, 303
thrombolytic and thrombotic pathways, 304–305

W

Western blot analysis
  AtSerpin1, 353
  GST–SCCA1 and SspB, 341
  SDS-PAGE resolved proteins, 340–341
  staphopain, 340–341
Whey acidic protein (WAP)-maspin transgenic mice
  experimental approach, 151–154
  generation
    F1 lines, 154
    microinjection, 153–154
    PCR, 154
    steps, 153
    transgene, 151–152
  isolation and purification, construct, 152–153
White gene, 251
Wild-type mouse models, PAI-1
  cancer, 87
  cardiovascular disease
    aging, venous thrombosis, 85
    arterial thrombosis, 86
    bleomycin induced pulmonary fibrosis, 87
    EIM, 84
    electrolyte stasis and mechanical injury, 84
    ferric chloride role, 86
    IVC stasis/ligation, 84, 85
    MA-33H1F7, 85–86
    monoclonal antibodies, 85–86
    NFκB activation and inflammation, 85
    PAI-1 role, 85
    photochemical injury, 84
    thrombosis, 83–84
    thrombotic occlusion, 86–87
  thrombosis, 83
  vessel size and blood flow, 83
Wolbachia
  description, 218–219
  Wsp primers, 218–219

Y

Yeast nitrogen base (YNB) solution, 67–68, 69–71

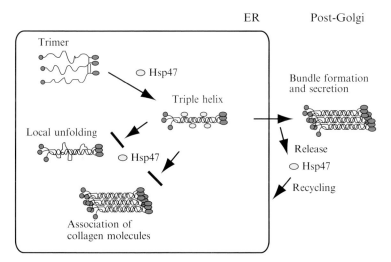

Yoshihito Ishida and Kazuhiro Nagata, Figure 9.3 A possible role in Hsp47 for collagen maturation. Newly synthesized collagen is inserted into the ER, and trimer and triple helix formation occurs. Hsp47 preferentially binds to triple helical procollagen. One hypothesis is that Hsp47 binds to triple helical procollagen to prevent association of collagen molecule and formation of aggregation in the ER since collagens have a characteristic to associate and form aggregate under the neutral condition. Another hypothesis is that Hsp47 hampers the local unfolding of triple helical procollagen.

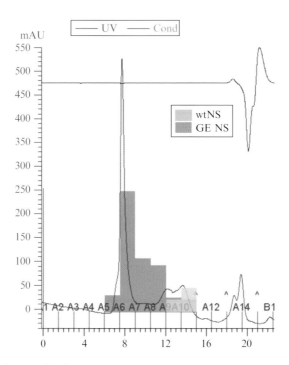

Thomas R. Jahn *et al.*, Figure 12.2 Size exclusion chromatography demonstrates partitioning of mutant neuroserpin into high molecular mass species in fly brain protein extracts. S200 size exclusion chromatography using ÄKTA equipment (GE Healthcare, UK) permits the fractionation of brain proteins from flies expressing wild-type human neuroserpin (wtNS, green bars) and the G392E variant (GE-NS, purple bars). ELISA quantitation of neuroserpin content shows that the polymerogenic serpins elute earlier, indicating a higher molecular mass.

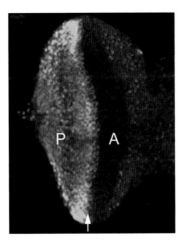

Thomas R. Jahn et al., Figure 12.8 Immunolabeling of the eye imaginal disc. The eye imaginal disc costained for a positive control protein (elav, green) and a *GMR-GAL4* protein of interest (red). Red staining is seen at the morphogenetic furrow (arrow) and posterior (P) to this band of cells. Anterior (A) to the furrow expression has not yet occurred.

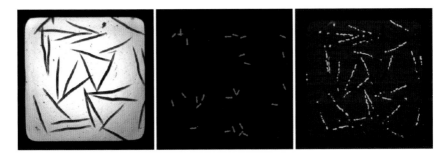

Olivia S. Long et al., Figure 13.5 Image acquisition using the ArrayScanV^{TI}. Brightfield (*left*), mCherry (*center*), and GFP (*right*) images of transgenic worms expressing an aggregation-prone serpin–GFP fusion. The SpotDetector BioApplication is used to automatically quantify the number, area, and intensity of GFP-positive, serpin aggregates (green spots) within each worm. Images obtained (in part) from Gosai *et al.*, 2010.

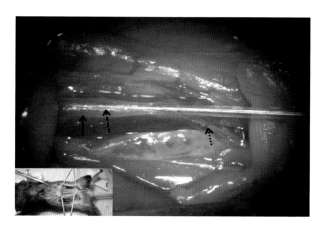

Hao Chen et al., Figure 15.3 Picture of balloon angioplasty in mouse. Lower left inserted picture shows abdominal incision and surgical exposure of the aorta for angioplasty balloon insertion and injury. Mag 10×. ••••▶, represents the right iliac artery into which the balloon is inserted; ⟶▶, represents the abdominal aorta; – –▶, represents the balloon inserted into the right iliac artery and advanced into the abdominal aorta where it is inflated with saline and dragged back and forth in the aorta to stretch the aorta and to induce damage.